EXPLORING GEOGRAPHY
& GLOBAL ISSUES

Richard G. Boehm, Ph.D.

Gary E. Clayton, Ph.D.

Nafees M. Khan, Ph.D.

Peter Levine, Ph.D.

Emily M. Schell, Ed.D.

About the Cover

 Fitz Roy is a mountain peak in Patagonia, located on the border between Argentina and Chile.

 Ships carrying containers filled with traded goods travel around the world.

 Commuters in cars, buses, and on foot travel throughout New York City.

 A man from Ghana makes traditional kente cloth on a hand loom.

 A compass is a tool that helps determine direction.

 The Allée Des Nations, or Avenue of the Nations, leads to the United Nations Office in Geneva, Switzerland.

mheducation.com/prek-12

 McGraw Hill

 A female biologist gathers water samples in Slovenia.

Send all inquiries to:
McGraw Hill
8787 Orion Place
Columbus, OH 43240

ISBN: 978-0-07-902284-4
MHID: 0-07-902284-7

Printed in the United States of America.

1 2 3 4 5 6 7 8 9 LKV 29 28 27 26 25 24 23 22

Authors

Richard G. Boehm, Ph.D., is one of the original authors of the *Guidelines for Geographic Education*, in which the Five Themes of Geography were first articulated. Dr. Boehm has received many honors, including Distinguished Geography Educator by the National Geographic Society (1990), the George J. Miller Award from the National Council for Geographic Education (NCGE) for distinguished service to geographic education (1991), Gilbert Grosvenor Honors in geographic education from the American Association of Geographers (2002), and the NCGE's Distinguished Mentor Award (2010). In 2020, he was named a Fellow of the American Association of Geographers. He served as president of the NCGE and also received the NCGE's Distinguished Teaching Achievement award. Presently, Dr. Boehm holds the Jesse H. Jones Distinguished Chair in Geographic Education at Texas State University in San Marcos, Texas, where he serves as Director of The Gilbert M. Grosvenor Center for Geographic Education. His most current project includes the development of "Powerful Geography," a new method of teaching and learning that acknowledges the diverse classroom and how geography can prepare students to fulfill their career goals in life.

Gary E. Clayton, Ph.D., is Professor and Chair of the Economics and Finance Department at Northern Kentucky University. He received his Ph.D. in economics from the University of Utah and an honorary doctorate from the People's Friendship University of Russia in Moscow. Dr. Clayton has authored several best-selling textbooks and a number of articles, has appeared on numerous radio and television programs, and was a guest commentator for economic statistics on NPR's Marketplace. Dr. Clayton won the Freedoms Foundation Leavey Award for Excellence in Private Enterprise Education in 2000. Other awards include a national teaching award from the National Council on Economic Education (NCEE), NKU's 2005 Frank Sinton Milburn Outstanding Professor Award, and the Excellence in Financial Literacy Education Award from the National Institute for Financial Literacy® in 2009. Dr. Clayton has taught international business and economics to students in England, Austria, and Australia. In 2006 he helped organize a microloan development project in Uganda.

Nafees M. Khan, Ph.D., is Assistant Professor of Social Foundations at Clemson University. Nafees holds a Ph.D. in Educational Studies from Emory University and a B.A. in Sociology with a minor in History from Tufts University. His doctoral work was on how the history of slavery was presented in secondary U.S. and Brazilian history textbooks. He serves on the Operational Committee for the *Slave Voyages Consortium* (www.slavevoyages.org) and has led numerous presentations on this digital humanities resource at conferences and workshops for teachers around the country. In addition, he is on the planning and advisory committee of the African Diaspora Consortium (www.adcexchange.org), wherein he is one of the developers of a new Advanced Placement (AP) Seminar course with African Diaspora Content with the College Board. His current research interests incorporate the legacies of slavery as related to education and the experiences of Afro-Brazilians, African Americans, and other diaspora communities.

Peter Levine, Ph.D., is the Associate Dean of Academic Affairs and Lincoln Filene Professor of Citizenship & Public Affairs in Tufts University's Jonathan Tisch College of Civic Life. He is also a full professor of Political Science at Tufts. In the domain of civic education, Levine was a co-organizer and co-author of *The Civic Mission of Schools* (2003), *The College, Career & Citizenship Framework for State Social Studies Standards* (2013), and *The Educating for American Democracy Roadmap* (2021). He has published eleven books, including *What Should We Do? A Theory of Civic Life* (Oxford University Press, 2022). He formerly directed CIRCLE (The Center for Information & Research on Civic Learning & Engagement) and has served on the boards of such civic organizations as AmericaSpeaks, Street Law Inc., the Newspaper Association of America Foundation, the Campaign for the Civic Mission of Schools, Discovering Justice, the Charles F. Kettering Foundation, the American Bar Association Committees for Public Education, and Everyday Democracy.

Emily M. Schell, Ed.D., is Executive Director of the California Global Education Project in the School of Leadership and Education Sciences at the University of San Diego. She received her Ed.D. in Education Leadership at the University of San Diego, M.S. in Journalism from Northwestern University, and B.A. in Diversified Liberal Arts at the University of San Diego. Emily was a teacher, district Social Studies Resource Teacher, and school principal in San Diego Unified, K-12 History-Social Science Coordinator at the San Diego County Office of Education, and liaison for the National Geographic Society Education Foundation. She was Teacher Education faculty at San Diego State University teaching Social Studies Methods for 20 years. She has authored numerous articles and two books, *Teaching Social Studies: A Literacy-based Approach* and *Social Studies Matters: Teaching and Learning with Authenticity*. She serves in statewide leadership roles to promote History-Social Science, Geography and Global Education, and Environmental Literacy in California schools. Her research and current work is based in professional learning for K-12 teachers.

Contributing Author

Douglas Fisher, Ph.D., is Professor and Chair of Educational Leadership at San Diego State University and a leader at Health Sciences High & Middle College, having been an early intervention teacher and elementary school educator. He is the recipient of an International Reading Association William S. Grey citation of merit, an Exemplary Leader award from the Conference on English Leadership of NCTE, as well as a Christa McAuliffe award for excellence in teacher education. He has published numerous articles on reading and literacy, differentiated instruction, and curriculum design as well as books, such as *The Distance Learning Playbook, PLC+: Better Decisions and Greater Impact by Design,* and *Visible Learning for Social Studies.*

Contributing Writer

David J. Rutherford, Ph.D., is Associate Professor in the Department of Public Policy Leadership at the University of Mississippi where he also serves as Executive Director of the Mississippi Geographic Alliance. His undergraduate and Masters degrees are in geography, and he earned his Ph.D. in geographic education from Texas State University. Rutherford's broad background in research and teaching in geography covers physical, human, regional, techniques, and educational components of the discipline. The current focus of his teaching and research is on major processes of change in the contemporary world.

Program Consultants

Timothy M. Dove, M.A.
Secondary Social Studies Educator
Founding staff member of Phoenix Middle School
Worthington, Ohio

Linda Keane, M.Ed.
Special Education Resource Teacher
Merrimack Middle School
Merrimack, New Hampshire

Meena Srinivasan, MA, NBCT
Executive Director
Transformative Educational Leadership (TEL)

Dinah Zike, M.Ed.
Creator of Foldables™
Dinah Zike Academy
Author, Speaker, Educator

Academic Consultants

Pedro Amaral, Ph.D.
Associate Professor of Economics
California State University

David Berger, Ph.D.
Ruth and I. Lewis Gordon Professor of Jewish History
Dean, Bernard Revel Graduate School
Yeshiva University, New York, New York

Carmen P. Brysch, Ph.D.
Assistant Professor of Geography
Texas A&M University

Xi Chen, Ph.D.
Assistant Professor of Geography
University of Cincinnati

Kimberly P. Code, Ph.D.
Director, Institute for Talent Development
and Gifted Studies
Northern Kentucky University

Seife Dendir, Ph.D.
Professor of Economics
Radford University, Radford, Virginia

Thomas Herman, Ph.D.
Director, California Geographic Alliance
Project Director, YESS Research Center
San Diego State University

Doug Hurt, Ph.D.
Assistant Professor of Geography
University of Missouri

Richard McCluskey, Ph.D.
Associate Professor of Geography
Aquinas College, Grand Rapids, Michigan

Osvaldo Muñiz-Solari, Ph.D.
Professor of Geography
Texas State University

Kwame Adovor Tsikudo, Ph.D.
Visiting Assistant Professor of Geography
Augustana College, Rock Island, Illinois

Jamie Wagner, Ph.D.
Assistant Professor of Economics
Director of UNO Center for Economic Education
University of Nebraska

Table of Contents

Included Sections

Topics	v
Primary and Secondary Sources	xii
Maps	xv
Charts, Graphs, and Diagrams	xvi
Topic Activities	xviii
Scavenger Hunt	xix

TOPIC 1
The World in Spatial Terms

INTRODUCTION LESSON

01 Introducing The World in Spatial Terms — G2

LEARN THE CONCEPTS LESSON

02 Thinking Like a Geographer — G7

INQUIRY ACTIVITY LESSON

03 Analyzing Sources: The Meaning of Place — G11

TAKE INFORMED ACTION — G16

LEARN THE CONCEPTS LESSONS

04 The Geographer's Tools — G17

05 Maps and Projections — G21

06 Geospatial Technologies — G25

INQUIRY ACTIVITY LESSON

07 Analyzing Sources: Acquiring and Using Geographic Information — G29

TAKE INFORMED ACTION — G34

REVIEW AND APPLY LESSON

08 Reviewing The World in Spatial Terms — C35

TOPIC 2
Places and Regions

INTRODUCTION LESSON

01 Introducing Places and Regions · G40

LEARN THE CONCEPTS LESSON

02 The Concept of Place · G45

INQUIRY ACTIVITY LESSON

03 Understanding Multiple Perspectives: Places and Identity · G49

    TAKE INFORMED ACTION · G52

LEARN THE CONCEPTS LESSONS

04 How Are Regions Defined · G53

05 Connections Between Places and Regions · G57

INQUIRY ACTIVITY LESSON

06 Global Issues: Connections Between Places · G61

    TAKE INFORMED ACTION · G66

REVIEW AND APPLY LESSON

07 Reviewing Places and Regions · G67

TOPIC 3
Physical Geography

INTRODUCTION LESSON

01 Introducing Physical Geography · G72

LEARN THE CONCEPTS LESSONS

02 Planet Earth · G77

03 Forces of Change · G83

INQUIRY ACTIVITY LESSON

04 Analyzing Sources: Forces Shaping the Earth's Crust · G89

    TAKE INFORMED ACTION · G94

LEARN THE CONCEPTS LESSONS

05 Factors Affecting Climate · G95

06 Climates and Biomes of the World · G99

INQUIRY ACTIVITY LESSON

07 Analyzing Sources: Geography and Climate · G103

    TAKE INFORMED ACTION · G108

REVIEW AND APPLY LESSON

08 Reviewing Physical Geography · G109

TOPIC 4
Population Geography

INTRODUCTION LESSON

01 Introducing Population Geography   G114

LEARN THE CONCEPTS LESSONS

02 Population Growth   G119

03 Population Patterns   G123

04 Population Movement   G127

05 The Growth of Cities   G131

INQUIRY ACTIVITY LESSON

06 Analyzing Sources: Population Change   G135

    TAKE INFORMED ACTION   G140

07 Global Issues: Causes of Population Change   G141

    TAKE INFORMED ACTION   G146

REVIEW AND APPLY LESSON

08 Reviewing Population Geography   G147

TOPIC 5
Cultural Geography

INTRODUCTION LESSON

01 Introducing Cultural Geography   G152

LEARN THE CONCEPTS LESSONS

02 Elements of Culture   G157

03 Expressions of Culture   G161

INQUIRY ACTIVITY LESSON

04 Analyzing Sources: Culture and Identity   G165

    TAKE INFORMED ACTION   G170

LEARN THE CONCEPTS LESSON

05 Causes and Effects of Cultural Change   G171

INQUIRY ACTIVITY LESSON

06 Global Issues: Continuity and Change in Culture   G175

    TAKE INFORMED ACTION   G180

REVIEW AND APPLY LESSON

07 Reviewing Cultural Geography   G181

TOPIC 6
Economic Geography

INTRODUCTION LESSON

01   Introducing Economic Geography   G186

LEARN THE CONCEPTS LESSON

02   Economic Systems and Activities   G191

INQUIRY ACTIVITY LESSON

03   Global Issues: The Location
of Economic Activities   G197

     TAKE INFORMED ACTION   G202

LEARN THE CONCEPTS LESSONS

04   Economic Development   G203

05   Economies and World Trade   G209

INQUIRY ACTIVITY LESSON

06   Understanding Multiple Perspectives:
Participation in Local and Global
Economic Systems   G213

     TAKE INFORMED ACTION   G218

REVIEW AND APPLY LESSON

07   Reviewing Economic Geography   G219

TOPIC 7
Political Geography

INTRODUCTION LESSON

01   Introducing Political Geography   G224

LEARN THE CONCEPTS LESSON

02   Features of Government   G229

INQUIRY ACTIVITY LESSON

03   Global Issues: Political Borders   G235

     TAKE INFORMED ACTION   G240

LEARN THE CONCEPTS LESSON

04   Conflict and Cooperation   G241

INQUIRY ACTIVITY LESSON

05   Analyzing Sources: Conflict and
Cooperation   G247

     TAKE INFORMED ACTION   G252

REVIEW AND APPLY LESSON

06   Reviewing Economic Geography   G253

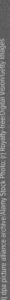

TOPIC 8
Human-Environment Interaction

INTRODUCTION LESSON

01 Introducing Human-Environment Interaction G258

LEARN THE CONCEPTS LESSONS

02 The Environment and Human Settlement G263

03 Natural Resources and Human Activities G269

INQUIRY ACTIVITY LESSON

04 Analyzing Sources: Human-Environment Interactions Throughout History G273

TAKE INFORMED ACTION G278

LEARN THE CONCEPTS LESSON

05 Human Impact on the Environment G279

INQUIRY ACTIVITY LESSON

06 Global Issues: Adapting to Environmental Change G285

TAKE INFORMED ACTION G290

REVIEW AND APPLY LESSON

07 Reviewing Human-Environment Interaction G291

TOPIC 9
What Is Economics?

INTRODUCTION LESSON

01 Introducing What Is Economics? E2

LEARN THE CONCEPTS LESSONS

02 Scarcity and Economics Decisions E7

03 Economic Systems E13

INQUIRY ACTIVITY LESSON

04 Analyzing Sources: Early Economic Systems E19

TAKE INFORMED ACTION E24

LEARN THE CONCEPTS LESSONS

05 Demand and Supply in a Market Economy E25

06 Prices in a Market Economy E31

07 Economic Flow and Economic Growth E37

INQUIRY ACTIVITY LESSON

08 Turning Point: The Industrial Revolution E41

TAKE INFORMED ACTION E46

REVIEW AND APPLY LESSON

09 Reviewing What Is Economics? E47

TOPIC 10
Markets, Money, and Businesses

INTRODUCTION LESSON

01   Introducing Markets, Money, and Businesses   E52

LEARN THE CONCEPTS LESSONS

02   Capitalism and Free Enterprise   E57

03   Business Organization   E63

04   Roles and Responsibilities of Businesses   E69

05   Employment and Unions   E73

06   Money and Financial Institutions   E79

INQUIRY ACTIVITY LESSON

07   Analyzing Sources: A Cashless Society?   E85

    TAKE INFORMED ACTION   E90

LEARN THE CONCEPTS LESSONS

08   Government and Business   E91

09   Income Inequality   E95

INQUIRY ACTIVITY LESSON

10   Multiple Perspectives: The Minimum Wage   E99

    TAKE INFORMED ACTION   E104

REVIEW AND APPLY LESSON

11   Reviewing Markets, Money, and Businesses   E105

TOPIC 11
Government and the Economy

INTRODUCTION LESSON

01   Introducing Government and the Economy   E110

LEARN THE CONCEPTS LESSONS

02   Gross Domestic Product   E115

03   Measuring the Economy   E119

04   The Federal Reserve System and Monetary Policy   E125

05   Financing the Government   E131

INQUIRY ACTIVITY LESSON

06   Multiple Perspectives: Taxes   E137

    TAKE INFORMED ACTION   E142

LEARN THE CONCEPTS LESSON

07   Fiscal Policy   E143

INQUIRY ACTIVITY LESSON

08   Multiple Perspectives: The Federal Debt   E149

    TAKE INFORMED ACTION   E154

REVIEW AND APPLY LESSON

09   Reviewing Government and the Economy   E155

TOPIC 12
The Global Economy

INTRODUCTION LESSON

01 Introducing The Global Economy   E160

LEARN THE CONCEPTS LESSONS

02 Why Nations Trade   E165

03 Exchange Rates and Trade Balances   E169

04 Global Trade Alliances and Issues   E173

05 The Wealth of Nations   E179

INQUIRY ACTIVITY LESSON

06 Analyzing Sources: Environmental Balance   E185

     TAKE INFORMED ACTION   E190

REVIEW AND APPLY LESSON

07 Reviewing The Global Economy   E191

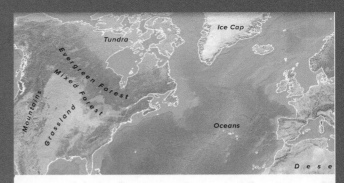

Appendix

World Religions Handbook   A3

Reference Atlas   A25

Glossaries/Glosarios

   Geography   A74

   Economics   A89

Indexes

   Geography   A100

   Economics   A110

Primary and Secondary Sources

TOPIC 1

President Barack Obama, from address to National Geographic Bee participants, 2012 .... G2

"Tivoli," *Fodor's Essential Scandinavia,* 2009 ...G15

"Foundation Document Overview: Frederick Douglass National Historic Site," National Park Service, 2021.......................... G16

TOPIC 2

Alastair Bonnett, from *Beyond the Map,* 2018 .... G40

"A Sacred Site to American Indians," National Park Service website, 2016 .... G50

"A New African American Identity: The Harlem Renaissance," Smithsonian National Museum of African American History and Culture website, 2021 .... G51

Kenneth R. Fletcher, from "Tangier Island and the Way of the Watermen," *Smithsonian Magazine,* 2009 .... G52

Daniel A. Gross, from "The Economics of Chocolate," *Smithsonian Magazine,* 2015..... G63

Trevor Williams, from "Examining How K-pop Helps Drive Korea's Global Growth," *Global Atlanta,* 2021 .... G65

TOPIC 3

"The Himalayas: Two continents collide," U.S. Geological Survey website, 2015 .... G90

Dr. Parker D. Trask, from "The Mexican Volcano Parícutin," Address before the Geologic Section of the New York Academy of Sciences in New York, October 4, 1943........G91

Jennifer Billock, from "Shimmy Through the World's Most Spectacular Slot Canyons," *Smithsonian Magazine,* 2017............... G92

"12 Things You Didn't Know About Great Sand Dunes National Park and Preserve," U.S. Department of the Interior website, 2020....................... G93

Saul Price, from "Climate of Hawai'i," National Weather Service website, 2021............. G104

Steve Linn, from "Rainfall in Mountainous Areas," ARM Climate Research Facility newsletter, U.S. Department of Energy, 2011................G107

"Climate of Rwanda," World Climate Guide/ Climatestotravel.com, 2015 .... G108

TOPIC 4

Lester R. Brown, from *For the Beauty of the Earth: A Christian Vision for Creation Care,* 2010 .... G114

Rasmenia Massoud, from *Broken Abroad,* 2013 .... G115

Guillaume Vandenbroucke, from "The Link between Fertility and Income," Federal Reserve Bank of St. Louis website, 2016 .... G142

Joseph J. Bish, from "Population Growth in Africa: Grasping the Scale of the Challenge," *The Guardian,* 2016....................... G143

Denise Hruby, from "How to Slow Down the World's Fastest-Shrinking Country," BBC.com, 2019 .... G144

William H. Frey, from "What the 2020 Census Will Reveal About America: Stagnating Growth, an Aging Population, and Youthful Diversity," The Brookings Institution, 2021............G145

"Population Growth in Canada: From 1851 to 2061," Statistics Canada, 2018..............G146

TOPIC 5

Mohandas K. Gandhi, from *My Non-violence,* 2015 ....G152

Sir Edward B. Tylor, from *Culture and Language Use,* 2009 ....G152

Wangari Maathai, from *Taking Root: The Vision of Wangari Maathai,* 2008........G152

Edward T. Hall, from *The SAGE Encyclopedia of Intercultural Competence,* 2015 ....G152

Thomas L. McPhail, from *Global Communication: Theories, Stakeholders, and Trends,* 2011.....................G153

Bobbie Kalman, from *What is Culture?,* 2009.....................................G153

Walter Lippmann, from *The Essential Lippmann: A Political Philosophy for Liberal Democracy,* 1982 ....G153

Matthew Arnold, from *Literature and Dogma: An Essay Towards a Better Apprehension of the Bible, 1883*.............G153

Rebecca Gross, from "The Quilts of Gee's Bend: A Slideshow," *National Endowment for the Arts blog,* 2015.......................G167

Mimi Kirk, from "A Hip Tradition," *Smithsonian Magazine,* 2007...........................G176

Wendi Maloney, from "Japanese-America's Pastime: Baseball," *Library of Congress Blog,* May 25, 2018 ....G179

Jennifer Vintzileos, from "9to5: The Story of a Movement," *Starry Constellation Magazine,* July 9, 2020 .... G180

TOPIC 6

William J. Bernstein, from *A Splendid Exchange: How Trade Shaped the World,* 2020.....................................G187

Max Opray, from "How Australia's 'White Gold' Could Power the Global Electric Vehicle Revolution," *The Guardian,* 2020 .... G198

Abby Wendle, from "Massive Corn Belt Crops Form Backbone of Meat Industry," KCUR/NPR, Harvest Public Media, 2015 .... G199

John McQuaid, from "The Secrets Behind Your Flowers," *Smithsonian Magazine,* 2011 ...... G200

Jonathan Nash, from "Eco-Tourism: Encouraging Conservation or Adding to Exploitation?," Population Reference Bureau website, 2001 .... G202

Stanley F. Stevens, from "From Tibet Trading to the Tourist Trade," *Claiming the High Ground,* 1993 ....G214

Janet L. Yellen, from "The History of Women's Work and Wages and How It Has Created Success For Us All," Brookings Institute, 2020.................................... G215

Katy Askew, from "Local Brands are Winning Hearts and Minds," FoodNavigator website, 2018 .... G216

Shamel Rolle Sands, Kenchera Ingraham, Bukola Oladunni Salami, from "Caribbean Nurse Migration—A Scoping Review," *Human Resources for Health,* 2020................ G218

TOPIC 7

The United Nations Guide to Model UN, "Preamble to the UN Charter," 2020 ....G225

Gregory Pappas, from "Pakistan and Water: New Pressures on Global Security and Human Health," *American Journal of Public Health,* 2011 .... G237

President Andrew Johnson, "Navajo Treaty of 1868," University of Groningen, Netherlands G238

Feargus O'Sullivan, from "When Borders Melt," from Bloomberg CityLab, 2017........ G239

U.S. Department of State website, "U.S.-Soviet Alliance, 1941–1945," National Archives and Records Administration G248

Imperial War Museums, from "What Caused the Division of the Island of Cyprus?" 2021...... G249

"Schengen Area—The World's Largest Visa Free Zone," SchengenVisaInfo.com, 2021......... G250

Rachel Will, from "China's Stadium Diplomacy," *World Policy Journal,* 2012 ....G251

Sylvie Corbet, "France to announce sanctions amid fishing dispute with UK," Associated Press News, 2021....................... G252

TOPIC 8

Joesph Romm, from *Climate Change: What Everyone Needs to Know,* 2016 .... G259

"History of Agriculture," Center for a Livable Future, Johns Hopkins Bloomberg School of Public Health website, 2021 .... G274

"Dunhuang," Silk Roads Programme, UNESCO website, 2021 .... G275

"Cities and the Fall Line," Virginia Studies website, 2019 .... G276

Benjamin Wilson Mountford and Stephen Tuffnell, from "How gold rushes helped make the modern world," The Conversation, 2018..... G277

"Coal," Alberta Culture and Tourism website, 2022................................... G278

Arys Aditya, from "Indonesia Sets 2024 Deadline to Move Its New Capital to Borneo," *Bloomberg News,* 2021............................... G286

Chris Arsenault, "In Canada, climate change could open new farmland to the plow," Reuters, 2017 .... G287

Mike Corder and Aleksandar Furtula, "Dutch reinforce major dike as sea rises, cliamte changes," AP News, 2019 .... G288

"Avoiding a water crisis: how Cape Town avoided 'Day Zero,'" *Resilience News,* Global Resilience Institute at Northeastern University, 2019.... G289

Luke Runyon, "The Bountiful Benefits of Bringing Back the Beavers," NPR, 2018............. G290

TOPIC 9

Miguel Barral, "The Silk Road: The Route for Technological Exchange that Shaped the Modern World," www.bbvaopenmind.com..... E21

Laura LaHaye, "Mercantilism," Library of Economics and Liberty.....................E23

Adam Smith, *An Inquiry into the Nature and Causes of the Wealth Of Nations,* Edinburgh: Tomas Nelson, 1827........................E24

Samuel Smiles, *Men of Invention and Industry,* 1895 ....E42

Daily Evening Traveller, September 1, 1880 . . . . . . E43

Henry Ford, *My Life and Work,* 1922 . . . . E44

Article 16, *Cuba's Constitution of 1976 with Amendments through 2002.* E49

President Bush Discusses Financial Markets and World Economy, November 13, 2008 . . . . . E49

TOPIC 10

Coral Ouellette, "Online Shopping Statistics You Need to Know in 2021," January, 2021 . . . . . E87

www.federalreserve.gov/paymentsystems/2019-December-The-Federal-Reserve-Payments-Study.htm, Trends in Noncash Payments, by Number. E88

Amy Fontinelle, "10 Reasons to Use Your Credit Card," Investopedia.com, 2021 . . . . E89

World Cash Report by Cash Essentials, Government of Japan, Percentage of Cash Payments in Selected Countries, 2018 . . . . E90

U.S. Bureau of Labor Statistics: U.S. Department of Labor, Federal Minimum Hourly Wage 2000–2020 . . . . E100

www.dol.gov/agencies/whd/minimum-wage/state, State Minimum Wages Per Hour E101

Economic Policy Institute Fact Sheet, "Why the U.S. needs a $15 minimum wage," 2021 . . . . . . E102

Congressional Budget Office, "How Increasing the Federal Minimum Wage Could Affect Employment and Family Income," 2021 . . . . E103

President Lyndon B. Johnson, Annual Message to the Congress on the State of the Union, January 8, 1964 . . . . E107

Benjamin Franklin, "On the Price of Corn, and Management of the Poor," *The London Chronicle,* 1766 . . . . E107

TOPIC 11

Bradfordtaxinstitute.com, "History of the Federal Income Tax Rates: 1913–2021" E138

Congressional Budget Office, Income Before and After Transfers and Taxes, 2020 . . . . E139

David Wessel, "Who Are the Rich And How Might We Tax Them More?" Brookings, 2019 . . . . E140

Adam Michel, "In 1 Chart, How Much the Rich Pay in Taxes," www.heritage.org, 2021 . . . . E141

Internal Revenue Service, Income Earned and Federal Income Taxes Paid by Income Group, 2018. . . . . E141

Emily Stewart, "Seriously, Just Tax the Rich," vox.com, 2021 . . . . E142

White House, Office of Management and Budget, Government Deficits Become Debt E150

President Franklin D. Roosevelt, "Address Before the American Retail Federation, Washington, D.C.," 1939 . . . . **E151**

Louise Gaille, "12 Key Balanced Budget Amendment Pros and Cons," vittana.org, 2018 . . . . E152

Peter Wehner and Ian Tufts, "Does the Debt Matter?" National Affairs, 2020. . . . . E153

U.S. Treasury, Federal Government Spending. E154

Paul Krugman, "An Interview with Paul Krugman," *The Washington Post,* 2012 . . . . E157

James Surowiecki, "The Stimulus Strategy," *The New Yorker,* 2008 . . . . E157

TOPIC 12

Tucker Davey, "Developing Countries Can't Afford Climate Change," futureoflife.org, 2016 . . . . E186

Blog.constellation.com, "How Much Energy Do Game Consoles Really Use?" E187

Gary Clayton, *Principles of Economics,* McGraw Hill, 2024 . . . . E188

Daniel Griswold, "Why We Have Nothing to Fear from Foreign Outsourcing," Free Trade Bulletin #10, Cato Institute, 2004 . . . . E193

Lou Dobbs, "Exporting America: False Choices," CNN Money, CNN, March 10, 2004 . . . . E193

Maps

TOPIC 1

Ushuaia, Argentina............................G12
Canada's Provinces and Territories..............G13
Earth's Hemispheres.........................G17
The Global Grid.............................G19
Common Map Projections.....................G22
Political Map of the United States..............G23
Airline Routes..............................G30
Rainfall in Africa............................G31
Lyme Disease..............................G33

TOPIC 2

Regions of the National Football League.........G41
Regions of the United States..................G43
Paris, France..............................G46
Perceptual Regions of the United States........G54
Commuting to Work.........................G62

TOPIC 3

Latitude and Temperature.....................G73
Plates and Plate Movement...................G85
The Himalaya..............................G90
Global Wind Patterns........................G97
World Climate Regions......................G100
World Biomes.............................G101
Boreal Forests............................G105
Rainfall in Texas..........................G106

TOPIC 4

World Population Density....................G116
World Population Distribution.................G125
Global Migration Patterns...................G128
Population Change of U.S. States.............G138

TOPIC 5

Cultural Change and the Columbian Exchange..G155
World Languages...........................G158
World Culture Regions......................G162
Generic Names for Soft Drinks...............G169
World Culture Hearths......................G171
Blues and the Great Migration...............G178

TOPIC 6

World Trade in Goods......................G189
Agriculture and Economic Development.......G204
Industry and Economic Development.........G205
Services and Economic Development........G206
Global Trade of Oil........................G210
Texas–Mexico Border......................G217

TOPIC 7

Tigris and Euphrates Rivers.................G225
Political Map of North America..............G226
Antarctic Claims..........................G227
Governments Around the World.............G231
Melbourne, Australia......................G233
The Rhine River: A Natural Boundary.........G234
Headwaters of the Indus River..............G237
The Treaty of Guadalupe Hidalgo...........G241
Conflict Over Kashmir.....................G242
The Great Lakes..........................G244
The Schengen Area.......................G250

TOPIC 8

The Fall Line in the United States............G265
Geography and the Industrial Revolution......G266
Land from the Sea........................G267
Natural Resources Around the World.........G270
Arctic Resources.........................G271
Desertification Vulnerability.................G280

TOPIC 10

Right-to-Work States......................E76

TOPIC 11

The Federal Reserve System...............E127

TOPIC 12

Export Partners: Selected Nations...........E166
Top Ten Trading Partners of the United States,
    2020.................................E168
European Union..........................E175
GDP Per Capita–Selected Countries.........E180

Charts, Graphs, and Diagrams

TOPIC 1

Studying the World in Spatial Terms............. G6
Skills for Thinking Like a Geographer............ G8
The Inca Trail Marathon....................... G34

TOPIC 2

Studying Places and Regions.................. G44
Types of Regions G53
Exporting Waste G64
International Travel Destinations.............. G66

TOPIC 3

Studying Physical Geography G76
Water, Land, and Air....................... G77
The Continental Shelf G79
The Water Cycle G82
Inside Earth G83
Earth's Seasons........................... G96
The Rain Shadow Effect G98, G107

TOPIC 4

World Urban Population G115
Studying Population Geography G118
World Population G119
Stages of Growth: Population Pyramid:
    South Sudan, 2020 (rapid growth) G120
Stages of Growth: Population Pyramid:
    United States, 2020 (slow growth) G121
States of Growth: Population Pyramid:
    Italy, 2020 (decline) G121
Where is the World's Population G124
Top Five Countries to Receive Remittances,
    2021 .... G130
Central Place Theory G132
World Population Growth, 1750–2100 ....G136
Birth Rates in Decline......................G137
Population Pyramid: Japan, 1965 ....G139
Population Pyramid: Japan, 2015 ....G139
Race-Ethnic Profile for U.S. Population,
    2000 and 2019........................ G140

Relationship Between Fertility and Income......G142
Emigration of Bulgarians to OECD Countries G144
U.S. Annual Population Growth, 1900 to
    2020................................G145

TOPIC 5

Studying Cultural Geography................. G156
Major World Religions G159

TOPIC 6

Studying Economic Geography G190
Economic Choices G193
Exports and Imports, Leading Countries
    in 2020.............................. G209

TOPIC 7

Studying Political Geography................ G228
Functions of Government G229
Comparing Democratic and Authoritarian
    Governments G232

TOPIC 8

Studying Human-Environment Interaction...... G262
Classifying Resources G269
Climate Change........................... G283

TOPIC 9

Economic Connections E6
Benefit-Cost Analysis........................E11
Choices All Societies Face E12, E47
Price Signals in a Market Economy E16
Illustration of a Market...................... E20
Illustration of a Manor E22
Demand Schedule and Graph for Fruit
    Smoothies E26
Supply Schedule and Graph for Fruit
    Smoothies E27
Change in Demand for Video Games..........E29
Change in Supply for Video Games E30
Market for Fruit Smoothies E34

Demand and Supply for Oil. E36
Circular Flow of Economic Activity E37
Demand, Supply, and Equilibrium E47

TOPIC 10

Characteristics of Free Enterprise Capitalism. . . . E56
American Free Enterprise Capitalism E58
Most Popular Social Networks, 2021 . . . . E60
Forms of Business Organization E64
Corporate Structure E67
Population and the Civilian Labor Force,
    2020. . . . . .E73
Employment by Sector, 2020. . . . . .E74
Types of Unions E75
Features of U.S. Currency. E80
Average Revenue Per Online Shopper. E87
Trends in Noncash Payments, by Number. E88
Percentage of Cash Payments in Selected
    Countries. E90
Selected U.S. Government Regulatory
    Agencies E94
Weekly Earnings by Level of Education, 2020 . . . E96
Poverty Guidelines for the 48 Contiguous
    States and Washington, D.C., 2021. . . . . .E97
Federal Minimum Hourly Wage 2000–2020 . . . .E100
State Minimum Wages Per Hour E101

TOPIC 11

Estimating Total Annual Output (GDP) E116
Real GDP 1950–2020 (in constant 2012 prices) . . E118
The Business Cycle E119
U.S. Real GDP 1975–2021. . . . . E121
U.S. Unemployment Rate 1975–2021 . . . . .E122
Changes in Prices, January 2021. . . . . .E123
Monetary Policy and Interest Rates. E128
The Federal Budget Process E131
The Federal Budget, Fiscal Year 2020. . . . . .E133
State and Local Government Revenue and
    Expenses. E135
Income Before and After Transfers and
    Taxes, 2020. . . . . .E139
Income Earned and Federal Income Taxes
    Paid by Income Group, 2018 . . . . E141
Government Deficits Become Debt E147, E150
Federal Government Spending E154
The Business Cycle E155

TOPIC 12

Leading Exporters and Importers E165
International Value of the U.S. Dollar E170
A Strong Dollar vs. a Weak Dollar E172
Country Classification Levels E179
Comparing Economies. E182

Topic Activities

Debating Globalization

Taking a Position...........................G222

Exploring Careers

Writing an Employment Advertisement G37

Geographic Reasoning

Finding Locations Using Coordinates G37
Gaining a New Perspective G69
Contrasting Elevations.......................G111
Drawing Three Dot Density Maps G149
Drawing an Illustrated Map of Two Nations.....G184
Comparing and Contrasting GDP G221
Reimagining a Border G255

Geography and History

Making a Display...........................G70

Making Connections to Today

Imagining Alternate History G38
Designing a Museum Exhibit G70
Creating a Multimedia Presentation About the
    Creation of a Place by Physical Geography ...G112
Presenting a News Report G150
Creating a Multimedia Presentation About an
    Advance in Technology G184
Tracking a Resource........................G222
Identifying Influences........................G256
Tracking Price Changes on a Chart and Graph...E50
Creating a New Product E108
Being an Active Citizen E158
Being an Active Citizen E194

Solving Conflicts

Advising a Leader..........................G256

Understanding Causes

Explaining Increasing Life Expectancies G150

Understanding Economic Concepts

Applying Economic Concepts to a Graph E50
Finding Evidence of a Market in a Photograph...E108
Applying Evidence in a Photograph
    to the Federal Budget...................E158
Creating a Chart Showing Comparative
    Advantage..............................E194

Understanding Markets

Learning From History G221

Understanding Multiple Perspectives

Making a Display...........................G38
Creating an Identity Flow Chart...............G69
Creating a Travel Brochure...................G111
Analyzing Points...........................G183
Taking a Position..........................G255
Comparing and Contrasting Perspectives
    on Economic Systems...................E49
Comparing and Contrasting Viewpoints on
    Government Assistance E107
Using Evidence in an Essay About
    Fiscal Policy E157
Using Sources to Write About Outsourcing......E193

Understanding Urban Sprawl

Taking a Position..........................G149

Writing a Persuasive Argument

Writing an Argument About a
    Cultural Concept G183
Writing a Report About an Economic Choice E49
Writing About a Trade Policy for
    Athletic Shoes..........................E193

Writing an Informative Essay

Writing an Essay About Natural Disaster
    Preparedness G112

Writing an Informative Report

Writing About a Company's Charitable
    Giving.................................E107
Writing About the Federal Budget.............E157

Scavenger Hunt

This book contains a wealth of information. The trick is to know where to look to access all the information.

ACTIVITY Complete this scavenger hunt exercise with your teachers or parents. You will see how the textbook is organized and how to get the most out of your reading and studying time. Let's get started!

1. How many lessons are in Topic 8?

2. What is the title of Topic 12?

3. How many Inquiry Activity Lessons are in Topic 2?

4. What is the Compelling Question for Topic 3, Lesson 4?

5. What is the title of the map found in Topic 5, Lesson 2?

6. You want to quickly find a map in the book about the world. Where do you look?

7. In which two places can you find the contents and page numbers for a topic?

8. Where can you find information on what you will learn in a particular topic?

9. If you needed to know the definition for *biome,* where would you look?

10. Where in the back of the book can you find page numbers for information about Gross Domestic Product?

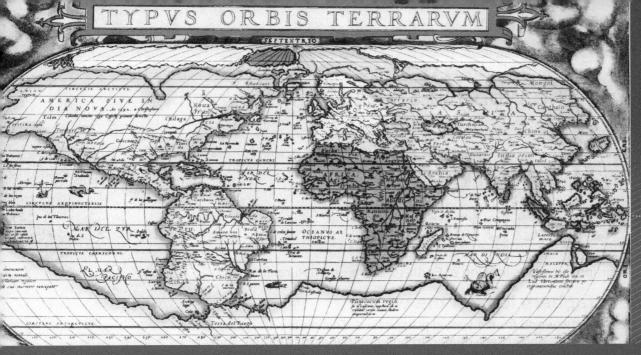

SEPTENTRIO

TYPVS ORBIS TERRARVM

Typus Orbis Terrarum, Latin for "map of the world," is from 1570. It is the first map of the world to appear in a standard atlas.

The World in Spatial Terms

INTRODUCTION LESSON

01 Introducing The World in Spatial Terms   G2

LEARN THE CONCEPTS LESSONS

02 Thinking Like a Geographer   G7

04 The Geographer's Tools   G17

05 Maps and Map Projections   G21

06 Geospatial Technologies   G25

INQUIRY ACTIVITY LESSONS

03 Analyzing Sources: The Meaning of Place   G11

07 Analyzing Sources: Acquiring and Using Geographic Information   G29

REVIEW AND APPLY LESSON

08 Reviewing The World in Spatial Terms   G35

Introducing The World in Spatial Terms

Thinking Like a Geographer

The study of geography involves looking at Earth's people and natural features.

How do geographers look at the world?

Geographers examine the world from a spatial perspective. This means that they focus on how people, places, and objects are related to one another across Earth's surface.

MAP ILLUSTRATING NORTH AMERICAN TRANS-CONTINENTAL RAILWAYS.

The U.S. railroad system is an example of the interactions within and between the human and physical aspects of places. Geographers might ask: How did physical features affect the routes of the railroads? What role did the expanding railroads play in opening up the middle of the country? The West coast?

> 66 The study of geography is about more than just memorizing places on a map. It's about understanding the complexity of our world, appreciating the diversity of cultures that exists across continents. And in the end, it's about using all that knowledge to help bridge divides and bring people together. 99

—President Barack Obama, address to 2012 National Geographic Bee participants

Holi, a spring festival, is an important celebration for Hindus around the world. Geographers are interested in peoples, cultures, and their interactions.

PHOTO: (t) North Wind Picture Archives/Alamy Stock Photo, (bl) Library of Congress Prints and Photographs Division [LC-DIG-ppbd-00358], (br) powerofforever/Getty Images; TEXT: Obama, Barack. Comments at 2012 National Geographic Bee. In Parkinson, Alan. Why Study Geography? London: London Publishing Partnership, 2020.

The Geographer's Tools

Geographers have many things to help them learn about the world. They rely on other subjects like history and economics. And of course, they use globes and maps. Different kinds of maps show countless kinds of information. Maps can show

- physical features like mountains, rivers, and lakes;

- human features like countries, states, and cities;

- populated areas, climate regions, and land use.

Technology is also a part of the geographer's tool kit. In fact, some of the same tools may be part of your daily life. Your cell phone has a feature that allows you to find a specific location like a house address. This technology is part of the same global positioning system (GPS) geographers use to determine the location of various things they study. Computer mapping programs and satellites are also important ways of gathering, organizing, and analyzing geographic information.

Perhaps you participate in geocaching, using a GPS device or your cellphone to find the location of hidden treasures.

Geographers use satellite images and photographs to gather information about human and environmental situations. This photograph of a highway overpass may help geographers understand the transportation network in a city.

Getting Ready to Learn About . . . The World in Spatial Terms

A Geographer's Perspective

Geographers think spatially. That is, they study things in terms of their location in the spaces on planet Earth. In their work, geographers consider

- where things are located;
- their size and direction;
- how far apart they are;
- what characteristics they have;
- why they are located where they are;
- what relationships they have with other things.

Thinking spatially begins with the study of the location or the size of things. But it is more than just that. Thinking spatially means looking at the characteristics of Earth's features and the relationships among them. For example, geographers look at the interaction of people with all kinds of business locations, such as shopping centers and restaurants. They study the layout of cities and think about how easy or difficult it is for people to move around in them. Geographers look to other subjects like history, economics, and government to help them describe the distribution of peoples, places, and environments.

Geographers are interested in the location of Cape Town, South Africa. Its location at the southern tip of the African continent makes it important for water-based trade. The city developed around the area's physical features like Table Mountain, highlighting the relationship between humans and the environment.

Maps and technology help geographers gather and analyze information about the world.

Geographers Use Maps

Maps are one of the most important tools that geographers use. Maps show many types of features and objects on Earth, including those in the physical environment and those created by humans. They show the locations of these features and objects in order to help us identify the relationships among them. Information gathered from maps can even help us understand processes like the expansion of cities, the movement of people, and the impact on natural resources involved in those relationships.

Maps provide a lot of information to interpret as we improve our ability to read them. As you will learn in this topic, spatial information from maps is used in a variety of ways—to plan new communities, to monitor agricultural activities, and to track the spread of disease.

Geographers Use Technology

The word *technology* simply means any way that scientific discoveries are applied to practical use. Geographers use many types of technologies to provide practical information about the physical and human aspects of Earth. You are probably already familiar with some types of these technologies—like GPS and Google Earth—found on computers and even on smartphones. They are so present in our lives that we often take them for granted. But we can learn to use them more effectively as we come to know more about them.

Technology is always changing. Geographers and others who use technology to study our world can gather what seems like endless amounts of information as new and improved ways allow them to do so.

Looking Ahead

You will learn about the essential elements of geography, the spatial perspective, and how geographers view the world. You will learn about globes and maps, as well as the parts of a map. You will learn about map projections and different types of maps. You will learn about geospatial technologies and their limitations. You will examine Compelling Questions and develop your own questions about the world in spatial terms in the Inquiry Activity Lessons. You can see some of the ways in which you might already use the spatial perspective in your life by completing the activity.

What Will You Learn?

In these lessons about the world in spatial terms, you will learn

- that geographers look at people and the world in terms of space and place;
- that geographers examine physical features and how humans interact with them;
- the six elements of geography;
- that geography is used to understand historical patterns, economies, politics, and the impact of societies and cultures on the landscape;
- that geographers use map projections to view the round Earth on a flat map;
- how lines of latitude and longitude create a grid system on Earth's surface that is used to determine location;
- that there are many parts to a map, including a title, a map key, and a compass rose;
- that maps are either large-scale or small-scale to convey distances on Earth;
- that different information is conveyed on either general-purpose or thematic maps;
- how a mental map describes an individual's perception of features on Earth's surface;
- that geospatial technologies include global positioning systems (GPS), geographic information systems (GIS), aerial photographs, and remote sensing from satellites;
- that geospatial technologies can help identify and navigate information, but geographers still need to ask and answer these questions to better understand the world.

 COMPELLING QUESTIONS IN THE INQUIRY ACTIVITY LESSONS

- **How do you define a *place*?**
- **How can maps and other geographic representations help us?**

Studying The World in Spatial Terms

Thinking Like a Geographer	• the spatial perspective • skills for thinking like a geographer • the essential elements of geography • geography and other subjects
The Geographer's Tools	• globes and maps • absolute location • relative location
Maps and Map Projections	• parts of a map • map projections • types of maps
Geospatial Technologies	• global positioning systems (GPS) • geographic informatin systems (GIS) • satellites • limitations of technology

Thinking Like a Geographer

READING STRATEGY

Analyzing Key Ideas and Details As you read, take notes about how geographers consider each of the essential elements of geography in their work.

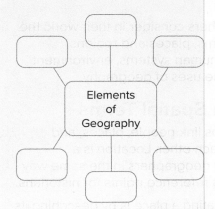

Geographers Think Spatially

GUIDING QUESTION

What does it mean to think like a geographer?

Information from many subjects helps us understand the world. For example, biology is the study of how living things survive and relate to one another. History is the study of events that have occurred over time and how those events are connected. **Geography** is the study of Earth and its peoples, places, and environments. This includes where people live on the surface of Earth, why they live there, and how they interact with each other and the physical environment.

Geography, then, emphasizes the **spatial** aspects of Earth's physical and human features. These are examined in terms of their locations, their shapes, their relationships to one another, and the patterns that they form. Features in the physical world such as mountains and lakes can be located on a map. These features can be described in terms of height, width, depth, and their distances and directions to other features. The human world also has spatial dimensions that can be located on a map. Geographers describe cities, states, and countries. They measure how close or far apart these human features are to one another. They think about the relationships between human features and physical features. In addition, geographers consider the scale of any feature or issue. They ask if the scale is local, regional, state, national, global, or some combination of these.

But thinking spatially includes more than just the location or size of things. It also means studying the characteristics of Earth's features.

geography the study of Earth and its peoples, places, and environments

spatial Earth's physical and human features in terms of their places, shapes, relationships to one another, and the patterns they form

Today, what we know about the world continues to grow as geographers study the world's environments.

Skills for Thinking Like a Geographer
Asking Geographic Questions helps you pose questions about your surroundings.
Acquiring Geographic Information helps you answer geographic questions.
Organizing Geographic Information helps you analyze and interpret information you have collected.
Analyzing Geographic Information helps you look for patterns, relationships, and connections.
Answering Geographic Questions helps you apply information to real-life situations and problem solving.

The five skills described in the table are key to geographic understanding.

Geographers ask what mountains in different locations are made of. They examine what kinds of fish live in different lakes. They study the layout of cities and how easy or difficult it is for people to move around in them. Geographers identify and study the processes by which physical and human features form and change. The goal is to explain as well as describe.

One of the most important tools a geographer needs is the ability to think geographically and spatially about the world. There are five basic skills that are key to such thinking. These are asking, acquiring, organizing, analyzing, and answering geographic questions. Asking geographic questions provides information that can be used to better understand one's surroundings. As knowledge is **acquired**, it can be applied to recognize patterns and relationships that will help in real-life situations.

✓ **CHECK FOR UNDERSTANDING**

Explaining How is geography related to history?

The Essential Elements of Geography

GUIDING QUESTION

What are the essential elements of geography?

Geographers study interactions between peoples, places, and environments. They explain why and how spatial patterns occur. There are six essential

elements geographers consider in their work: the world in spatial terms, places and regions, physical systems, human systems, environment and society, and the uses of geography.

The World in Spatial Terms

Spatial relationships link people, places, and environments to each other. Location is a reference point for geographers in the same way that dates serve as reference points for historians.

One way of locating a place is by describing its **absolute location**. This is the exact spot at which the place is found on Earth. To determine absolute location, geographers use the system of latitude and longitude. We also identify a place based on **relative location**—a place's location in relation to other places. For example, New Orleans is located near the mouth of the Mississippi River. The broad definition of location based on absolute or relative location also considers a place's site and situation. *Site* is the characteristics specific a place, including its physical setting. *Situation* is the geographic position of a place in relation to the other places or features.

Places and Regions

A **place** is what a location is like. Its distinguishing characteristics include both physical and human qualities. Geographers study the similarities and differences between places. They group places with similar characteristics into regions. A **region** is defined by its shared physical or human characteristics. Physical characteristics include

---

acquire to gain

absolute location the exact spot at which a place is found on Earth

relative location a place's location in relation to other places

place what a specific location is like

region a group of places with similar characteristics

climate, landforms, soils, vegetation, animal life, and natural resources. Human characteristics include language, religion, culture, and political or economic systems.

Physical and Human Systems

Geography covers a broad range of topics, so geographers focus on either physical geography or human geography. Physical geography looks at climate, land, water, vegetation, and animal life. It examines how these combine to form ecosystems and biomes. Geographers also study human geography, the processes by which people operate across Earth's surface. This includes how they settle, earn their livings, form societies and cultures, and interact with one another.

Physical and human geography are broad in their focus. They can be further divided into subject areas. For example, climatology is the study of climate and atmospheric conditions. It analyzes their impact on the environment and society. Historical geography is the study of places and human activities over time based on the geographic factors that have shaped them.

Environment and Society

The relationship between people and their physical environment is an important area of study for geographers. One aspect of this relationship involves ways that people use and change their surroundings. Geographers identify these ways and examine the consequences that result. Pollution, construction, population growth,

conservation of parks, and reintroduction of species into the wild are a few of the ways humans change the physical environment.

Another aspect of the relationship between people and the physical environment involves ways the physical environment can affect humans. For example, physical features such as rivers, mountains, and deserts can limit human movement and settlement. Natural **phenomena** such as hurricanes, earthquakes, and heavy storms or droughts force humans to adapt to the changing environment. Understanding how the Earth's physical features and processes shape and are shaped by human activity helps societies make informed decisions.

The Uses of Geography

Geography provides an understanding into how physical and human systems developed in the past. It also considers current trends about those systems to plan for the future. For example, planning and policy making must consider interactions between humans and the natural environment. Data related to physical features can highlight suitable sites for certain types of economic activities. Urban planners analyze trends in human growth in a certain region. They use this information to determine what things—schools, roads, public services, and businesses—are necessary for supporting a growing population and where they can be located.

---

phenomena a rare or significant event

Rice paddy fields in China reflect both human systems and physical systems at work.

1. **Evaluating** What are the human systems at work in this photograph?
2. **Explaining** What are the physical systems at work in this photograph?

Geography skills are useful in many different situations. Geographers have many different job titles. Geographers work in a variety of jobs in government, business, and education. They often combine the study of geography with other areas of study. For example, a historian must know the geographic characteristics of a place or region so that they can explain how and why people settled in a particular way. A travel agent must know the physical and human geography of places in order to plan trips for clients.

Knowledge of geography is not only something people apply in a wide variety of jobs. Knowing geography helps people understand the causes, meanings, and effects of physical and human events. It also provides insight into what is likely to occur in the future. Geographic knowledge plays an important role in developing effective, ethical, and lasting solutions to the difficult challenges and issues we all face. Geography can help us deal with issues such as climate change, environmental hazards, transportation planning, international trade, and conflict or cooperation.

☑ **CHECK FOR UNDERSTANDING**

1. **Describing** Describe two ways of locating a place.
2. **Contrasting** What is the difference between a place and a region?

Geography and Other Subjects

GUIDING QUESTION

How is geography related to other subjects?

Geography has important relationships to other subjects. Geographers use a spatial perspective to understand patterns in history, economics, and political systems. They study the relationships between societies, cultures, and the environment.

To visualize what a place could have looked like in the past, geographers consider historical perspectives of that place. For example, to gather information about how a city has changed over time, geographers collect data from historical sources. This may include information about population growth, economic activity, natural disasters, and disease. Such data can address questions concerning how human activities have changed the natural vegetation, or how waterways are different today than in the past. Historical perspectives can show how to avoid repeating decisions that created conflicts between human growth and the environment.

Geographers analyze historical and current political patterns to learn about changing boundaries and government systems. They study how the natural environment has influenced political decisions and how governments change natural environments. For example, in the 1960s the Egyptian government built the massive Aswān High Dam on the Nile River to help irrigate the land. The dam required the relocation of people and resulted in decreased soil fertility.

Human geographers use ideas from sociology and anthropology. Sociology is the study of societies and social relationships. Anthropology is the study of human origins and cultures. These subjects help geographers study human trends and past cultures and their influence on contemporary societies. Because people come from diverse cultural backgrounds, their interpretations of information and experiences differ. For example, people in different neighborhoods may define boundaries based on different locations of their activities, such as the stores they frequent.

Geographers study economies to understand how the locations of resources affect the ways people make, transport, and use goods. The study of economies also helps geographers understand how and where services are provided. Geographers are interested in how locations are chosen for economic activities. Where and what is produced and consumed depend on a variety of factors. These include the location of natural resources, fertile soil, suitable climates, and closeness to transport routes.

☑ **CHECK FOR UNDERSTANDING**

Summarizing How is geography related to sociology?

LESSON ACTIVITIES

1. **Explanatory Writing** Write a paragraph explaining how geography helps interpret the past, understand the present, and plan for the future.

2. **Interpreting Information** Work with a group to learn how your town or city looked 100 years ago. Find historical photos or maps. Compare those to the way the community looks today. Use your findings to create a slide show. Be sure to include captions explaining how the community has changed.

03

Analyzing Sources: The Meaning of Place

? COMPELLING QUESTION

How do you define a *place*?

Plan Your Inquiry

DEVELOPING QUESTIONS

Think about different types of places you are familiar with—hometown, school, a frequent vacation destination. Consider why a place is described in a certain way. Then read the Compelling Question for this lesson. What questions can you ask to help you answer this Compelling Question? Create a graphic organizer like the one below. Write these Supporting Questions in your graphic organizer.

Supporting Questions	Source	What this source tells me about the meaning of *place*	Questions the source leaves unanswered
	A		
	B		
	C		
	D		
	E		

ANALYZING SOURCES

Next, examine the sources in this lesson. Analyze each source by answering the questions that follow it. How does each source help you answer each Supporting Question you created? What questions do you still have? Write these in your graphic organizer.

After you analyze the sources, you will
- use the evidence from the sources;
- communicate your conclusions;
- take informed action.

Background Information

Geographers use the term *place* to refer to a location with unique physical and human characteristics. People describe these characteristics based on their experiences and perceptions. Some places are large, such as the country of Russia or the Sahara Desert. Some places are small, such as a school campus or Liberty Island, where the Statue of Liberty stands.

The term *place* is flexible in part to be useful at different scales. You might have a favorite place in your house, and you can also describe a large city as the place that you are from.

» A sign on an Australian highway identifies distances (in kilometers) to a list of places.

Ushuaia, Argentina

Ushuaia (OO·SWAI·UH) is a city in Argentina, located on the southern coast of South America. While it is a relatively small city, Ushuaia is well known to adventure travelers as being the southernmost city in the world. On this map, the location of Ushuaia is labeled.

PRIMARY SOURCE: MAP

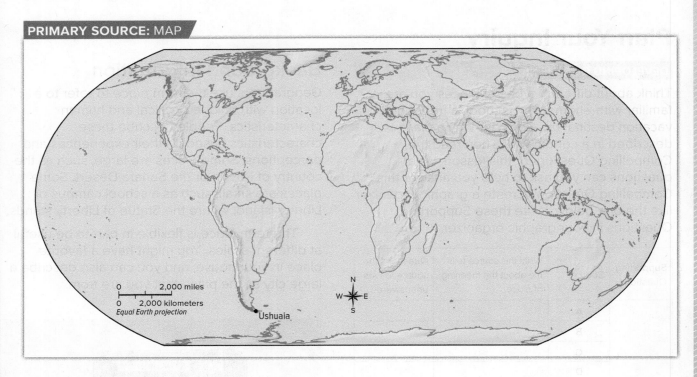

EXAMINE THE SOURCE

1. **Calculating** Using the concept of relative location, how would you verify Ushuaia's claim to being the world's southernmost city?

2. **Inferring** Why do you think Ushuaia has been an important location for adventurers traveling to Antarctica?

Canada's Provinces and Territories

Like many countries, Canada is made up of several political regions. In Canada these are known as provinces and territories. Each province and territory has its own government that is responsible for establishing and administering, or carrying out, its own laws. For example, in the province of Quebec, French is one of two official languages. Provinces have more powers than territories. The territory of Nunavut was separated from the Northwest Territories in 1999 so that the **Inuit** people living there could establish their own government.

PRIMARY SOURCE: MAP

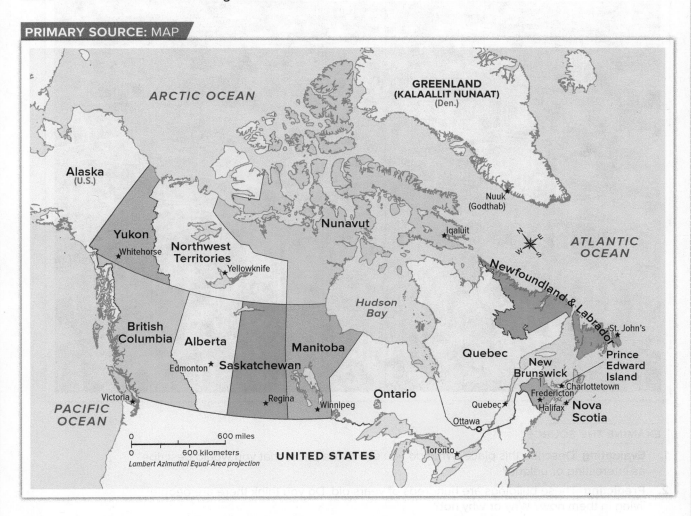

Inuit an indigenous people of northern Canada

EXAMINE THE SOURCE

1. **Identifying** How are the provinces and territories of Canada legally distinguished from each other?

2. **Contrasting** Describe some differences between the provinces of Saskatchewan and Prince Edward Island.

The Old City of Sanaa, Yemen

Yemen is a country on the Arabian Peninsula, and Sanaa is its largest city. Like many cities in the region, Sanaa has been inhabited for over a thousand years. Many people are interested in preserving a section of Sanaa known as the Old City, shown in this photograph.

PRIMARY SOURCE: PHOTOGRAPH

EXAMINE THE SOURCE

1. **Evaluating** Describe this place. What do you see in the photo that you would describe as interesting or unique?

2. **Predicting** These buildings are hundreds of years old. Do you think there are people living in them now? Why or why not?

Tivoli Gardens

Tivoli Gardens is a 20-acre attraction in Denmark's capital city of Copenhagen. It was created for the enjoyment of people. Specific features were included in the design to create a unique type of environment. This article talks about why Tivoli Gardens has drawn visitors for over 175 years.

Tivoli Gardens features numerous rides and attractions.

PRIMARY SOURCE: ARTICLE

66 Copenhagen's best-known attraction, conveniently next to its main train station, attracts an astounding 4 million people from mid-April to mid-September. Tivoli is more sophisticated than a mere amusement park: among its attractions are a **pantomime** theater, an open-air stage, 38 restaurants (some of them very elegant), and frequent concerts, which cover the spectrum from classical to rock to jazz. Fantastic flower exhibits color the lush gardens and float on the swan-filled ponds.

The park was established in the 1840s, when Danish architect George Carstensen persuaded a worried King Christian VIII to let him build an amusement park on the edge of the city's fortifications, rationalizing that "when people amuse themselves, they forget politics." The Tivoli Guard, a youth version of the Queen's Royal Guard, performs every day. Try to see Tivoli at least once by night, when 100,000 colored lanterns illuminate the Chinese pagoda and the main fountain. Some evenings there are also fireworks displays. 99

—from *Fodor's Essential Scandinavia*, 2009

pantomime a performance that tells a story without words by using body movements and facial expressions

EXAMINE THE SOURCE

1. **Analyzing Points of View** What kinds of experiences do you think visitors to Tivoli are seeking?

2. **Summarizing** What information does the source provide regarding the site and situation of Tivoli Gardens?

Frederick Douglass National Historic Site

The National Park Service preserves natural and cultural resources in the United States. The mission of that federal department is to help protect the **heritage** of the country. This article from the National Park Service describes Frederick Douglass National Historic Site in Washington, D.C.

PRIMARY SOURCE: ARTICLE

" Frederick Douglass National Historic Site was established by Congress on September 5, 1962... This site at Cedar Hill, where Douglass lived from 1877 until his death in 1895, honors the life and legacy of Frederick Douglass, the most recognizable African American public spokesman of his time...

Douglass spent his career as an advocate not only for the rights of minorities and the **abolition** movement, but also for women's **suffrage** and **temperance**. His life triumphs were many: abolitionist, women's rights activist, author, owner-editor of antislavery newspapers, United States Minister to Haiti, and the most respected African American orator of the 1800s.

The Frederick Douglass Home, Cedar Hill, is a 14-room house with associated outbuildings that was constructed in the 1850s. The home is situated on a 14-acre tract on a hill with a commanding view of the capital city. "

---

heritage something of value to a group or society that can be passed to future generations
abolition the end of slavery
suffrage voting rights
temperance avoidance of alcoholic drinks

EXAMINE THE SOURCE

1. **Explaining** What action ensured that Frederick Douglass's home and the property on which it sits would be preserved?
2. **Contrasting** Why was this place more significant in 1962 than it was in 1850?

Complete Your Inquiry

EVALUATE SOURCES AND USE EVIDENCE

Refer back to the Compelling Question and the Supporting Questions you developed at the beginning of the lesson.

1. **Determining Context** Which source describes the largest place, by area?
2. **Making Connections** Which sources in this lesson provide information about a place that is located within a city?
3. **Gathering Sources** Which sources helped you answer the Supporting Questions and the Compelling Question? Which sources, if any, challenged what you thought you knew when you first created your Supporting Questions? What information do you still need in order to answer your questions? What other viewpoints would you like to investigate? Where would you find that information?
4. **Evaluating Sources** Identify the sources that helped answer your Supporting Questions. How reliable is the source? How would you verify the reliability of the source?

COMMUNICATE CONCLUSIONS

5. **Collaborating** Share with a partner which of these sources you found most useful in answering your Supporting Questions. Discuss how a place you know well is similar to one of the places introduced in the sources. How do these sources support your thinking? Use the graphic organizer that you created at the beginning of the lesson to help you. Share your conclusions with the class.

TAKE INFORMED ACTION

Creating an Informative Video Think about the area where you live and identify a specific place that is special to you for some reason. This could be a place that is named and known to others or a new place that you want to identify and bring to others' attention. Create a 30-second commercial that can be shared on social media or a school website to educate others about the importance of this place.

National Park Service. Foundation Document Overview: Frederick Douglass National Historic Site. 2021.

04
The Geographer's Tools

Analyzing Key Ideas and Details As you read, take notes about how geographers use globes, maps, and location to learn about the world.

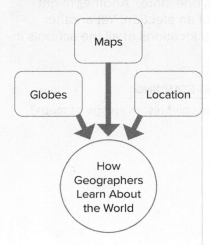

GUIDING QUESTION
What is the difference between globes and maps?

Making and using maps is a big part of geography. Of course, geographers make maps that have many parts. Their maps are more detailed than your mental map. Still, paper maps are essentially the same as your mental map. Both are a way to picture the world and show where things are located.

The most accurate way to show places on Earth is with a globe. Globes are the most accurate because globes, like Earth, are spheres; that is, they are shaped like a ball. As a result, globes most closely represent the relative sizes and shapes of land and bodies of water. They also show distances and directions between places more accurately than flat images of Earth.

The Equator and the Prime Meridian each divide Earth in half. Each half of Earth is called a **hemisphere**. The Equator divides Earth into two halves called the Northern and Southern Hemispheres. The Prime Meridian, together with the International Date Line, splits Earth into two halves called the Eastern and Western Hemispheres.

hemisphere each half of Earth

Earth's Hemispheres
A set of imaginary lines divides Earth into hemispheres.

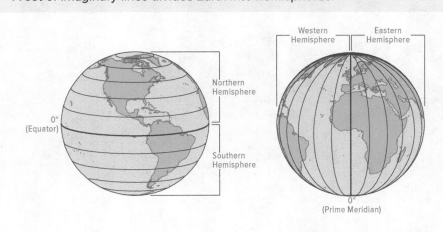

EXAMINE THE MAP

1. **Spatial Thinking** What line divides Earth into Eastern and Western Hemispheres?

2. **Spatial Thinking** What line divides Earth into Northern and Southern Hemispheres?

Maps are not round like globes. Instead, maps are flat representations of the round Earth. They might be sketched on a piece of paper, printed in a book or poster, or displayed on a computer screen.

Maps **convert**, or change, a round surface into a flat surface. As a result, maps **distort** physical reality, or change it so that it is no longer accurate. This is why maps are not as accurate as globes are, especially maps that show large areas or the whole world.

Despite this distortion problem, maps have several advantages over globes. Globes may be used to show the whole planet. Maps, though, often show only a part of it, such as one country, one city, or one mountain range. As a result, maps can provide more detail than globes can. Think about how large a globe would have to be to show the streets of a city. You could certainly never carry such a globe around with you. Maps are especially useful if you want to study a small area. They can focus on just that area, and they are quite accurate. In addition, they are easy to store and carry.

Maps show more kinds of information than globes. Globes generally show major physical and human features, such as landmasses, bodies of water, the countries of the world, and the largest cities. They cannot show much else without becoming too difficult to read or too large. Some maps show these same features, but maps can also be specialized. One map might illustrate a single mountain range or focus on the railroad networks in one state. Another might display the results of an election. Yet another map could show the locations of all the schools in a certain city.

✓ **CHECK FOR UNDERSTANDING**

Analyzing What is the chief disadvantage of maps?

---

convert to change from one thing to another
distort to pull or change something so it is no longer accurate

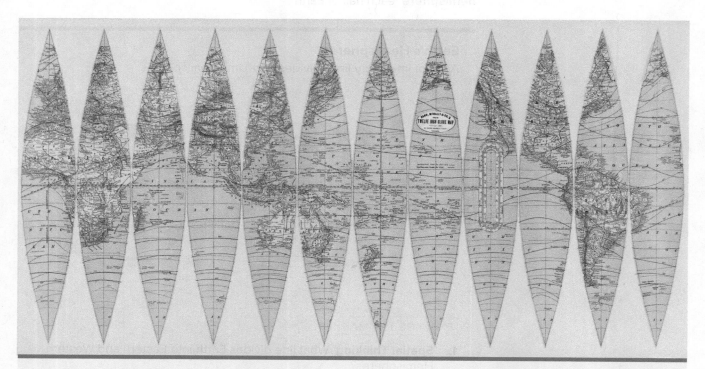

This image shows the flattened pieces of a globe. Because they have been cut, the map is interrupted. To create maps that are not interrupted, mapmakers use mathematical formulas to transfer information from the three-dimensional globe to a two-dimensional map.

The Global Grid

Lines of latitude and longitude create a grid system on Earth's surface.

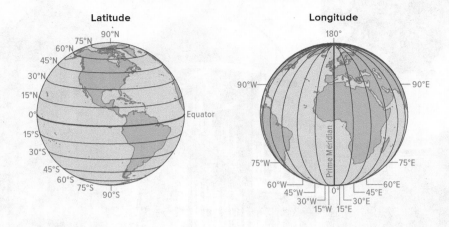

Latitude

Longitude

EXAMINE THE MAP

1. **Spatial Thinking** What kinds of lines circle Earth parallel to the Equator and measure the distance north or south of the Equator in degrees?
2. **Spatial Thinking** What kinds of lines circle the Earth from the North Pole to the South Pole and measure distances east or west of the Prime Meridian?

Determining Location

GUIDING QUESTION

How is location determined?

Geography addresses the question of *where*. To answer this question, a geographer identifies a location. Location in geography is where something is found on Earth. There are two types of location. **Relative location** describes where a place is compared to other places. This approach often uses the cardinal directions—north, south, east, and west. A school might be on the east side of town. Relative location can also tell us about the characteristics of a place. For example, knowing that New Orleans is near the mouth of the Mississippi River helps us understand why the city became an important trading port.

Absolute location is the exact location of something. An address like 123 Main Street is an absolute location. Geographers identify the absolute location of places using a system of imaginary lines called latitude and longitude.

Those lines form a grid for locating a place precisely. For example, the absolute location of Tokyo, Japan, is near 36° N, 140° E. Both globes and maps use a grid system in order to form a pattern of lines that cross one another.

Lines of **latitude**, also known as parallels, run east to west, but they measure distance on Earth in a north-to-south direction. One of these lines, the Equator, circles the middle of Earth. This line is equally distant from the North Pole and the South Pole. Other lines of latitude circle the Earth parallel to the Equator. These lines between the Equator and the North Pole and Equator and the South Pole are assigned a number from 0° to 90°. The higher the number, the farther the line is from the Equator. The Equator is 0° latitude. The North Pole is at 90° north latitude (90° N), and the South Pole is at 90° south latitude (90° S).

Lines of **longitude** run from north to south, but they measure distance on Earth in an

---

relative location location in relation to other places

absolute location the exact position of a place on Earth's surface, often determined by latitude and longitude

---

latitude the lines on a map or globe that run east to west but measure distance on Earth in a north to south direction

longitude the lines on a map or globe that run north to south but measure distance on Earth in an east to west direction

The Royal Greenwich Observatory was founded in 1675 by King Charles II of England. Its purpose was to aid in navigation. The Greenwich meridian was adopted as Earth's Prime Meridian (0° longitude) in 1884.

east-to-west direction. They go from the North Pole to the South Pole. These lines are also called *meridians*. The Prime Meridian is the starting point for measuring longitude. It runs through Greenwich, England, and has the value of 0° longitude. There are 180 lines of longitude to the east of the Prime Meridian and 180 lines to the west. They meet at the meridian 180, which is called the International Date Line. All countries on Earth have agreed that this is the line that separates one day from the next.

Geographers use the global grid of latitude and longitude to locate anything on Earth. In stating absolute location using this system, geographers always list latitude first. For example, the absolute location of Washington, D.C., is 39° N, 77° W.

✓ CHECK FOR UNDERSTANDING

Explaining How are latitude and longitude are used to determine absolute location?

LESSON ACTIVITIES

1. **Informative/Explanatory Writing** Conduct research to find out more about the International Date Line. Then write a brief essay explaining its significance and its relationship to time zones.

2. **Collaborating** Work with a partner to create an outline of this lesson to be used as a study guide. Your outline should include main ideas and supporting details. You may want to add sketches, maps, or photos to help present some of the information covered in the lesson.

Glow Images

05
Maps and Map Projections

READING STRATEGY

Analyzing Key Ideas and Details As you read, take notes about the parts of a map and explain what each part shows.

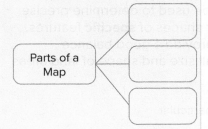

All About Maps

GUIDING QUESTION

How do maps work?

You can find maps in many places. You see them in a subway station, where subway maps indicate the routes each train takes. In a textbook, a map might show new regions that were added to the United States at different times in history. At a company's website, a map can display the locations of all its stores in a city. The map of a state park tells visitors what activities they can enjoy in each area of the park. Each of these maps is different from the others, but they have some traits in common.

Parts of a Map

Maps have several important elements, or parts. These elements are tools that convey important information. The *map title* tells what area the map will cover. It also identifies what kind of information the map presents about that area. The *key* unlocks the meaning of the map by explaining its symbols, colors, and lines. The *scale bar* is an important part of the map. It tells how a measured distance on the map corresponds to actual distance on Earth. For example, by using the scale bar, you can determine how many miles in the real world each inch on the map represents. The *compass rose* shows direction. North, south, east, and west are the four cardinal directions. The intermediate directions—northeast, northwest, southeast, and southwest—may also be shown. Some maps include insets that show more detail for smaller areas, such as cities on a state map. Many maps show latitude and longitude lines to help you locate places.

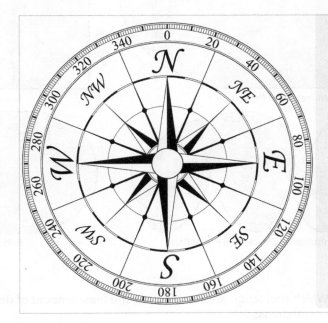

The compass rose indicates direction or orientation of a map. The compass rose looks like intersecting arrows or points of a star.

Volgarud/Shutterstock

Map Projections

To convert the round Earth to a flat map, geographers use map projections. A **map projection** distorts some aspects of the round Earth in order to represent other aspects as correctly as possible on a flat map. Some projections show the correct size of areas in relation to one another. Other map projections emphasize making the shapes of areas as **accurate** as possible.

Mapmakers, known as *cartographers*, choose which projection to use based on the purpose of the map. In addition to distorting shapes and sizes, each projection distorts the distances between things on some parts of the globe more or less than other parts. Cartographers think about what part of Earth they are drawing and how large an area they want to cover.

Some projections, like Goode's Interrupted Equal-Area projection, break apart the world's oceans. By doing so, these maps show land areas more accurately. However, distances between land features are generally distorted.

A Mercator projection is most accurate at the Equator because shapes and distances are increasingly distorted moving toward the Poles. However, a Mercator projection displays true direction, and this makes it useful for navigation on the sea by ships and in the air by airplanes.

Many world maps used for general reference utilize the Winkel Tripel projection. This map projection cannot be used to determine precise distances, sizes, or shapes of **specific** features. It does, however, provide a good balance between the overall size and shape of land areas shown.

map projection the method used to represent the round Earth on a flat map

accurate without mistakes or error

specific special or particular

Common Map Projections

Each type of map projection has its advantages as well as some degree of inaccuracy.

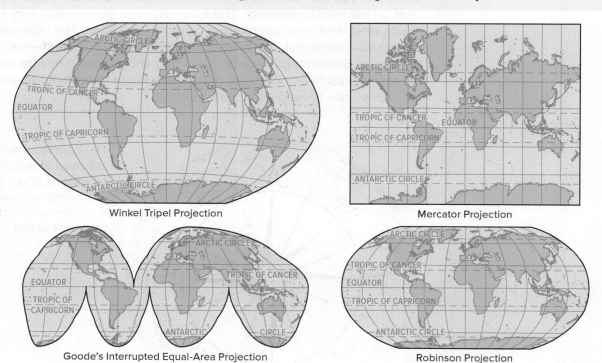

Winkel Tripel Projection

Mercator Projection

Goode's Interrupted Equal-Area Projection

Robinson Projection

EXAMINE THE MAP

1. **Spatial Thinking** Which projection appears to have the least amount of distortion of distances and size of landmasses?

2. **Spatial Thinking** Which projection is preferred for sea navigation? Why?

Political Map of the United States

This political map of the United States shows the boundaries and locations of political units such as states and cities.

EXAMINE THE MAP

1. **Exploring Location** How are state capitals depicted on this map?
2. **Spatial Thinking** How can you tell that a river forms the boundary between Indiana and Kentucky but not between Ohio and Indiana?

The Robinson projection looks similar to a Winkel Tripel projection, although its lines of latitude run in a straight line. The Robinson projection produces minor distortions, especially in the polar areas. The sizes and shapes near the eastern and western edges of the map are accurate, and outlines of the continents appear much as they do on the three-dimensional globe.

The Gall-Peters projection shows all areas according to their true sizes. This makes it easy for accurate comparisons of size. However, this projection stretches the shape of land masses.

Map Scale

Scale is another important feature of maps. **Scale** is the relationship between distances on the map

and on Earth. As you learned, the scale bar relates distances on the map to actual distances on Earth. The scale bar is based on the scale at which the map is drawn.

Maps are either *large scale* or *small scale*. A large-scale map focuses on a smaller area. An inch on a large-scale map might correspond to 10 miles (16 kilometers) on the ground. A small-scale map shows a relatively larger area. An inch on a small-scale map might be the same as 1,000 miles (1,609 kilometers) on the ground.

Each type of scale has benefits and drawbacks. Which scale to use depends on the map's purpose. Do you want to map your school and the streets and buildings near it? Then you need a large-scale map to show this small area in

---

scale the relationship between distances on the map and on Earth

great detail. Do you want to show the entire United States? In that case, you need a small-scale map that shows that larger area but with less detail.

✓ **CHECK FOR UNDERSTANDING**

Contrasting What is the difference between large-scale maps and small-scale maps?

Types of Maps

GUIDING QUESTION

What are the main types of maps?

There are three main types of maps: general purpose, thematic, and mental. The type depends on what kind of information is on the map. General-purpose maps are often referred to as reference maps. They show a wide range of information about an area. They generally show either the human-made features of an area or its physical features.

Political maps are one common type of general-purpose map that shows human-made features. They show the boundaries of countries or the divisions within them, like the states of the United States. They can also show the locations and names of cities. Physical maps are another common type of general-purpose map. They display natural features such as mountains, valleys, rivers, and lakes. They show the location, size, and shape of these features. Many physical maps show **elevation**, or the height of land above or below sea level. Maps often use colors to present this information. A key on the map explains what height above or below sea level each color represents.

Some physical maps show the **topography**, or shape of the land. This involves the **relief** of the land, or the vertical elevation changes that exist in a landscape. Shape and relief are generally shown by using *contour lines* that represent the elevation of mountains, hills, valleys, plains, and the like. Topographic maps often show human-made features such as buildings, cities and towns, roads, railroads, canals, dams, bridges, tunnels, and parks.

Thematic maps are a type of map that shows more specialized information. A **thematic map** emphasizes a particular theme or subject. For example, a map might indicate the kinds of plants that grow in different areas. That kind of map is a vegetation map. Another could show where farming, ranching, or mining takes place. That is called a land-use map. Road maps show people how to travel from one place to another by car. Just about any physical or human feature can be displayed on a thematic map.

A third type of map is a **mental map**. These are maps that people have in their minds. They are based on the understandings and perceptions people have of features on Earth's surface. Mental maps exist at local, regional, and global scales. They can include information from the point of view of a person within a classroom or a house. They can also apply to the view one has from an airplane. Mental maps can change according to a person's experiences and perceptions of people and places.

✓ **CHECK FOR UNDERSTANDING**

Explaining What is a general-purpose map?

LESSON ACTIVITIES

1. **Argumentative Writing** Which map projection discussed in this lesson is best suited for a general reference map of the world's countries? Write a paragraph supporting your answer using details from the lesson.

2. **Interpreting Information** Work with a partner or small group to find three examples of general-purpose maps and three examples of thematic maps. Study the maps to identify the type of information on each. Then create a table describing each map and explaining one situation in which the map could be used to solve a problem.

---

elevation the height of a land surface above or below sea level

topography the shape of the land on Earth's surface

relief the vertical elevation change that exists in a landscape

thematic map a map that shows specialized information

mental map a map that people have in their minds that is based on the understandings and perceptions they have of features on Earth's surface

Geospatial Technologies

READING STRATEGY

Analyzing Key Ideas and Details As you read, take notes about the types and uses of geospatial technologies.

Technology	Uses

What Are Geospatial Technologies?

GUIDING QUESTION

How are geospatial technologies used to learn about the world?

Have you seen maps on smartphones? On GPS devices in cars? Have you seen satellite images on television? These electronic maps and images are examples of geospatial technologies. **Technology** is any way that scientific discoveries are applied to practical use. Geospatial technologies help us think spatially. They provide information about the locations and interactions of physical and human features.

Global Positioning Systems

A **global positioning system** (GPS) is used to determine the exact, or absolute, location of something on Earth. Made up of a network of satellites and receiving instruments, GPS provides an accurate location with respect to latitude and longitude. GPS technology in the United States relies on a system of 24 satellites. The European Union, as well as some individual countries such as Russia and China, have their own satellites that support GPS systems. This U.S. network was built by the government. Some parts of it can be used only by the U.S. armed forces. Other parts, though, can be used by ordinary people all over the world. The GPS has three elements.

technology any way that scientific discoveries are applied to practical use

global positioning system a navigational system that can determine absolute location by using satellites and receivers on Earth

Biologists use GPS technology to track the movement and behavior of animals in the wild.

1. **Drawing Conclusions** How might wildlife biologists use GPS technology to protect endangered species such as the African lion?

2. **Inferring** What other fields, or lines of work, use GPS technology to gather information?

The first **element** of this network consists of satellites that orbit Earth constantly. The U.S. government launched the satellites into space and maintains them. The satellites send out radio signals. Almost any spot on Earth can be reached by signals from at least four satellites at all times.

The second part of the network is the control system. Workers around the world track the satellites to make sure they are working properly and are on course. The workers reset the clocks on the satellites when needed.

The third part of the GPS system consists of GPS devices on Earth. These devices receive the signals sent by the satellites. By combining the signals from different satellites, the receiving device calculates its location on Earth in terms of latitude and longitude. The more satellite signals the device receives at any time, the more accurately it can determine its location. Because satellites have accurate clocks, the GPS device also displays the correct time.

GPS is used in many ways. It is used to track the exact location and course of airplanes. That information helps ensure the safety of flights. Farmers use it to help them work their fields. Businesses use it to guide truck drivers. Cell phone companies use GPS to provide services. And, of course, GPS in cars helps guide us to our destinations.

Geographic Information Systems

Advances in technology have changed the way maps are made. An important tool in mapmaking today involves computers with software programs called **geographic information systems** (GIS). But more than simply making maps, GIS can be used to perform advanced spatial and geographical analysis.

A GIS is a powerful tool because it links data about all kinds of physical and human features with the locations of those features. Because computers can store and process large amounts of data, the GIS is accurate and detailed. Maps can be created—and changed—quickly and easily to display various types of information with a single map.

People select what features they want to study using the GIS. Then they can combine different features on the same map and analyze the patterns. For instance, a farmer might want to compare the amount of moisture in the soil to the health of the plants. At the same time, he or she could add soil types around the farm to the comparison. The farmer could then use the results of the analysis to answer all kinds of questions: What plants should I plant in different locations? How much irrigation water should I use? How can I drive the tractor most efficiently?

---

element a particular part of something

geographic information systems computer programs that process and organize details about places on Earth

GIS helps scientists at the National Oceanic and Atmospheric Administration (NOAA) forecast weather, issue severe-storm warnings, and monitor climate.

This satellite image of Earth at night reveals some stark differences among the regions of the world.

1. **Evaluating Sources** What does this image tell you about North America? about Africa?
2. **Analyzing** How might the information from this image be used by geographers?

GIS is used in a range of fields, including environmental and urban planning, the military, marketing products, establishing stores and other retail businesses, environmental protection, and many other professions.

Satellites

Since the 1970s, satellites have gthered data about Earth's surface. They do so using remote sensing. **Remote sensing** simply refers to acquiring information about an object without being in full contact with it. Aerial photographs were originally used to get information about Earth, and they are still used today. But satellites have become very important tools in the remote-sensing process. Most early satellite sensors were used to gather information about the weather. Weather satellites helped save lives during disasters by providing warnings about approaching storms. Before satellites, tropical storms were often missed because they could not be tracked over open water.

Satellites gather information in different ways. They may use powerful cameras to take photographs of the land. They can also pick up other kinds of information, such as the amount of moisture in the soil, the amount of heat the soil holds, or the types of vegetation that are present. In the early 2000s, scientists used satellites and GIS technology to help conserve the plants and animals that lived in many places on Earth, including the Amazon rainforest. Using the technology, scientists can compare data gathered on the ground to data taken from the satellite images. Land-use planners use this information to help local people make good decisions about how to use the land. These activities help prevent degradation or destruction of environments such as the rainforest.

Some satellites gather information regularly for every spot in the world. That way, scientists can compare the information from one year to another. They look for changes in the shape of the land or in its makeup, spot problems, and take steps to fix them.

✓ **CHECK FOR UNDERSTANDING**

1. **Summarizing** What are the three elements of a GPS?
2. **Analyzing** How does GIS allow maps to be created and changed quickly and easily?

remote sensing the method of getting information from far away, usually with aircraft or satellites

Limits of Geospatial Technology

What can geospatial technology provide?

Geospatial technologies are excellent sources of information because they provide actual images and data related to a location and can provide a great amount of detail. While scientists can use observational and historical data to gather information about a place, geospatial technology acts as a primary source for compiling raw data.

Geospatial technologies are a relatively new innovation in comparison to traditional forms of mapmaking. They are constantly changing, and they will continue to improve with the advancement of computer, aerospace, and Internet-based technology. Continued development in geospatial technology offers possibilities for its use by government, private industry, scientists, and the general public. Because the economic, cultural, and political activities of the world's regions have become increasingly interconnected, information related to the world's physical and human systems needs to be readily available, consistent, and up to date. The combination of mental mapping with GPS, GIS, and aerial imagery can create a very detailed picture of places and regions.

Geospatial technologies allow **access** to a wealth of information about the features and objects in the world and where those features and objects are located. This information can be helpful for identifying and navigating. By itself, however, such information does not help much in answering questions about why features and objects are located where they are or about "why we should care" if we can identify the locations of features and objects. But these questions lie at the heart of understanding and making decisions about the world in which we live. It is important to go beyond the information provided by geospatial technologies and build understanding of peoples, places, and environments and the connections among them. Only then is the information from the geospatial technologies most useful.

This pair of images show the New Jersey coast before (bottom) and after (top) Hurricane Sandy made landfall in 2012. Geospatial technology helps authorities assess the damage after a natural disaster.

✓ **CHECK FOR UNDERSTANDING**

Describing What are the limitations of geospatial technology?

LESSON ACTIVITIES

1. **Informative/Explanatory Writing** How could GIS help business owners make better decisions? Write a brief essay describing at least two ways in which a person could use GIS to improve their business.

2. **Analyzing Information** NASA is an excellent source for research. Work with a small group to find NASA weather satellite images online. Choose two images and discuss the data and information the images provide. Evaluate the importance of the information. Then share your images and analyses with the class.

access a way to get information

07

Analyzing Sources: Acquiring and Using Geographic Information

? COMPELLING QUESTION

How do geographic tools help us make decisions?

Plan Your Inquiry

DEVELOPING QUESTIONS

Think about the reasons you might have for using a map or another source of geographic information. Then read the Compelling Question for this lesson. What questions can you ask to help you answer this Compelling Question? Create a graphic organizer like the one below. Write these Supporting Questions in your graphic organizer.

Supporting Questions	Source	What this source tells me about how geographic tools help us make decisions	Questions the source leaves unanswered
	A		
	B		
	C		
	D		
	E		

ANALYZING SOURCES

Next, examine the sources in this lesson. Analyze each source by answering the questions that follow it. How does each source help you answer each Supporting Question you created? What questions do you still have? Write these in your graphic organizer.

After you analyze the sources, you will
- use the evidence from the sources;
- communicate your conclusions;
- take informed action.

Background Information

Geographers develop techniques to collect, organize, and communicate information about the world. The information may be about the natural environment, how people change and interact with the environment, or even about cultures themselves. Geographers collect information using different methods of observation. They can also access information others have gathered and shared through maps, stories, and other types of geographic representations.

People analyze geographic information to help them understand the world around them and make decisions. Governments, businesses, organizations, and individuals all use geographic information in different ways.

» *The Geographer*, painted by Dutch artist Johannes Vermeer in 1668–69, depicts a scholar studying maps with a globe in the background.

Airline Route Map

Airlines are businesses that provide transportation as a service. They maintain a fleet of airplanes and operate scheduled flights between airports. This map shows the routes of one airline, Western Airways.

PRIMARY SOURCE: MAP

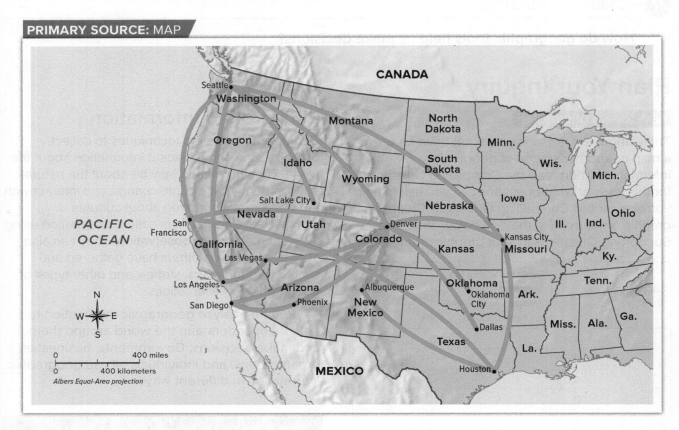

EXAMINE THE SOURCE

1. **Identifying** Name a city that is connected to only one other city. Name the city that serves as the major hub for the airline, where the greatest number of routes meet.

2. **Making Connections** How could you travel on this airline between Salt Lake City and Oklahoma City? List the cities visited in order from first flight to last flight.

Rainfall in Africa

Rainfall is an important source of water for drinking and for growing crops. The amount of rainfall in a location helps determine the type of vegetation that can grow there. The average amount of rainfall for an area indicates that area's long-term climate pattern. However, an area can have a single year in which rainfall is either far above or far below average. The map below shows the average rainfall across the continent of Africa.

SECONDARY SOURCE: MAP

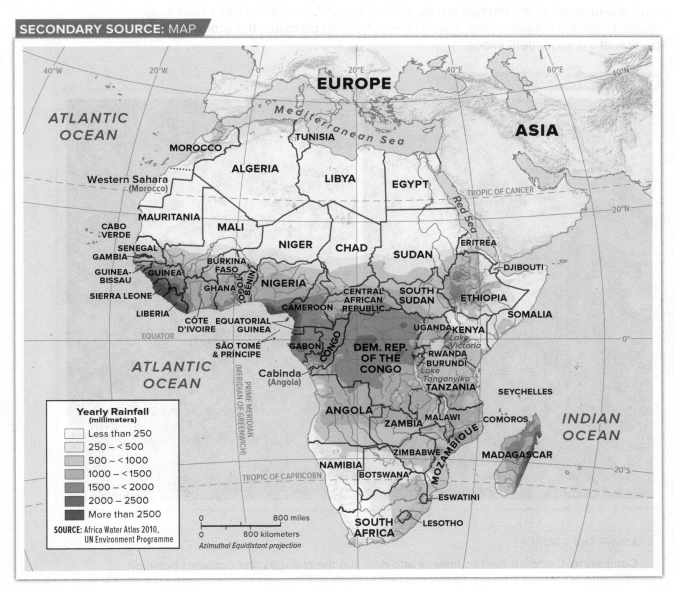

EXAMINE THE SOURCE

1. **Identifying** In which parts of Africa would you likely find drifting sand dunes with very little plant life?

2. **Drawing Conclusions** In which parts of Africa would it make sense to plant crops that need a lot of water to grow?

Remote Sensing

Remote sensing is a scientific technique in which geographers make observations of Earth from above. They use an airplane or satellite to capture images of Earth. Remote sensing involves photography or other kinds of technology such as radar that provide information about conditions on the land. These two images of the same area of Brazil were captured by satellite at two different times, in 1986 and 2001. The dark-green areas of the images are dense rain forest. Land where trees have been cleared appears a lighter color. A small area of clouds obscures the ground in a portion of the older (left) image.

PRIMARY SOURCE: PHOTOGRAPH

1986   2001

EXAMINE THE SOURCE

1. **Comparing** Compare the two images and describe the changes that occurred between 1986 and 2001.

2. **Speculating** In what portion of the area represented in the 2001 photo could you establish a preserve to protect species native to the rain forest?

Lyme Disease

Lyme disease is the most common **vector-borne** disease in the United States. It is transmitted to humans through the bite of infected ticks. Most cases of Lyme disease can be treated successfully with antibiotics, but serious health problems can occur if it is left untreated. This map showing reported cases of Lyme disease in one year is from the Centers for Disease Control and Prevention, a United States federal agency.

PRIMARY SOURCE: MAP

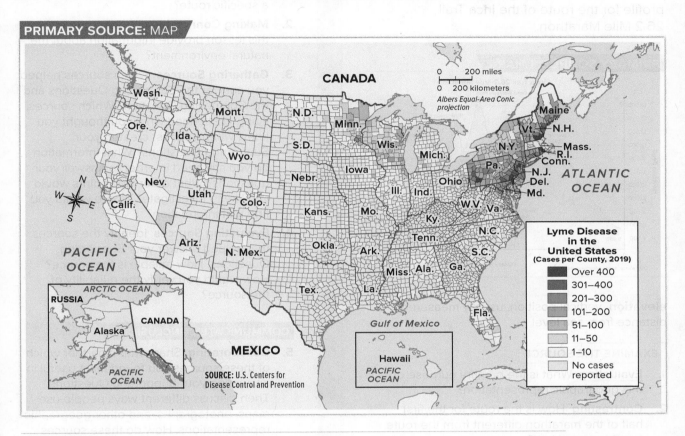

vector-borne spread to humans through insect bites

EXAMINE THE SOURCE

1. **Assessing credibility** Do you think this map accurately reflects the threat of Lyme disease for the whole country? Why or why not?

2. **Explaining** How could you use this map to choose a good location for a new center to train doctors to identify the signs of Lyme disease?

The Inca Trail Marathon

A marathon is a foot race that covers 26.2 miles (42.1 km). Professional and amateur athletes both participate in this type of race. Marathons are held in many parts of the world. The chart below shows an **elevation** profile for the route of the Inca Trail 26.2 Mile Marathon.

PRIMARY SOURCE: CHART

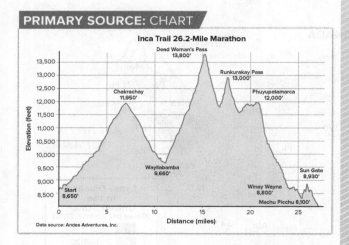

Inca Trail 26.2-Mile Marathon

Data source: Andes Adventures, Inc.

elevation vertical position, usually measured in distance from sea level

EXAMINE THE SOURCE

1. **Evaluating** What is the primary purpose of this geographic representation?

2. **Contrasting** How is the route for the first half of the marathon different from the route for the second half?

Complete Your Inquiry

EVALUATE SOURCES AND USE EVIDENCE

Refer back to the Compelling Question and the Supporting Questions you developed at the beginning of the lesson.

1. **Determining Context** Which source(s) would help someone plan for travel along a specific route?

2. **Making Connections** Which source(s) in this lesson provide information about the natural environment?

3. **Gathering Sources** Which sources helped you answer the Supporting Questions and the Compelling Question? Which sources, if any, challenged what you thought you knew when you first created your Supporting Questions? What information do you still need in order to answer your questions? What other viewpoints would you like to investigate? Where would you find that information?

4. **Evaluating Sources** Identify the sources that helped answer your Supporting Questions. How reliable is the source? How would you verify the reliability of the source?

COMMUNICATE CONCLUSIONS

5. **Collaborating** Share with a partner which of these sources you found most useful in answering your Supporting Questions. Then discuss different ways people use maps, photographs, and other geographic representations. How do these sources support your thinking? Use the graphic organizer that you created at the beginning of the lesson to help you. Share your conclusions with the class.

TAKE INFORMED ACTION

Mapping Your School Think about how the map of your school grounds could provide useful information for students who are new to the school. What features at your school should be included? Create a printed or digital map of the school campus or of one specific section of the campus or building. Share the map with others at your school to help new students make decisions about how to navigate around the school.

08

Reviewing The World in Spatial Terms

Summary

Thinking Like a Geographer

Geographers study the location and relationships of Earth's physical and living features. They think spatially, looking for links and patterns between people, places, and environments. Geographers consider six elements in their work—the world in spatial terms, places and regions, human systems, physical systems, environment and society, and the uses of geography.

The Geographer's Tools

- The Equator divides Earth into the Northern and Southern Hemispheres.

- The Prime Meridian, together with the International Date Line, divides Earth into the Eastern and Western Hemispheres.

- Maps distort shape, size, and distance, but they can show more types of information than globes.

- Globes usually show major physical and human features, such as landmasses, bodies of water, the countries of the world, and the largest cities.

- Absolute location is determined using latitude and longitude.

- Relative location describes where a place is compared to another place.

Maps and Map Projections

- The parts of a map include the title, key, scale bar, compass rose, and latitude and longitude lines.

- Map projections distort some aspects of the round Earth in order to represent other aspects as correctly as possible on a flat map.

- Common map projections are Goode's Interrupted Equal-Area, Mercator, Winkel Tripel, Robinson, and Gall-Peters.

- Scale is the relationship between distances on the map and on Earth.

- The two main types of maps are general-purpose and thematic.

- A mental map is a map that people have in their minds.

Geospatial Technologies

- Geospatial technologies provide information about the locations of physical and human features.

- Geospatial technologies include global positioning systems (GPS), geographic information systems (GIS), and remote sensing.

- A GPS is used to determine the absolute location of something on Earth. A network of satellites provides a location using latitude and longitude.

- A GIS is a computer program that processes, organizes, and maps details about places on Earth.

- Remote sensing is the method of getting information from far away, usually with aircraft or satellites.

- There are limits to geospatial technologies. Information does not help in answering questions about "why" features and objects are located where they are or about "why care."

Checking For Understanding

Answer the questions to see if you understood the topic content.

IDENTIFY AND EXPLAIN

1. Identify each of these terms as they relate to maps.

   A. map projection
   B. scale
   C. elevation
   D. topography
   E. relief
   F. thematic map
   G. mental map
   H. latitude
   I. longitude

REVIEWING KEY FACTS

2. **Identifying** What is the difference between absolute and relative location?

3. **Describing** List and describe the four common map projections.

4. **Comparing** How are globes and maps related?

5. **Explaining** How is absolute location determined?

6. **Analyzing** How are geospatial technologies used to learn about the world?

7. **Determining Meaning** What is a spatial perspective?

8. **Identifying** What are the essential elements of geography?

9. **Evaluating** How is geography related to other subjects?

CRITICAL THINKING

10. **Describing** Describe the problems that occur when the curves of a globe become straight lines on a map.

11. **Comparing and Contrasting** Explain the similarities and differences between the Winkel Tripel projections and the Mercator projection.

12. **Analyzing** Why is scale important when reading a map?

13. **Identifying** List three examples of things that a map can show.

14. **Evaluating** What are the different types of geospatial technologies? How has this advanced technology improved the way maps are created?

15. **Contrasting** How is mental mapping different from other forms of mapping?

16. **Describing** Describe three ways that geographic knowledge and skills are used in jobs other than geographer.

17. **Determining Central Ideas** Why do geographers study more than a place's location and dimensions?

18. **Describing** Why do maps distort the way Earth's surface really looks?

19. **Summarizing** What are the parts of a map? How do they help you read a map?

20. **Contrasting** How do the two main types of maps differ?

NEED EXTRA HELP?

If You've Missed Question	1	2	3	4	5	6	7	8	9	10
Review Lesson	4, 5	4	5	4	2	6	2	2	2	4

If You've Missed Question	11	12	13	14	15	16	17	18	19	20
Review Lesson	5	4	5	6	5	2	2	4, 5	5	5

Apply What You Have Learned

A Exploring Careers

The career path of a geographer can head in many directions. Geographers study the physical and human characteristics of Earth, on scales from local to global. This professional skill is useful in a wide variety of workplaces, from government agencies to educational institutions to private corporations.

ACTIVITY Writing an Employment Advertisement Create an advertisement for a company or organization that is seeking a geographer. Have the advertisement specify the type of geographic knowledge and the skills and experience the job seeker must have. Alternatively, create an advertisement for a position with another job title, specifying how that position requires skills related to those of a geographer. The advertisement should be 150–200 words in length.

B Geographic Reasoning

Geographers and cartographers use a grid to give each place on Earth an absolute location, using lines of latitude and longitude. Every point has coordinates on that grid. Using this system, if you have the latitude and longitude coordinates of a place, you can easily find that place on a map. For example, the city of St. Louis, Missouri, is located at 38.62° N (north of the Equator), 90.18° W (west of the Prime Meridian).

ACTIVITY Finding Locations Using Coordinates Display a map of Earth showing political boundaries and major cities as well as physical features. Working with a partner, use a graphic organizer like the one below to create a list of five features like cities, mountains, lakes, archaeological sites, or other points on the map. For each item, list its latitude and longitude coordinates on a second piece of paper. Then exchange latitude/longitude lists with another pair of students. Find the location of each place on the list and write the answer on the list. After everyone has finished the search, return the lists to the original partners and check answers.

Geographic Feature	Latitude & Longitude

C Understanding Multiple Perspectives

Map projections display the three-dimensional Earth in a two-dimensional format. All map projections distort the appearance of their displayed areas in some way. On world maps, the areas that most often are distorted are regions near polar areas.

ACTIVITY **Making a Display** Research several map projections that are used to show large areas of Earth. Using these projections, make a display showing a state, country, or landmass in a polar region as it appears on the different projections. Possible areas include Greenland, Alaska, Scandinavia, or Antarctica. Be sure to show both the shape of the place and the scale of its area in your display.

D Making Connections to Today

Global positioning system (GPS) technology is essential to many industries, governments, and researchers. The information it provides makes it possible to access up-to-the-minute information about weather, changes to the environment, human activities, and much more. Had it been available to leaders and explorers of past generations, such information could have changed the course of history in many ways.

ACTIVITY **Imagining Alternate History** Research an event in history that might have turned out differently had today's GPS technology been available. Possibilities include exploration voyages, military campaigns, migrations, or searches for people or objects. Write a short essay explaining how the outcome of the chosen event would have been altered by the use of some component of GPS technology. Think creatively to imagine both short-term and long-term changes to history.

A county fair is a place that has meaning for the local community.

Places and Regions

INTRODUCTION LESSON

01 Introducing Places and Regions   G40

LEARN THE CONCEPTS LESSONS

02 The Concept of Place   G45

04 How Are Regions Defined?   G53

05 Connections Between Places and
    Regions   G57

INQUIRY ACTIVITY LESSONS

03 Understanding Multiple Perspectives:
    Places and Identity   G49

06 Global Issues: Connections Between
    Places   G61

REVIEW AND APPLY LESSON

07 Reviewing Places and Regions   G67

Jim West/Alamy Stock Photo

Introducing Places and Regions

How Do Geographers Organize the World?

Places and regions help us organize the world around us.

What is Place?

Place refers to all of the human and physical characteristics, or traits, that give an area its own special identity and meaning to people.

> **"** A place is a storied landscape, somewhere that has human meaning. . . . places aren't just about people . . . they reflect our attempt to grasp and make sense of the . . . land and its many inhabitants that are forever around and beyond us. **"**
>
> —Alastair Bonnett, *Beyond the Map*

The places you live and go to school are important to you. This sign in North Carolina includes names in the Cherokee language.

Larger places like your town, state, or country are places that have meaning for many people. Charleston, the capital city of West Virginia, has meaning for people across the state.

What is a Region?

A *region* is an area of Earth's surface that has similar physical or human characteristics. Regions may be defined differently depending on what is being studied. You encounter regions in your daily life.

Regions can be

- different sizes;
- defined as an area with similar physical features like mountains;
- defined as an area with similar human features like language.

Regions of the National Football League

This map shows that NFL teams are popular in larger areas than just the cities or states in which they are located. These groups of places are regions.

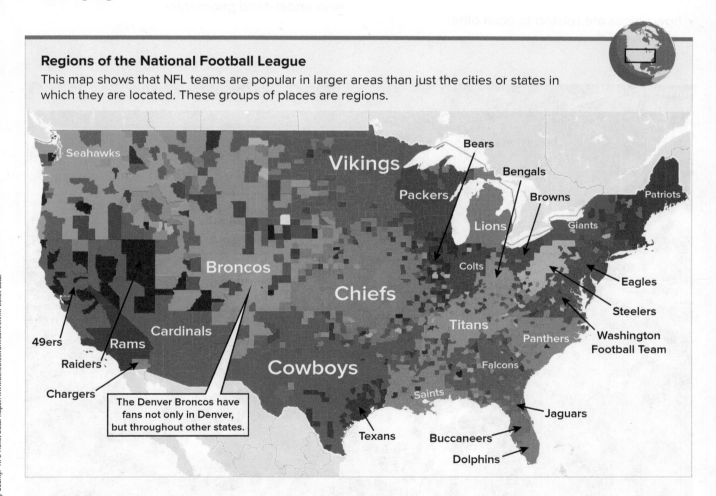

The Denver Broncos have fans not only in Denver, but throughout other states.

TicketNetwork. "Most Popular NFL Teams by County." NFL Ticket Data. https://www.ticketnetwork.com/en/nfl-ticket-data.

Getting Ready to Learn About . . .
Places and Regions

Studying People and Places

Geographers study people and places and their locations in space on Earth. They look at the relationships between people in different places to understand how each place is important. Geographers consider

- the activities of people in a place;

- the physical and human features of a place;

- the ways in which a place has meaning for the people there; and

- how places are related to each other.

Geographers also study how places change over time. They try to understand what causes change and what impact those changes have. For example, what factors made a city grow? Were they human factors? Physical factors? A combination of both? What effect did a growing city have on the people who lived there?

Changes to a place may also affect nearby locations. Geographers may ask: What effect did the city's growth have on nearby communities? On the land and water near it? On the wildlife near it? These are questions asked by people who understand geography.

The holy city of Varanasi, India, and the Ganges River that flows through the city have meaning for Hindus not only in India but around the world.

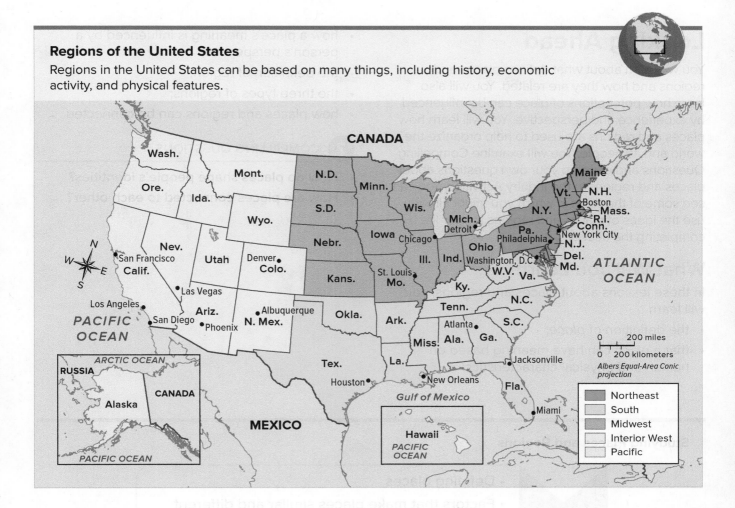

Regions of the United States

Regions in the United States can be based on many things, including history, economic activity, and physical features.

Legend:
- Northeast
- South
- Midwest
- Interior West
- Pacific

Defining Places

A place can be defined by the physical characteristics of its location. For example, the city of New Orleans is located near the mouth of the Mississippi River. Places can also be defined by human characteristics such as history, language, and economic activities. The city of New Orleans has a history influenced by French settlers, Native Americans, and enslaved Africans. The city is an economic hub for the Gulf Coast of the United States.

Places can be defined by people's perceptions, meaning the ways in which they think about or understand a place. Perceptions can come from visiting the place, reading about it, or seeing it in a movie or television show. As a result, one person's perception of a place might be different from someone else's. For example, a person who lives in Orlando, Florida, understands and defines the city differently than someone who is there on vacation.

Identifying Regions

The definition of a region can vary depending on what is being studied. For example, the United States is often divided into regions to examine broad patterns such as trade, population distribution, language, religion, and natural resources. You might be familiar with U.S. regions like the Northeast or the Midwest. Similarly, the world is organized into regions.

Since regions can be defined in a variety of ways, you may see one list of world regions in a geography book but a slightly different list in a history book. This is because regions are being used to organize the world when studying different things and ideas. As mentioned above, geographers may be studying trade or natural resources. Historians may be studying the development and spread of civilizations. Whatever the need, regions allow us to group areas with common characteristics as a way to organize the world and its people.

Looking Ahead

You will learn about what defines places and regions and how they are related. You will also learn how perceptions of place can be influenced by experience and perspective. You will learn how places and regions are used to help organize the world and its people. You will examine Compelling Questions and develop your own questions about places and regions in the Inquiry Activities. You can see some of the ways in which you might already use the ideas of places and regions in your life by completing the activity.

What Will You Learn?

In these lessons about places and regions, you will learn

- the definition of *place*;
- that a place can have meaning based on human and physical characteristics;
- how a place's meaning is influenced by a person's perspective;
- the definition of *region*;
- the three types of regions;
- how places and regions can be connected.

? COMPELLING QUESTIONS

- **How do places shape people's identities?**
- **How are places connected to each other?**

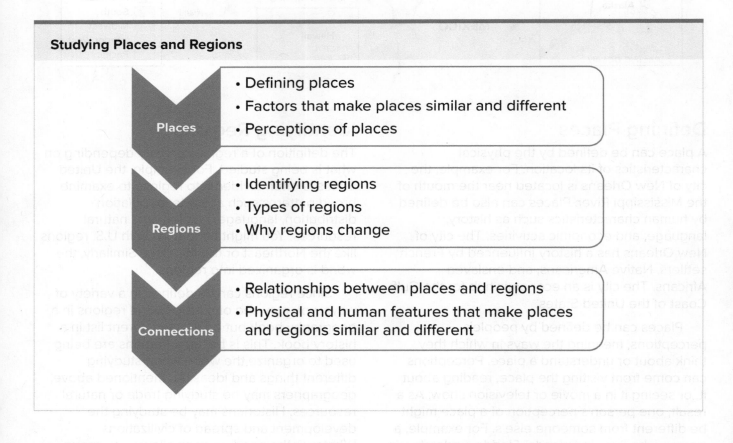

Studying Places and Regions

Places
- Defining places
- Factors that make places similar and different
- Perceptions of places

Regions
- Identifying regions
- Types of regions
- Why regions change

Connections
- Relationships between places and regions
- Physical and human features that make places and regions similar and different

The Concept of Place

READING STRATEGY

Analyzing Key Ideas and Details As you read, take notes about what influences how place is defined.

- Characteristics
- Meaning
- Identity

What Is Place?

GUIDING QUESTION

How is a place defined?

A **place** is a specific location with distinct characteristics. The characteristics of a place could be human or physical, or some combination of the two. Human characteristics include population, culture, settlement, economic activities, and political activities. Physical characteristics include landforms, lakes, rivers, and climate. Examples of places that have a combination of both are cattle herders in semi-arid Kenya and banana plantations in the humid tropics of Honduras. Every place on Earth has its own **unique** characteristics determined by the surrounding environment and the people who live there.

Places have meaning to people. We develop an attachment to a place by filling it with meaning and emotion. The places where we live, work, and go to school are important to us. Even small places such as our bedroom or a classroom can have a unique and special meaning. In the same way, larger locations, such as our hometown, a cemetery, a historical battlefield, a religious shrine, or a state or national park are places that have shared meanings.

place a location on Earth with distinctive characteristics that make it meaningful to people

unique unlike anything else

Places often have meaning because an important event occurred there. This sign marks the place that bluegrass music originated—the Ryman Auditorium in Nashville, Tennessee.

Identifying What places can you name that are the locations of important events?

A place provides one way that people gain an identity. This is because the characteristics of a place affect the lives of people who live there. Additionally, a place usually has a name, location, and character that many people recognize, even people not from that place. For example, "New Yorker," "Parisian," and "Texan" are common identities associated with those places. People may associate a place with a certain character or feel. When we experience and give meaning to places, we can have a feeling of "home" when we are in a familiar place or a feeling of adventure in an unfamiliar place.

One way that geographers learn about places is by studying landscapes. **Landscapes** are portions of Earth's surface that can be viewed at one time and from one location. They can be as small as the view from the front porch of your home, or they can be as large as the view from a tall building that includes the city and surrounding countryside.

Whether we visit a landscape or look at photographs of the landscape, it can tell us much about the people who live there. Geographers look at landscapes and try to explain their unique combinations of physical and human features. As you study geography, notice the great variety in the world's landscapes.

When we define a place, we often consider a place's site and situation. **Site** refers to the characteristics specific to a place, including its physical setting. For example, the site of San Francisco is its location at the end of a peninsula in central California. **Situation** refers a place's geographic position in relation to other places and its connections to other regions. San Francisco's situation is a port city on the Pacific coast, close to California's agricultural lands.

The characteristics of a place change over time. Although the physical environment may seem unchanging, it is not. Some change is due to natural processes. Islands form through volcanic activity. Islands disappear as a result of a rise in sea levels. Coastal areas are reshaped by hurricanes. Glaciers form and then melt as temperatures change.

landscape a portion of Earth's surface that can be viewed at one time

site the characteristics of a place, including its physical setting

situation the geographic position of a place in relation to other places of a larger region

Edwin Remsberg/VWPics/Alamy Stock Photo

Paris's site as a large city on the Seine River is shown in the photo. The map shows Paris's situation in north central France, southwest of Belgium.

Making Connections Describe the site and situation of your hometown.

Other changes to the physical landscape are the result of human activities. As populations grow, the landscape is changed by adding human characteristics like cities and towns, roads, farms, and factories. Land is cleared and leveled to build houses and businesses. People cut down trees to supply the timber industry. The land is mined for coal and minerals, sometimes leaving open holes in Earth's surface. Farmers clear land for agriculture. Roads and railroads are built to transport people and the goods they produce and trade.

Geographers are concerned with a place at given moments in time. But to understand the development and importance of a place, they must study the forces and events that shaped that place. Characteristics of places today are the result of continuous change in the physical and human landscape. Geographers ask, "How did it come to be this way?" when studying a place.

✓ CHECK FOR UNDERSTANDING

Summarizing How does a place help give a person an identity?

Perception and Place

GUIDING QUESTION

How might perception affect a person's idea of a place?

The cultures and experiences that people have influence their **perceptions** of places.

A perception is the way you think about or understand something. Your perception of a place is often informed by what you know and what is familiar to you. It gives you a **context**, or setting, within which to put something new. Perceptions are helpful for understanding a place's location, characteristics, and importance. For example, a person who grew up in coastal Florida has a much different understanding of the role of water in life than someone who grew up in the desert area of Arizona. The abundance and scarcity of water in each of these two places has a major effect on the culture of each place.

People's knowledge and perceptions of places can come from direct experiences like living in or traveling to places. You might have developed a perception of life in a small town because you live in one, or because you visit friends or family members who live in a small town. People's knowledge and perceptions about a place can also come from indirect experiences. People often develop ideas about places they have never visited. These perceptions are created through the media, books, and movies, as well as from stories and pictures shared by family and friends who have visited places. You may have a perception of living in a large city in a high-rise apartment building from watching a television show or movie. The media coverage of wildfires in California may have shaped your perception of the dangers of living in a dry area with little rainfall.

perception the way you think about something

context the situation in which something happens

Change in a place can be dramatic. The photo on the left shows Miami, Florida, in 1913 as just a small settlement on the banks of the Miami River. The photo on the right shows Miami in 2021 and reflects how much it has grown and changed the physical landscape.

People learn about places through direct and indirect experiences.

Analyzing Perspectives What do you know about Times Square in New York City? Is what you know about this place from direct or indirect experiences?

Have you ever stopped to think about how you know what you know about a place? People often don't realize how their own cultures and experiences influence their perceptions of other places. This can often lead to the formation of geographic stereotypes, or unfair and untrue beliefs about all people or places with a certain characteristic. For example, some people have a perception of Australia as a hot, dry, barren landscape full of kangaroos and koalas. Others imagine a place where everyone is a surfer. While these characteristics are true for some parts of Australia, they do not represent all of Australia. The country is home to modern cities full of people who work at a variety of jobs. The natural environment offers beaches, mountains, and rain forests to explore.

You or someone you know might have formed some stereotypes of a place nearby. What do you know about a neighboring town or state? Think about what you hear people say

about that town or state. Could any of it be a stereotype? As we learn about people, places, and other cultures, we must be aware of where we get our information. We should be careful to question how our cultures and experiences might affect our perceptions.

✓ CHECK FOR UNDERSTANDING

Explaining What is a geographic stereotype?

LESSON ACTIVITIES

1. **Narrative Writing** Write a paragraph describing why your hometown gives you a "sense of home" when you're there.

2. **Analyzing Information** Work with a partner to analyze travel writing. Research to find a book or online blog from a travel writer. Analyze samples from one writer to help you determine whether the descriptions of a place are stereotypes.

03

Understanding Multiple Perspectives: Places and Identity

? COMPELLING QUESTION

How do places shape people's identities?

Plan Your Inquiry

DEVELOPING QUESTIONS

Think about different places with which you are familiar. Consider how those places might shape the identities of people. Then read the Compelling Question for this lesson. What questions can you ask to help you answer this Compelling Question? Create a graphic organizer like the one below. Write these Supporting Questions in your graphic organizer.

Supporting Questions	Source	What this source tells me about how places shape people's identities	Questions the source leaves unanswered
	A		
	B		
	C		

ANALYZING SOURCES

Next, examine the sources in this lesson. Analyze each source by answering the questions that follow it. How does each source help you answer each Supporting Question you created? What questions do you still have? Write these in your graphic organizer.

After you analyze the sources, you will
- use the evidence from the sources;
- communicate your conclusions;
- take informed action.

Background Information

Places can influence people's lives in many ways. Sometimes people live in the same place for many generations. Other people move to different places where they find new opportunities and have new experiences. Places can also be important to people who do not live there. For example, people who like country music may identify with the city of Nashville, Tennessee, while others who like rap music may identify with Atlanta, Georgia.

» Some residents of southern California have lifestyles and identities centered on the beach.

Devils Tower National Monument, Wyoming

There are 574 Native American groups that are recognized by the U.S. federal government. Not all of these groups have control of their historic homelands today. For example, a prominent mountain in Wyoming that was important to several groups is now a national monument managed by the National Park Service. Devils Tower National Monument was created to preserve natural and cultural resources. Visitors from around the world now come to enjoy this physical feature and to participate in recreational activities like hiking and rock climbing. The National Monument's website explains the connection between this place and Native Americans.

SECONDARY SOURCE: WEBSITE

66 The connections which tie American Indian culture to the place known as Devils Tower are both ancient and modern. Oral histories and sacred narratives explain not only the creation of the Tower, but also its significance to American Indians. They detail peoples' relationships with the natural world, and establish those relationships through literal and symbolic language....

» Devils Tower is a physical feature that rises up out of the surrounding prairie.

Modern connections are maintained through personal and group ceremonies. Sweat lodges, sun dances, and others are still practiced at the monument today. The most common ritual that takes place at the Tower are prayer offerings. Colorful cloths or bundles are placed near the Tower—commonly seen along the park's trails—and represent a personal connection to the site. They are similar to ceremonial objects from other religions, and may represent a person making an offering, a request, or simply in remembrance of a person or place....

It is important to note a key difference between American Indian religions and many other contemporary religions...a sense of place dominates the religion of American Indians, as opposed to the sense of time that dominates many western religions. 99

—"A Sacred Site to American Indians," National Park Service website

EXAMINE THE SOURCES

1. **Analyzing Perspectives** What are some of the reasons that a member of a Native American group might visit Devils Tower?

2. **Inferring** Why do you think the National Park Service wants to inform park visitors about the significance of this place to Native Americans?

The Harlem Renaissance

During the Great Migration from 1910 to 1970, approximately 6 million African Americans left agricultural regions of the South. They sought economic opportunities and escape from discriminatory Jim Crow laws. They moved to industrial cities in the Northeast and Midwest to take jobs in factories. This website excerpt explains how Harlem in New York City was affected by the Great Migration.

SECONDARY SOURCE: WEBSITE

66 By the turn of the 20th century, the Great Migration was underway as hundreds of thousands of African Americans relocated to cities like Chicago, Los Angeles, Detroit, Philadelphia, and New York. The Harlem section of Manhattan, which covers just three square miles, drew nearly 175,000 African Americans, giving the neighborhood the largest concentration of black people in the world. Harlem became a destination for African Americans of all backgrounds. From unskilled laborers to an educated middle-class, they shared common experiences of slavery, emancipation, and racial oppression, as well as a determination to forge a new identity as free people.

The Great Migration drew to Harlem some of the greatest minds and brightest talents of the day, an astonishing array of African American artists and scholars. Between the end of World War I and the mid-1930s, they produced one of the most significant eras of cultural expression in the nation's history—the Harlem Renaissance....

The Harlem Renaissance encompassed poetry and prose, painting and sculpture, jazz and swing, opera and dance. What united these diverse art forms was their realistic presentation of what it meant to be black in America, what writer Langston Hughes called an "expression of our individual dark-skinned selves," as well as a new militancy in asserting their civil and political rights. 99

—"A New African American Identity: The Harlem Renaissance," website of the Smithsonian National Museum of African American History and Culture

EXAMINE THE SOURCES

1. **Identifying** What made Harlem, a neighborhood in New York, unique among the other places where African Americans were moving during the Great Migration?

2. **Comparing** How were the African Americans who moved to Harlem similar to those who moved to other cities?

Tangier Island, Virginia

The land along the Chesapeake Bay in Maryland and Virginia was settled by English colonists beginning in the 1600s. Tangier Island is a small island in Chesapeake Bay. The island has a distinctive local culture that developed because of its location.

PRIMARY SOURCE: ARTICLE

Tangier Island is an isolated patch of Virginia marshland in the middle of the Chesapeake Bay, just south of the Maryland line. For centuries the island has been a community of watermen, the Chesapeake term for people who harvest the crabs, oysters and fish in the bay... .

Houses line narrow streets that follow patches of high ground in the town of Tangier, population 535. With no bridge to the mainland, supplies and people arrive on the daily mail boat from Crisfield, Maryland, 12 miles away. Most people get around the 3-mile-long island by foot, golf cart or bicycle.

Residents speak with an accent so distinctive that after a quick listen they can easily tell if someone is from Tangier or another nearby harbor. And the island has its own vocabulary, prompting a resident to compile an extensive dictionary of local terms (including "mug-up" for hearty snack, "cunge" for deep cough). Conversations are peppered with expressions like "yorn" for yours and "onliest" for only.

—"Tangier Island and the Way of the Watermen," Kenneth R. Fletcher, *Smithsonian Magazine*, March 31, 2009

EXAMINE THE SOURCE

1. **Identifying Cause and Effect** What is the reason that the identities of people who live on Tangier Island are very different from those who live in Richmond, Virginia, or Washington, D.C.?

2. **Speculating** What other kind of environment, besides an island, could encourage the development of a unique local identity?

Complete Your Inquiry

EVALUATE SOURCES AND USE EVIDENCE

Refer back to the Compelling Question and the Supporting Questions you developed at the beginning of the lesson.

1. **Determining Context** Which source provides evidence of how a feature of the natural landscape can be important to someone's identity?

2. **Identifying Themes** Each source describes a connection between a different place and the identity of a specific group of people. For each source, list the place and the group of people whose identity is connected to that place.

3. **Gathering Sources** Which sources helped you answer the Supporting Questions and the Compelling Question? Which sources, if any, challenged what you thought you knew when you first created your Supporting Questions? What information do you still need in order to answer your questions? What other viewpoints would you like to investigate? Where would you find that information?

4. **Evaluating Sources** Identify the sources that helped answer your Supporting Questions. How reliable is the source? How would you verify the reliability of the source?

COMMUNICATE CONCLUSIONS

5. **Collaborating** Share with a partner which of these sources you found most useful in answering your Supporting Questions. Discuss how a place can be special to a group of people. Think of examples of places that have shaped the identities of people you know. How do these sources support your thinking? Use the graphic organizer that you created at the beginning of the lesson to help you. Share your conclusions with the class.

TAKE INFORMED ACTION

Design a Community Monument Think about what defines the identity of the community in which you live: your shared history, culture, symbols, and experiences. Design a monument that reflects its identity. Create a drawing of the monument and a short message explaining why it would be a positive addition to the community. Send the message and design to a local leader.

TEXT: "Tangier Island and the Way of the Watermen." Smithsonian Magazine. Kenneth R. Fletcher. https://www.smithsonianmag.com/science-nature/tangier-island-and-the-way-of-the-watermen-117890294/.

04

How Are Regions Defined?

Analyzing Key Ideas and Details As you read, take notes about the three types of regions that geographers study.

Region	Description

Defining Regions

GUIDING QUESTION

What is a region?

Areas on Earth's surface that share similar characteristics belong to the same **region**. A region can be defined by physical traits such as climate, landforms, soils, vegetation, animal life, and natural resources. For example, Los Angeles and San Diego are located in southern California. They have some features in common, such as nearness to the ocean. Both cities are also mostly sunny throughout the year. In the case of those two cities, the region is defined using physical characteristics.

A region can be defined by human characteristics such as language, religion, political or economic systems, and population distribution. For instance, one reason the countries of North Africa are often grouped into a region is that most of the people living in these countries follow the same religion, Islam.

region an area of Earth's surface that has similar physical or human characteristics

Types of Regions

To interpret the Earth's complexity, geographers have developed the concept of regions.

Types of Regions	Examples	Why It's Important
Formal (or uniform) Region: An area defined by common human or physical characteristics.	Human: Political boundaries define the state of Ohio. Physical: The Rocky Mountains define a region in the American West.	Certain assumptions may be made about life in these regions.
Functional Region: A central place surrounded by an area in which people look to the central place for jobs, entertainment, education and shopping.	The city of Atlanta, Georgia, and the surrounding area is a functional region. A school district is a functional region.	Understanding how humans interact and connect with each other is important for political and economic reasons.
Perceptual Region: An area defined by a common understanding that developed over time.	Examples are the Bible Belt, the South, Frontier Texas, the West, or Cajun Country.	Think of it as an introduction to a region, how people in that area live and work.

Geographers study regions to identify the broad patterns of areas. They compare and contrast features in one region with those in another. The study of regions is important for understanding how the characteristics of each region affect the lives of the people who live there.

Regions occur at different **scales**, or sizes of the areas being studied. Smaller regions are often included in larger or more important regions. For example, counties and cities are within states and provinces, which are within countries.

Geographers identify three types of regions: formal, functional, and perceptual. A **formal region**, or uniform region, features a unifying characteristic. Formal regions can be states or countries. Other might produce a specific product. For example, the Corn Belt is a band of farmland stretching from Ohio to Nebraska in the United States. It is a formal region because corn is its primary crop. Another example of a formal region is an *ethnic enclave*. This is a geographic area where an ethnic group is clustered together for safety and support. Italian immigrants arriving in New York City in the 1880s set up Italian-style markets and restaurants and celebrated Italian holidays in an area of Manhattan that became known as Little Italy.

A **functional region** incorporates a central point and a surrounding area that is connected to the point by some defined function. For example, a cell

scale the geographic size of the area being studied

formal region a region defined by a common characteristic

functional region a central place and the surrounding territory linked to it

Perceptual Regions of the United States

Some regions are not defined by data or specific boundary lines but instead by culture and speech, such as in this map showing perceptual regions of the United States.

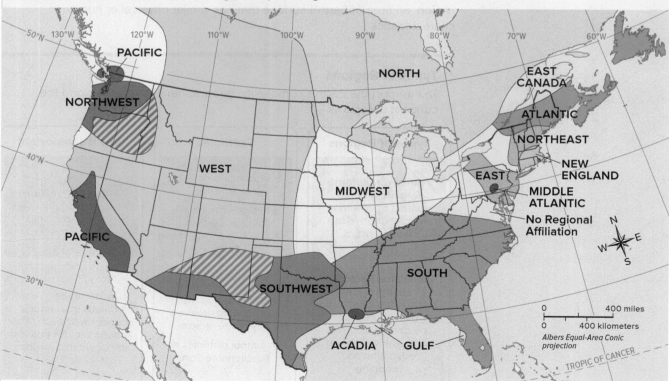

EXAMINE THE MAP

1. **Exploring Regions** According to the map, in which region do you live?

2. **Exploring Regions** What aspects of culture do you think define a perceptual region in the United States?

tower provides the central point for a surrounding area in which cell-phone users can obtain phone reception. A city and its surrounding suburbs are another example of a functional region.

A **perceptual region** is descriptive and based on a shared understanding that has developed over time. For example, the term "heartland" refers to a central area in the United States in which traditional values of family and hospitality are believed to prevail. Another example of a perceptual region is the South. Some people frequently refer to the South as a place with cultural and economic features perceived to be quite different from the rest of the United States.

✓ CHECK FOR UNDERSTANDING

Summarizing What kinds of characteristics are used to define regions?

Regional Change

GUIDING QUESTION

What causes regions to change?

All regions experience change. They change as the physical and human characteristics of Earth change. Physical processes cause earthquakes and volcanoes that alter the natural landscape. Wind and ocean waves change the shape and size of coastal areas. Severe weather events like hurricanes, typhoons, and tornadoes bring rains and high winds that cause flooding and destroy houses, roads, and even whole towns.

Climate change can affect the physical characteristics of a region in many ways. Warming temperatures result in the expansion of dry areas, leading to increasing wildfires. For example, the "wildfire season" in the western United States used to be about four months long. Today fire season lasts about six to eight months of the year. According to the U.S. Forest Service, wildfires are starting earlier, burning more intensely, and burning larger areas of land than ever before.

In addition to the expansion of dry areas, climate change has affected other regions of the world. In the polar regions and in high mountains, there has been increased melting of ice. This has led to changes in the availability of water for plant and animal populations. There has been disruption to weather patterns in many regions. For example,

The shrinking ice in the Arctic region makes it more difficult for polar bears to find and hunt for food.

Making Connections Identify changes to the physical or human characteristics of the region in which you live. What do you think might be the cause of these changes?

more-intense rainfall in the midwestern United States has resulted in spring flooding. This flooding can destroy homes and interferes with planting schedules for farmers in the region.

Regions also change as Earth's human characteristics change over time. These causes of regional change include wars and conflict, shifts in land use, population growth and decline, the expansion of cities, and the development of transportation and communications networks. For example, the regional boundaries and names of African countries have changed over time. In some instances, these changes were the result of European colonization. Africa was divided up into colonies by European powers—mainly Britain, France, Portugal, Belgium, and Germany. They competed to **expand** their empires and to protect their trade routes. In less than 40 years, Africa was divided into multiple colonial divisions with

---

perceptual region a region defined through common understanding developed over time

expand to increase in size

These people are leaving their home country of Afghanistan after the fall of the national government and takeover by the Taliban in 2021. This change in the region pushed some Afghan people to flee to regions such as North America and Europe.

borders that ignored traditional ethnic territories. Now those countries have become independent, though languages and forms of government introduced by Europeans remain.

Regions experience change as people **migrate**, or move from one place to another. People move to a new place in search of new opportunities such as jobs, a desirable climate and environment, political and religious freedom, and better living conditions. People may also move to a new place to get away from a bad situation. In the nineteenth century, for example, many Irish immigrants came to the United States to escape the Irish potato famine of 1845 to 1852. During this same period, a diverse group of immigrants was arriving in California to participate in the Gold Rush. This influx of new people and cultures resulted in changes to the names, boundaries, and characteristics of the places and regions in which these immigrants settled.

Economic activity brings change to a region. In the 1960s, as South Korea **industrialized**, or became more focused on manufacturing, more people moved to cities. Now, most South Koreans live in urban areas. In Southeast Asia, industries provide an economic boost to the region, but they sometimes **exploit** and harm the environment. Tin mining has created huge wastelands in Malaysia, Thailand, and Indonesia. Commercial logging and farming destroy tropical forests at an alarming rate.

The development of transportation networks can lead to regional change. The Pan-American Highway, for example, runs from Argentina to Panama. A trans-Andean highway connects cities in Chile and Argentina. Peru and Brazil are building the Transoceanic Highway. Eventually, this road will link Amazon River ports in Brazil with Peruvian ports on the Pacific Ocean. Increased connections between countries or cities in a region link them economically and make it easier for people to move throughout the region.

✓ **CHECK FOR UNDERSTANDING**

Explaining Explain the things that cause changes to the human characteristics of a region.

LESSON ACTIVITIES

1. **Narrative Writing** Look at the map of U.S. perceptual regions earlier in this lesson. Learn more about the perceptual region in which you live. Why does the region have this name? Write a paragraph describing how it feels to live in this region.

2. **Using Multimedia** Work with a partner to create a video news story about regional change anywhere in the world. Be sure your news story identifies the cause of the change and explains its effects. Include maps or photographs to display in the news video.

---

migrate to move from one place to another

industrialize to build factories and engage in manufacturing

exploit to get value or use from

05
Connections Between Places and Regions

READING STRATEGY

Analyzing Key Ideas and Details As you read, take notes about how spatial interaction leads to connections.

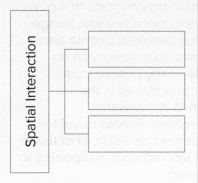

Spatial Interaction

GUIDING QUESTION

What is spatial interaction?

Geographers use the term **spatial interaction** when talking about all kinds of movement and flows related to human activity. For example, shipment of raw materials or finished products, traveling to work or for a vacation, shopping trips, telephone and Internet communications, and electronic banking are all forms of spatial interaction.

Spatial interaction occurs when people in one place want or need to access resources, goods and services, or people from another place. In our interconnected world, a vast number of products move from place to place. Fruits such as grapes, blueberries, and cherries from Chile move to supermarkets in the United States. Poultry from Alabama is shipped to Latin American countries. Oil from Saudi Arabia powers cars and trucks across Europe and the United States. All this movement relies on transportation systems that use ships, railroads, airplanes, and trucks.

Look at the labels in your clothes or on other products you buy. How many different country names can you find? How and why do we get goods from far across the world?

spatial interaction the movement and flow of people, products, and ideas related to human activity

The European Union is an example of countries cooperating to provide common social, economic, and security policies throughout the continent.

Trade is the business of buying, selling, or bartering. When you buy something at a store, you are trading money for a product. On a much bigger scale, countries trade with each other, and companies spread their operations across multiple countries. Countries have different resources. Resources can include raw materials, such as iron ore. Expertise and labor are also resources. One country may have workers with high levels of training, while labor may be cheaper in another country where workers earn lower wages. As a result, labor-intensive activities such as manufacturing and software development can be done more cost effectively in some countries.

Trade can benefit countries. One country can **export** a product or service that it is able to produce. Another country **imports** that product from the exporting country to meet the needs of its people. In some instances, a group of countries may work together to improve trade. For example, the European Union (EU) was created by several countries in Europe to unite much of the region into one trading community. Trade, both within Europe and between Europe and the rest of the world, has changed as a result. The EU has one of the largest levels of trade in the world.

Trade can also lead to harm in countries. Some countries only have one or two products to export. The countries lose income and can enter recession, or a period of reduced economic activity, when prices for those products decline. In addition, the income from exports in some countries only benefits a small number of people who produce the exports and does not benefit most of the people in the country.

When people move and interact with people in another place, spatial interaction is occurring. Needs are being met. People may migrate to a new place in search of better opportunities. Or people may migrate to a new place to escape bad situations like war, hunger, and conflict. Interaction between people also happens when we travel to a new place on vacation. For example, many people from Sweden take vacations in places in the Mediterranean region to enjoy warm temperatures and sunshine. People travel for work and to make business deals. In all of these instances, the spatial interaction is happening to help meet peoples' needs.

Information can move and therefore interact at an even faster pace than products and people. Communications systems, such as telephone, television, radio, and the Internet, carry ideas and

export to send a product from one country to another for purposes of trade

import to bring a product into one country from another country

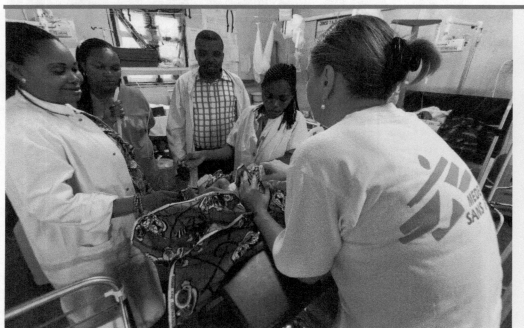

Aid workers may travel to a country to help people in need. In this image, members of Doctors Without Borders work with local doctors to help children in need of medical care in Central Africa.

Making Connections Describe examples in which people interact in order to meet economic or social needs.

information all around the Earth. Remote villagers on the island of Borneo watch American television shows and learn about life in the United States. Political protesters in the United States use text messaging and social-networking sites to coordinate their activities.

Spatial interaction also occurs through movements and flows related to physical geography. For example, erupting volcanoes emit smoke and ash that settle on places far away. The process of erosion carries pieces of rock and soil by water or wind and then deposits them in other places such as river deltas or sand dunes. Earthquakes under the ocean can generate tsunami waves that impact shorelines thousands of miles away. It is important to consider spatial interaction in both physical and human geography.

✓ **CHECK FOR UNDERSTANDING**

Summarizing Why does spatial interaction occur?

Connections

GUIDING QUESTION

How are places connected?

Places are connected within and across regions. However, the strength of their connections can be influenced by distance, **function**, and ease of movement between places. Barriers make it difficult for people, products, and ideas to move between places. They can be physical, such as oceans or mountains. For example, the Andes mountains in South America are a physical barrier making travel across the continent difficult. As a result, many indigenous communities developed as isolated groups. Some mountain villages exhibit centuries-old social customs. Other barriers can be cultural, like language and traditions. Communication between places where people speak different languages or have different ways of life can make establishing and maintaining connections difficult.

Distance affects connections between places. Generally, interaction decreases as distance increases. This relationship is called *distance decay*. Distance has a strong influence on how places are connected. For example, if a new grocery store opens on your street, you are more likely to shop there than at the grocery store on the other side of town.

Chicago (top) is the largest city in Illinois. It is a major economic and railroad hub for the Midwest region and for the country. The smaller city of Decatur, Illinois (bottom) is home to manufacturing and agriculture. Though connected through economic and transportation networks, each city serves a different function.

Another factor in the connection between places is the function, or purpose, of each place. People tend to travel short distances to obtain goods and services that they need frequently, like groceries and gasoline. These are called lower-order goods and services, and are available in many places. Larger places also provide higher-order goods and services such as specialty furniture stores, live entertainment, and airports. They serve as regional centers for some functions and provide links to other places outside of the region. This **interdependence** with smaller cities is important for people living in both places.

The ease of movement, or *accessibility,* between places is also a factor in how places are connected. Accessibility refers to how easy or difficult is it to overcome the barriers separating

interdependence a condition in which people or groups rely on each other, rather than only relying on themselves

function purpose

The World in Spatial Terms **G59**

places. For example, distance isolated North America from Europe until the development of ships and airplanes that reduced the travel time between the continents. Today, technology like the Internet and cell phones has removed or reduced many of the barriers to interaction between people in places that are far away from each other.

✓ **CHECK FOR UNDERSTANDING**

Describing What is accessibility?

Globalization

GUIDING QUESTION

What is globalization?

Globalization is the process by which the peoples, places, and things around the world have experienced greater connectivity and interdependence. This is especially true in the areas of economics, politics, and culture. Advances in transportation and communications have driven globalization by making spatial interaction more efficient across greater distances. For example, commercial airlines now allow people to travel to the other side of the world in a matter of hours. In contrast, just over 100 years ago the same trip would have taken days or weeks on a ship. The Internet allows for instantaneous communication between any two locations that are connected to it.

There is much debate about the benefits and challenges of globalization. Some beliefs about the issue depend on perspective, such as whether one is from a more developed country or from a less developed country. Globalization affects different people and places in different ways.

Some people argue that the benefits of globalization outweigh the challenges. Examples of benefits of globalization include

- an increased level of wealth, goods, and necessities available for people in developing countries;
- a larger selection of goods and services;
- the promotion of innovation through competition;
- technology transfer. Less-developed countries can enjoy the benefits of new technologies developed by more developed countries;

- cooperation and awareness. Governments are better able to work together toward common goals;
- greater access to other cultures.

Others believe that globalization is more harmful. Examples of challenges of globalization include

- job loss in wealthier countries. Moving manufacturing of products for **domestic** use or sale to countries where labor is cheaper can create jobs for people in one country but also takes away jobs from skilled workers in the home country;
- an increasing gap between the rich and the poor. Globalization helps wealthier countries profit from the resources from poorer countries. Many countries obtain loans from wealthier countries but are often unable to repay them;
- loss of culture. Although globalization encourages the sharing and awareness of cultures, the spread of Western culture results in a narrowing of beliefs and values. The unique characteristics of individual cultures begin to fade;
- the spread of diseases across a large area or even the world;
- environmental degradation. Globalization has resulted in increased consumption of goods, as well as an increase in the extraction and use of raw materials such as fossil fuels and timber.

✓ **CHECK FOR UNDERSTANDING**

Explaining How is globalization related to connectivity?

LESSON ACTIVITIES

1. **Informative/Explanatory Writing** The movement of ideas creates connections between people and places. Research to identify and learn about an event that occurred as a result of the spread of ideas. Write an essay to discuss your findings.

2. **Interpreting Information** Work in a group to learn more about the cities of Chicago and Decatur. How are they connected? What types of goods and services does each city offer? How does this influence their connection? Share your findings with the class.

---

globalization increasing interconnection of economic, political, and cultural processes to the point that they become global in scale and impact

domestic relating to a home country

06

Global Issues: Connections Between Places

"Figure 6: A Commuter Flow-Based Regionalization of the United States: From Commutes to Megaregions. PLoS ONE 11, no. 11 (2016): e0166083. https://doi. org/10.1371/journal.pone.0166083

? COMPELLING QUESTION

How are places connected to each other?

Plan Your Inquiry

DEVELOPING QUESTIONS

Think about different places with which you are familiar. Consider how those places are related to the place where you live, as well as to each other. Then read the Compelling Question for this lesson. What questions can you ask to help you answer this Compelling Question? Create a graphic organizer like the one below. Write these Supporting Questions in your graphic organizer.

Supporting Questions	Source	What this source tells me about how places are connected	Questions the source leaves unanswered
	A		
	B		
	C		
	D		
	E		

ANALYZING SOURCES

Next, examine the sources in this lesson. Analyze each source by answering the questions that follow it. How does each source help you answer each Supporting Question you created? What questions do you still have? Write these in your graphic organizer.

After you analyze the sources, you will
- use the evidence from the sources;
- communicate your conclusions;
- take informed action.

Background Information

There are many ways in which places can be connected to each other. Spatial interaction is a term that geographers use to describe connections between places. Distance is one major factor affecting interaction between places. The time and energy it takes to travel or transport goods between locations is a cost of interaction. That cost must be balanced against the benefit that is received from the interaction.

Over time, advancements in transportation and communications have increased people's ability to interact across distances. These technologies have made possible an era of globalization in which the world's economies, cultures, and populations are increasingly interconnected.

» Container ships like the one pictured here have made it possible to easily transport large quantities of consumer goods around the world.

Commuting to Work

The trip from where people live to where they work is called a *commute*. People make their commutes using different types of transportation, including cars, trains, or bicycles. This map traces the commutes of American workers in many different cities. The routes are color-coded by city. For example, the routes taken by people who work in Des Moines, Iowa are shown in green.

SECONDARY SOURCE: MAP

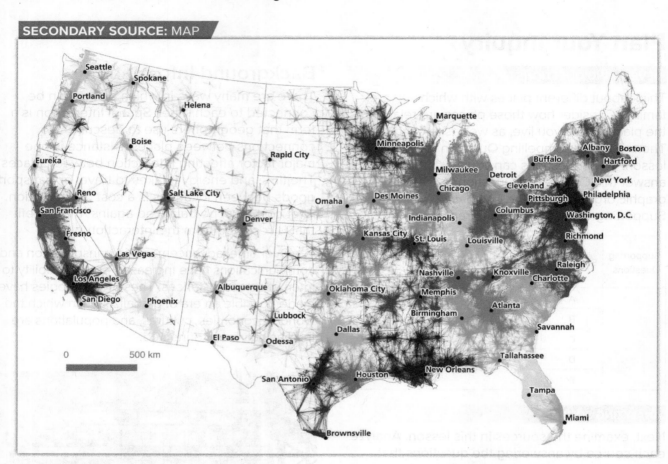

"Figure 6: A Commuter Flow-Based Regionalization of the United States." In Dash Nelson G. and A. Rae, An Economic Geography of the United States: From Commutes to Megaregions. PLoS ONE 11, no. 11 (2016): e0166083. https://doi.org/10.1371/journal.pone.0166083

EXAMINE THE SOURCE

1. **Interpreting** What information does this source provide about where many Americans live and work?

2. **Speculating** If a person lives in a small town, what are some reasons other than work that might encourage them to travel into a large city?

B

The Economics of Chocolate

Chocolate is a popular treat in many parts of the world. Cocoa, which is grown on trees, is one of the main ingredients in chocolate. This article explains how the production of chocolate creates connections between the places where it is grown and the places where it is eaten.

PRIMARY SOURCE: ARTICLE

Cocoa makes a long and winding journey from bean to bar. The crop starts in the farms of tropical nations, especially in West Africa, and travels through ports, shipping containers and processing plants. [. . .]

» A farmer harvests cocoa on a farm in Cameroon, Africa.

Cocoa requires [a] tropical climate and shady conditions, which means that cocoa farms don't look much like wheat fields or orange orchards. Trees are grown under a canopy of taller trees, so many farms look like cultivated rainforest. On average, cocoa farms are small operations, around 4 hectares—the size of just 8 football fields. (The average farm in the US, by contrast, is around 95 hectares.) Though cocoa farms can generate relatively big profits, the long-term survival of some farms is in question: Recent climate change predictions have made producers nervous, and the world's largest chocolate manufacturers are at work breeding heat- and drought-resistant trees. [. . .]

Almost all of the world's cocoa is grown in **developing** countries and consumed by **industrialized** countries. The top four producers—Ivory Coast, Nigeria, Ghana and Indonesia—are all in the bottom half of nations by **per-capita GDP**. More strikingly, the top ten countries ranked by chocolate consumed are all in the top 15 percent. Nine of those countries are in Europe. (In 2012, the United States was ranked 15th.)

—"The Economics of Chocolate," Daniel A. Gross, *Smithsonian Magazine*, February 11, 2015

---

developing having an economy mainly based on agriculture and a low level of wealth

industrialized having an advanced economy and a relatively high level of wealth

per-capita GDP the average value of goods and services produced by one person

EXAMINE THE SOURCE

1. **Explaining** Describe the conditions required to grow cocoa.
2. **Contrasting** How are the top chocolate-consuming countries different from the top chocolate-producing countries?

Exporting Waste

Managing waste products is essential for maintaining a healthy environment. But where does all of our trash go? Waste management systems collect waste and move it to designated locations for disposal. A *landfill* is a place where trash is accumulated, compacted, and covered with soil. Recycling programs capture and divert usable materials. This helps reduce the need for new landfills. This graphic shows the countries to which plastic scrap materials from the United Kingdom and United States were being shipped for disposal or recycling in 2018.

PRIMARY SOURCE: INFOGRAPHIC

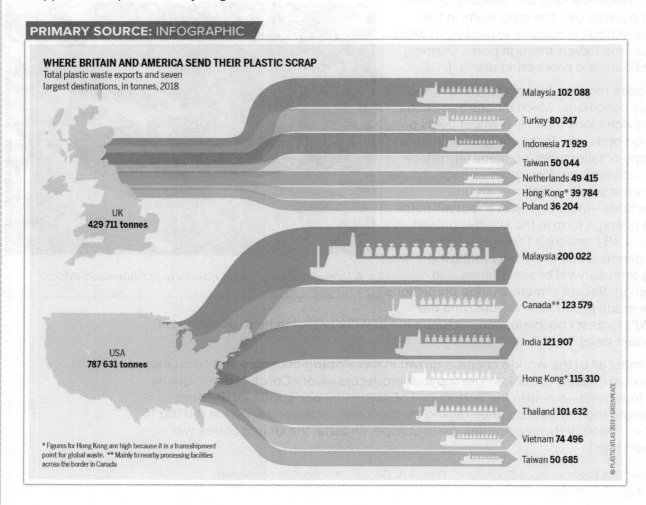

WHERE BRITAIN AND AMERICA SEND THEIR PLASTIC SCRAP
Total plastic waste exports and seven largest destinations, in tonnes, 2018

UK
429 711 tonnes

Malaysia **102 088**
Turkey **80 247**
Indonesia **71 929**
Taiwan **50 044**
Netherlands **49 415**
Hong Kong* **39 784**
Poland **36 204**

USA
787 631 tonnes

Malaysia **200 022**
Canada** **123 579**
India **121 907**
Hong Kong* **115 310**
Thailand **101 632**
Vietnam **74 496**
Taiwan **50 685**

* Figures for Hong Kong are high because it is a transshipment point for global waste. ** Mainly to nearby processing facilities across the border in Canada

© PLASTIC ATLAS 2019 / GREENPEACE

EXAMINE THE SOURCE

1. **Interpreting** Which country receives the most plastic scrap from both the United Kingdom and the United States? Which other countries also receive plastic scrap from both the United Kingdom and the United States?

2. **Speculating** Why do you think countries are willing to accept this scrap material?

Globalization of Korean Pop Culture

Communications technologies like television and the Internet make it possible for cultural influences to extend around the world. For example, many people throughout the world have learned about the United States by hearing American music styles such as jazz and hip-hop. Japanese animation has gained fans worldwide. This article explains how pop culture has become an important export for South Korea.

PRIMARY SOURCE: ARTICLE

"K-pop groups are now . . . accustomed to amassing huge numbers of views [online] and selling out concert venues on extensive worldwide tours. But it wasn't always this way, and it took a blend of government backing, creative serendipity and a well-placed diaspora over the decades to turn this trickle of cultural exports into what has become a bona fide "Korean wave". . . [Last year] Bong Joon-ho's "Parasite" became the first non-English film to take home the Oscar for best picture. Soon after, mega K-pop group BTS became the first all-Korean band to top the Billboard Hot 100 chart with its English-language hit "Dynamite". . .

» K-pop band BTS served as special presidential envoys for South Korea during a meeting of the United Nations General Assembly in September 2021.

Jenny Wang Medina, an assistant professor at Emory University's Department of Russian and East Asian Languages and Cultures, said K-pop was certainly a beneficiary of government efforts to push more Korean creative content into the world — supporting the translation of Korean literature for instance. But the **genre** also became a force for ushering in a larger, more diverse media landscape.

"While there is an economic incentive behind this. . .there is also a real desire on the part of the Korean government as well as Korean people to have their culture recognized," said Dr. Wang Medina."

—"Examining How K-pop Helps Drive Korea's Global Growth," Trevor Williams, *Global Atlanta*, 2021

genre style

EXAMINE THE SOURCE

1. **Identifying** What events showed the popularity of Korean pop music in the U.S.?
2. **Interpreting** Why do you think the government of South Korea provides generous support for the K-pop industry?

International Travel Destinations

Every year, millions of Americans travel outside of the United States. People may travel for work or for personal reasons, such as to visit family or to enjoy a vacation. This infographic shows the destinations to which Americans traveled in one year.

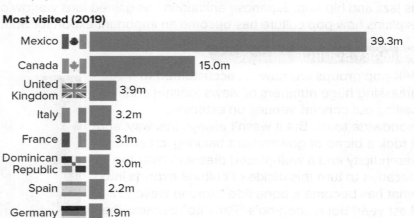

The Most Popular International Travel Destinations of Americans

Most popular destinations of U.S. resident travelers

Most visited (2019)

Mexico	39.3m
Canada	15.0m
United Kingdom	3.9m
Italy	3.2m
France	3.1m
Dominican Republic	3.0m
Spain	2.2m
Germany	1.9m

Stays of one or more nights, airline travel only except to Mexico and Canada
Source: National Travel and Tourism Office

EXAMINE THE SOURCE

1. **Calculating** According to the source, how many Americans traveled to a country in Europe in 2019?

2. **Identifying Cause and Effect** How does the data presented in this source reflect the impact of distance on spatial interaction?

Complete Your Inquiry

EVALUATE SOURCES AND USE EVIDENCE

Refer back to the Compelling Question and the Supporting Questions you developed at the beginning of the lesson.

1. **Evaluating** Which source(s) describe how the movement of people creates connections between places?

2. **Making Connections** Which sources show evidence of globalization?

3. **Gathering Sources** Which sources helped you answer the Supporting Questions and the Compelling Question? Which sources, if any, challenged what you thought you knew when you first created your Supporting Questions? What information do you still need in order to answer your questions? What other viewpoints would you like to investigate? Where would you find that information?

4. **Evaluating Sources** Identify the sources that helped answer your Supporting Questions. How reliable is each source? How would you verify the reliability of the source?

COMMUNICATE CONCLUSIONS

5. **Collaborating** Share with a partner which of these sources you found most useful in answering your Supporting Questions. Take turns identifying some ways that your community is connected to other places that are both close and far away. How do these sources support your thinking? Use the graphic organizer that you created at the beginning of the lesson to help you. Share your conclusions with the class.

TAKE INFORMED ACTION

Writing a Letter to a Local Newspaper Research the ways in which your local community is connected to other places around the world. You may want to interview community members or conduct an Internet search to gather information. Then write a letter to the editor of a local newspaper or magazine. Express your opinion about why it is important for people in your community to be aware of how they are connected to other parts of the world. Send the letter and ask that it be published in the newspaper or magazine.

Reviewing Places and Regions

Summary

The Concept of Place

- A *place* is a specific location with distinct characteristics.
- Places can have human or physical characteristics, or a combination of the two.
- Places have meaning to people.
- People's culture and experiences influence their perception of places.
- A perception of a place can also come from the media, books, movies, and other people.

How Are Regions Defined?

- A region is an area with similar characteristics.
- A region can have physical or human significance, or both.
- Regions can occur at different scales.
- There are three types of regions: formal (uniform), functional, and perceptual.
- Regions may change as the physical and human characteristics of Earth change.

Connections Between Places and Regions

- Spatial interaction is the movement and flows associated with human activity.
- Spatial interaction occurs when people in one place want or need to access resources, goods and services, or people from another place.
- The connections between places and regions are influenced by distance, function, the ease of moving between places, and modern forms of communication.
- Globalization is the process by which places around the world have experienced greater degrees of interaction and connection.
- There are both benefits and challenges associated with globalization.

Checking For Understanding

Answer the questions to see if you understood the topic content.

IDENTIFY AND EXPLAIN

1. Identify each of these terms as they relate to places and regions.

   A. place
   B. landscape
   C. region
   D. formal (uniform) region
   E. functional region
   F. perceptual region
   G. spatial interaction
   H. interdependence
   I. globalization

REVIEWING KEY FACTS

2. **Explaining** What kinds of characteristics define a place?

3. **Making Connections** How do places come to have meaning for people?

4. **Drawing Conclusions** How do places provide people with an identity?

5. **Summarizing** Describe the relationship between places and landscapes.

6. **Contrasting** What is the difference between site and situation?

7. **Describing** What influences people's perception of places? Distinguish between direct and indirect experiences.

8. **Explaining** What kinds of characteristics can be used to define regions?

9. **Determining Meaning** What do geographers mean when they talk about scale?

CRITICAL THINKING

10. **Contrasting** How is a formal region different than a perceptual region?

11. **Identifying** Describe an example of a functional region.

12. **Drawing Conclusions** What types of physical characteristics can influence regional change?

13. **Making Connections** How is migration related to regional change?

14. **Explaining** Why does spatial interaction occur?

15. **Identifying** Name some examples of spatial interaction.

16. **Evaluating** How does distance influence the connections between places?

17. **Describing** How do physical and human barriers affect the connections between places?

18. **Analyzing** What does *function* mean when talking about interdependence between places?

19. **Determining Meaning** How does connectivity relate to globalization?

20. **Identifying** Explain one benefit and one challenge of globalization.

NEED EXTRA HELP?

If You've Missed Question	1	2	3	4	5	6	7	8	9	10
Review Lesson	2, 4, 5	2	2	2	2	2	2	4	4	4

If You've Missed Question	11	12	13	14	15	16	17	18	19	20
Review Lesson	4	4	4	6	6	6	6	6	6	6

Apply What You Have Learned

 ## A Understanding Multiple Perspectives

You live in a place with its own unique characteristics. Your current home is part of a region that is defined by economic, political, cultural, and other characteristics as well as physical features. How important each of these characteristics is to you may differ from the importance that your family members or friends place on the same characteristics.

ACTIVITY **Creating an Identity Flow Chart** Think about what features are associated with the place in which you live. Then draw a flow chart. At the top, name either the place you currently live, a place you have lived before, or a place from which your family moved in an earlier generation. Below the place name, draw arrows to at least four boxes underneath. Each box should contain one part of your life that is affected by the place at the top of the chart. For example, your family's place of origin might affect what foods you eat, or your current home determines what outdoor activities you can enjoy. Explain your flow chart in a short class presentation.

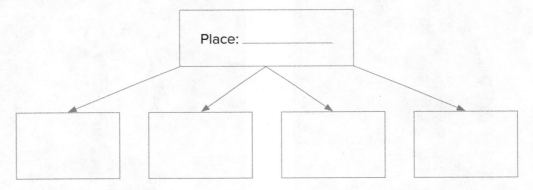

 ## B Geographic Reasoning

The Information Revolution that began at the end of the 1900s made it possible to instantly communicate and share content across international borders. It is no longer necessary to physically travel overseas to meet people in other parts of the world. Thanks to social media apps, discussion groups, and online gaming, it is more and more common for people to have faraway acquaintances. In addition, people who in past generations may have been seen only on special occasions, such as family members living in other regions, can now talk to you using online video platforms.

ACTIVITY **Gaining a New Perspective** Think of one feature of the place where you live that might be attractive to someone in another part of the world. This might be a vacation spot, a unique geographic attraction, a festival or other regular event, or high-profile sports events or other entertainment. Then use an e-mail or other electronic message to promote that feature to someone who is physically as far from you as possible, either in another country or another region of this country. Ask the person to share an equivalent feature that might make you want to visit the place where that person lives. In a short presentation, compare and contrast the two features and their attractiveness for travelers.

C Geography and History

Throughout history, decisions made by leaders and governments have had a major impact on regional environments. Elements such as forms of government or social systems have driven those decisions. Such decisions always attempt to find a positive balance between costs and benefits. For example, the Aral Sea, once the fourth-largest lake in the world, was almost completely drained for agricultural purposes by the Soviet Union beginning in the 1960s. However, the diversion of water has had a positive economic effect for some in central Asia.

ACTIVITY **Making a Display** Research a decision by a leader or a government that had an impact on the regional environment. The impact could be immediate or long-term, and could be beneficial, harmful, or both. Then create a display that illustrates what motivated the decision and how it changed the environment. Include information about what cultural elements may have influenced the decision. Explain what the environment in the affected region was like both before and after the decision.

» The Aral Sea in the year 2000

» The Aral Sea in the year 2017

D Making Connections to Today

Globalization has defined recent history for most people. As more and more societies become connected, they gain advantages such as improved technology, cheaper goods, healthier foods, and diversity in the arts. However, globalization has also led to new attention to the preservation of peoples' own traditions and identities.

ACTIVITY **Designing a Museum Exhibit** Choose a cultural element from the place where you live or a place where you previously have lived. Consider how globalization could change, or may have already changed, that element or its use. Then design a small museum exhibit showcasing up to three artifacts related to that cultural element. Think about what is most important about that element that you would want a museum visitor far into the future to see and understand. Finalize your design either on paper or on a computer.

Mount Etna in Italy is one of the most active volcanoes in the world.

Physical Geography

INTRODUCTION LESSON

01 Introducing Physical Geography   G72

LEARN THE CONCEPTS LESSONS

02 Planet Earth   G77

03 Forces of Change   G83

05 Factors Affecting Climate   G95

06 Climates and Biomes of the World   G99

INQUIRY ACTIVITY LESSONS

04 Analyzing Sources: Forces Shaping the Earth's Crust   G89

07 Analyzing Sources: Geography and Climate   G103

REVIEW AND APPLY LESSON

08 Reviewing Physical Geography   G109

01

Introducing Physical Geography

What Shapes Our World?

Earth has air, land, and water that make it suitable for plant, animal, and human life. Natural forces inside and outside of Earth shape its surface.

Earth's Structure

The surface of Earth is made up of water and land. We breathe air from a layer of gases that extends above Earth's surface.

Earth's water is found in oceans as well as underground and in lakes, rivers, and streams. Together the waterfall, cliffs, and clouds shown here surrounding the village of Gásadalur, Faroe Islands, Denmark, represent the part of Earth that supports life.

Did You Know?

Natural forces create the physical features you see on Earth. The movement of water and wind can change the shape of these features.

- mountains
- valleys
- cliffs
- lakes
- rivers

The waves at the beach on the coast of Ireland have made once-jagged rocks smooth and rounded.

Earth's Climates

There are many things that influence climate, depending on the characteristics of the location. Climates are organized into climate regions, each with its own plant and animal life.

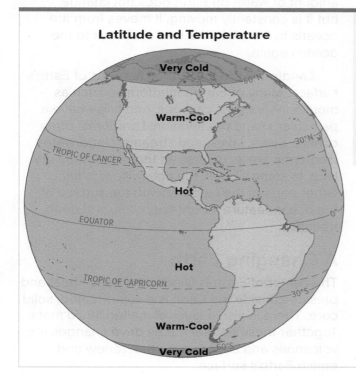

Latitude and Temperature

Very Cold

Warm-Cool

TROPIC OF CANCER

Hot

EQUATOR

Hot

TROPIC OF CAPRICORN

Warm-Cool

Very Cold

60°N

30°N

0°

30°S

60°S

Latitude plays a major role in temperature and climate. The farther away you get from the Equator, the cooler the climate.

Think about Alaska. Although it is part of the United States, it is located close to the North Pole. It is cold and snowy with cool summers.

Now think about Florida, which is located closer to the Equator. Its climate is warm most of the year.

Earth's climate zones range from tropical to high latitude. Tropical areas are located near the Equator. They are hot and often covered in large grasslands. The tropical grasslands of eastern Africa (left) are home to such wildlife as zebras and wildebeests. In contrast, areas farther away from the Equator have colder climates. Some parts of these high-latitude climate regions, like the Canadian Rockies (right), have evergreen trees.

Getting Ready to Learn About . . .
Physical Geography

Water, Land, and Air

The surface of Earth is made up of water and land. Oceans, lakes, and other bodies of water cover most of Earth's surface. The rest of Earth's surface is land, including the seven continents and numerous islands. The air we breathe is part of the layer of gases extending above Earth. Together, Earth's water, land, and air make it possible for people, animals, and plants to live.

Almost all of the water on Earth, about 97 percent, is salt water found in oceans, seas, and a few large salt water lakes. The rest of Earth's water is freshwater found in lakes, streams, rivers, and water in the ground. The total amount of water on Earth does not change, but it is constantly moving. It moves from the oceans to the air to the land and back to the oceans again.

Landforms are the natural features of Earth's surface. Many of Earth's landforms—such as mountains, valleys, canyons, and hills—have a particular shape or elevation. Landforms often contain rivers, lakes, and streams. Underwater landforms are as diverse as those found on dry land. In some places, the ocean floor is flat. Other parts of the land beneath the surface of the ocean feature mountains, cliffs, and deep trenches.

A Changing Earth

The center of Earth is filled with intense heat and pressure. Inside the Earth is a superheated, solid core. There is also a layer of melted liquid metal. Together these natural forces drive changes like volcanoes and earthquakes that renew and enrich Earth's surface.

Earth's surface has changed greatly over time. Some of these changes come from forces related to the movement of the large slabs of rock that cover Earth. Scientists estimate that these slabs of rock, or plates, have been moving slowly around the globe shaping Earth's surface for 2.5 to 4 billion years. This plate movement has produced Earth's largest features—not only continents, but also oceans and mountain ranges. Their movement is slow, so we don't feel it. Plates may crash into each other, pull apart, or grind and slide past each other. They push up mountains, create volcanoes, and produce earthquakes.

In addition to plate movement, forces such as wind, water, and temperature changes also transform Earth's surface. The movement of wind, water, and ice breaks down rocks and wears away Earth's surface. For example, pounding waves wear rocks into sandy beaches. Sometimes these processes carry dust, sand, and soil from one place to another. In some instances, the dust carried by wind forms large deposits of rich soil.

The stunning rock formations at Arches National Park in Utah were created as water seeped through cracks in the rock. Ice formed, expanding the cracks and weakening the rock. Years and years of wind and water wore through the rock, and pieces began to drop away.

Nature Picture Library / Alamy Stock Photo

Earth, Sun, and Climate

Daily life on Earth is influenced by the dynamic relationship between Earth and the sun. The amount of direct sunlight reaching Earth's surface plays an important role in affecting the temperature of different places. The Earth's rotation, which occurs once every 24 hours, determines when we receive sunlight, giving us day or night. The Earth's tilt and its revolution around the sun each year result in the four seasons—winter, spring, summer, and fall—we experience.

The climate of a particular place may have extreme conditions and temperature ranges that are caused by geographic features. Both the latitude and the elevation of a place, along with wind and ocean currents, influence its climate. Landforms and bodies of water can also affect the climate of a place. The Earth's atmosphere influences climate because it filters energy from the sun.

During Earth's annual revolution around the sun, the sun's direct rays fall on the planet in a regular pattern. This pattern can be associated with bands, or zones of latitude, to describe climate regions. Within each latitude zone, climate follows general patterns. Within each broad zone, geographers further divide the zones into smaller regions.

The low latitudes lie between about 30° N to 30° S latitude. Most of the low latitudes are located in the Tropics between the Tropic of Cancer (23.5° N) and the Tropic of Capricorn (23.5° S). Parts of the low latitudes receive the direct rays of the sun year-round. Places in the low latitudes have warm to hot climates.

The midlatitudes are located between about 30° N and 60° N in the Northern Hemisphere and between about 30° S and 60° S in the Southern Hemisphere. The midlatitudes generally have a climate that ranges from fairly hot to fairly cold.

The high latitudes stretch from about 60° N to 90° N in the Northern Hemisphere, and from about 60° S to 90° S in the Southern Hemisphere. These are often called the polar areas because they include the North Pole and the South Pole. Freezing temperatures are common throughout the year because these areas never receive the direct rays of the sun. As a result, the amount and variety of plant life are limited.

Abundant rainfall supports the growth of forests in some midlatitude climates like that of New York in the United States (top). High latitude climates like the Andes region of South America (bottom) have sparse vegetation because of the lack of sunlight and cold temperatures.

Looking Ahead

You will learn about Earth's structure and patterns in the environment that are part of the biosphere. You will learn about the forces of change that shape Earth's surface. You will learn about the factors that influence the world's climate regions. You will examine Compelling Questions and develop your own questions about physical geography in the Inquiry Activity Lessons. You will see some of the ways in which you might already use the ideas of physical geography in your life by completing these lessons.

What Will You Learn?

In these lessons about places and regions, you will learn

- what forms the biosphere;
- that landforms are grouped by their characteristics;
- that water on Earth's surface can be freshwater or salt water;
- how the water cycle works;

- how acid rain is formed, and how it upsets the balance of the ecosystem;
- that Earth is made up of the core, the crust, and the mantle;
- that Earth's rigid crust is made up of pieces called tectonic plates;
- how the movement of plates changes Earth's surface features;
- how weathering and erosion change things on Earth's surface;
- the difference between weather and climate;
- the factors that influence climate;
- how climate is classified;
- that each climate region has its own unique types of natural vegetation and animal life;
- the indicators of climate change.

 COMPELLING QUESTIONS

- **Why is Earth's surface constantly changing?**
- **Why does the Earth have a great diversity of climates and ecosystems?**

Studying Physical Geography

Planet Earth
Earth's physical systems consist of four major subsystems: the hydrosphere, the lithosphere, the atmosphere, and the biosphere.

Forces of Change
Powerful processes operate below and on top of the surface of Earth to create the physical environment

Factors Affecting Climate
Elevation, wind and ocean currents, weather, and landforms influence climate.

Climates of the World
Climates in the world are organized into four zones: tropical, dry, midlatitude, and high latitude climates.

02
Planet Earth

Getting to Know Earth

GUIDING QUESTION

What are Earth's four major physical systems?

Powerful **processes** operate in the physical systems on the surface of Earth. Earth's physical systems include the hydrosphere, the lithosphere, the atmosphere, and the biosphere.

The *hydrosphere* is the system that consists of Earth's water. Water is found in the ocean, seas, lakes, ponds, rivers, groundwater, ice, and as water vapor in the air. About 71 percent of Earth's surface is water, but only 3 percent of the water on Earth is freshwater. About 29 percent of Earth's surface is land above sea level. Land makes up the part of Earth called the *lithosphere*. Landforms are the shapes that occur on Earth's surface. Landforms include plains, hills, plateaus, and mountains. Land also occurs in the ocean basins, the land beneath the ocean.

The air we breathe is part of the *atmosphere*, the thin layer of gases that envelop Earth. The atmosphere is made up of about 78 percent nitrogen, 21 percent oxygen, and small amounts of other gases. The atmosphere is thickest at Earth's surface and gets thinner higher up. About 98 percent of the atmosphere is found within 16 miles

process a series of actions that produce something

Water, Land, and Air
The atmosphere, hydrosphere, and lithosphere contain the biosphere, the life that exists on Earth.

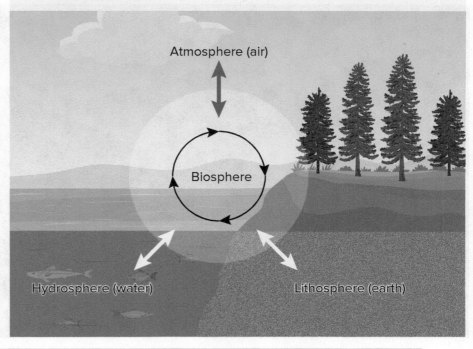

Atmosphere (air)

Biosphere

Hydrosphere (water)

Lithosphere (earth)

(26 km) of Earth's surface. Outer space begins at 100 miles (161 km) above Earth, where the atmosphere ends.

The *biosphere* is made up of all that is living on Earth's surface, close to the surface, or in the atmosphere. All people, animals, and plants form the biosphere.

Landforms

Earth has many different landforms. When scientists study landforms, they find it useful to group them by characteristics. One characteristic that is often used is elevation.

Elevation describes how far above sea level a landform or a location is. Low-lying areas, such as ocean coasts and deep valleys, may be just a few feet above sea level. Mountains and highland areas can be thousands of feet above sea level. Even flat areas of land can have high elevations, especially when they are located far inland from ocean shores.

Plateaus are flat areas that lie at high elevations. A steep cliff often forms at least one side of a plateau. *Plains* are also flat areas but they are found at lower elevations. Some plains have a gentle roll. They are often found along coastlines but can also be far inland. Some plains are home to grazing animals, such as horses and

elevation the height of a land surface above the level of the sea

antelope. Farmers and ranchers use plains to raise crops and livestock.

A *valley* is a lowland area between higher elevation sides. Some valleys are small, level places surrounded by hills or mountains. Other valleys are huge expanses of land with highlands or mountain ranges on their sides. Because they are often supplied with water and topsoil that run off from the higher lands around them, many valleys have rich soil and are used for farming and grazing livestock.

Another way to classify some landforms is to describe them in relation to bodies of water. Some types of landforms are surrounded by water. *Continents* are the largest of all landmasses, although most continents are bordered by land as well as water. Australia and Antarctica are completely surrounded by water. *Islands* are landmasses that are surrounded by water, but they are much smaller than continents.

A **peninsula** is a long, narrow area of land that extends into a river, lake, or ocean. Peninsulas are connected to a larger landmass at one end. An **isthmus** is a narrow strip of land that connects two larger land areas and has water on both sides. One well-known isthmus is the Central American country of Panama. Panama connects

peninsula a portion of land nearly surrounded by water

isthmus a narrow strip of land that connects two larger land areas

The fertile Punakha valley in Bhutan, South Asia, is famous for rice farming.

Analyzing Visuals How can you tell from this image that farming is an important activity?

The continental shelf is an extension of the coastal plain, the low-lying land next to the ocean.

Summarizing What other types of landforms can be found on the ocean floor?

the continents of North America and South America, and has the Pacific and Atlantic Oceans on either side. Because it is the narrowest place in the Americas, the Isthmus of Panama is the location of the Panama Canal, a human-made canal that connects the two oceans.

The Ocean Floor

In many ways, the ocean floor and land are similar. If you could see an ocean without its water, you would see plains, valleys, mountains, hills, and plateaus. Some landforms were shaped by the same forces.

One type of landform underneath the surface of the ocean is the continental shelf. A **continental shelf** is an underwater plain that borders a continent. Continental shelves usually end at cliffs or downward slopes to the deep ocean floor. When scientists explore oceans, they sometimes find **enormous** underwater cliffs that drop off into total darkness. These cliffs extend downward for hundreds or even thousands of feet. The deepest location on Earth is the Mariana Trench in the Pacific Ocean. A **trench** is a long, narrow, steep-sided cut in the ground or on the ocean floor. At its deepest point, the

Mariana Trench is more than 35,000 feet (10,668 m) below the ocean surface.

Other landforms on the ocean floor include volcanoes and mountains. When underwater volcanoes erupt, islands can form because layers of lava build up until they reach the ocean's surface. Mountains on the ocean floor can be taller than Mount Everest and can also form ranges. The Mid-Atlantic Ridge, the longest underwater mountain range, is longer than any mountain range on land.

✓ **CHECK FOR UNDERSTANDING**

Determining Meaning How is a valley similar to an ocean trench?

Earth's Water

GUIDING QUESTION

What types of water are found on Earth's surface?

Water exists in different forms. Water in each of the three states of matter—solid, liquid, and gas—can be found all over the world. Glaciers, polar ice caps, and ice sheets are large masses

continental shelf the part of a continent that extends out underneath the ocean and is fairly flat, but then drops sharply to the ocean floor

enormous very great in size

trench a long, narrow, steep-sided cut on the ocean floor

of water in solid form. Rivers, lakes, and oceans contain liquid water. The atmosphere contains water vapor, which is water in the form of a gas.

Two Kinds of Water

Water at Earth's surface can be freshwater or salt water. Salt water is water that contains a large percentage of salt and other dissolved minerals. About 97 percent of the planet's water is salt water. Salt water makes up the world's oceans and also a few lakes and seas, such as the Great Salt Lake and the Dead Sea.

Salt water supports a huge variety of plant and animal life, such as marine mammals like whales and dolphins, fish of many types, and other sea creatures. Because of its high concentration of minerals, humans and most animals cannot drink salt water. Humans have

Many people around the world, like these women in Kiniero, Guinea, in West Africa, draw groundwater from wells for drinking and cooking.

developed a way to remove minerals from salt water. **Desalination** is a process that separates most of the dissolved chemical elements to produce water that is safe to drink. People who live in dry regions of the world use desalination to process seawater into drinking water. But this process is expensive.

Freshwater makes up the remaining 3 percent of water on Earth. Most freshwater stays frozen in the ice caps of the Arctic and Antarctic. Only about 1 percent of all water on Earth is the liquid freshwater that humans and other living organisms use. Liquid freshwater is found in lakes, rivers, ponds, swamps, and marshes, as well as in the rocks and soil underground.

Water underground inside Earth's crust is called **groundwater**. Groundwater often gathers in *aquifers*. These are underground layers of rock that allow water to flow. When humans dig wells down into rocks and soil, groundwater flows from the surrounding area and fills the well. Groundwater is an important source of drinking water, and it is used to irrigate crops. It also flows naturally into rivers, lakes, and oceans.

Bodies of Water

You are probably familiar with some of the different kinds of bodies of water. Some bodies of water contain salt water, and others hold freshwater. The world's largest bodies of water are its five vast, salt water oceans.

From largest to smallest, the oceans are named the Pacific, Atlantic, Indian, Southern, and Arctic. The Pacific Ocean covers more area than all of Earth's land combined. The Southern Ocean surrounds the continent of Antarctica. Although it is convenient to name the different oceans, it is important to remember that these water bodies are actually connected and form one global ocean. Things that happen in one part of the ocean can affect the ocean all around the world.

When oceans and lakes meet landmasses, unique land features and bodies of water form. A coastal area where the water is partially surrounded by land is called a *bay*. Bays are protected from rough waves by the surrounding land, making them useful for docking ships, fishing, and boating. Larger areas of ocean

desalination a process that makes salt water safe to drink

groundwater the water contained inside Earth's crust

waters partially surrounded by landmasses are called *gulfs*. Gulfs have many of the features of oceans but are smaller and are affected by the landmasses around them.

Bodies of water such as lakes, rivers, streams, and ponds usually hold freshwater. Freshwater contains some dissolved minerals, but only a small percentage. The fish, plants, and other life forms that live in freshwater cannot live in salty ocean water. Freshwater rivers are found all over the world. Rivers begin at a source where water feeds into them. Some rivers begin where two other rivers meet; their waters flow together to form a larger river. Other rivers are fed by sources such as lakes, natural springs, melting snow flowing down from higher ground, and rainfall.

A river's end point is called the *mouth* of the river. Rivers end where they empty into other bodies of water. A river can empty into a lake, bay, gulf, river, or ocean. A *delta* is an area where sand, silt, clay, or gravel is deposited, often at the mouth of a river. But some deltas are found on the land. They are called inland deltas, and they enrich the soil with the nutrients they deposit. River deltas can be huge areas with their own ecosystems.

Another ecosystem near the mouths of rivers that drain into the ocean is called an *estuary*. An estuary is where freshwater and salt water meet. They contain a **gradient** of saltiness from purely fresh to salty ocean, which makes for a wide range of fish, shellfish, and other creatures.

Bodies of water of all kinds affect the lives of people who live near them. Water provides food, work, transportation, and recreation to people in many parts of the world. People get food by fishing in rivers, lakes, and oceans. The ocean floor is mined for minerals and drilled for oil. All types of waters have been used for transportation for thousands of years. People also use water for sports and recreation, such as swimming, sailing, fishing, and scuba diving. Water is vital to human culture and survival.

✓ **CHECK FOR UNDERSTANDING**

Describing Describe three ways in which water affects your life.

The Water Cycle

GUIDING QUESTION

What is the water cycle?

The total amount of water on Earth does not change, but it is constantly moving—from the oceans to the air to the land and finally back to the oceans. This regular movement of water is called the **water cycle**. The sun drives the water cycle by evaporating water from bodies of water and from plants. *Evaporation* is the changing of liquid water into vapor, or gas. The sun's energy causes evaporation by heating the water and moisture from plants. Water vapor rising from bodies of water and plants is gathered in the air. The amount of water vapor the air holds depends on its temperature. Warm air can hold more water vapor than cool air.

When warm air cools, it cannot retain all of its water vapor, so the excess water vapor changes into liquid water—a process called *condensation*. Tiny droplets of water come together to form clouds. When clouds gather more water than they can hold, they release moisture, which falls to the Earth as *precipitation*—rain, snow, hail, or sleet, depending on the air temperature and wind conditions. This precipitation sinks into the ground or collects in streams and lakes as it returns to the oceans. Soon most of it evaporates, and the cycle begins again.

The amount of water that evaporates is approximately the same amount that falls back to Earth. This amount varies little from year to year. Thus, the total volume of water in the water cycle is fairly constant.

Human actions have done damage to the world's water. Waste from factories and runoff from toxic chemicals used on lawns and farm fields have polluted rivers, lakes, oceans, and groundwater. Chemicals like pesticides and fertilizers seep into wells that hold drinking water, poisoning the water and causing deadly diseases.

Some of the fuels we burn release poisonous gases into the atmosphere. These gases combine with water vapor in the air to create toxic acids. These acids then fall to Earth as a

---

gradient an increase or decrease in the amount of something

water cycle the process in which water is used and reused on Earth, including precipitation, collection, evaporation, and condensation

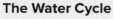

The Water Cycle

Water is constantly moving—from the oceans to the air to the ground and finally back to the oceans.

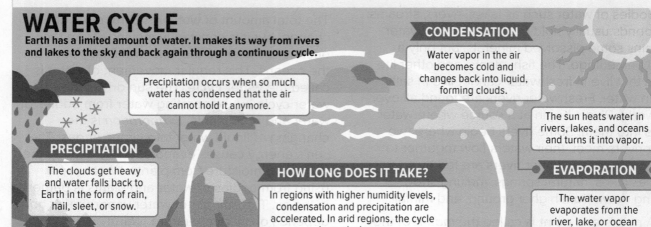

WATER CYCLE

Earth has a limited amount of water. It makes its way from rivers and lakes to the sky and back again through a continuous cycle.

CONDENSATION
Water vapor in the air becomes cold and changes back into liquid, forming clouds.

Precipitation occurs when so much water has condensed that the air cannot hold it anymore.

The sun heats water in rivers, lakes, and oceans and turns it into vapor.

PRECIPITATION
The clouds get heavy and water falls back to Earth in the form of rain, hail, sleet, or snow.

HOW LONG DOES IT TAKE?
In regions with higher humidity levels, condensation and precipitation are accelerated. In arid regions, the cycle is much slower.

EVAPORATION
The water vapor evaporates from the river, lake, or ocean and goes into the air.

When water falls back to Earth as precipitation, it may fall back in oceans, lakes, and rivers or end up on land.

When water ends up on land, it will either soak in and become part of the groundwater, or it may run over the soil and collect in oceans, lakes, or rivers where the cycle starts again.

COLLECTION

Source: http://www.kidzone.ws/water/

EXAMINE THE DIAGRAM

1. **Identifying** What are the four types of precipitation? According to the water cycle, where might this precipitation end up?

2. **Speculating** How might contaminated water end up affecting people even if they live far away from its source?

---

deadly mixture called **acid rain**. Acid rain damages the environment in several ways. It pollutes the water humans and animals drink. The acids damage trees and other plants. As acid rain flows over the land and into waterways, it kills plant and animal life in bodies of water. This upsets the balance of the ecosystems.

✓ **CHECK FOR UNDERSTANDING**

Explaining What causes evaporation?

---

acid rain rain that contains harmful amounts of poisons due to pollution

LESSON ACTIVITIES

1. **Informative/Explanatory Writing** Many cities and towns develop near sources of water. Write a paragraph describing the sources of water used by your community.

2. **Analyzing Information** Work with a partner and consider the ratio of water and land on Earth—71 percent water and 29 percent land. Discuss how life might be different if the proportions were reversed. Create a list to capture your answers.

Forces of Change

Analyzing Key Ideas and Details As you read, take notes about plate movements and how they change Earth.

Type of Plate Movement	Landforms Created

Inside Earth

What are the three layers of the Earth?

Thousands of miles beneath your feet, Earth's heat has turned metal into liquid. Such forces as this operate even though you do not feel them. What lies inside Earth **affects** what lies on top. Mountains, plateaus, and other landscapes were formed over time by forces acting below Earth's surface—and those forces are still changing the landscape.

The inside of Earth is made up of three layers: the core, the mantle, and the crust. The center of Earth—the **core**—has a solid inner core and an outer core of melted liquid metal. Surrounding the outer core is a thick layer of hot, dense rock called the **mantle**. Scientists estimate that the mantle is about 1,793 miles (2,886 km) thick.

affect to act on and cause a change

core the innermost layer of Earth

mantle a thick layer of hot, dense rock surrounding the Earth's core

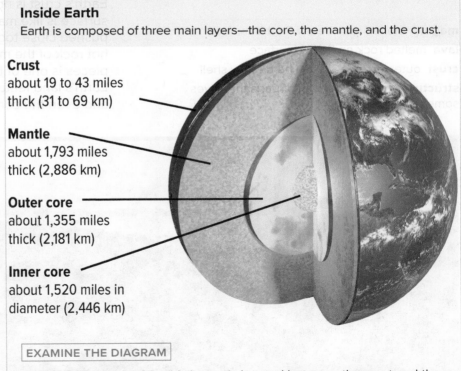

Inside Earth

Earth is composed of three main layers—the core, the mantle, and the crust.

Crust
about 19 to 43 miles
thick (31 to 69 km)

Mantle
about 1,793 miles
thick (2,886 km)

Outer core
about 1,355 miles
thick (2,181 km)

Inner core
about 1,520 miles in
diameter (2,446 km)

Identifying Which of Earth's layers is located between the crust and the outer core?

When volcanoes erupt, the glowing-hot lava that flows from the mouth of the volcano comes from magma in Earth's mantle. **Magma** is melted rock below Earth's surface, but **lava** is melted rock at Earth's surface. The outer layer of Earth is the **crust**, a rocky shell forming the surface of Earth. The crust is thin, ranging from about 19 miles (31 km) thick under the ocean to about 43 miles (69 km) thick under mountains.

The deepest hole ever drilled into Earth is about 8 miles (13 km) deep. That is still within Earth's crust. The farthest any human has traveled down into Earth's crust is only about 2.5 miles (4 km). Still, scientists have developed an accurate picture of the layers that form Earth's **structure**. One important way that scientists do this is to study vibrations that travel deep within Earth. The vibrations are caused by earthquakes, underground explosions, and by large "thumper trucks" that pound on the ground. From observations produced by these sources, scientists have learned what materials are inside Earth and have estimated the thickness and temperature of Earth's layers.

✓ **CHECK FOR UNDERSTANDING**

Describing What makes up the core of Earth?

magma melted rock below Earth's surface

lava melted rock on Earth's surface

crust outer layer of Earth, a hard rocky shell

structure the arrangement of parts that gives something its basic form

Plate Movements

GUIDING QUESTION
How was the surface of Earth formed?

Since Earth was formed, the surface of the planet has been in constant motion. Landmasses have shifted and moved over time. Landforms have been created and destroyed. The way Earth looks from space has changed many times because of the movement of continents.

Earth's Surface

A continent is a large mass of land. Continents are part of Earth's crust. Earth has seven continents: Asia, Africa, North America, South America, Europe, Antarctica, and Australia. The region around the North Pole is not a continent because it is made of a huge mass of dense ice that lies on the surface of the Arctic Ocean, not land. Greenland might seem as big as a continent, but it is classified as the world's largest island. Each continent has features that make it unique.

Even though you usually cannot feel it, the land beneath you is moving. This is because Earth's crust is not a solid sheet of rock. Earth's surface is like many massive puzzle pieces pushed close together and floating on the very hot rock of the mantle. The movement of these pieces is one of the major forces that create Earth's land features. New mountains grow taller and even the continents move.

This photo shows a fault, or crack, in Earth's crust in Thingvellir National Park, Iceland. This is where the Eurasian and the North American tectonic plates meet.

Plates and Plate Movement
Tectonic plates make up Earth's crust.

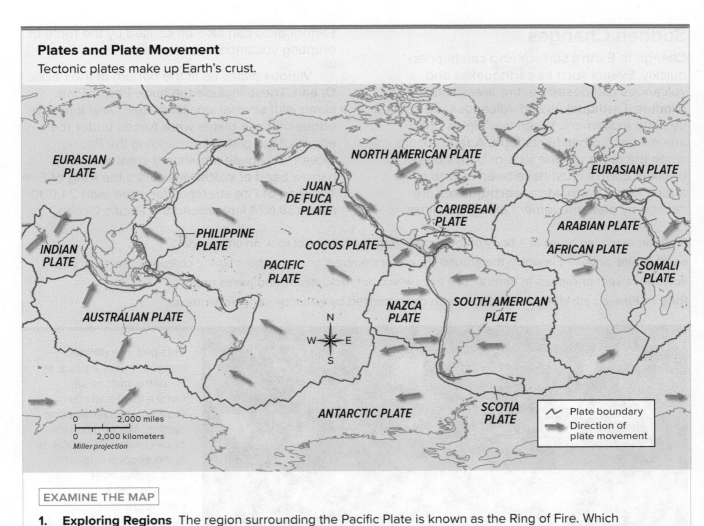

EURASIAN PLATE

NORTH AMERICAN PLATE

EURASIAN PLATE

JUAN DE FUCA PLATE

CARIBBEAN PLATE

ARABIAN PLATE

PHILIPPINE PLATE

COCOS PLATE

AFRICAN PLATE

INDIAN PLATE

PACIFIC PLATE

SOMALI PLATE

AUSTRALIAN PLATE

NAZCA PLATE

SOUTH AMERICAN PLATE

ANTARCTIC PLATE

SCOTIA PLATE

N
W E
S

0   2,000 miles
0   2,000 kilometers
Miller projection

~ Plate boundary
→ Direction of plate movement

EXAMINE THE MAP

1. **Exploring Regions** The region surrounding the Pacific Plate is known as the Ring of Fire. Which tectonic plates border the Pacific Plate, making them part of this chain of volcanic activity?

2. **Exploring Location** Which tectonic plates are moving toward each other along colliding boundaries?

Tectonic Plates

The many enormous puzzle pieces that make up Earth's rigid crust are called **tectonic plates**. These plates vary in size and shape. They also vary in the amount they move over the more flexible layer of the hot mantle below. Heat from deep within the planet causes the plates to move. This movement generally happens so slowly that humans do not feel it. But some of Earth's plates move as much as a few inches each year. This might not seem like much, but over millions of years, it causes the plates to move thousands of miles.

This movement of the plates changes Earth's surface features very slowly. It takes millions of years for plates to move enough to create landforms. Some land features form when plates

are crushed together. At times, forces within Earth push the edge of one plate up over the edge of a plate beside it. This dramatic movement along these *colliding boundaries* can create mountains, volcanoes, and deep trenches in the ocean floor.

At other times, the colliding plates are crushed together in a way that causes the edges of both plates to crumble and break. This event can form jagged mountain ranges. If plates on the ocean floor move apart, known as *spreading boundaries*, the space between them widens into a giant crack in Earth's crust. Magma rises through the crack and forms new crust as it hardens and cools. If enough cooled magma builds up that it reaches the surface of the ocean, an island will begin to form. Sometimes plates grind past one another along what are called *strike-slip boundaries*.

---

tectonic plate one of the large pieces of Earth's crust

Sudden Changes

Change to Earth's surface also can happen quickly. Events such as earthquakes and volcanoes can destroy entire areas within minutes. Earthquakes and volcanoes are caused by plate movement. When two plates grind against each other, faults form. A **fault** results when the rocks on one side or both sides of a break in Earth's crust have been moved by forces within Earth. Many **earthquakes** are caused by plate movement along fault lines.

Earthquakes can also be caused by the force of erupting volcanoes.

Various plates lie at the bottom of the Pacific Ocean. These include the huge Pacific Plate along with several smaller plates. Over time, the edges of these plates were forced under the edges of the plates surrounding the Pacific Ocean. This plate movement created a long, narrow band of **volcanoes** called the **Ring of Fire**. The Ring of Fire stretches for more than 24,000 miles (38,624 km) around the Pacific Ocean.

fault an extended break in a body of rock, marked by movement of rock on either side

earthquake an event in which the ground shakes or trembles, brought about by the collision of tectonic plates

volcano a vent or rupture in Earth's crust from which hot rock, steam, and gases erupt

Ring of Fire a path along the Pacific Ocean characterized by volcanoes and earthquakes

This pair of satellite images shows a place in Sumatra, Indonesia, before (top) and after (bottom) the devastating earthquake and tsunami of December 26, 2004. It is the second-largest earthquake ever recorded by a seismograph. Seismographs are used to measure the strength and duration of earthquakes.

Science History Images/Alamy Stock Photo

The Colorado River has been shaping the main gorge of the Grand Canyon in the United States for millions of years.

Summarizing Explain how weathering and erosion helped form the Grand Canyon.

The intense vibrations caused by earthquakes and erupting volcanoes can transfer energy to Earth's surface. When this energy travels through ocean waters, it can generate a wave that builds up to enormous height when it nears land. Such an ocean wave is called a **tsunami**. It is caused by volcanic eruptions or movement of Earth under the ocean floor. Tsunamis have caused terrible flooding and damage to coastal areas. The forces of these mighty waves can level entire coastlines.

✓ **CHECK FOR UNDERSTANDING**

1. **Describing** What causes plates to move?
2. **Explaining** What causes earthquakes?

Other Forces at Work

GUIDING QUESTION

How can wind and water change Earth's surface?

What happens when the tide comes in and washes over a sand castle on the beach? The water breaks down the sand castle. Similar changes take place on a larger scale across Earth's lithosphere. These changes happen much slower—over hundreds, thousands, or even millions of years.

Weathering

Some landforms are created when materials such as rock and soil build up on Earth's surface. Other landforms take shape as rocks and soil break down and wear away over time. **Weathering** is a process by which Earth's surface is worn away by forces such as wind, rain, chemicals, the change of temperature, and the movement of ice and flowing water. Even plants can cause weathering. Plant roots and small seeds can grow into tiny cracks in rock, gradually splitting the rock apart as the roots expand and the seeds grow.

You may have seen the effects of weathering on an old building or statue made of stone. The edges become chipped and worn, and features such as raised lettering are smoothed down. Landforms such as mountains are affected by weathering, too. The Appalachian Mountains in the eastern United States have become rounded and crumbled after millions of years of weathering by natural forces.

Erosion

Erosion is a process that works with weathering to change surface features of Earth. Erosion occurs when weathered bits of rock are moved elsewhere by water, wind, or ice. Rain and moving water can weather and erode even the hardest stone over time. When material is broken down by weathering, it can easily be carried away by the action of erosion. For example, the Grand

tsunami a sea wave caused by an undersea earthquake that gets larger when it nears land

weathering the process of wearing away or breaking down rocks into smaller pieces

erosion the process by which weathered bits of rock are moved elsewhere by water, wind, or ice

Canyon was formed by weathering and erosion that occurred by flowing water and blowing winds. Water flowed over the region for millions of years, weakening the surface of the rock. The moving water carried away tiny bits of rock. Over time, weathering and erosion carved a deep canyon into the rock. Erosion by wind and chemicals caused the Grand Canyon to widen until it became the amazing landform we see today.

Weathering and erosion cause different materials to break down at different speeds. Soft, porous rocks, such as sandstone and limestone, wear away faster than dense rocks like granite. The spectacular rock formations in Utah's Bryce Canyon were formed as different types of minerals within the rocks were worn away by erosion, some more quickly than others. The result is landforms with jagged, rough surfaces and unusual shapes.

The Okpilak River in Alaska shows how rivers deposit soil where they empty into larger bodies of water. This creates coastal plains and wetland areas.

Buildup and Movement

The buildup of materials creates landforms such as beaches, islands, and plains. Ocean waves pound coastal rocks into smaller and smaller pieces until they are tiny grains of sand. Over time, waves and ocean currents deposit sand along coastlines, forming sandy beaches. Sand and other materials carried by ocean currents build up on mounds of volcanic rock in the ocean, forming islands. Rivers deposit soil where they empty into larger bodies of water, creating coastal plains and wetland ecosystems.

Entire valleys and plains can be formed by the incredible force and weight of large masses of ice and snow. These masses are often classified by size as glaciers, polar ice caps, or ice sheets. A **glacier**, the smallest of the ice masses, moves slowly over time, sometimes spreading outward on a land surface. There were many glaciers during the Ice Age, but they can still be found on Earth today.

Ice caps are high-altitude ice masses. Ice sheets, extending more than 20,000 square miles (5,800 sq. km), are the largest ice masses. Ice sheets cover most of Greenland and Antarctica.

✓ **CHECK FOR UNDERSTANDING**

1. **Contrasting** How does weathering differ from erosion?

2. **Identifying Cause and Effect** How does weathering contribute to erosion?

LESSON ACTIVITIES

1. **Informative/Explanatory Writing** Write a paragraph describing the plate movement of one Earth's continents. Identify the tectonic plates associated with the continent and whether or not the continent is moving away from or toward another continent.

2. **Analyzing Information** There are many online resources dealing with plate tectonics. Work with a partner to discover which sites are the most helpful in explaining these processes. Before you begin, create a list of the criteria you will use to judge each site. Share your findings with the class.

glacier a large body of ice that moves slowly across land

Analyzing Sources: Forces Shaping the Earth's Crust

 COMPELLING QUESTION

Why is Earth's surface constantly changing?

Plan Your Inquiry

DEVELOPING QUESTIONS

Think about the kinds of landforms you know about on the Earth's surface, or crust. Consider what forces might have been involved in shaping those features of the Earth's crust. Then read the Compelling Question for this lesson. What questions can you ask to help you answer this Compelling Question? Create a graphic organizer like the one below. Write these Supporting Questions in your graphic organizer.

Supporting Questions	Source	What this source tells me about why the Earth's surface is constantly changing	Questions the source leaves unanswered
	A		
	B		
	C		
	D		
	E		

ANALYZING SOURCES

Next, examine the sources in this lesson. Analyze each source by answering the questions that follow it. How does each source help you answer each Supporting Question you created? What questions do you still have? Write these in your graphic organizer.

After you analyze the sources, you will
- use the evidence from the sources;
- communicate your conclusions;
- take informed action.

Background Information

Humans, like all other land-based species, live on the outer layer of the Earth's crust. The shape of the land's surface can have great impact on habitats and influences what kinds of life are possible. For example, high mountains can influence rainfall patterns, and broad river valleys have deep and rich soil to support plant life.

The land's surface is also constantly changing. Forces are continuously at work. They combine to create diverse features in locations all over Earth. New features are created while others change or disappear.

» The Earth's outer crust takes on extreme shapes in the Patagonian section of the Andes mountains in Argentina.

GO ONLINE Explore the Student Edition eBook and find interactive maps, charts, graphs, and tools.

G89

The Himalaya

The Himalaya mountain ranges (Himalayas) contain the highest mountains in the world, including Mount Everest, Earth's highest peak at 29,032 feet (8,849 m) above sea level. The mountain ranges extend for approximately 1,500 miles (2,414 km) and pass through India, Pakistan, Afghanistan, China, Bhutan, and Nepal. The Himalaya are the source of several major river systems in Asia. Perhaps surprisingly given their massive size, the Himalaya are among the youngest mountains in the world. A website created by the U.S. Geological Survey explains how the Himalaya were created as a result of plate tectonics.

SECONDARY SOURCE: WEBSITE

The Himalayas: Two Continents Collide

66 Among the most dramatic and visible creations of plate-tectonic forces are the lofty Himalayas, which stretch 2,900 kilometers along the border between India and Tibet. This immense mountain range began to form between 40 and 50 million years ago, when two large landmasses, India and Eurasia, driven by plate movement, collided. Because both these continental landmasses have about the same rock density, one plate could not be **subducted** under the other. The pressure of the **impinging** plates could only be relieved by thrusting skyward, contorting the collision zone, and forming the jagged Himalayan peaks...

About 80 million years ago, India was located roughly 6,400 kilometers south of the Asian continent, moving northward at a rate of about 9 meters a century. When India rammed into Asia about 40 to 50 million years ago, its northward advance slowed by about half...In just 50 million years, peaks such as Mt. Everest have risen to heights of more than 9 kilometers...The Himalayas continue to rise more than 1 cm a year—a growth rate of 10 kilometers in a million years! 99

—"The Himalayas: Two continents collide," U.S. Geological Survey website, 2015.

subducted forced under another layer of crust
impinging striking against

The path of the landmass of India leading to its collision with the Eurasian Plate about 40 to 50 million years ago.

EXAMINE THE SOURCE

1. **Identifying Cause and Effect** What is the force that caused the dramatic uplift of the mountain range?

2. **Inferring** Mount Everest is Earth's highest peak. It was measured at 29,032 feet (8,849 m) tall in 2020. Would you expect Mt. Everest to be taller when measured again in 100 years?

United States Geological Survey, https://pubs.usgs.gov/gip/dynamic/himalaya.html. 2015.

The Mexican Volcano Parícutin

Sometimes natural forces can create rapid change on Earth's surface. In 1943, a farmer in Michoacán, Mexico, made a discovery in his cornfield. The phenomenon he observed would last for nine years and lead to the destruction of a town. Later, the location became a tourist attraction. This article from 1943 explains why a cornfield in Mexico suddenly became a place of great interest to geologists.

PRIMARY SOURCE: ARTICLE

Parícutin in October 1943, during the first year of an eruption that lasted nine years.

❝ The new volcano in Mexico, El Parícutin (pronounced pah-ree-koo-teen) is a unique geological phenomenon; for, before our very eyes, it has sprung into existence and has grown to a very respectable height of 1,500 feet, all within a period of 8 months. It lies within a region in which no previous volcanic activity has been known within the memory of man, though in 1759 the volcano El Jorullo, some 50 miles to the southeast, likewise suddenly was born, grew to a height of more than 1,000 feet within 5 months, and then quieted down, never more to erupt violently. Will Parícutin do likewise? That remains to be seen, for at present it is still going strong.

For the first time in their lives geologists have been able to observe in a single volcano all stages of its history. Parícutin exhibits many of the features of other volcanoes; but other volcanoes have been encountered by geologists after they have been in existence for some time, and their early history is unknown. The early history of Parícutin therefore fills important gaps in our understanding of volcanism.

To me the most outstanding aspect of this volcano is the incredible rapidity with which it grew. Within one week it was 550 feet high and within 10 weeks it was 1,100 feet in altitude. Up to this time, all the material in its cone had come from fragments that had been blown into the air from the volcano. No lava came from the cone until nearly four months after the eruption started; and then, contrary to some popular reports, it did not flow over the lip of the crater. Instead, it broke through the sides of the cone, undermining the overlying fragmental material. Lava appeared within two days of the first explosion, but it issued quietly from a fissure about 1,000 feet north of the explosive vent. ❞

— Address presented before the Geologic Section of the New York Academy of Sciences in New York, October 4, 1943. Published by permission of The Director, Geological Survey, U.S. Department of the Interior.

EXAMINE THE SOURCE

1. **Determining Context** What makes Parícutin an important volcano for geologists to study?

2. **Making Connections** What information in the source suggests that the region around Parícutin might continue to be volcanically active in the future?

Slot Canyons

Slot canyons are features that are found throughout the world, wherever the conditions for their formation exist. Hikers can explore the narrow passageways of slot canyons that wind through solid rock. This article explains how slot canyons are created.

PRIMARY SOURCE: ARTICLE

Travelers must travel through a slot canyon to reach the ancient city of Petra, Jordan.

" It begins with a simple crack in the rock on the ground. But add a few million years and that crack opens into a deep winding gorge in the earth, with a narrow path and sheer sides. The crack has become a slot canyon.

Slot canyons—the narrow, tall channels through otherwise solid rock—can be found anywhere in the world but are particularly numerous in the southwestern U.S. and Australia, where the perfect canyon-forming combination of soft rock and extreme climate collide. It happens like this: the initial crack is covered by a flash flood from heavy rain pooling in a natural wash. The water seeps into the crack, bringing with it rocks, sediment, and other debris that carve a little bit away from the inside edges of the crack. Rain, flood, repeat. Sandstone is most susceptible to this kind of earth carving, but slot canyons can also form out of limestone, granite, basalt and other types of rock.

Once formed, careful hikers can trek through the base of these otherworldly canyons, shimmying through tapered sections, bracing themselves against both walls in the narrowest portions and beholding scenery unlike just about anything else in the world. Intrigued? Be sure to plan carefully or take a guide as flash floods and extreme conditions can make these canyons as dangerous as they are beautiful. "

—"Shimmy Through the World's Most Spectacular Slot Canyons," Jennifer Billock, *Smithsonian Magazine*, 2017

EXAMINE THE SOURCE

1. **Identifying** What force is responsible for carving slot canyons?
2. **Explaining** How does the process that creates slot canyons create a potential danger for people who explore them?

PHOTO: Jan Wlodaczyk/Alamy Stock Photo; TEXT: Jennifer Billock. "Shimmy Through the World's Most Spectacular Slot Canyons". Smithsonian Magazine. https://www.smithsonianmag.com/travel/most-beautiful-slot-canyons-180962270/.

Great Sand Dunes

Great Sand Dunes National Park in Colorado is home to the tallest sand dunes in North America. A website created by the Department of the Interior helps visitors understand this impressive feature.

SECONDARY SOURCE: WEBSITE

> At Great Sand Dunes, you can witness geology coming alive. Sand is dormant, but the forces of wind and water that move geological structures are very much active. The traditional Ute word for the Great Sand Dunes is *Saa waap maa nache*, "sand that moves." Though the locations of the larger dune forms have remained fairly constant for centuries, some smaller dunes may grow and migrate across grasslands. How do dunes grow? Each day, sand erodes from surrounding mountains and is carried by wind and water to the dunes. After thousands of years, these tiny fragments of rock begin to add up, gradually forming the largest dunes in North America.

—"12 Things You Didn't Know About Great Sand Dunes National Park and Preserve," U.S. Department of the Interior website, 2020

Elk stand in front of the Great Sand Dunes near the Sangre de Cristo Mountains.

EXAMINE THE SOURCE

1. **Explaining** What is the source of the sand that forms the dunes?
2. **Identifying** What moves the sand?

Mississippi River Delta

The Mississippi River is the longest river in North America. It collects drainage waters from approximately one-eighth of the continent and deposits them into the Gulf of Mexico. Near the mouth of the river, a new landform has been created. Sediment and plant matter have built up in layers over time to create land. New land that forms at the mouth of a river is called a delta. The photo, a satellite image of the Mississippi Delta taken in 2001, shows land that has been created in this way. Note how the river branches out in several directions as the course of the river has shifted to travel through and around each new land formation.

PRIMARY SOURCE: PHOTO

EXAMINE THE SOURCE

1. **Summarizing** Where does the sediment come from that forms new land in the delta?
2. **Speculating** How might the delta be affected by the construction of dams along the Mississippi that trap sediment upriver?

Complete Your Inquiry

EVALUATE SOURCES AND USE EVIDENCE

Refer back to the Compelling Question and the Supporting Questions you developed at the beginning of the lesson.

1. **Drawing Conclusions** Which sources provide evidence that water is significant in changing the surface of the land?
2. **Making Connections** Which two sources describe how plate tectonics change the surface of Earth?
3. **Gathering Sources** Which sources helped you answer the Supporting Questions and the Compelling Question? Which sources, if any, challenged what you thought you knew when you first created your Supporting Questions? What information do you still need in order to answer your questions? What other viewpoints would you like to investigate? Where would you find that information?
4. **Evaluating Sources** Identify the sources that helped answer your Supporting Questions. How reliable is the source? How would you verify its reliability?

COMMUNICATE CONCLUSIONS

5. **Collaborating** Share with a partner which of these sources you found most useful in answering your Supporting Questions. Discuss how the sources help you understand how the land's surface is shaped into landforms. Take turns relating how processes described in the sources might explain how features you are aware of in your local area were formed. How do these sources support your thinking? Use the graphic organizer that you created at the beginning of the lesson to help you. Share your conclusions with the class.

TAKE INFORMED ACTION

Showing Love to a Local Landform
Think about the features that define the physical geography of your local area. Identify one landform or feature that you think is important and research the forces that have shaped it. Consult someone with knowledge of the area if needed. Then create a poster to draw attention to this landform or feature. Include an image and a brief explanation of how the feature was formed. Display the poster in your school or community.

05

Factors Affecting Climate

READING STRATEGY

Analyzing Key Ideas and Details As you read, take notes about the different factors that affect climate.

Factor	How It Affects Climate
Elevation	
Wind & Ocean Currents	
Atmosphere	
Landforms	

Weather and Climate

GUIDING QUESTION

How does the relationship between Earth and the sun influence climate?

Weather is the state of the atmosphere at a given time, such as during a week, a day, or an afternoon. Weather refers to conditions such as hot or cold, wet or dry, calm or stormy, or cloudy or clear. Weather is what you can observe at any time by going outside or looking out a window. **Climate** is the average weather conditions in a region or an area over a longer period. One useful measure for comparing climates is the average daily temperature. This is the average of the highest and lowest temperatures that occur in a 24-hour period. In addition to the average temperature, climate includes the typical wind conditions and rainfall or snowfall that occur in an area year after year.

While the Earth rotates on its **axis**, it also revolves around the sun. It takes the Earth one year, approximately 365 days, to complete one **revolution** around the sun. The Earth's revolution, combined with its tilted axis, affects the amount of sunlight that reaches different locations on Earth at different times of the year.

Earth is tilted 23.5 degrees on its axis. If you look at a globe that is attached to a stand, you will see what the tilt looks like. Because of the tilt, not all places on Earth receive the same amount of direct sunlight at the same time.

As Earth revolves around the sun, it stays tilted. This means that one-half of the planet is usually tilted toward the sun, while the other half is tilted away. As a result, Earth's Northern and Southern Hemispheres experience the warmer and cooler seasons at different times of the year.

On about June 21, the North Pole is tilted toward the sun. The Northern Hemisphere is receiving the direct rays of the sun. The sun appears directly overhead at noon on the Tropic of Cancer. This day is the summer **solstice**, or the beginning of summer, in the Northern Hemisphere. It is the day of the year when the Northern Hemisphere experiences the most hours of sunlight during Earth's 24-hour rotation. At the same time, the day is shortest in the Southern Hemisphere.

Six months later—on about December 22—the North Pole is tilted away from the sun. The sun's direct rays strike the Tropic of Capricorn. This is called the winter solstice in the Northern Hemisphere because it

weather condition of the atmosphere in one place during a short time

climate weather patterns typical for an area over a long period of time

axis an imaginary line that runs through the center of Earth between the Poles

revolution a complete trip of Earth around the sun

solstice one of two days of the year when the sun reaches its northernmost or southernmost point

is when winter starts here. But it is when summer begins in the Southern Hemisphere. The days are short in the Northern Hemisphere but long in the Southern Hemisphere.

Midway between the two solstices, about September 23 and March 21, the sun's rays are directly overhead at the Equator. These are **equinoxes**, when day and night in the Northern and Southern hemispheres are of equal length—12 hours of daylight and 12 hours of nighttime everywhere on Earth.

 CHECK FOR UNDERSTANDING

Identifying When it is winter in the Southern Hemisphere, what season is it in the Northern Hemisphere?

What Influences Climate?

GUIDING QUESTION
How do elevation, wind and ocean currents, weather, and landforms influence climate?

The climate of a particular place may have extreme weather and temperature ranges that are caused by several geographic features. Both the latitude and the elevation of a place, along with wind and ocean currents, influence its climate. Two sides of a mountain range may also

---

equinox one of two days each year when the sun is directly overhead at the Equator

---

have two different climates. While one side receives more **precipitation** as air rises, the opposite side has drier, warmer air as the air descends.

Elevation

Elevation is an important influence on climate. Although warmer temperatures are found in the low latitudes and cooler temperatures in the high latitudes, elevation influences climate at all latitudes. This is because Earth's atmosphere thins as altitude increases. Thinner air **retains** less heat. As elevation increases, temperatures decrease by about 3.5°F (1.9°C) for every 1,000 feet (305 m). For example, if the temperature averages 70°F (21.1°C) at sea level, the average temperature at 5,000 feet (1,524 m) is only 53°F (11.7°C). A high elevation will be colder than lower elevations at the same latitude.

Wind and Ocean Currents

In addition to latitude and elevation, the movement of air and water helps create Earth's climates. Moving air and water help circulate the sun's heat around the globe.

Movements of air are called winds. Winds are the result of changes in air pressure caused by uneven heating of Earth's surface. Warmer air

---

precipitation the water that falls on the ground as rain, snow, sleet, hail, or mist

retain to keep or hold

Earth's Seasons
The tilt of Earth as it revolves around the sun is what causes the seasons to change.

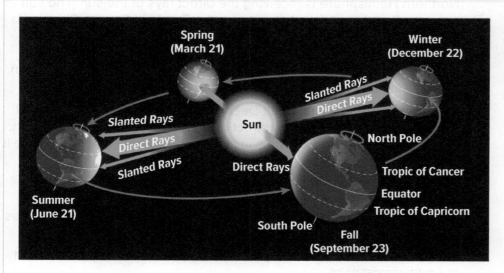

Spring
(March 21)

Winter
(December 22)

Slanted Rays
Direct Rays

Slanted Rays
Direct Rays

Sun

Slanted Rays

North Pole

Direct Rays

Tropic of Cancer

Summer
(June 21)

Equator

Direct Rays

Tropic of Capricorn

South Pole

Fall
(September 23)

EXAMINE THE DIAGRAM

Analyzing Why are the seasons reversed in the Northern and Southern Hemispheres?

Global Wind Patterns

Wind patterns and zones of latitude both influence Earth's weather and climate.

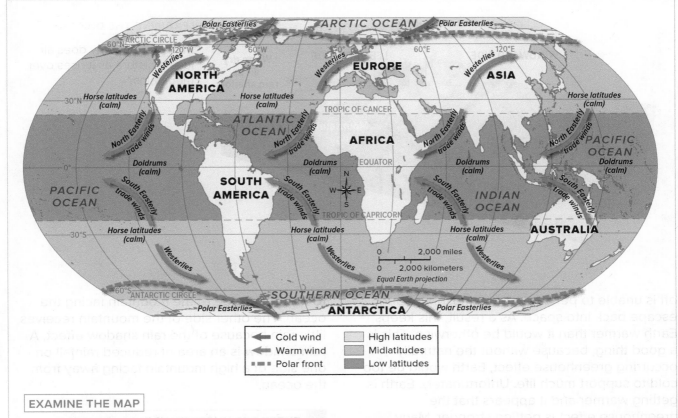

EXAMINE THE MAP

1. **Patterns and Movement** In what general directions do winds blow over Africa?
2. **Exploring Regions** What wind currents blow over the midlatitudes?

---

tends to rise higher in Earth's atmosphere, which creates low pressure at the surface.

Winds are created as air is drawn across the surface of Earth toward the low-pressure areas. The Equator is **constantly** warmed by the sun, so warm air masses tend to form near the Equator. This air rises, which creates low pressure at the surface, and then cooler, high-pressure air rushes in, causing wind. This helps balance Earth's temperature as it creates winds that follow prevailing, or typical, patterns.

Just as winds move in patterns, cold and warm streams of water, known as currents, circulate through the oceans. Warm water moves away from the Equator, transferring heat energy from the equatorial region to higher latitudes. Cold water from the polar regions moves toward the Equator, also helping to balance the temperature of the planet.

constant happening all the time

The Atmosphere

Earth's atmosphere has a strong influence on climate because it filters energy. The sun sends out many kinds of energy, including X-rays and ultraviolet rays as well as visible light. Some of that radiation would be harmful to humans and other animals, but the atmosphere stops it from reaching the surface of Earth.

The atmosphere also regulates energy through the naturally occurring *greenhouse effect*. Even as the atmosphere stops harmful radiation, it allows visible light to enter. You have probably noticed that a bright, sunny day is warmer than a dark, cloudy day, or than night time. This is because the light from the sun warms Earth's atmosphere. But the light also warms the land of the lithosphere and the water of the hydrosphere. The land and water absorb the energy from the light and convert it into heat. Most of the heat energy the land and water give

The Rain Shadow Effect

A rain shadow affects the amount of rain a region receives.

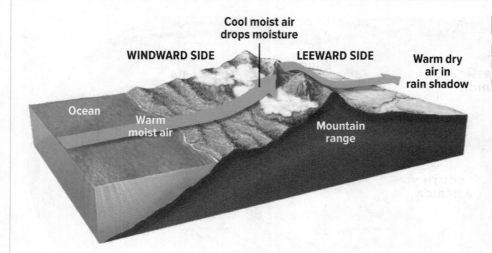

Cool moist air drops moisture

WINDWARD SIDE **LEEWARD SIDE**

Warm dry air in rain shadow

Ocean

Warm moist air

Mountain range

EXAMINE THE DIAGRAM

Interpreting Why does air lose moisture as it rises over mountains?

off is unable to penetrate the atmosphere and escape back into space. As a result, this keeps Earth warmer than it would be otherwise. This is a good thing, because without the naturally occurring greenhouse effect, Earth would be too cold to support much life. Unfortunately, Earth is getting warmer and it appears that the greenhouse effect is getting stronger. Many people think this is occurring because of the emissions that come as humans burn fossil fuels like coal, oil, and natural gas. Many countries have worked together to create plans to slow down the human contribution to climate change.

Landforms

It might seem strange to think that landforms such as mountains can affect weather and climate, but landforms and landmasses change the strength, speed, and direction of wind and ocean currents. These flows carry heat and precipitation, which shape weather and climate. The sun warms the land and the surface of the world's oceans at different rates, causing differences in air pressure. As winds blow inland from the oceans, they carry moist air with them. As the land rises in elevation, the atmosphere cools. When masses of moist air approach mountains, the air rises and cools, causing rain

to fall on the side of the mountain facing the ocean. The other side of the mountain receives little rain because of the rain shadow effect. A **rain shadow** is an area of reduced rainfall on one side of a high mountain facing away from the ocean.

✓ **CHECK FOR UNDERSTANDING**

1. **Explaining** What happens to temperature as elevation increases?

2. **Describing** What causes winds?

LESSON ACTIVITIES

1. **Narrative Writing** Imagine you live on a planet that is not tilted on its axis. What might this planet's seasons be like? How might life on this planet differ from life on Earth? Write a fictional narrative essay that addresses these questions.

2. **Using Multimedia** Working with a partner, research to find information about one of the four seasons—winter, spring, summer, and autumn. Create a slide show about the season you chose. Explain how the movement of Earth creates this season. Be sure to describe the typical seasonal weather where you live.

rain shadow an area that receives reduced rainfall because it is on the side of a mountain facing away from the ocean

06

Climates and Biomes of the World

READING STRATEGY

Analyzing Key Ideas and Details As you read, take notes about Earth's four climate zones.

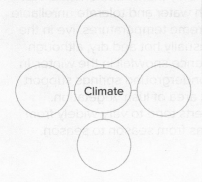

Climate

Climate Regions

GUIDING QUESTION

How are Earth's climates organized?

Climates in the world are organized into four zones. These are tropical, dry, midlatitude, and high-latitude climates. Some climates can vary within these broad zones, so geographers further divide the major climate zones into smaller regions. Different climates support different kinds of biomes. A **biome** is a major type of ecological community defined primarily by distinctive **natural vegetation** and animal groups.

Tropical Climates

Tropical climates are found in or near low latitudes in areas otherwise referred to as the Tropics. The two most widespread kinds of tropical climate regions are wet climates and wet/dry climates.

Tropical wet climates are often called *tropical rain forest* climates. They have an average daily temperature of 80°F (27°C) and almost daily rain year-round. Annual rainfall averages from 50 to 260 inches (125 to 660 cm). Tall trees form a canopy over shorter trees and bushes, and shade-loving plants grow on the completely shaded forest floor. The world's largest tropical rain forest is in the Amazon River basin. **Similar** climate and vegetation exist in other parts of South America, the Caribbean, Asia, and Africa.

---

biome a type of large ecosystem with similar life-forms and climates

natural vegetation plant life that grows in a certain area if people have not changed the natural environment

similar comparable

Due to the vast amount of plant food, wildlife like this crested macaque in Indonesia is abundant in tropical rain forests. Scientists estimate that more than half of all of the plant and animal species exist in this climate.

Tropical wet/dry climates have pronounced dry and wet seasons. Temperatures are high year-round. These regions, also called **savannas**, have fewer plants and animals than the tropical rain forest climates. One distinguishing characteristic of a tropical wet/dry climate is that sunlight is not blocked by trees and is able to reach much of the ground surface. Tropical savannas are found in Africa, Central and South America, Asia, and Australia.

Dry Climates

The two main types of dry climates are semi-arid (or steppe) and arid (or desert). Both of these climates occur in both low latitudes and midlatitudes. Geographers distinguish these dry climates based on the amount of rainfall and the vegetation in each.

Steppes are usually located away from oceans or large bodies of water. They are therefore less humid. However, they do receive an average of 10 to 30 inches (25 to 76 cm) of rainfall per year. Steppes experience warm summers and harshly cold winters. Some steppes have heavy snowfall, while others are susceptible to droughts and violent winds. Steppes are found on almost every continent and are home to a **diverse** variety of grasses.

Deserts are extremely dry areas that receive less than about 10 inches (25 cm) of rainfall per year. This climate supports a very small amount of plant and animal life. Only plants and animals that can live without much water and tolerate unreliable precipitation and extreme temperatures live in the desert. Deserts are usually hot and dry, although some deserts experience snowfall in the winter. In some desert areas, underground springs support oases. An oasis is an area of lush vegetation. Temperatures in deserts tend to vary widely from day to night, as well as from season to season.

savanna a large area of land with grass and scattered trees

diverse different

World Climate Regions

Climate patterns vary from region to region around the world.

EXAMINE THE MAP

1. **Exploring Regions** How can you tell whether Australia or South America receives more rain?
2. **Exploring Place** Which two climate regions occur on and around the Equator?

World Biomes

Biomes are determined by climate and physical geography.

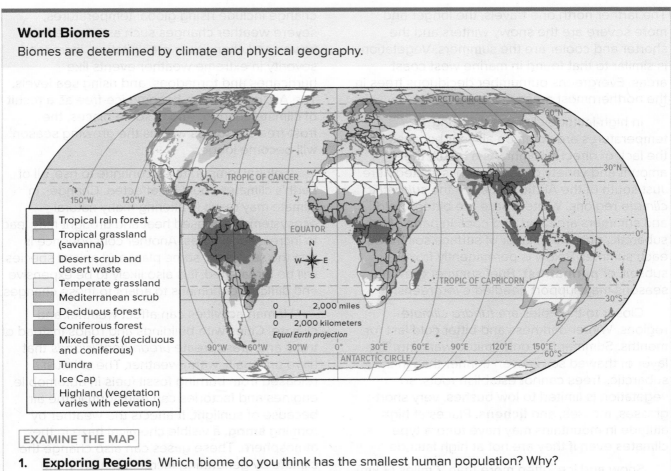

Legend:
- Tropical rain forest
- Tropical grassland (savanna)
- Desert scrub and desert waste
- Temperate grassland
- Mediterranean scrub
- Deciduous forest
- Coniferous forest
- Mixed forest (deciduous and coniferous)
- Tundra
- Ice Cap
- Highland (vegetation varies with elevation)

ARCTIC CIRCLE · 60°N · TROPIC OF CANCER · 30°N · EQUATOR · 0° · TROPIC OF CAPRICORN · 30°S · ANTARCTIC CIRCLE · 60°S

150°W · 120°W · 90°W · 60°W · 30°W · 0° · 30°E · 60°E · 90°E · 120°E · 150°E

0 2,000 miles
0 2,000 kilometers
Equal Earth projection

EXAMINE THE MAP

1. **Exploring Regions** Which biome do you think has the smallest human population? Why?
2. **Exploring Place** In which type of biome is your town or city?

Midlatitude and High-Latitude Climates

Humid subtropical climates include the southeastern United States as well as parts of Brazil, China, Japan, Australia, and India. They are characterized by short, mild winters and nearly year-round rain. The wind patterns and high pressure from nearby oceans keep **humidity** levels high. Vegetation consists of **prairies** and forests with evergreen and deciduous trees. **Coniferous** trees, most of which are evergreens, have cones. **Deciduous** trees, most of which have broad leaves, change color and drop their leaves in autumn.

Marine west coast climates include the southern coast of Chile, parts of Australia, the British Isles, and parts of the Pacific coast of North America. They are mainly between the latitudes of about 30° N and 60° N and about 30° S and 60° S. Ocean winds bring cool summers and cool, damp winters. Abundant rainfall supports both coniferous and deciduous trees.

Lands surrounding the Mediterranean Sea, in addition to the southwestern coast of Australia and central California, have mild, rainy winters and hot, dry summers. The natural vegetation includes thickets of woody bushes and short trees. Geographers classify any such coastal midlatitude area with similar climate and vegetation as a *Mediterranean* climate.

In some midlatitude regions of the Northern Hemisphere, landforms influence climate more than winds, precipitation, or ocean temperatures do. *Humid continental* climate regions do not experience the moderating effect of ocean winds because of their northerly inland locations.

humidity moisture in the air

prairie an inland grassland area with no trees except along rivers

coniferous evergreen trees that produce cones to hold seeds and that have needles instead of leaves

deciduous trees that shed their leaves in the autumn

The farther north one travels, the longer and more severe are the snowy winters and the shorter and cooler are the summers. Vegetation is similar to that found in marine west coast areas. Evergreens outnumber deciduous trees in the northernmost areas.

In high-latitude climates, freezing temperatures are common all year because of the lack of direct sunlight. As a result, the amount and variety of vegetation is limited here. Just south of the Arctic Circle are the *subarctic climate* regions. Winters here are bitterly cold, and summers are short and cool. In parts of the subarctic, only a thin layer of surface soil thaws each summer. Below is permanently frozen subsoil, or **permafrost**. Brief summer growing seasons may support needled evergreens.

Closer to the Poles are *tundra climate* regions. Winter darkness and bitter cold last for months. Summer has only limited warming. The layer of thawed soil is even thinner than in the subarctic. Trees cannot establish roots, so vegetation is limited to low bushes, very short grasses, mosses, and **lichens**. Places at high altitude in mountains may have tundra type climates even if they are not at high latitude.

Snow and ice, often more than 2 miles (3 km) thick, constantly cover the surfaces of ice-cap regions. Lichens are the only form of vegetation that can survive in these areas where monthly temperatures average below freezing.

✓ **CHECK FOR UNDERSTANDING**

1. **Making Connections** Why do high-latitude climates have limited vegetation?
2. **Contrasting** How do midlatitude and high-latitude climates differ?

Climate Change

GUIDING QUESTION

What causes climates to change over time?

Climate change refers to major changes in the factors used to measure climate over time. For example, scientists have found that the average global temperature has increased by 1.4°F (0.8°C) over the last century. Some indicators of climate change include rising global temperatures, severe weather changes such as intense heat waves and changes in precipitation, increased severity in extreme weather events like hurricanes and tornadoes, and rising sea levels. The Arctic is likely to become ice-free as a result of climate change. And in some places, the frost-free season as well as the growing season will become longer.

If global temperatures continue to rise, all of Earth's climates could be affected. Changes in climate may result in altering many natural ecosystems. Increased heat and drought may lead to increased wildfires. Another consequence is that the survival of some plant and animal species will be threatened. It is also likely to be expensive and difficult for humans to adapt to these changes.

Human activities can affect weather and climate. Cities with buildings and roads instead of trees and grass create *urban heat islands* that have unusually warm weather. The exhaust released from burning fossil fuels in automobile engines and factories can transform in the air because of sunlight. It affects the weather by forming **smog**, a visible chemical haze in the atmosphere. These gases can also change the climate by generating increased heat in the atmosphere. Fewer forests may also result in climate change when deforestation occurs as humans cut down trees for lumber, paper, and livestock grazing.

✓ **CHECK FOR UNDERSTANDING**

Identifying What are some indicators of climate change?

LESSON ACTIVITIES

1. **Informative/Explanatory Writing** Write a paragraph detailing how the four major climate zones are related to the three zones of latitude.
2. **Presenting** Working with a group, consider these biomes: rain forest, desert, grassland, and tundra. Create a slide show that highlights major features of the climate of each biome, and names some of the plants and animals that live in each.

---

permafrost the permanently frozen, lower layers of soil found in the tundra and subarctic climate zones
lichen tiny sturdy plants that grow in rocky areas
smog haze caused by chemical fumes from automobile exhausts and other pollution sources

07

Analyzing Sources: Geography and Climate

 COMPELLING QUESTION

Why does the Earth have a great diversity of climates and ecosystems?

Plan Your Inquiry

DEVELOPING QUESTIONS

Think about the climate and the natural environment of places with which you are familiar. Consider how climate varies from place to place and the different types of ecosystems that exist. Then read the Compelling Question for this lesson. What questions can you ask to help you answer this Compelling Question? Create a graphic organizer like the one below. Write these Supporting Questions in your graphic organizer.

Supporting Questions	Source	What this source tells me about why there is diversity in Earth's climate and ecosystems	Questions the source leaves unanswered
	A		
	B		
	C		
	D		
	E		

ANALYZING SOURCES

Next, examine the sources in this lesson. Analyze each source by answering the questions that follow it. How does each source help you answer each Supporting Question you created? What questions do you still have? Write these in your graphic organizer.

After you analyze the sources, you will
- use the evidence from the sources;
- communicate your conclusions;
- take informed action.

Background Information

Earth is home to a wide range of climates and a tremendous diversity of natural ecosystems. *Climate* is a term that describes long-term weather patterns and conditions. The average temperatures and amounts of precipitation experienced at different locations can vary significantly. Climate in turn influences the community of plants and animals that exists within a particular place. Warm temperatures and high rainfall may support a dense rain forest in one location. In another location, cold and dry temperatures mean that only a small number of well-adapted plants and animals can survive there.

» Antarctica's harsh climate means that very few plants and animals can live there.

The Climate of Hawaii

Hawaii is unique among the states of the U.S. as the only state not on the continent of North America. Many people from around the world travel to Hawaii to enjoy its climate and natural environment. The website of the National Weather Service provides an introduction to the climate of Hawaii.

PRIMARY SOURCE: WEBSITE

> " The outstanding features of Hawaii's climate include mild temperatures throughout the year, moderate humidity, persistence of northeasterly trade winds, significant differences in rainfall within short distances, and infrequent severe storms. For most of Hawaii, there are only two seasons: "summer," between May and October, and "winter," between October and April.

The Na Pali coast on the island of Kauai, Hawaii

Latitude

Hawaii is in the tropics, where the length of day and temperature are relatively uniform throughout the year. Hawaii's longest and shortest days are about 13½ hours and 11 hours, respectively, compared with 14½ and 10 hours for Southern California and 15½ hours and 8½ hours for Maine. Uniform day lengths result in small seasonal variations in incoming solar radiation and, therefore, temperature.

The Surrounding Ocean

The ocean supplies moisture to the air and acts as a giant thermostat, since its own temperature varies little compared with that of large land masses. The seasonal range of sea surface temperatures near Hawaii is only about 6 degrees, from a low of 73 or 74 degrees between late February and March to a high near 80 degrees in late September or early October. The variation from night to day is one or two degrees.

Hawaii is more than 2,000 miles from the nearest continental land mass. Therefore, air that reaches it, regardless of source, spends enough time over the ocean to moderate its initial harsher properties. Arctic air that reaches Hawaii, during the winter, may have a temperature increase by as much as 100 degrees during its passage over the waters of the North Pacific. Hawaii's warmest months are not June and July, but August and September. Its coolest months are not December and January, but February and March, reflecting the seasonal lag in the ocean's temperature. "

—"Climate of Hawai`i," Saul Price, National Weather Service website

EXAMINE THE SOURCE

1. **Analyzing** How does latitude influence the amount of daylight a location receives?
2. **Drawing Conclusions** What factors account for the small amount of variation in temperatures that are experienced in Hawaii throughout the day and throughout the year?

Climate and Forests

Forests cover over one-third of the Earth's land surface. Forest ecosystems have different characteristics depending on climate. Every forest fits into one of four categories based on the type of climate where it is found: tropical, subtropical, temperate, or boreal. Boreal forests are made up primarily of **coniferous** trees. This map shows the locations of boreal forests worldwide.

SECONDARY SOURCE: MAP

Worldwide distribution of the boreal forest is shown in green.

---

coniferous a cone-bearing tree with needle- or scale-like leaves; evergreen tree

EXAMINE THE SOURCE

1. **Summarizing** In what countries are most of the world's boreal forests?
2. **Inferring** Based on their locations, to what kind of climate are the plants and animals of the boreal forest likely able to adapt?

Rainfall in Texas

Texas is the second-largest state in the United States, over 268,000 square miles (432,000 sq km) in area. The state contains several different climate zones with widely varying temperatures and precipitation. This map shows the average amount of rainfall received throughout the state over a period of several decades.

SECONDARY SOURCE: MAP

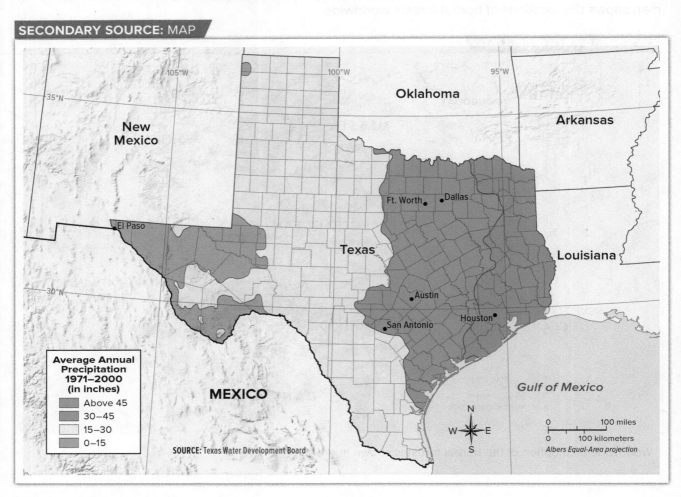

Average Annual Precipitation 1971–2000 (in inches)

- Above 45
- 30–45
- 15–30
- 0–15

SOURCE: Texas Water Development Board

Albers Equal-Area projection

EXAMINE THE SOURCE

1. **Identifying** Which parts of Texas received the highest average amounts of precipitation per year in this period?

2. **Interpreting** The locations of the largest cities in Texas are shown on this map. What do most of those cities have in common with regard to precipitation?

Rainfall in the State of Washington

Seattle, Washington is a city that typically receives more rainfall than most other parts of the continental United States. But the climate in Seattle is vastly different than the climate of some other parts of the state. This article, written by someone who lives in another part of the state, explains the reason for this difference.

PRIMARY SOURCE: ARTICLE

❝ It's an age-old question: So, where do you live?

My answer: "Washington—the STATE, that is." Inevitably, the next remark I hear is, "Oh, it rains a lot there. I bet you have webbed feet!" [...]

My new friend is amazed to learn that while Washington is home to a rainforest (yes, there are non-tropical rainforests), and plenty of rain on its western half, there is an area covering roughly half the state, on the eastern side of the Cascade Mountains, that receives very little rainfall.

As air over the ocean is forced over land, mountain slopes facing the wind (windward) experience the orographic effect while the slopes facing away from the direction of the wind (leeward) experience the rain shadow effect.

On the eastern side of the Cascade Mountains, a rain shadow effect prevails, and much of the region is semi-arid, with less than 10 inches of rain per year. By definition, a rain shadow is a region of relatively low rainfall that occurs downwind of a mountainside or mountain range facing away from the direction of the wind. The mountains block the passage of rain-producing weather systems, casting a "shadow" of dryness behind them.

There are two basic effects of precipitation caused by mountains. There is the "orographic" effect and the "rain shadow" effect...The orographic effect happens on the windward side of a mountain. Winds carry air masses up and over the mountain range, and as the air is driven upward over the mountain, falling temperatures cause the air to lose much of its moisture as precipitation.

The other effect is the rain shadow effect. The rain shadow effect is where precipitation amounts drop significantly on the leeward side of a mountain. ❞

"Rainfall in Mountainous Areas," Steve Linn, ARM Climate Research Facility newsletter, 2011

EXAMINE THE SOURCE

1. **Identifying Cause and Effect** If the eastern side of the Cascade Mountains is drier because of the rain shadow effect, from which direction is the wind coming?

2. **Predicting** Based on the information in the source, on which side of the Cascade Mountains would you expect to find more plants growing?

Climate of Rwanda

Rwanda is a small nation on the continent of Africa. It is a landlocked country, which means it does not have any coastline. This article was written to help visitors prepare for the weather.

PRIMARY SOURCE: WEBSITE

❝ In Rwanda, the "land of a thousand hills," located just south of the Equator, the climate is pleasantly warm all year round, with cool nights, because of the altitude. In fact, most of the country is located on a plateau, around 1,500 meters (5,000 feet) above sea level. The altitude decreases below a thousand meters (3,300 feet) only in the westernmost part, along the Rusizi River. . .which is therefore the only area where it can get hot, and the temperature can sometimes reach 35° C (95° F).

Precipitation ranges from 1,000 to 1,400 millimeters (40 to 55 inches) per year depending on area. There's a dry season from June to August (with July as the driest month) and a rainy season from September to May.[...] [T]he rainiest months are usually April and May, when precipitation reaches 150 millimeters (6 inches) per month, or even more. [...] [T]here is not much sun, in fact, the sky is often cloudy. . .

In Rwanda, there are also mountainous areas. . . . Along the mountain slopes, the sky is often covered with clouds. At middle-low altitudes, a cloud forest grows, while above 3,000 meters (9,800 feet), the climate is colder, the vegetation is sparser and the forest is replaced by shrubs, and the temperature at night can drop below freezing (0° C or 32° F). Finally, at high altitude, we find a cold desert. ❞

—"Climate of Rwanda," Climatestotravel.com

EXAMINE THE SOURCE

1. **Explaining** Why doesn't Rwanda have a very hot climate like other places that are close to the Equator? What part of the country is described as a cold desert?

2. **Calculating** Is the dry season or the rainy season longer in Rwanda?

Complete Your Inquiry

EVALUATE SOURCES AND USE EVIDENCE

Refer back to the Compelling Question and the Supporting Questions you developed at the beginning of the lesson.

1. **Making Generalizations** Which sources provided information about how latitude relates to climate and ecosystems?

2. **Identifying Themes** Which sources provide information about how mountains can affect climate and ecosystems?

3. **Gathering Sources** Which sources helped you answer the Supporting Questions and the Compelling Question? Which sources, if any, challenged what you thought you knew when you first created your Supporting Questions? What information do you still need in order to answer your questions? What other viewpoints would you like to investigate? Where would you find that information?

4. **Evaluating Sources** Identify the sources that helped answer your Supporting Questions. How reliable is the source? How would you verify its reliability?

COMMUNICATE CONCLUSIONS

5. **Collaborating** Share with a partner which of these sources you found most useful in answering your Supporting Questions. Discuss the ways in which Earth's climate and ecosystems are diverse. Discuss the characteristics of the climate and ecosystem of a place you know about. Take turns sharing evidence you found associated with one of your supporting questions. How do these sources support your thinking? Use the graphic organizer that you created at the beginning of the lesson to help you. Share your conclusions with the class.

TAKE INFORMED ACTION

Creating a Climate Information Record Conduct research on the climate of the place where you live or another place of your choosing. Find data and descriptive information that can help you summarize the climate for the location, such as average temperature and annual rainfall, as well as latitude, longitude, and elevation. Also collect information about extreme variations from normal climate. Create an informative video report.

08

Reviewing Physical Geography

Summary

Planet Earth

- The part of Earth that supports life is the biosphere. It includes the hydrosphere (water), the lithosphere (land), and the atmosphere (air).

- Landforms can be classified by elevation. They can also be described in relation to bodies of water.

- Earth's surface has two types of water: freshwater and salt water. Groundwater is freshwater below Earth's surface.

- The total amount of water on Earth does not change. It is constantly moving. This is called the water cycle.

Forces of Change

- Earth is made up of three main layers—the core, the mantle, and the crust.

- Earth's crust is broken up into great slabs of rock called plates. These plates float on the mantle, carrying Earth's oceans and continents.

- Tectonic plates moving around the globe have produced Earth's largest features like mountain ranges and oceans. The movement of plates can create earthquakes and volcanoes.

- Weathering and erosion also shape Earth's surface.

PHYSICAL GEOGRAPHY

Factors Affecting Climate

- Weather is the condition of the atmosphere at a given time. Climate is the average weather conditions over a longer period of time.

- Earth's tilt and revolution cause changes in the amount of direct sunlight that reach different locations. It is what causes seasons.

- Latitude plays a major role in climate—the farther one gets from the Equator, the cooler the climate.

- High elevations are generally cooler than the surrounding landscape.

- Other factors that help determine climate are wind and water currents, landforms, and the atmosphere.

Climates of the World

- Geographers organize the world into major climate zones: tropical, dry, midlatitude, and high-latitude climates. Each of these can be broken down into smaller regions.

- These climates support different biomes, each with its own characteristic natural vegetation and animals. The characteristics of biomes may overlap with each other.

- Climate change involves major changes in the factors used to measure climate over time. Some indicators of climate change are increasing global temperatures, severe weather changes, rising sea levels, and an increase in severe weather events such as hurricanes and tornadoes.

Checking For Understanding

Answer the questions to see if you understood the topic content.

IDENTIFY AND EXPLAIN

1. Identify each of these terms as it relates to forces of change in physical geography.

    A. core   **F.** earthquake

    B. mantle   **G.** volcano

    C. magma   **H.** weathering

    D. crust   **I.** erosion

    E. fault

REVIEWING KEY FACTS

2. **Making Connections** How is the term *revolution* related to the term *solstice*?

3. **Describing** How does Earth's tilted axis and revolution around the sun cause the seasons we know as winter, spring, summer, and fall?

4. **Determining Meaning** Do the terms *weather* and *climate* mean the same thing? Explain.

5. **Explaining** What causes the erosion of rocks on Earth's surface?

6. **Determining Central Ideas** Earth's tectonic plates are moving. Why don't we feel the ground moving under us?

7. **Analyzing** How is the ocean floor similar to the surface of dry land?

8. **Identifying** Identify these bodies of water as either freshwater or salt water: lake, river, ocean, pond, sea, gulf, groundwater.

9. **Determining Meaning** Explain how the processes of evaporation and condensation are both similar and different.

10. **Analyzing** How does water enter the air during the water cycle?

CRITICAL THINKING

11. **Describing** How does acid rain affect animal and plant life?

12. **Determining Meaning** What is a *rain shadow*?

13. **Explaining** How do landforms cause the formation of a rain shadow?

14. **Summarizing** What pattern occurs in ocean currents? What is the effect of this pattern? How is this pattern like that of wind currents?

15. **Drawing Conclusions** What causes climates to change?

16. **Explaining** How does the burning of fossil fuels create smog?

17. **Identifying** Which types of climates are located in the low-latitude zones?

18. **Analyzing** Why are freezing temperatures common all year in high-latitude climates?

19. **Identifying** Which midlatitude climate does not experience the moderating effects of ocean winds?

20. **Making Connections** How are permafrost and lichens related to high-latitude climates?

NEED EXTRA HELP?

If You've Missed Question	1	2	3	4	5	6	7	8	9	10
Review Lesson	3	5	5	5	3	3	2	2	2	2

If You've Missed Question	11	12	13	14	15	16	17	18	19	20
Review Lesson	2	5	5	5	6	6	6	6	6	6

Apply What You Have Learned

 ## A Understanding Multiple Perspectives

When studying physical geography, it is both more informative and more fun to examine places with climates and ecosystems that are very different from your own. In doing so, it helps to imagine how your life would be different if you lived in that environment. Your home, clothing, food, and more would likely change as you adapted to that new place.

ACTIVITY **Creating a Travel Brochure** Choose a location that features a climate as different as possible from the climate in which you live. Using paper and art supplies or a digital-publishing program, create a colorful brochure (a foldable pamphlet or a small poster) advertising tourism to that location. The focus of the brochure should be the natural features of that place's environment, displaying and labeling key natural features and landmarks and explaining what a visit to that location would be like for the traveler. Present the finished brochure to your class.

 ## B Geographic Reasoning

Elevation is one factor that helps determine the climate of a place. It can play a part in determining an area's precipitation patterns, air temperatures, amounts of wind, and more. On the other hand, there are places at similar elevations with drastically different climates due to other factors such as latitude or proximity to bodies of water.

ACTIVITY **Contrasting Elevations** Choose two places that are located at significantly different elevations; for example, Lhasa, Tibet (China) is at 12,000 feet (3,658 m) above sea level, while Amsterdam, Netherlands is at 7 feet (2.13 m) below sea level. Describe how the two places are different in terms of climate, natural vegetation, and other characteristics. Then determine which of these characteristics is, or is not, the result of elevation. Summarize your findings in a five-minute class presentation.

Writing an Informative Essay

When a large event in Earth's crust, such as an earthquake or a volcanic eruption, occurs underneath the ocean's waters, a tsunami can occur. This type of ocean wave can spread for thousands of miles, causing great destruction when they reach land. The Indian Ocean tsunami that occurred on December 26, 2004, was one of the deadliest disasters in history, impacting people in 14 countries from Indonesia to Somalia.

ACTIVITY **Write an Essay about Natural Disaster Preparedness** Research ways in which people in different coastal areas around the world have attempted to prepare for tsunamis, both before and after catastrophic occurrences. Then write a short essay explaining these efforts to protect people and property and how they have been tested.

 ## Making Connections to Today

In this topic, you have learned about a wide variety of processes and events that have created the world we know today. Volcanic islands, mountain ranges, and deep canyons can all trace their origins to geographic forces that have acted on the environment over time.

ACTIVITY **Creating a Multimedia Presentation About the Creation of a Place by Physical Geography** Choose a place on Earth that was created by erosion, volcanic activity, or the interaction of tectonic plates. Using photographs, drawings, video, and text, explain how that place gained its current form. Describe both the process that created the place and the effect that place has on its surrounding climate (or the effect that the climate has on the place). Share your presentation with the class.

Population tends to cluster in large cities like New York City. It is the largest city in the United States in terms of population.

Population Geography

INTRODUCTION LESSON

01 Introducing Population Geography   G114

LEARN THE CONCEPTS LESSONS

02 Population Growth   G119

03 Population Patterns   G123

04 Population Movement   G127

05 The Growth of Cities   G131

INQUIRY ACTIVITY LESSONS

06 Analyzing Sources: Population Change   G135

07 Global Issues: Causes of Population Change   G141

REVIEW AND APPLY LESSON

08 Reviewing Population Geography   G147

Introducing Population Geography

The World's People

Earth's human population is dynamic—it is growing, and it is scattered around the globe in small and large settlements. People are on the move. Migration has fueled the growth of cities.

Where People Live

People are distributed unevenly on Earth's surface. Population distribution is related to physical geography.

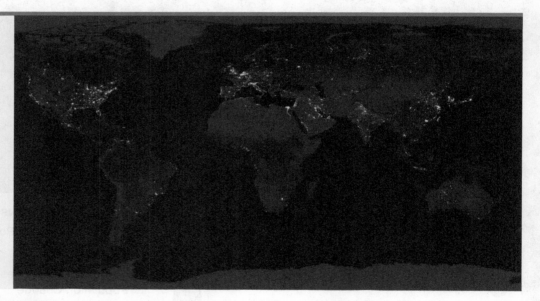

This image shows the world at night with lighted areas quite visible. The world's population is increasing, and people live on much of Earth's surface. Some people live in heavily populated areas, such as large cities. Some people live in areas where they may not have access to electricity or running water.

Population growth and the ways people live impact the environment. People are using resources more quickly than ever before. In this image, big tracts of forest are being cleared so the land can be used for large-scale agriculture.

Migration and Cities

People move from one place to another to leave negative situations or because they are drawn to more attractive situations in a new place. Some of this migration results in the growth of cities as people move from the countryside to cities and towns in search of ways to make a living.

» People migrate for different reasons. These men migrated from Africa to southern Italy for work in agriculture. And these young Chinese women moved to Australia to attend college.

> " A city isn't so unlike a person. They both have the marks to show they have many stories to tell. They see many faces. They tear things down and make new again. "
>
> —Rasmenia Massoud, *Broken Abroad*

World Urban Population

More people live in cities today than ever before.

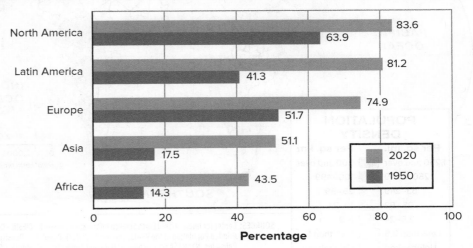

Region	2020	1950
North America	83.6	63.9
Latin America	81.2	41.3
Europe	74.9	51.7
Asia	51.1	17.5
Africa	43.5	14.3

Percentage

Source: UN Population Division

Getting Ready to Learn About . . . Population Geography

Growth and Patterns

Geographers study the world's growing population in order to understand the reasons for its changes over time. When examining population growth, geographers consider

- what changes lead to growth in populations;

- how economic and technological factors influence birth rates, death rates, and migration;

- which areas of Earth are growing in population most quickly;

- the impacts of population growth on Earth and its resources.

The study of populations requires geographers to use various tools and methods to identify patterns in population change. The places in which people choose to live or are forced to live, the living conditions in those places, and the connections between those places can all be seen as spatial patterns. For example, people around the world are more likely to live in an area with regular rainfall than in a desert. Cities located along an important trade route are more likely to grow than those that are more isolated. Regions in which people rapidly consume natural resources might grow quickly and then decline just as quickly as they grew as resources are depleted.

World Population Density

The world's people are unevenly distributed across the globe.

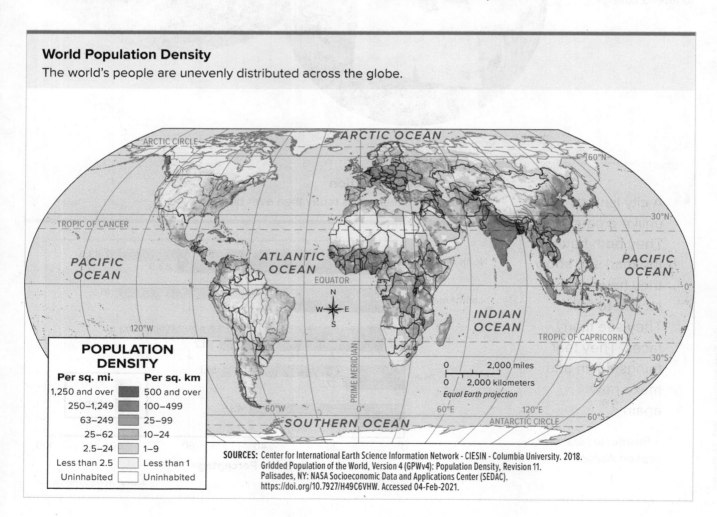

POPULATION DENSITY

Per sq. mi.	Per sq. km
1,250 and over	500 and over
250–1,249	100–499
63–249	25–99
25–62	10–24
2.5–24	1–9
Less than 2.5	Less than 1
Uninhabited	Uninhabited

SOURCES: Center for International Earth Science Information Network - CIESIN - Columbia University. 2018. Gridded Population of the World, Version 4 (GPWv4): Population Density, Revision 11. Palisades, NY: NASA Socioeconomic Data and Applications Center (SEDAC). https://doi.org/10.7927/H49C6VHW. Accessed 04-Feb-2021.

Mexico City is one of the largest cities in the world. As it grows, it continues to spread outward, taking up undeveloped land near the city.

Migration

Throughout history, people around the world have moved from place to place, establishing cities and countries that have grown and declined over time. Over the past century, advances in transportation have made it possible for more people than ever before to migrate, or move, over long distances. The development of new forms of communication in recent decades has improved the ability of people to learn about opportunities in other regions, influencing their decisions about where to live.

Migration has led to dramatic changes in the makeup of some cities and regions. For example, the World Bank estimates that almost 90 percent of the current population of the United Arab Emirates consists of migrants from other countries. Within the United States, many towns in agricultural areas have seen a significant drop in population, while there has been rapid growth in some cities. The population of Austin, Texas, has nearly tripled in the last 40 years, while in neighboring Arkansas the city of Pine Bluff has lost about 30 percent of its population during the same amount of time.

Geographers look for the reasons behind such differences. Often a combination of factors—economic, political, environmental, and social—drives people to leave one home for another. Some of those factors are voluntary, such as relocating to a city where a new workplace has been built. A new distribution center for a shipping company, for example, may offer better pay to its employees than businesses in nearby regions. Other factors are involuntary. A region experiencing warfare will often be too dangerous for many people to remain, leading them to seek both physical and economic security in another country. Such migration can impact the areas that migrants leave, the areas in which they make new homes, and the areas through which they must travel to reach their destinations.

Looking Ahead

You will learn about Earth's human population. You will learn about its growth, distribution and density, and the factors that affect each. You will learn about the causes and effects of migration. You will learn about the growth of cities. You will examine Compelling Questions and develop your own questions about population geography in the Inquiry Activity Lessons. You will see some of the ways in which you might already use the ideas of population geography in your life by completing these lessons.

What Will You Learn?

In these lessons about population geography, you will learn

- how birthrates, death rates, and migration contribute to population changes;
- how population pyramids are used to study population structure and growth;
- the effects of population growth;
- why population growth varies throughout the world;
- what population distribution and population density can tell us about a country;
- what influences where people settle;
- the causes and effects of migration;
- the causes and effects of urban growth;
- how urban areas interact and influence surrounding areas.

 COMPELLING QUESTIONS IN THE INQUIRY ACTIVITY LESSONS

- **How do populations change over time?**
- **Why do populations change over time?**

Studying Population Geography

Growth
- Birthrates
- Death rates
- Natural increase
- Doubling time
- Population pyramids

Patterns
- Population distribution
- Urban areas
- Rural areas
- Population density

POPULATION

Movement
- Migration
- Push factors
- Pull factors
- Cultural diffusion
- Brain drain
- Remittances

Cities
- Urbanization
- Megalopolis
- World city
- Infrastructure
- Urban sprawl

Population Growth

Analyzing Key Ideas and Details As you read, take notes about the causes and effects of population growth.

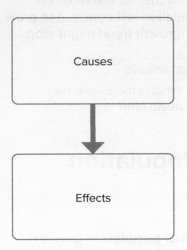

Causes

Effects

Causes of Population Growth

GUIDING QUESTION

What factors influence population growth?

Earth's population increased dramatically during the nineteenth century and much of the twentieth century. Although that growth has begun to slow in recent decades, there are far more people on Earth today than ever before. But there are also places on Earth where population is declining. The result is that issues of population growth or decline are crucial to all countries. Geographers play an important role in examining ways to plan for the future and solve problems of tomorrow.

Nearly 8 billion people now live on Earth today. Global population continues to grow and is expected to reach just under 10 billion by the year 2050. Such rapid growth was not always the case. From the year 1000 until 1800, the world's population increased slowly. Then the number of people on Earth more than doubled between 1800 and 1950, from 1 billion to over 2.6 billion. It doubled again between 1950 and about 2000, from 3 billion to more than 6 billion.

One factor involved with population growth is a falling **death rate**. The death rate is the number of people who die per year for every 1,000 people. The death rate decreases for many reasons. Better health care, more food, and cleaner water have helped more people—young and old—live longer, healthier lives.

---

death rate the number of deaths per year for every 1,000 people

---

World Population

World population is projected to surpass 10 billion people by the year 2100.

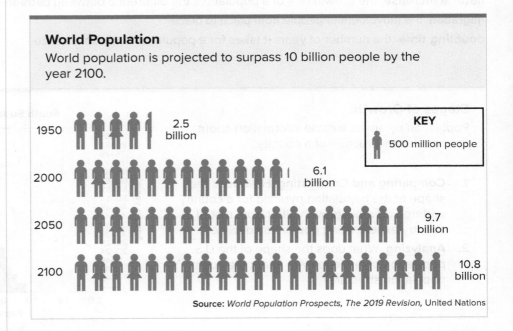

1950	2.5 billion
2000	6.1 billion
2050	9.7 billion
2100	10.8 billion

KEY
500 million people

Source: *World Population Prospects, The 2019 Revision,* United Nations

Another factor is the **birth rate**. It is the number of babies born per year for every 1,000 people. In time, many of the babies born today will **mature** and have children and grandchildren of their own.

Since about 1950, the birth rate of Earth as a whole has been decreasing slowly, although in some areas of Asia, Africa, and Latin America, birth rates have been slow to decline. One reason for this is that families in these regions traditionally are large because of cultural beliefs about marriage, family, and the value of children. A second reason is that children are valuable for working agricultural lands that support many of these families.

This difference between the birth rate and the death rate is called **natural increase**, or the growth rate of an existing population. For Earth as a whole, the birth rate is still higher than the death rate, which is the cause of population growth. **Migration** is also considered when examining population growth in a particular place. People moving into an area increase the population in that area. People migrating out of an area can make population growth decline or even cause the overall population to decrease.

In some countries, a high number of births has combined with low death rates to greatly increase population growth. As a result, **doubling time**, or the number of years it takes a population to double in size, is relatively short. In some parts of Asia and Africa, for example, the doubling time is only about 50 years or less. In contrast, the average doubling time of countries with slow growth rates, such as Japan, can be nearly 200 years.

Despite the fact that the global population is growing, the rate of growth is gradually slowing. The United Nations predicts that the world's population will pass 10 billion by the year 2100. After that, the population is projected to begin decreasing. This means that for the next few decades, Earth's population will continue to grow. In time, however, this growth trend might stop.

✓ **CHECK FOR UNDERSTANDING**

Determining Meaning What is the difference between birth rate and death rate?

Effects of Population Growth

GUIDING QUESTION

What are the effects of population growth?

When human populations grow, the places people inhabit can become crowded. In many parts of the

birth rate the number of births per year for every 1,000 people

mature to age; to grow up

natural increase the growth rate of a population; the difference between birthrate and death rate

migration the movement of people from place to place

doubling time the number of years it takes for a population to double in size

Stages of Growth
Population pyramids include information about the population structure of a country.

1. **Comparing and Contrasting** How does the shape of the population pyramid for a country undergoing rapid growth differ from that of a country experiencing population decline?

2. **Analyzing** What does the shape of the U.S. population pyramid tell you about the country's population structure?

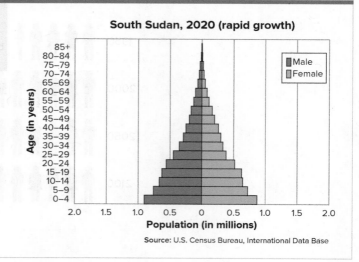

South Sudan, 2020 (rapid growth)

Age (in years): 85+, 80–84, 75–79, 70–74, 65–69, 60–64, 55–59, 50–54, 45–49, 40–44, 35–39, 30–34, 25–29, 20–24, 15–19, 10–14, 5–9, 0–4

Population (in millions): 2.0 1.5 1.0 0.5 0 0.5 1.0 1.5 2.0

Male / Female

Source: U.S. Census Bureau, International Data Base

world, cities, towns, and villages have grown and expanded beyond a comfortable capacity. Some cities are now so filled with people that they are becoming dangerously overcrowded.

When the population of an already crowded area continues to grow, serious problems can arise. For example, diseases spread quickly in crowded environments. Sometimes there is not enough work for everyone, and many households live in **ongoing** poverty. Where many people share tight living spaces, crime can be a serious problem and pollution can increase.

Effects on Society

One concern is that the world's population is unevenly distributed by age, with the majority of some countries' populations being infants and very young children who cannot contribute to food production. This is called the *population structure*, and is represented with a **population pyramid**. Population pyramids show a country's population organized by age and gender. A growing population will be wider at the bottom. This represents a large number of young people. A declining population will be wider at the top. This means the country's population is mostly elderly. A stable population is represented by bars of similar length over several age groups.

In places with the largest and fastest-growing populations, the need for resources, jobs, health care, and education is great. When millions of people living in a small area need food, water, and housing, there is sometimes not enough for everyone. Water supplies in crowded cities are often polluted with waste that can cause diseases. Some areas do not have enough land resources and materials for people to build safe, sturdy homes.

Children in crowded places are often affected by extreme poverty. They may live in crowded neighborhoods called *slums*. Slums surround some of the world's cities. These places are frequently dirty and unsafe. In areas with job shortages, people are forced to live on low incomes. Governments and organizations such as the United Nations work to make these areas safer, healthier places to live. Because populations grow at different rates, some areas experience more severe problems such as lack of security and education.

In the late 1900s, some countries in Europe began to experience a trend called *negative population growth*, in which the annual death rate is greater than the annual birthrate. Hungary, for example, shows a change rate of about −0.3. This situation has economic consequences different from, but just as serious as, those caused by high growth rates. In countries with negative population growth, it is difficult to find enough workers to keep the economy going. Labor must be recruited from other countries, often by encouraging immigration or granting

---

ongoing continuing

population pyramid a diagram that shows the distribution of a population by age and gender

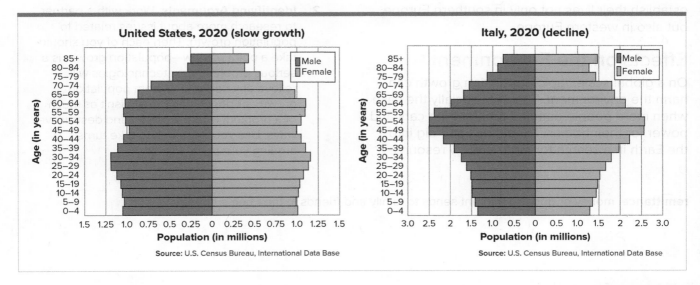

United States, 2020 (slow growth)

Age (in years): 85+, 80–84, 75–79, 70–74, 65–69, 60–64, 55–59, 50–54, 45–49, 40–44, 35–39, 30–34, 25–29, 20–24, 15–19, 10–14, 5–9, 0–4

Population (in millions): 1.5, 1.25, 1.0, 0.75, 0.5, 0.25, 0, 0.25, 0.5, 0.75, 1.0, 1.25, 1.5

Male / Female

Source: U.S. Census Bureau, International Data Base

Italy, 2020 (decline)

Age (in years): 85+, 80–84, 75–79, 70–74, 65–69, 60–64, 55–59, 50–54, 45–49, 40–44, 35–39, 30–34, 25–29, 20–24, 15–19, 10–14, 5–9, 0–4

Population (in millions): 3.0, 2.5, 2, 1.5, 1.0, 0.5, 0, 0.5, 1.0, 1.5, 2, 2.5, 3.0

Male / Female

Source: U.S. Census Bureau, International Data Base

temporary work permits. The use of foreign labor has helped countries with negative population growth rates maintain their levels of economic activity. But it also can create tensions between the "host" population and the communities of newcomers.

For example, many countries in southern Europe like Italy and Spain have aging populations and negative growth population growth rates. This creates increased economic opportunities for young people from North Africa and Southwest Asia, giving them a reason to migrate to southern Europe. Young workers are needed to fill the spaces left as older people retire and stop working. There are also factors that push many people to leave North Africa and Southwest Asia. These include high unemployment, declining wages, political instability, and conflict.

Once employed in southern Europe, many foreign workers send **remittances** or portions of their pay home to their families in North Africa and Southwest Asia. These increased funds have reduced poverty in some places. And the incoming funds usually mean that there is more money to invest in health and education in countries where remittances are sent.

Besides providing labor, North African and Southwest Asian migrants have influenced the cultural development of the countries of Europe. The migrants have enriched the cultural landscape of their new countries. They have brought their religious beliefs, social customs, arts, cuisine, and dress to southern Europe. In some places, the change has come at the expense of the migrants. The arrival of migrants from North Africa and Southwest Asia has led to tensions and sometimes conflict as they try to establish their lives not only in southern Europe but also in western Europe.

Effects on the Environment

On a global scale, rapid population growth can harm the environment. This is especially the case when more people demand fuel for their cars and power for their homes. Miners drill and dig into the Earth in a search for more energy resources.

Forests are cut down to make farms that produce food to feed growing populations and to raise livestock to meet demands for more meat.

Factories build cars, computers, and appliances for the growing population. Most of those factories generate pollution, and some of them dump chemicals into waterways and vent poisonous smoke into the air.

Over time, and with tens of thousands of factories all over the world, chemical wastes have polluted Earth's atmosphere. Many groups and individuals are working to clean up the environment and restore once-polluted areas.

Humans have many methods of finding and using the resources we need for survival. Some of these methods are wasteful and destructive. More impact on the environment occurs as more people use these wasteful and destructive ways. However, humans can change the ways they live. People are creative in solving modern problems. People in all parts of the world have invented new ways to produce power and harvest resources. Wind, solar, and geothermal energy do not pollute the environment as much as coal and gas. Humans are rising to the challenge of finding new methods of using these natural resources.

✓ **CHECK FOR UNDERSTANDING**

Summarizing What are the social effects of population growth?

LESSON ACTIVITIES

1. **Explanatory Writing** Write a paragraph summarizing the factors that have contributed to Earth's growing population.

2. **Identifying Arguments** Work with a partner to research more about issues related to population growth. Then each of you should take a point of view—population growth is a serious problem, new technologies will relieve any bad effects from population growth, or population growth isn't as much of a problem as are wasteful and destructive ways. Use your research to take turns arguing your point of view.

---

remittance money or goods a migrant sends to family and friends in their home country

03
Population Patterns

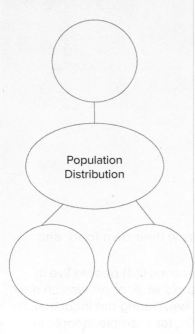

READING STRATEGY

Analyzing Key Ideas and Details As you read, take notes about the things that affect where people live.

Population Distribution

Population Distribution

GUIDING QUESTION

What influences the distribution of people and where they live?

Not only do population growth rates vary among Earth's regions, but the pattern of human settlement, or **population distribution**, is uneven as well. Some areas of the world have almost no people living in them, other areas are sparsely settled, and still other areas have large numbers of inhabitants.

Population distribution is strongly related to Earth's physical geography. Only about 29 percent of Earth's surface is made up of land, and much of that land is unwelcoming. High mountain peaks, barren deserts, and frozen tundra make human activity difficult in many places. Most of the people on Earth lives on a relatively small **portion** of the planet's land—a little less than one-third.

Most people live where fertile soil, available water, and a climate without harsh extremes make it more likely for humans to thrive. People tend to gather in lowland areas such as river valleys and coastal plains. Coastal areas are often the most-populated areas in the world. As elevation increases, the number of people generally decreases. This is because higher elevations have lower temperatures and a shorter growing season that can make it difficult to live.

population distribution the geographic pattern of where people live
portion a part of the whole

One of the things that makes the village of Woolacombe, England, a desirable place to live is its coastal location.

Where is the World's Population?

The world's population is unevenly distributed. Most of the total global population lives in Asia.

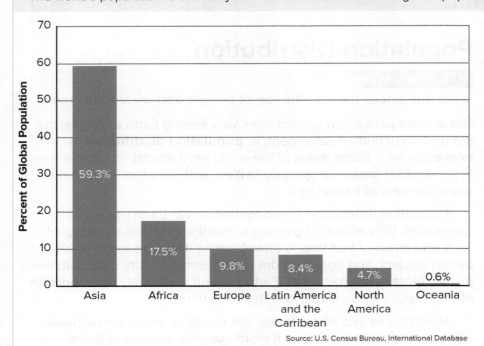

Percent of Global Population

- Asia: 59.3%
- Africa: 17.5%
- Europe: 9.8%
- Latin America and the Carribean: 8.4%
- North America: 4.7%
- Oceania: 0.6%

Source: U.S. Census Bureau, International Database

Analyzing What part of the world has the next-highest percentage of people after Asia?

One of the main reasons people settle in some areas and not in others is their need for resources. People live where their basic needs can be met. People need shelter, food, water, and a way to earn a living. Many people throughout the world live in **urban** areas—cities and their surrounding areas—where there are many places to live and work and there is greater access to needed resources. Populations in these places are highly concentrated. Today, most people in Europe, North America, South America, and Australia live in or around urban areas.

Unlike urban areas, rural areas are sparsely populated. A **rural** area is an open stretch of land that has few homes or other buildings, and not many people. Some rural areas are not very desirable to live in because of extreme climate conditions. In contrast, other rural areas are important for agriculture because of fertile land and abundant rainfall. In some parts of the world, people make their homes on open grasslands in rural areas where they build their own shelters, grow their own food, and raise livestock.

There are other reasons that people live in certain places. A person's work or profession may influence where they live. During the Industrial Revolution in the 1800s, for example, people in **more developed countries** moved from rural areas to cities because farm labor was not needed as much as work in the growing factories. Today, people in **less developed countries** are also moving from rural areas to cities, but their reasons for moving are different. There is no longer enough land in the rural areas for them. They hope to get work in the cities, but the factories are not growing as fast as they did in the past.

People gather in some places because those places hold some kind of cultural significance. For example, Philadelphia, Pennsylvania, is considered the birthplace of the United States. It is where the founders discussed, debated, and formed a new country. Religion is a part of culture that also causes people to gather in certain places. One

urban an area that is densely populated
rural an area that is lightly populated

more developed country a country with a highly developed economy and advanced technological infrastructure
less developed country a country that shows the lowest indicators of social and economic development

such place is the Ganges River in India, sacred to the Hindu people. People might also choose to live in places because the places are government or transportation centers. Houston, Texas, for example, is home to one of the world's largest ports. It is vital to national and international trade.

Of all the continents, Asia is the most heavily populated. Asia alone contains about 60 percent of the world's people. Study the map of world population distribution. Can you see the major clusters of population around the world? The first is in East Asia, primarily in eastern China but also in Japan and Korea. China's population is clustered in urban areas on the coastal plain as well as along the Chang Jiang and Huang He river valleys.

The second major population cluster is in South Asia. It is centered in India, but extends into Bangladesh, Pakistan, and the island of Sri Lanka. As in East Asia, many people in South Asia tend to cluster in major cities, on the coasts, and along rivers like the Ganges and Indus rivers. However, many people also live in rural areas.

Physical geography creates the boundaries of the South Asian population cluster—the Himalaya mountains to the north and the dry desert west of the Indus River in Pakistan.

The third major world population cluster is Europe. An area of high population stretches from Ireland and Great Britain into Russia. It includes large parts of Germany, Poland, Ukraine, Belarus, Netherlands, France, Belgium, and Italy. Population in Europe is mostly clustered in cities and towns.

✓ CHECK FOR UNDERSTANDING

Analyzing How does physical geography influence where people live?

Population Density

GUIDING QUESTION

What is population density?

One way to look at population is by measuring **population density**—the number of people living

population density the average number of people living in a square mile or a square kilometer

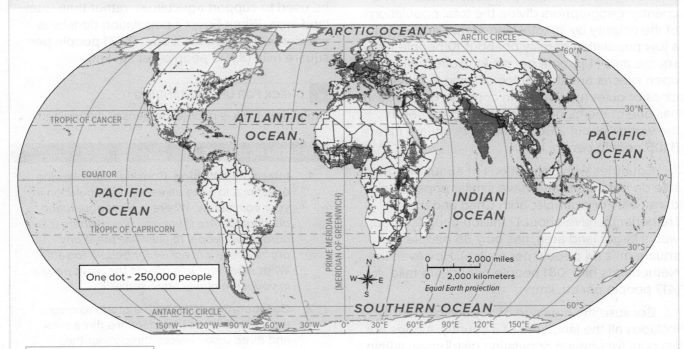

World Population Distribution

Population distribution maps show where heavy concentrations of population are located.

One dot = 250,000 people

EXAMINE THE MAP

1. **Human Population** Describe the pattern of population distribution in the United States.
2. **Exploring Regions** Which parts of Africa are the most densely populated?

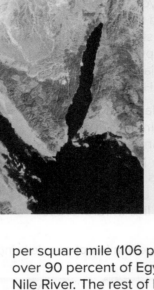

The Nile River and delta in Egypt are clearly reflected by the green of natural vegetation and agricultural fields in this satellite image.

Drawing Conclusions
What does the pattern of vegetation in this image tell you about Egypt's population density?

on a square mile or a square kilometer of land. To say an area is *densely populated* means that a large number of people live within that area.

Population density **varies** widely from country to country. To determine population density in a country, geographers divide the total population of the country by its total land area. Canada, with a low population density of about 10 people per square mile (4 people per sq. km), offers wide-open spaces and the choice of living in thriving cities or quiet rural areas. In contrast, Bangladesh has one of the highest population densities in the world—about 2,863 people per square mile (1,105 people per sq. km).

Countries with populations of about the same size do not necessarily have similar population densities. For example, both Chad and the Netherlands have about 17 million people. Chad, with a larger land area, has only 35 people per square mile (14 people per sq. km). However, the Netherlands has 1081 people per square mile (417 people per sq. km).

Because the measure of population density includes all the land area of a country, it does not account for uneven population distribution within a country. In Egypt, for example, the overall population density of the country is 275 people per square mile (106 people per sq. km). In reality, over 90 percent of Egypt's people live along the Nile River. The rest of Egypt is desert. Thus, some geographers describe a country's population density in terms of *arable* land, or land that can be used to support agriculture, rather than total land area. When Egypt's population density is measured this way, it is about 9,831 people per square mile (3,796 people per sq. km).

✓ **CHECK FOR UNDERSTANDING**

Explaining How is population density determined?

LESSON ACTIVITIES

1. **Informative Writing** Consider the following questions to write an essay on population in your community. *Where is most population distributed? Are there any areas where it is harder for people to live? Are there any areas where it is easier for people to settle? What other features affects the ways people in your community have settled?*

2. **Collaborating** Work with a group to create a two-column chart that identifies three rural and three urban places throughout the world. Use what you already know, your travel experiences, or research to help you complete your chart. Then share your chart with the class.

vary to make different

Population Movement

READING STRATEGY

Analyzing Key Ideas and Details As you read, take notes about the causes and effects of migration.

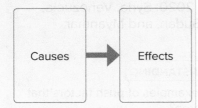

Causes → Effects

Causes of Migration

GUIDING QUESTION

What are the causes of migration?

The populations of different regions change as people move from one region to another. When many people leave an area, that area's population decreases. When large numbers of people move into a city, a state, or a country, the population of that area increases. Moving from one place to another is called **migration**.

Migration is one of the main causes of population shifts in the world today. Some people migrate permanently, while other people may move for a certain period of time and then move back home. What causes people to leave their home and migrate to different parts of the world?

To **emigrate** means "to leave one's home to live in another place." Emigration can happen within the same country, such as when people move from a village to a city inside the same country. Often, emigration happens when people move from one country to another. For example, millions of people have emigrated from countries in Europe, Asia, and Africa to start new lives in the United States. The term *immigrate* is closely related to emigrate, but it does not mean the same thing. To **immigrate** means "to enter and live in a new place or country."

migration the movement of people from one place to another

emigrate to leave one's home to live in another place

immigrate to enter and live in a new place or country

The Fulani people are scattered throughout West Africa. Some Fulani are pastoral, meaning that they migrate with their animal herds. Their migrations are often with the seasons.

People migrate for many reasons. The reasons for leaving one area and going to another are interpreted as *push-pull factors*. Positive situations like better social and economic conditions and religious or political freedoms can draw people to a place. Some people move to new places to be with friends or family members. Many young people move to cities or countries to attend universities or other schools. Some people relocate in search of better jobs. Families sometimes move to places where their children will be able to attend better schools. These **pull factors** attract people to an area.

Negative situations such as wars, persecution, and famines can motivate people to migrate away from a place. Sometimes people emigrate from an area after a natural disaster such as a flood, an earthquake, or a tsunami has destroyed their homes and land. If the economy of a place becomes so weak that little or no work is available, people emigrate to seek new opportunities. Such **push factors** drive people from an area.

In most cases, push-pull factors result in voluntary migrations. But migrations can also be forced. One instance of forced migration occurred when enslaved Africans were brought to the Americas. Some people are forced to flee their country because of violence, war, food shortages, or persecution. They are called **refugees**. According to the UN Refugee Agency, more than two-thirds of all refugees came from just five countries in 2020: Syria, Venezuela, Afghanistan, South Sudan, and Myanmar.

✓ **CHECK FOR UNDERSTANDING**

Identifying What are examples of push factors that drive people from an area?

pull factor factors that attract people to a place

push factor factors that drive people from a place

refugee a person who flees a country to escape war, persecution, or natural disaster

Global Migration Patterns

This map shows the average annual population gains and losses as a result of migration.

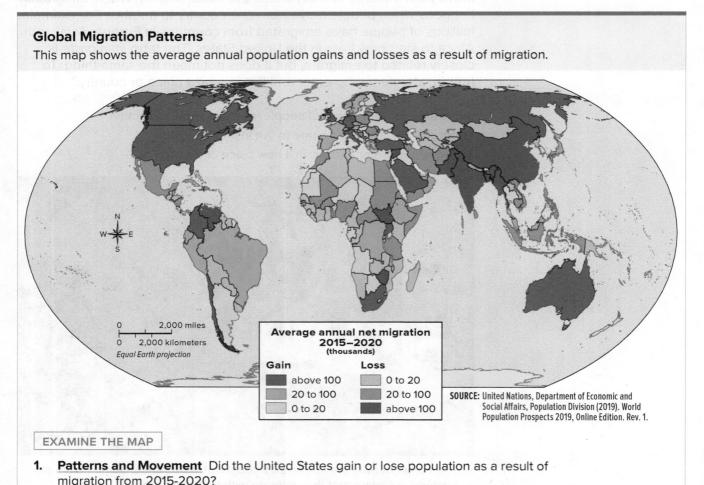

Average annual net migration 2015–2020 (thousands)

Gain
- above 100
- 20 to 100
- 0 to 20

Loss
- 0 to 20
- 20 to 100
- above 100

0   2,000 miles
0   2,000 kilometers
Equal Earth projection

SOURCE: United Nations, Department of Economic and Social Affairs, Population Division (2019). World Population Prospects 2019, Online Edition. Rev. 1.

EXAMINE THE MAP

1. **Patterns and Movement** Did the United States gain or lose population as a result of migration from 2015-2020?

2. **Exploring Regions** Which regions mainly experienced population loss from migration?

Effects of Migration

GUIDING QUESTION

What are the effects of migration?

The movement of people to and from different parts of the world can affect the culture, population, resources, and economy of an area. One effect of migration is **cultural diffusion**, or the spread of new ideas. As a result, there can be *cultural convergence,* in which cultures become more alike over time. When people from various diverse cultures migrate to the same place and live close together, their cultures can become more alike. Their cultures can also become mixed and blended. This blending creates new, unique cultures and ways of life. Artwork and music created in diverse urban areas is often an interesting mixture of styles and rhythms from around the world. Food, clothing, and languages spoken in urban areas change when people migrate into those areas and bring new influences.

Cultural divergence can also occur. In such instances, cultures separate from one another and develop in different ways. Some families and cultural groups work to preserve their original cultures even when they migrate. These people want to keep their cultural traditions alive so they can be passed down to future generations. For example, the traditional Chinese New Year is an important celebration for many Chinese American families. Chinese Americans can be part of a blended American culture but still celebrate Chinese holidays and enjoy traditional Chinese foods, music, and arts. It is possible to adapt to a local culture while maintaining strong ties to a home culture.

Migration can affect the population structure and population makeup of an area. Remember that when we refer to population structure, we are talking about the age and gender of people within a population. New people coming into an area from another place may include a large number of young people. This creates an increased need for teachers and schools. Incoming immigrants may be older, past the age of working, and not able to **contribute** economically. They may have additional or more frequent healthcare needs. In addition, migration affects the ethnic makeup of a

A Bolivian folk-dancing group performs in Arlington, Virginia. Arlington has a large Bolivian immigrant community.

population. For example, Hispanic and Asian groups represent a growing portion of the U.S. population due to changes in immigration into the country. Beginning in the 1960s, immigration from Europe decreased while immigration from Latin America and Asia increased. At the same time, overall immigration into the United States increased considerably. These trends have grown stronger ever since and include migration today. These changes have generated ethnic, political, and cultural tensions and may even include security concerns.

As new people migrate to an area, the population increases and there is need for additional resources such as housing, jobs, medical care, and education. In some places,

---

cultural diffusion the spread of new ideas from a source point

contribute to give

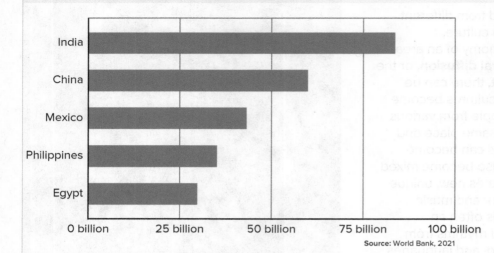

Top Five Countries to Receive Remittances, 2021

For the home country, remittances from citizens working in another country are an important source of much-needed money.

India

China

Mexico

Philippines

Egypt

0 billion   25 billion   50 billion   75 billion   100 billion

Source: World Bank, 2021

EXAMINE THE GRAPH

Calculating About how much more does India receive in remittances than Mexico?

this puts a strain on the local community. Additionally, an influx of people to an area can impact the natural environment. This may be in the form of increased pollution, greater consumption of water, the creation of more trash and waste, and the conversion of land to new uses.

Another effect of migration is the growth of urban areas. *Urbanization* occurs when cities grow and spread into surrounding areas. As more people migrate to cities, populations increase and urban areas become larger and more crowded. Farmland is bought by developers to build houses, apartments, factories, offices, schools, and stores to provide for the growing number of people. The loss of farmland means that food must be grown farther from cities, resulting in additional shipping costs and related pollution.

Migration also has economic consequences. In some instances, skilled professionals like scientists and lawyers emigrate away from their home country in search of higher salaries and a better life. They take skills that their host country needs. For example, in the United States many rural areas and small towns rely on foreign-trained doctors. However, the home country of such workers experiences **brain drain**. This exit of educated skilled, and professional people can leave the host country without the ability to progress.

Remittances are another economic effect of migration. These are money or goods that migrants send back to family and friends in their home country. As immigrants in a new country work, they often send a portion of their earnings back to friends and family to help with essential needs. The money is spent in the home country on things such as food and clothing. This **promotes** growth in the economy of the home country and helps individual families rise out of poverty.

✓ **CHECK FOR UNDERSTANDING**

Making Connections How is cultural diffusion related to migration?

LESSON ACTIVITIES

1. **Informative Writing** Conduct research to learn about refugee flows in the world today. Select one group or refugee movement and write an essay about that group. Discuss the reasons why the group's refugees had to leave their home. Explain where they migrated to and how their lives are different.

2. **Analyzing Information** Working with a partner, think about how push and pull factors affect immigration and population. Create a cause-and-effect chart to capture your thinking.

brain drain migration of skilled or talented people out of a place

remittance money or goods a migrant sends to family and friends in their home country

promote to help

05
The Growth of Cities

Analyzing Key Ideas and Details As you read, take notes about what influences the location and growth of cities.

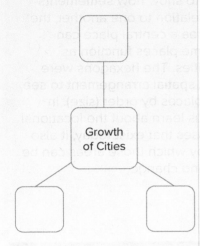

Patterns of Urbanization

What influences the location and growth of cities?

The Earth's population is moving in great numbers. People are moving from city to city or from rural villages to cities. The resulting growth of city populations brought about by such migration is called **urbanization**. The world's urban population is growing at a much faster rate than that of the rural population.

This urbanization is a major way that today's world is changing, especially in less developed countries. Over half of the world's people now live in cities. About 82 percent of people in the United States live in urban areas, and more than two thirds of the people in Europe and Latin America do as well. Much of this growth is due to increased *connectivity*—the directness of routes and communication linking places.

A main cause of urbanization is the desire of rural people to find jobs and a better life in more prosperous urban areas. Rural populations have certainly grown in some countries, but the amount of farmland has not increased to meet the growing number of people. As a result, many people migrate from rural areas to cities in search of jobs in manufacturing and service industries.

Urbanization is occurring rapidly. Between 1950 and 2021, the population of Mexico City increased from about 3.4 million to nearly 22 million.

urbanization the growth of cities as people migrate from rural areas to urban areas

Urban growth in Thailand is dominated by the capital city of Bangkok. It is the only city in the country that has more than 1,000,000 people.

Other cities in Latin America, as well as in Asia and Africa, have seen similar growth. A huge city or a cluster of cities is called a **megalopolis**.

Some factors that led to the growth of cities in the past are still important today. The growth of American cities, for example, began in the late 1700s. People went where waterways could be crossed, such as at river crossings or narrow straits. They also went where there was fertile land such as deltas and river valleys. Protection from enemies was another important consideration. The basics of survival are food and water sources and security from enemies. Most cities began where these factors were present. Although some of these factors are no longer as important as in the past, the cities remain located in these places.

In addition to factors that promote population increase in urban areas, there are factors that can shrink population in a region. For example, if an industry is no longer needed, the city's population will move on in order to seek new opportunities. Ghost towns are a prime example of this. Railroads in the United States provided the transportation for people to get to towns in California that were thriving during the Gold Rush. But when the gold mines were depleted, the people moved on and there was no longer any reason for these towns to exist.

Cities display patterns of human settlements—cities, towns, and villages—that are arranged in a region. Often these patterns are similar. The *central place theory* is a spatial model in geography that helps explain reasons for these distribution patterns. It observes the size and number of cities and towns that are present. It attempts to show how settlements locate and grow in relation to one another, the amount of market area a central place can control, and why some places function as villages, towns, or cities. The hexagons were considered the best spatial arrangement to see the organization of places by order (size). In addition to helping us learn about the locational patterns and processes that exist today, it also provides a context by which those areas can be studied for growth and change.

megalopolis a huge city or cluster of cities with an extremely large population

Central Place Theory

Central place theory is used to examine the distribution patterns, size, and number of cities and towns in a region. The hexagon is the best shape that minimizes overlaps between circles.

Legend:
- City
- Town
- Market Town
- Village
- Boundaries

EXAMINE THE DIAGRAM

Describing Describe the location of villages in relation to the city, towns, and market towns.

The city of İstanbul, Turkey, straddles the Bosporus Strait. This location at the crossroads of trade routes by land and by sea has made the city a center of regional trade for centuries.

Making Connections Can you name other cities that are important because of their location on the water?

All cities serve a variety of functions. For example, manufacturing, retail, and service centers are often located in urban areas. These functions are the economic base of a city, generating employment and wealth. The larger a city is, the more numerous and highly specialized its functions are likely to be. Smaller cities and towns have fewer functions, which tend to be of a more general nature. In the field of health care, for example, clinics are found in a wide range of places, but specialized teaching hospitals tend to be located only in larger cities.

A **world city** is a city that plays an important role in the global economic system. World cities have such features as international and diverse cultures, interactions in world affairs, a large population, a major international airport, and an advanced transportation system. One important world city that has resulted from the physical geography of its location is İstanbul. It straddles a strait, thus placing it on vital land and sea trade routes. Farming is good because of its fertile soil. In addition, it can easily defend itself against enemies because of its location by the water. İstanbul has been attacked many times throughout history. Its location is the very thing that has made it desirable.

Cities also tend to be centers of culture and creativity. Artists, musicians, architects, scientists, philosophers, and writers are attracted to cities where there are patrons, communities of other artists, universities, clients, and a skilled workforce.

There are several reasons cities can support a variety of functions. The large population of a city means there are plenty of workers available to support different types of industries. The large population also means there is a large market of consumers to buy goods and services.

world city a city generally considered to play an important role in the global economic system

Infrastructure provides systems that help the production of goods and services, and also the distribution of finished products to markets. Roads provide adequate transportation, and safe buildings provide secure housing. It also includes services such as schools and hospitals.

✓ CHECK FOR UNDERSTANDING

Contrasting In economic terms, how are the functions of larger cities different than those of smaller ones?

Challenges of Urban Growth

GUIDING QUESTION

What problems do urban areas face?

The urbanization process refers to much more than simple population growth. It involves changes in the economic, social, and political structures of a region. Rapid urban growth is responsible for many environmental and social changes, and its effects are related to issues of pollution and economics. These changes are not always positive. The rapid growth of cities strains their capacity to provide services such as energy, education, health care, transportation, sanitation, and physical security.

The more developed countries experienced urbanization during the nineteenth and twentieth centuries along with the Industrial Revolution. As the focus shifted from agricultural production to industrial production, people began moving to cities in large numbers. During this time, urbanization resulted from and contributed to industrialization. New job opportunities in the cities motivated people to migrate from rural areas to cities. At the same time, migrants provided cheap, plentiful labor for the emerging factories. As more people moved to cities, the physical size of cities also began to grow.

Today, the circumstances are rather different in less developed countries. People are forced out of rural areas because of insufficient land on which to grow subsistence crops. Meanwhile, there are not enough jobs to accommodate the many migrants looking for employment, creating a large surplus labor force. This influx keeps wages low and can lead to poverty in many urban areas.

Modern cities all over the world face many of the same problems: poor housing, homelessness, pollution, and social problems such as addiction, crime, and gang violence. Some areas of these modern cities are characterized by inadequate city services, low real estate values, almost absent businesses, and high unemployment. In addition, cars and industries pollute city air and water. Some cities continue to face conflicts between different cultural groups.

One of the major effects of rapid urban growth is **urban sprawl**. This is spreading of urban areas onto undeveloped land near cities. Urban sprawl increases traffic, drains local resources, and destroys open space. It is responsible for changes in the physical environment and can also **diminish** the local character of the community. Small local businesses find it difficult to compete with larger stores and restaurants. Some cities are trying to be proactive and establish measures aimed at fighting urban sprawl by limiting construction and using innovative land-use planning techniques or community cooperation.

✓ CHECK FOR UNDERSTANDING

Summarizing In what ways does rapid growth strain cities?

LESSON ACTIVITIES

1. **Argumentative Writing** Consider the negative aspects of urban sprawl. What are some possible solutions a government could provide to boost infrastructure and services? Develop an argument to support your ideas. Write an essay defending your argument.

2. **Presenting** Work with a group to gather information about a city challenged by urban sprawl. Your research should focus on the past, present, and future of the city. Create a multimedia presentation outlining that city's experiences along the road to urbanization.

---

infrastructure the set of systems that affect how well a place or organization operates, such as telephone or transportation systems, within a country

urban sprawl spreading of urban developments on undeveloped land near a city

diminish to make or become less

06
Analyzing Sources: Population Change

 COMPELLING QUESTION

How do populations change over time?

Plan Your Inquiry

DEVELOPING QUESTIONS

Think about all the people who make up the population of a place, such as your town, state, or country. Consider the ways in which the population of that place might change over time. Then read the Compelling Question for this lesson. What questions can you ask to help you answer this Compelling Question? Create a graphic organizer like the one below. Write these Supporting Questions in your graphic organizer.

Supporting Questions	Source	What this source tells me about how populations change over time	Questions the source leaves unanswered
	A		
	B		
	C		
	D		
	E		

ANALYZING SOURCES

Next, examine the sources in this lesson. Analyze each source by answering the questions that follow it. How does each source help you answer each Supporting Question you created? What questions do you still have? Write these in your graphic organizer.

After you analyze the sources, you will
- use the evidence from the sources;
- communicate your conclusions;
- take informed action.

Background Information

Population refers to the number of people living in an area. We can focus on the population of the planet, the population of a country, or the population of a more specific place. Geographers use statistics to describe populations and understand how they change over time. A *census* is an official count or survey of a population. Statistical information that describes populations and particular groups within populations is known as *demographics*.

» According to the U.S. Constitution, the population of the United States must be counted every ten years.

World Population

Population growth has been accelerating throughout modern history. Advances in healthcare and technology have helped people live longer lives and produce enough food to support a growing population. Some people worry about whether the world's environment can support the needs of the population in the future after so many years of population growth. This graph shows how the world's population has grown in the past and how it is expected to grow in the future. The graph shows the total population as well as the rate of growth of the population.

PRIMARY SOURCE: GRAPH

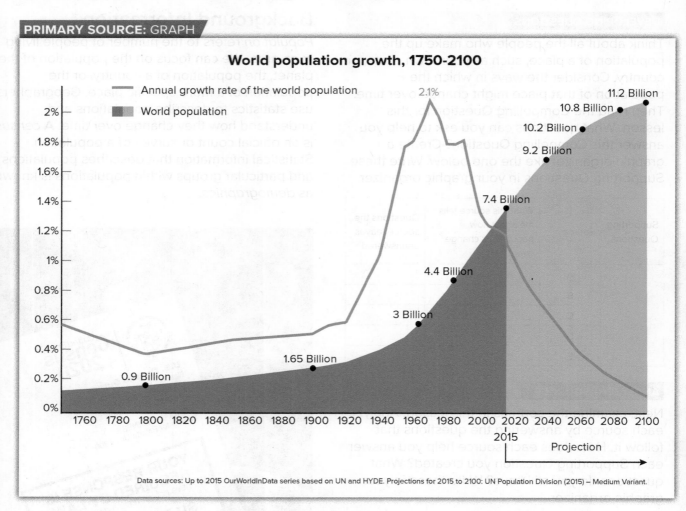

World population growth, 1750-2100

— Annual growth rate of the world population

World population

2.1%

11.2 Billion
10.8 Billion
10.2 Billion
9.2 Billion
7.4 Billion
4.4 Billion
3 Billion
1.65 Billion
0.9 Billion

1760 1780 1800 1820 1840 1860 1880 1900 1920 1940 1960 1980 2000 2020 2040 2060 2080 2100

2015

Projection

Data sources: Up to 2015 OurWorldInData series based on UN and HYDE. Projections for 2015 to 2100: UN Population Division (2015) – Medium Variant.

EXAMINE THE SOURCE

1. **Comparing and Contrasting** Compare the changes in the overall population to the changes in the growth rate of the population. When does the size of the population peak? When does the annual growth rate of the population peak?

2. **Calculating** How many people, on average, were added to the world population in each year between 1960 and 1980?

Birth Rates

One way that populations grow is through births. The birth rate of a population is determined by comparing the total number of births within a year to the size of the population. It is expressed as the number of births per 1,000 people. This measurement helps us easily compare birth rates between populations of different sizes. The graph below shows the birth rates for five different countries over the course of 60 years.

PRIMARY SOURCE: GRAPH

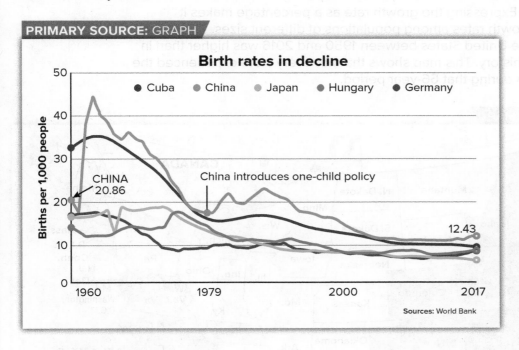

Birth rates in decline

● Cuba ● China ● Japan ● Hungary ● Germany

CHINA 20.86

China introduces one-child policy

12.43

Births per 1,000 people

Sources: World Bank

EXAMINE THE SOURCE

1. **Identifying** Rank the five countries from highest to lowest birth rates in each year: 1960, 1979, and 2017.

2. **Comparing and contrasting** How are changes in the birth rates of these five countries different from each other? Are there similarities between the countries?

Population Change of U.S. States

The Constitution requires that the population of the United States be counted every ten years. The population of the United States has increased in each decade since the country began counting its residents in the 1790 census. The Census Bureau calculates that the U.S. population currently adds one person every 29 seconds. Both immigration and births have contributed to the growth of the population over time. The population growth rate is calculated by comparing the population at two different points in time. Expressing the growth rate as a percentage makes it possible to compare growth rates among populations of different sizes. The rate of population growth in the United States between 1950 and 2016 was higher than in any previous period in history. This map shows that not every state experienced the same amount of growth during that 66-year period.

SECONDARY SOURCE: MAP

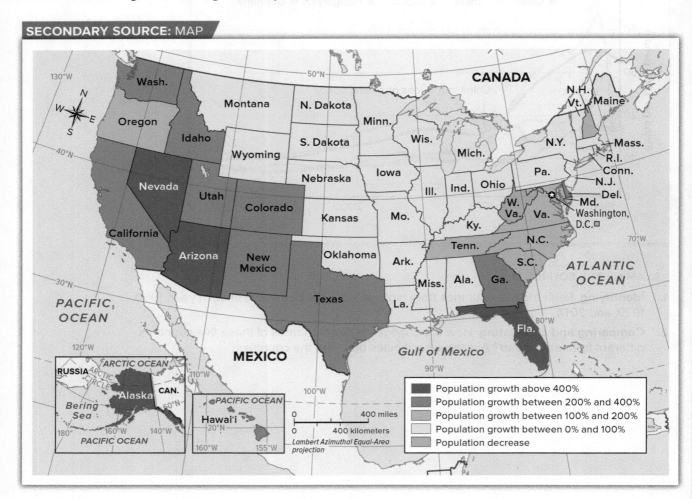

EXAMINE THE SOURCE

1. **Contrasting** Which states and regions of the U.S. experienced the most population growth from 1950 to 2016? In which states and regions did populations grow the least?

2. **Speculating** How might migration have impacted the differences in the growth rates experienced by the states?

The Population of Japan

A population is made up of people of different ages. A population pyramid is a special type of graph that shows how much of a population is made up of people of different ages and genders. One side of a population pyramid shows how many males there are in each age group. The other side of the pyramid shows how many females of each age are in the population. In this graphic, population pyramids are used to compare the population of Japan in 1965 with that of 2015.

PRIMARY SOURCE: GRAPH

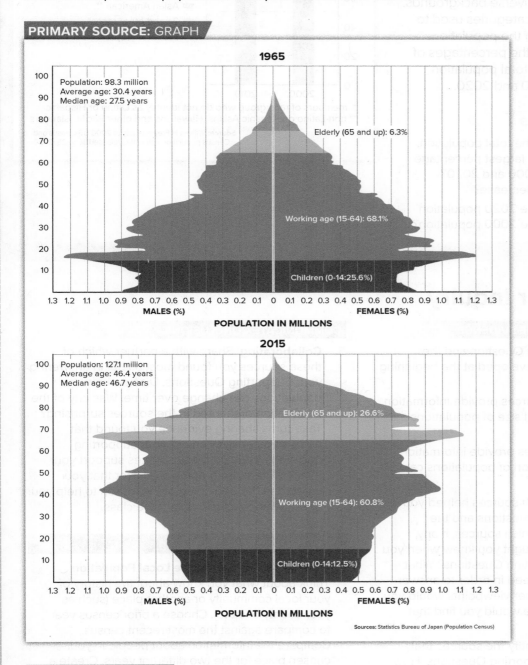

1965

Population: 98.3 million
Average age: 30.4 years
Median age: 27.5 years

Elderly (65 and up): 6.3%

Working age (15-64): 68.1%

Children (0-14:25.6%)

1.3 1.2 1.1 1.0 0.9 0.8 0.7 0.6 0.5 0.4 0.3 0.2 0.1 0 0.1 0.2 0.3 0.4 0.5 0.6 0.7 0.8 0.9 1.0 1.1 1.2 1.3
MALES (%)   **FEMALES (%)**
POPULATION IN MILLIONS

2015

Population: 127.1 million
Average age: 46.4 years
Median age: 46.7 years

Elderly (65 and up): 26.6%

Working age (15-64): 60.8%

Children (0-14:12.5%)

1.3 1.2 1.1 1.0 0.9 0.8 0.7 0.6 0.5 0.4 0.3 0.2 0.1 0 0.1 0.2 0.3 0.4 0.5 0.6 0.7 0.8 0.9 1.0 1.1 1.2 1.3
MALES (%)   **FEMALES (%)**
POPULATION IN MILLIONS

Sources: Statistics Bureau of Japan (Population Census)

EXAMINE THE SOURCE

1. **Analyzing** How did the population of Japan change between 1965 and 2015?
2. **Contrasting** In which age group do you see the biggest difference in the size of the male and female population? What does this mean?

Race and Ethnicity in the U.S. Population

The U.S. Census Bureau collects information about the racial and ethnic groups with which people identify. The United States has always had a population made up of people of diverse backgrounds. Race and ethnicity are categories used to describe the diversity of the population. This graph shows how the percentages of these categories in the total population changed between 2000 and 2020.

EXAMINE THE SOURCES

1. **Calculating** Within the total population, which group had the largest percentage increase between 2000 and 2020? Which group had a decrease?

2. **Analyzing** How is the 2020 population more diverse than the 2000 population?

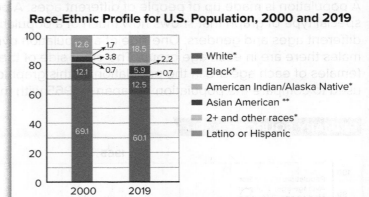

PRIMARY SOURCE: GRAPH

Race-Ethnic Profile for U.S. Population, 2000 and 2019

Legend:
- White*
- Black*
- American Indian/Alaska Native*
- Asian American **
- 2+ and other races*
- Latino or Hispanic

2000: 69.1, 12.1, 0.7, 3.8, 1.7, 12.6
2019: 60.1, 12.5, 5.9, 0.7, 2.2, 18.5

* members of race group who do not identify as Latino or Hispanic
** non-Latino or Hispanic Asians, Hawaiians and other Pacific Islanders

Source: William H Frey analysis of 2000 US Census and Census population estimates, released June 25, 2020

Complete Your Inquiry

EVALUATE SOURCES AND USE EVIDENCE

Refer back to the Compelling Question and the Supporting Questions you developed at the beginning of the lesson.

1. **Summarizing** Which sources provide information about how much the total size of populations can change over time?

2. **Explaining** Which sources provide information about how the composition of populations can change over time?

3. **Gathering Sources** Which sources helped you answer the Supporting Questions and the Compelling Question? Which sources, if any, challenged what you thought you knew when you first created your Supporting Questions? What information do you still need in order to answer your questions? What other viewpoints would you like to investigate? Where would you find that information?

4. **Evaluating Sources** Identify the sources that helped answer your Supporting Questions. How reliable is each source? How would you verify the reliability of the source?

COMMUNICATE CONCLUSIONS

5. **Collaborating** Share with a partner which of these sources you found most useful in answering your Supporting Questions. Discuss how populations can change over time. Was any of the information presented in the sources surprising? Take turns sharing evidence you found that helped you answer one of your supporting questions. How do these sources support your thinking? Use the graphic organizer that you created at the beginning of the lesson to help you. Share your conclusions with the class.

TAKE INFORMED ACTION

Visualizing Changes in the Local Population
Think about what you know about the history of your local community or another place (such as your county or state). Choose a prior census year to compare against the most recent census. Compare the information about people in your chosen place for the two different years. Create a poster that illustrates the most significant changes that occurred, such as how age ranges, genders, or ethnic groups have become larger or smaller parts of the population. Display the poster at school or in another location in your community.

07

Global Issues: Causes of Population Change

? COMPELLING QUESTION

Why do populations change over time?

Plan Your Inquiry

DEVELOPING QUESTIONS

Think about the population of a community, state, or country and the ways in which that population might have changed over time. The size of the population and the characteristics of the people who make up the population might have changed. Consider why these changes happened and why the population might change in the future. Then read the Compelling Question for this lesson. What questions can you ask to help you answer this Compelling Question? Create a graphic organizer like the one below. Write these Supporting Questions in your graphic organizer.

Supporting Questions	Source	What this source tells me about why populations change over time	Questions the source leaves unanswered
	A		
	B		
	C		
	D		
	E		

ANALYZING SOURCES

Next, examine the sources in this lesson. Analyze each source by answering the questions that follow it. How does each source help you answer each Supporting Question you created? What questions do you still have? Write these in your graphic organizer.

After you analyze the sources, you will:
- use the evidence from the sources;
- communicate your conclusions;
- take informed action.

Background Information

Populations are dynamic, which means they are constantly changing. Geographers and other professionals track population trends and try to understand why those trends occur. Understanding and anticipating changes in population helps governments plan to meet the needs of their citizens. For example, rapid population growth can put pressure on resources and **infrastructure**. On the other hand, a population that is not growing will gradually age and have different needs. Companies also track changes in populations to analyze markets for their goods and services.

infrastructure the set of systems that affect how well a place or organization operates, such as telephone or transportation systems, within a country

» Bengaluru, India, has experienced an average annual population growth rate of 22 percent per year since 1950. The city currently has over 12.7 million residents.

Total Fertility Rate

The number of children being born is one important factor in determining how a population will grow. The *total fertility rate* is the average number of children a woman has during her childbearing years (usually ages 15–49). A fertility rate of 2.1 is referred to as the *replacement level*. If each woman has an average of 2.1 children, the size of the total population will stay the same (setting aside the influence of migration). The graph and text compare the total fertility rates and income levels of different countries in the world. Income is measured here as gross domestic product (GDP), which is the value of the goods and services that a country produces. This comparison points to a trend in that connection between fertility rate and GDP.

SECONDARY SOURCE: WEBSITE

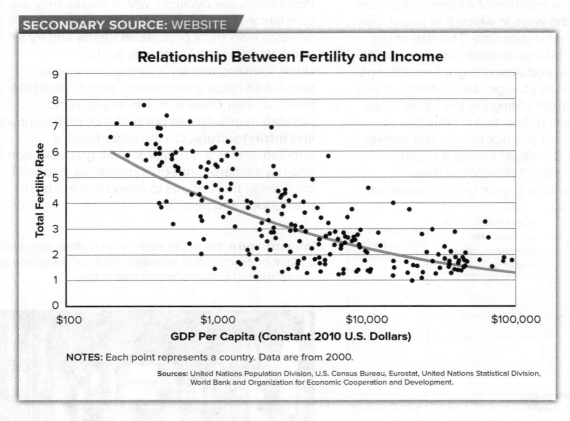

Relationship Between Fertility and Income

NOTES: Each point represents a country. Data are from 2000.

Sources: United Nations Population Division, U.S. Census Bureau, Eurostat, United Nations Statistical Division, World Bank and Organization for Economic Cooperation and Development.

> 66 The figure shows the relationship between fertility (more specifically, the total fertility rate) and gross domestic product (GDP) per capita (measured in 2010 U.S. dollars) across countries in 2000.
>
> Women tend to give birth to no fewer than three children in countries where GDP per capita is below $1,000 per year. In countries where GDP per capita is above $10,000 per year, women tend to give birth to no more than two children. 99

—"The Link between Fertility and Income," Guillaume Vandenbroucke, Federal Reserve Bank of St. Louis website, 2016

EXAMINE THE SOURCES

1. **Identifying** What are the maximum and minimum fertility rates and GDP per capita for the countries represented on the graph?

2. **Summarizing** How might the relationship between fertility and income be summarized based on the information in this graph?

Africa's Population

Population increases in regions of the world have taken place at very different rates. The reasons for these differences are many. This article examines why population growth in Africa has increased in recent years while that of other parts of the world has slowed or declined.

PRIMARY SOURCE: ARTICLE

❝ The last 100 years have seen an incredible increase in the planet's population. Some parts of the world are now seeing smaller increments of growth, and some, such as Japan, Germany, and Spain, are actually experiencing population decreases.

The continent of Africa, however, is not following this pattern. Now home to 1.2 billion (up from just 477 million in 1980), Africa is projected by the United Nations Population Division to see a slight acceleration of annual population growth in the immediate future.

In the past year the population of the African continent grew by 30 million. By the year 2050, annual increases will exceed 42 million people per year and total population will have doubled to 2.4 billion, according to the UN. This comes to 3.5 million more people per month, or 80 additional people per minute. At that point, African population growth would be able to re-fill an empty London five times a year.

The dynamics at play are straightforward. Since the middle of the last century, improvements in public health have led to an inspiring decrease in infant and child mortality rates. Overall life expectancy has also risen. The 12 million Africans born in 1955 could expect to live only until the age of 37. Encouragingly, the 42 million Africans born this year can expect to live to the age of 60.

Meanwhile, another key demographic variable — the number of children the average African woman is likely to have in her lifetime, or total fertility rate — remains elevated compared to global rates. The total fertility rate of Africa is 88% higher than the world standard (2.5 children per woman globally, 4.7 children per woman in Africa). ❞

—"Population Growth in Africa: Grasping the Scale of the Challenge," Joseph J. Bish, *The Guardian*, January 11, 2016

EXAMINE THE SOURCE

1. **Contrasting** How is population growth different in Africa than in Japan, Germany, and Spain?

2. **Identifying Cause and Effect** What factors contribute to the high rate of population growth in Africa?

Bulgaria's Population Decline

Bulgaria is a country in Eastern Europe. It shares borders with Romania, Turkey, and Greece and has coastline along the Black Sea. Bulgaria was part of the Soviet bloc (a group of Eastern European countries allied with the Soviet Union) until 1989. Its citizens were not able to easily move elsewhere. Since 2007, the country has been part of the European Union (EU), and its citizens can freely move to other EU countries. Populations are declining in many countries in Europe due to decreasing birth and fertility rates, but as this article explains, Bulgaria's situation is unique.

**PRIMARY SOURCE: ** ARTICLE

66 As natives of the poorest member of the European Union, Bulgarians have been leaving their home in droves, contributing to the world's fastest population decline. Bulgaria's population was around 9 million at the end of the 1980s, but it fell to fewer than 7 million in 2018, and is expected to fall below 6 million in 50 years. The UN Population Division projects that Bulgaria will lose 23% of its population by 2050—a projection so high that the country is neck-and-neck with Lithuania for the fastest shrinking population in the world.

Low birth rates are the biggest factor for such steep decline. But what sets Bulgaria apart from other declining European countries is its massive outbound migration.

The government does not keep reliable statistics but some economists...estimate that at least 60,000 Bulgarians leave each year. And even that estimate may be low, given that Germany alone says it welcomed 30,000 new Bulgarian residents in 2017. 99

—"How to Slow Down the World's Fastest-Shrinking Country," Denise Hruby, BBC.com, September 30, 2019

EXAMINE THE SOURCES

1. **Identifying** When did Bulgaria's population start to decline, and what other country is experiencing a similar decline?

2. **Explaining** What two factors are contributing to population decline in Bulgaria?

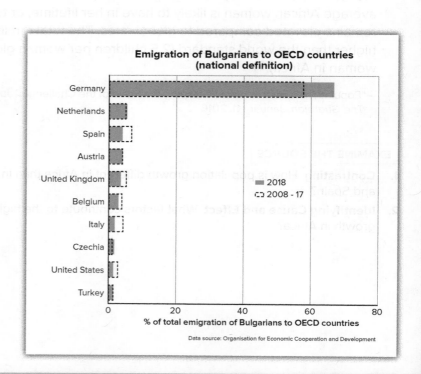

Emigration of Bulgarians to OECD countries (national definition)

Legend: 2018; 2008 - 17

% of total emigration of Bulgarians to OECD countries

Data source: Organisation for Economic Cooperation and Development

U.S. Population Growth

The population of the United States has grown throughout the country's history. Examining when the population growth rates increase and decrease can help reveal underlying causes for those changes. This report is about the most recent trends in the U.S. population. It also includes interesting historical information.

PRIMARY SOURCE: REPORT

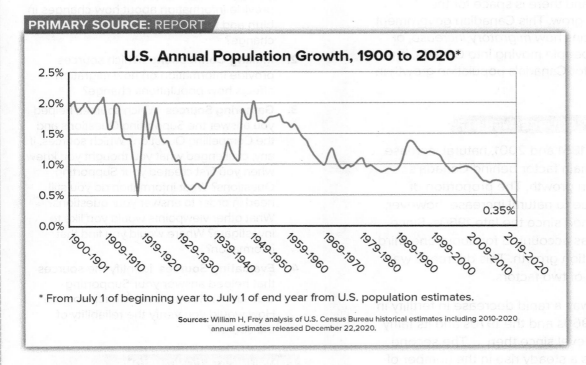

U.S. Annual Population Growth, 1900 to 2020*

0.35%

* From July 1 of beginning year to July 1 of end year from U.S. population estimates.

Sources: William H, Frey analysis of U.S. Census Bureau historical estimates including 2010-2020 annual estimates released December 22,2020.

" For much of the recent past, the U.S. has been one of the most rapidly growing countries in the industrialized world. This was especially true in the last half of the 20th century, due to the post-World War II baby boom and rising immigration in the 1980s and 1990s. . . .

Recently released Census Bureau population estimates show that from July 1, 2019 to July 1, 2020, the nation grew by just 0.35%. This is the lowest annual growth rate since at least 1900.

National population growth began to dip after 2000, especially after the Great Recession and, in recent years, due to new immigration restrictions. Yet the 2019-to-2020 rate is well below most growth rates over the past 102 years. . . . Part of this sharp decline can be attributed to the COVID-19 pandemic, which brought more deaths and further immigration restrictions. Still, the entire 2010s decade was one of fewer births, more deaths, and uneven immigration. "

—"What the 2020 Census Will Reveal About America: Stagnating Growth, an Aging Population, and Youthful Diversity," William H. Frey, The Brookings Institution, January 11, 2021

EXAMINE THE SOURCE

1. **Comparing** How did the growth rate of the U.S. population compare to the growth rate of other industrialized countries from 1950–2000?

2. **Identifying Cause and Effect** What are some of the events that have affected the growth rate of the U.S. population since 1900?

Population Change in Canada

Canada is the second-largest country in the world as measured by area, but it only has the thirty-ninth-largest population. This means its population density is low (about 10 persons per square mile) and there is space for the population to grow. This Canadian government website explains how *migratory increase*, or growth from people moving into Canada, is a large reason for Canada's population growth in recent times.

SECONDARY SOURCE: WEBSITE

❝ Between 1851 and 2001, natural increase was the main factor behind Canada's population growth. The proportion of growth due to natural increase, however, has declined since the late 1960s. Since 2001, it has accounted for about one-third of population growth. This decrease was the result of two factors.

The first was a rapid decrease in fertility in the late 1960s and the 1970s and its fairly constant level since then.... The second factor was a steady rise in the number of deaths. This was due in part to the aging of the population.... It was also due to population growth.

As a result, the numbers of births and deaths have converged since the end of the baby boom in Canada, and migratory increase has taken on an increasingly important role in recent Canadian population growth....

Without a sustained level of immigration or a substantial increase in fertility, Canada's population growth could, within 20 years, be close to zero. ❞

—"Population Growth in Canada: From 1851 to 2061," Statistics Canada, 2018

EXAMINE THE SOURCE

1. **Explaining** What is migratory increase?
2. **Identifying Cause and Effect** What factor other than rate of migratory increase determines Canada's overall population growth rate?

Complete Your Inquiry

EVALUATE SOURCES AND USE EVIDENCE

Refer back to the Compelling Question and the Supporting Questions you developed at the beginning of the lesson.

1. **Making Generalizations** Which sources provide information about how changes in birth and fertility rates relate to population change?

2. **Making Connections** Which sources provide information on how migration affects how populations change?

3. **Gathering Sources** Which sources helped you answer the Supporting Questions and the Compelling Question? Which sources, if any, challenged what you thought you knew when you first created your Supporting Questions? What information do you still need in order to answer your questions? What other viewpoints would you like to investigate? Where would you find that information?

4. **Evaluating Sources** Identify the sources that helped answer your Supporting Questions. How reliable is each source? How would you verify the reliability of the source?

COMMUNICATE CONCLUSIONS

5. **Collaborating** Share with a partner which of these sources you found most useful in answering your Supporting Questions. Discuss how the sources helped you understand why populations change. Take turns sharing evidence you found associated with your supporting questions. How do these sources support your thinking? Use the graphic organizer that you created at the beginning of the lesson to help you. Share your conclusions with the class.

TAKE INFORMED ACTION

Forecasting Population Changes Using U.S. Census data, work in pairs to try to predict how the population of your local area will change in the next twenty years, and what effects this population change will have on your community and its resources. Interview 2–3 community members to get their ideas about how the population might change in the future. Look for patterns that will help you decide what to predict. Discuss your predictions with your class.

Statistics Canada. "Population growth in Canada: From 1851 to 2061". https://www12.statcan.gc.ca/census-recensement/2011/as-sa/98-310-x/98-310-x2011003_1-eng.cfm. 2018.

Summary

Population Growth	• Earth's total human population continues to grow but has slowed in recent decades. • The global birth rate is higher than the global death rate, resulting in population growth. • Migration is also a factor in population growth. • Population pyramids show a country's population by age and gender. • The effects of population growth include crowded cities, poverty, and strains on natural resources.
Population Patterns	• Population distribution is the patterns of human settlement. • Population distribution is influenced by physical geography—nearness to resources, jobs, and places of cultural or religious significance. • The three main world population clusters are East Asia, South Asia, and Europe. • Population density refers to the number of people living on a square mile or square kilometer of land. • Population density varies from country to country depending on land area and total population.
Population Movement	• Migration occurs when people move from one place to another. It is one of the main causes for population shifts in the world today. • People migrate for many reasons. Push factors drive people from an area, while pull factors attract people to an area. • Some people, known as refugees, are forced to migrate to escape violence and war or other disasters. Enslaved people are subjected to forced migration. • Migration can affect the culture, population, and economy of an area.
The Growth of Cities	• Urbanization is the movement of people from rural areas to cities. One main cause of this is that people in rural areas want to find jobs and a better life in cities. • Factors that influence the growth of cities are nearness to water and a location that can be easily defended. • Central place theory helps explain the reasons for the distribution patterns, size, and number of cities and towns in an area. • Cities serve many functions—as centers of economic activity, as centers of government, and as centers of culture. • There are challenges to urban growth. These include homelessness, pollution, violence, and urban sprawl.

Checking For Understanding

Answer the questions to see if you understood the topic content.

IDENTIFY AND EXPLAIN

1. Identify each of these terms as they relate to population growth and population patterns.

   A. death rate
   B. birth rate
   C. natural increase
   D. doubling time
   E. population pyramid
   F. population distribution
   G. population density
   H. urban
   I. rural

REVIEWING KEY FACTS

2. **Explaining** Why might a refugee move to a new area?

3. **Identifying** What are examples of push factors to migration?

4. **Analyzing** What is the main reason people choose to settle in one area and not in another?

5. **Making Connections** Give one example of an urban area and one example of a rural area.

6. **Explaining** How is population density calculated?

7. **Describing** What is urban sprawl? What are its effects?

8. **Summarizing** What problems do urban areas face?

9. **Explaining** What does central place theory attempt to explain?

CRITICAL THINKING

10. **Identifying** Describe the three factors that have contributed to Earth's constantly growing population.

11. **Determining Central Ideas** Why do more people live in some parts of the world than in others?

12. **Identifying Cause and Effect** List three causes and three effects of human migration.

13. **Analyzing** A country has a growing population. Will its population pyramid be wider at the top or wider at the bottom? Why?

14. **Analyzing** What does a population pyramid that is wider at the top tell you about that country's population?

15. **Making Connections** What influences the growth of cities?

16. **Determining Meaning** Discuss the difference between the meanings of the words *emigrate* and *immigrate*.

17. **Contrasting** How does population density for a country differ when arable land area is considered rather than total land area?

18. **Identifying** What are the three major population clusters in the world?

19. **Making Connections** How are population growth rate and doubling time related?

20. **Describing** What is negative population growth? What impact can it have on a country?

NEED EXTRA HELP?

If You've Missed Question	1	2	3	4	5	6	7	8	9	10
Review Lesson	2, 3	4	4	3	4	4	5	5	5	2

If You've Missed Question	11	12	13	14	15	16	17	18	19	20
Review Lesson	3	4	2	2	5	4	3	3	2	2

Apply What You Have Learned

 Geographic Reasoning

Population density occurs at different scales. It can vary widely among those scales; for example, a very densely populated city can be located within a state or country that is otherwise made up of a sparsely populated mountainous area. In addition to physical geography, economic and political forces also determine patterns of population density in a specific area. Visual representations of density can help us understand these patterns.

> **ACTIVITY** **Drawing Three Dot Density Maps** Choose three geographic areas of different scales: one town or city, one state or province within a country, and one country. For each area, draw a simple map showing only the area's boundaries and dots that represent specific numbers of people. Present your three maps to the class, briefly explaining why you think the dots are distributed within each area as they are (reasons may include physical geography, economic activities, or other factors).

 Understanding Urban Sprawl

Urbanization has been a growing phenomenon worldwide for several decades, and even longer in more developed countries. When the movement of people to large cities happens more quickly than the growth of available housing within those cities, settlements of people begin to spread outward from city centers. The effects of that spread are many. They include positive changes (such as the building of homes that are more affordable than those inside city boundaries) and negative changes (such as the destruction of wildlife habitats and pollution from commuters' motor vehicles).

> **ACTIVITY** **Taking a Position** Research the effects of urban sprawl on one city and its surrounding area. Look for evidence of change in the affected communities and the impact of sprawl on the land around the city. Then present both positive and negative effects of the phenomenon in a panel discussion. Attempt to give equal time to both sides of the issue.

C Understanding Causes

Population growth around the world has happened in part due to falling death rates. Advances in medical technology, disease prevention, nutrition, and infrastructure have led to longer life spans in most regions, especially in recent decades. For example, in areas where smoking rates have declined, the number of deaths from smoking-related diseases such as lung cancer may have decreased.

ACTIVITY **Explaining Increasing Life Expectancies** Research online a single event, invention, lifestyle change, or other development that has contributed to lower death rates, either in one area or worldwide. Write a short essay explaining the connection between the development and peoples' life expectancies.

More widely available healthcare services for babies and children have had a dramatic impact on their life expectancies.

D Making Connections to Today

You have learned about how migration impacts an area's population growth and cultural changes. Migration has changed the distribution of people since prehistoric times, and it continues today. With the communication tools of the Internet, it is possible to observe the movement of groups of people around the world and learn about their reasons for migration.

ACTIVITY **Presenting a News Report** Research one or more online news sources to learn about the movement of a group of people that is happening today. This might be a migration for economic reasons, such as to cities or countries with employment opportunities, or for humanitarian reasons, such as to escape a military conflict or natural disaster. Then write a script for a five-minute news report about the event. Be sure to include the perspectives of the migrating people in your report. Present the report to the class, either in person or as a video presentation.

Street musicians and the music they play are an intricate part of Cuba's culture.

Cultural Geography

INTRODUCTION LESSON

01 Introducing Cultural Geography   G152

LEARN THE CONCEPTS LESSONS

02 Elements of Culture   G157

03 Expressions of Culture   G161

05 Causes and Effects of Cultural Change   G171

INQUIRY ACTIVITY LESSONS

04 Analyzing Sources: Culture and Identity   G165

06 Global Issues: Continuity and Change in Culture   G175

REVIEW AND APPLY LESSON

07 Reviewing Cultural Geography   G181

David Jensen / Alamy Stock Photo

What Is Culture?

66 A nation's culture resides in the hearts and in the soul of its people. 99

—Mohandas K. Gandhi, leader of India's nonviolent independence movement

66 Culture is "that complex whole which includes knowledge, belief, art, morals, law, custom and any other capabilities and habits acquired by man as a member of society. 99

—Sir Edward B. Tylor, English anthropologist

66 Culture is coded wisdom. . . wisdom that has been accumulated for thousands of years and generations. Some of that wisdom is coded in our ceremonies, it is coded in our values, it is coded in our songs, in our dances, in our plays. 99

—Wangari Maathai, Kenyan politician and environmental activist

66 One of the most effective ways to learn about oneself is by taking seriously the cultures of others. It forces you to pay attention to those details of life which differentiate them from you. 99

—Edward T. Hall, American anthropologist

> " Culture is basically an attitude; it is also learned. It is the learning of shared language and perceptions, which are incorporated in the mind through education, repetition, ritual, history, media, or mimicking. "

—Dr. Thomas L. McPhail, professor of media studies

> " Culture is the way we live. It is the clothes we wear, the foods we eat, the language we speak, the stories we tell, and the ways we celebrate. It is the way we show our imaginations through art, music, and writing. "

—Bobbie Kalman, teacher and author

> " Culture is the name for what people are interested in, their thoughts, their models, the books they read and the speeches they hear, their table-talk, gossip, controversies, historical sense and scientific training, the values they appreciate, the quality of life they admire. "

—Walter Lippmann, American journalist

> " Culture, the acquainting ourselves with the best that has been known and said in the world, and thus with the history of the human spirit. "

—Matthew Arnold, English poet and literary and social critic

Getting Ready to Learn About . . . Cultural Geography

Why Study Culture?

People around the world establish communities and develop ways of life. Evidence of culture includes buildings, farming patterns, language, types of government, ways of earning a living, and more—all parts of what geographers study.

To understand why groups of people have a certain way of life, geographers study culture from a spatial perspective. This means that they look at the ways in which culture is expressed in terms of location in the spaces that exist on Earth. They identify patterns and relationships among groups of people.

As the world becomes increasingly interconnected, cultures spread and are shared. Studying the relationships among groups helps geographers understand the factors associated with cultural change.

Elements and Expressions

As the quotations on the first two pages of this lesson illustrate, the definition of *culture* is broad and can include many different things. To help focus their work, geographers think about elements of culture and the numerous expressions of culture.

Elements of culture and expressions of culture together help define cultures. Elements of culture are basic components that are part of every culture. These are language, religion, social systems, type of government, and type of economy. A combination of these elements together form a unique culture. Expressions of culture are the ways in which individuals and groups express their beliefs and feelings. For example, music, literature, dance, art, and food are expressions of culture.

A member of the Meskwaki people performs a traditional hoop dance. In Native American culture, the hoop represents what the culture views as the never-ending circle of life. The hoop is often used in traditional healing ceremonies.

Cultural Change and the Columbian Exchange

The Columbian Exchange resulted in a dramatic transfer of people, animals, plants, culture, ideas, and diseases between the Eastern and Western Hemispheres.

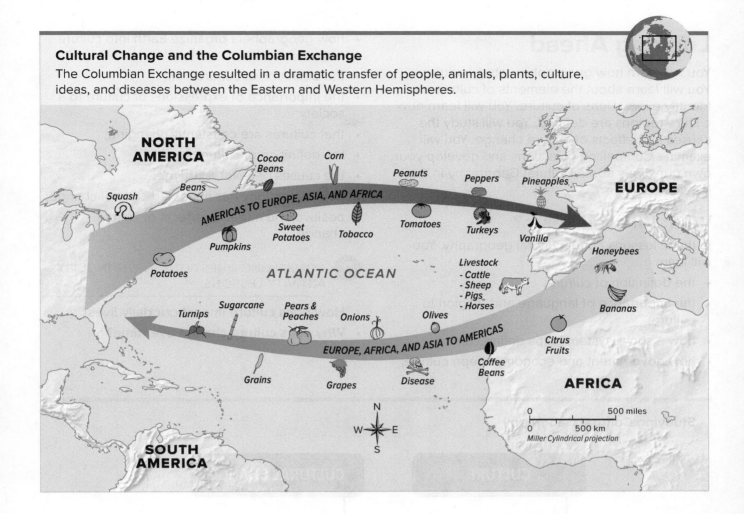

Cultural Change

The world and its people are always changing. Therefore, culture is in a state of change. Culture changes as a result of many things. For example, changes can spread from person to person within a place. This ultimately changes the whole culture of the place. Also, when people move from one place to another they often carry new ideas and lifestyles with them. Inventions and innovation also drive cultural change.

There are benefits to cultural change. For example, new foods, plants, and animals were exchanged between Europe and the Americas as a part of the Columbian Exchange. This altered the economies and ways of life of peoples in the Americas, Europe, and other continents. It created an economic trading situation.

There are also some drawbacks to cultural change. In some instances, change comes too quickly, and a culture is damaged or even

destroyed. In other instances, change is not wanted by the people in a culture. They worry that new ideas and ways of life could weaken their existing culture. During the Columbian Exchange, for example, the arrival of Spanish conquistadors in South America beginning in the late 1400s led to the crumbling of the Inca civilization. The Spanish, and later the British, French, and Dutch, brought diseases that caused the deaths of many of the Inca and other indigenous populations. The Europeans also brought new government and economic systems; this resulted in the decline of indigenous cultures.

Whatever the results, cultures today are coming into contact with one another more than ever before. Advances in technology and communications accelerate cultural change because of the speed with which new ideas can be shared between and among people. This has led to a considerable amount of cultural blending throughout the world.

Looking Ahead

You will learn how geographers study culture. You will learn about the elements of culture and identify expressions of culture. You will learn how culture regions are defined. You will study the causes and effects of culture change. You will examine Compelling Questions and develop your own questions about Cultural Geography in the Inquiry Activities.

What Will You Learn?

In these lessons about cultural geography. You will learn

- the definition of *culture*;
- the importance of language and religion to culture;
- how a social system helps define culture;
- how government and economy shape culture;
- how geographers organize Earth into culture regions;
- the definition of *culture trait*;
- the importance of expressions of culture to a society;
- that cultures are constantly changing;
- the definition of *culture hearth*;
- the causes of cultural change;
- how technology has impacted cultural change;
- positive and negative effects of cultural change.

 COMPELLING QUESTIONS IN THE INQUIRY ACTIVITY LESSONS

- **How does culture impact our daily lives?**
- **Why does culture change over time?**

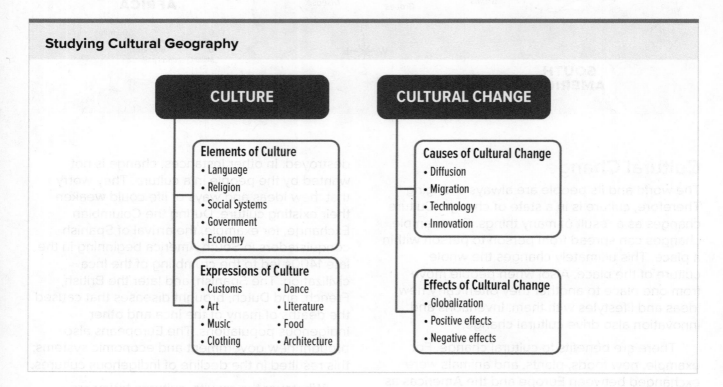

Studying Cultural Geography

CULTURE

Elements of Culture
- Language
- Religion
- Social Systems
- Government
- Economy

Expressions of Culture
- Customs
- Art
- Music
- Clothing
- Dance
- Literature
- Food
- Architecture

CULTURAL CHANGE

Causes of Cultural Change
- Diffusion
- Migration
- Technology
- Innovation

Effects of Cultural Change
- Globalization
- Positive effects
- Negative effects

02
Elements of Culture

Analyzing Key Ideas and Details As you read, take notes about the elements of culture.

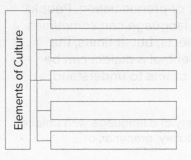

What Is Culture?

How is culture part of your life?

Geographers study **culture**, which is the way of life of a group of people who share similar beliefs and customs. **Customs** are actions or behaviors that are traditional among the people in a particular group or place. The term *culture* can also refer to the people of a certain culture. For example, saying "the Hindu culture" can mean the Hindu cultural traditions, the people who follow these traditions, or both. Culture is not something that people are born with. Instead, people learn the culture of the groups that they grow up in and within which they live their lives.

You might be part of more than one culture. If your family has strong ties to a culture, such as that of a religion or a country other than the United States, you might follow that cultural tradition at home. At the same time, you also might be part of a more mainstream American culture while at school and with friends.

A particular culture can be understood by looking at various elements: what languages are spoken, what religions are practiced, and what smaller groups form as parts of their society. Culture also includes how people govern their society and how they make their living.

✓ CHECK FOR UNDERSTANDING

Describing How would you define your culture?

culture the set of beliefs, behaviors, and traits shared by a group of people

custom an action or behavior that is traditional among the people in a particular group or place

Families that emigrated from Somalia to the United States might speak the Somali language and wear traditional Somali clothing but may enjoy playing sports that are popular in the United States.

Helen H. Richardson/The Denver Post/Getty Images

Language and Religion

GUIDING QUESTION

Why are language and religion important to a culture's development?

Language serves as a powerful form of communication. Through language, people communicate needs, information, and experience and pass on cultural beliefs and traditions. Thousands of different languages are spoken in the world. The world's languages have been organized into **language families**—large groups of languages having similar roots, or origins. Languages that may seem very different may belong to the same language family. For example, English, Spanish, Russian, and Hindi (spoken in India) are all members of the Indo-European language family. The map here shows the global distribution of language families.

Some languages have become world languages, or languages that are commonly spoken in many parts of the world. A language of this type is often called a **lingua franca** because it is a common language that meets the need for speakers of different languages to communicate with one another. Some languages are spoken differently in different regions or by different groups. A **dialect** is a regional variety of a language with unique features, such as vocabulary, grammar, or pronunciation. People who speak the same language can sometimes understand other dialects, but at times, the pronunciation, or accent, of a dialect can be nearly impossible for others to understand.

language family a group of related languages that have all developed from one earlier language

lingua franca a common language among speakers whose native languages are different

dialect a regional variety of a language with unique features, such as vocabulary, grammar, or pronunciation

World Languages

Most languages spoken around the world belong to one of 12 language families. Languages within a family have a common origin.

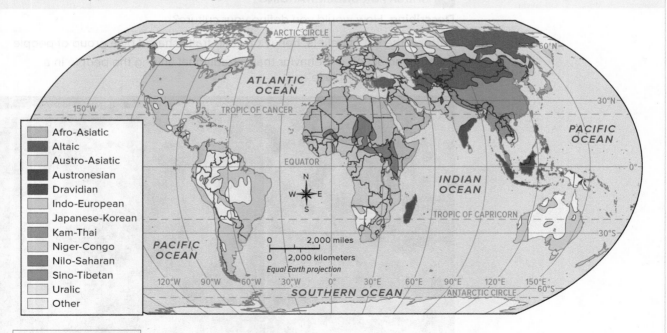

EXAMINE THE MAP

1. **Exploring Regions** What language group is spoken in the United States?
2. **Exploring Culture** Do people within the same language family always speak the same language? Explain your response.

Religion has a **major** influence on how people of a culture see the world. Religious beliefs are powerful. In many cultures, religion enables people to find a sense of identity. They view religion as the **foundation** and most important part of their life. In other cultures, people see their religion as only a tradition to follow during special occasions or holidays. Some people express no religion at all, and this also exerts a major influence on their culture.

Religious practices vary widely. Many cultures base their way of life on the spiritual teachings and laws of holy books. Throughout history, religious stories and symbols have influenced painting, sculpture, architecture, music, dance, and more. This continues into the present day. Some of the major world religions are Buddhism, Christianity, Hinduism, Islam, Judaism, and Sikhism. Those who express no religion are also a major group.

✓ **CHECK FOR UNDERSTANDING**

Summarizing How might religion influence the culture of a particular group?

major greater in importance or interest

foundation a basis on which something stands

Major World Religions

This table lists the leaders and beliefs of some of the major world religions. Religious beliefs influence cultures around the world.

Religion	Founder	Percent of World Population	Beliefs
Buddhism	Siddhārtha Gautama, the Buddha	5%	Suffering comes from attachment to earthly things, which are not lasting. People become free by following the Eightfold Path, rules of right thought and conduct. People who follow the Path achieve nirvana—a state of endless peace and joy.
Christianity	Jesus	31%	The one God is Father, Son, and Holy Spirit. God the Son became human as Jesus. Jesus died and rose to bring God's forgiving love to sinful humanity. Those who trust in Jesus receive eternal life with God.
Hinduism	No one founder	15%	One eternal spirit, Brahman, is represented as many deities. Every living thing has a soul that passes through many successive lives. Each soul's condition in a specific life is based on how the previous life was lived. When a soul reaches purity, it finally joins permanently with Brahman.
Islam	Muhammad	25%	The one God sent a series of prophets, including the final prophet Muhammad, to teach humanity. Islam's laws are based on the Quran, the holy book, and the Sunnah, examples from Muhammad's life. Believers are guided by the five pillars—belief, prayer, charity, fasting, and pilgrimage—to go to an eternal paradise.
Judaism	Abraham	<1%	The one God made an agreement through Abraham and later Moses with the people of Israel. God would bless them, and they would follow God's laws, applying God's will in all parts of their lives. The main laws and practices of Judaism are stated in the Torah, the first five books of the Hebrew Bible.
Non-religion	No one founder	16%	An absence of religion or a rejection of religion. An assumption or an active conviction that the natural world of matter and energy is all that people should think about, or even all that exists.
Sikhism	Guru Nanak	<1%	The one God made truth known through ten successive gurus, or teachers. God's will is that people should live honestly, work hard, and treat others fairly. The Sikh community, or Khalsa, bases its decisions on the principles of a sacred text, the Guru Granth Sahib.

Social Systems

GUIDING QUESTION

What is a social system?

Every culture includes a social system in which members of the society fall into various smaller groups. A social system develops to help the members of a culture work together to meet basic needs. In all cultures, the family is the most important group, although family structures vary somewhat from culture to culture.

Most cultures are also made up of social classes. These are groups of people organized according to ancestry, wealth, education, or other criteria. In addition, cultures may include people who belong to different **ethnic groups**. An ethnic group is made up of people who share a common language, history, place of origin, or combination of these things.

Members of the same Native American nation are an example of people of the same ethnic group. Other examples include the Maori of New Zealand and the Han Chinese. Large countries such as China can be home to hundreds of different ethnic groups. Some ethnic groups in a country are minority groups—people whose race or ethnic origin is different from that of the majority group. The largest ethnic minority groups in the United States are Hispanic Americans and African Americans.

Members of a culture might have special roles or positions as part of their cultural traditions. In some cultures, women are expected to care for and educate children. Most cultures expect men to earn money to support their families or to provide in other ways, such as by hunting and farming. Many cultures respect the elderly and value their wisdom. The leaders of older, traditional cultures are often elderly men or women who have leadership experience. Most cultures have clearly defined roles for their members. From an early age, young people learn what their culture expects of them.

✓ CHECK FOR UNDERSTANDING

Explaining How are ethnic groups related to culture?

ethnic group a group of people who share a common ancestry, language, religion, customs, or place of origin

Government and Economy

GUIDING QUESTION

Why is government important to culture?

Government is another element of culture. Despite differences, governments around the world share certain features. They maintain order within an area and provide protection from outside dangers. Governments also provide services to citizens, such as a legal system, education, and transportation infrastructure.

Different cultures have different ways of distributing power and making rules. Governments are organized according to levels of power—national, regional, and local. They also specify the type of authority found at the various levels—a single ruler, a small group of leaders, or a body of citizens or their representatives.

Economic activities also influence and shape a culture. People must make a living. In examining cultures, geographers look at economic activities. They study how a culture uses its natural resources to meet such human needs as food and shelter. They also analyze the ways in which people produce, obtain, use, and sell goods and services.

Some cultures have their own type of economy, but most follow the economy of the country or area in which they live. This allows people of different cultures living in an area to trade and conduct other types of business with one another. For example, many people in Benin, West Africa, sell goods in open-air markets. Some people bring items to the markets to trade for the goods they need because they don't live in a wage-earning culture. Other people pay for goods using the paper money and coins that they earn from employers.

✓ CHECK FOR UNDERSTANDING

Determining Central Ideas How are economic activities important in understanding a culture?

LESSON ACTIVITIES

1. **Informative Writing** Write a short essay describing one of the elements of culture. Explain its importance in defining a culture.

2. **Presenting** Work with a partner to create a slide show about the elements of culture. Include photographs showing examples of each element. Write captions explaining how geographers use the elements to learn about a culture.

03
Expressions of Culture

READING STRATEGY

Analyzing Key Ideas and Details As you read, take notes about culture regions and expressions of culture.

Culture Regions	Expressions of Culture

Defining Culture Regions

GUIDING QUESTION

What is a culture region?

Geographers organize the world into **culture regions** to help them understand cultural development. People in a culture region often share resources, have similar social groups, and work to keep their cultures and communities strong. A culture region includes different people and groups that have certain culture traits in common. A **culture trait** is a central characteristic of the culture that is shared by most members. Of course, the same trait—Christianity or the Spanish language, for example—can be part of more than one culture.

Culture regions may share similar economic systems, forms of government, and social groups. Their histories, religions, and art forms may share similar influences. The food, clothing, and housing of people in the culture region may all have common characteristics. Culture regions can be large or relatively small. For example, one of the world's largest culture regions stretches across North Africa and Southwest Asia. This culture region is home to hundreds of millions of people who practice the Islamic religion and share a Muslim culture. But culture regions can also be quite small and can exist within the larger world culture regions. An example of this is Spanish Harlem in New York City. This culture region is home to a large and growing Hispanic culture.

✓ **CHECK FOR UNDERSTANDING**

Identifying What types of things might a culture region have in common?

culture region section of the Earth in which people share a similar way of life, including language, religion, economic systems, and types of government

culture trait a characteristic of the culture that is shared by most members

Chinatown in Chicago is a culture region. As in other Chinatown neighborhoods in major cities like San Francisco, the Chicago Chinatown is home to a large Chinese population who share a common culture.

Bruce Yuanyue Bi/Getty Images

World Culture Regions

Why are world culture regions important?

When looking at maps, you can start to identify regional patterns of some culture traits like language, religion, or a common history. Geographers are interested in the spatial distribution of these individual traits, but they are also interested in the broader culture regions in which these traits exist. So they often organize the world according to these broad culture regions. The map here shows such regions.

The culture region of the United States and Canada stretches from the Pacific Ocean in the west to the Atlantic Ocean in the east. Mountains frame the eastern and western sides of the region, cradling a central area of vast plains that are important for agriculture. The region is a land of immigrants. Many people made this land their home by choice. Others were forced to come as exiles or enslaved workers. Together these groups have shaped the culture of the region.

The Latin America culture region includes Mexico, Central America, the Caribbean, and South America. The high peaks of the Andes to the lush rain forests of the Amazon have affected human settlement and cultural development here. Cultures have collided in Latin America. Native American civilizations originally flourished here. Then Europeans arrived and through the process of colonialism forced new laws, new languages, and a new religion onto the region's inhabitants. Today the region reflects this shared colonial history. There are also contrasts between urban and rural, rich and poor, and more developed and less developed.

Europe has several major peninsulas and many minor ones, with many pieces of land extending into the Atlantic Ocean and the Mediterranean Sea. Over the centuries, Europeans have taken advantage of this location by using the water as an avenue for trade and exploration. This has shaped the region's culture through interactions with other peoples and cultures. The Europe culture region has many different languages and culture groups.

World Culture Regions

This map shows culture regions of the modern world. It is just one of many possible ways of identifying and organizing the world into culture regions.

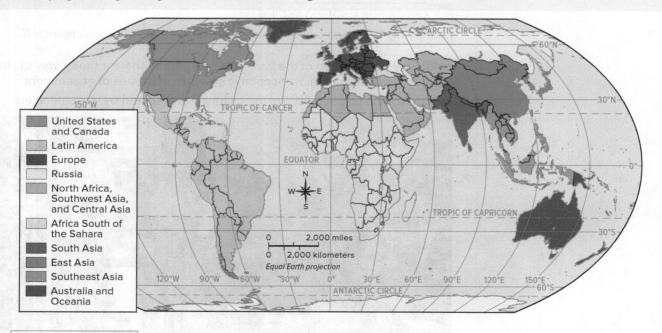

1. **Exploring Regions** Which culture regions include parts of Asia?
2. **Exploring Culture** Describe what you know about one of the culture regions on the map.

Throughout history, conflicts between competing nations have caused destruction in the region. In recent years, however, most of the countries of Europe have joined together in an economic and political union fostering peace and prosperity.

Russia extends from its boundary with Europe in the west to the Pacific Ocean in the east. Much of the region consists of plains and low plateaus. It lies at a high latitude, making the region cold and snowy. These factors have influenced settlement in the region. Over its history, Russia has established **empires** in which it has ruled over other countries or places. These empires have primarily seen expansion of Russia's influence over land areas because it has little access to the ocean. In the western part of the region, Russia has been influenced some from Europe, but most of the region is rather isolated.

Stretching from the Atlantic Ocean in the west to the borders of China in the east, the culture region of North Africa, Southwest Asia, and Central Asia is vast. The common physical geographic characteristic is that water is scarce. This has made two large river valleys—the Nile and the Tigris/Euphrates—vital to most aspects of life. Oil is also an important resource in parts of the region, which creates an income gap. There are bustling cities but also large expanses of uninhabited land. Southwest Asia is the birthplace of three world religions: Judaism, Christianity, and Islam. Muslims, followers of Islam, form the majority of the population in most of the region. The teachings and traditions of Islam are a major force in many parts of this culture region.

Africa south of the Sahara is a region of great diversity. There are areas of high and low elevation, and ecosystems that range from deserts to tropical rain forests. These features influence settlement and economic activities in the region. Culturally it is made up of several thousand ethnic groups with different languages and customs. Parts of this culture region have been shaped by Islamic influences and other parts by Christian influences. All of these mix with traditional African culture. Some parts of the region have been affected by French and British influences. Across this culture region, there is a blend of traditional cultures and the deep markings of a colonial past.

The South Asian landscape is one of contrasts—from lowlands to the Himalaya.

Monsoons not only cause wet and dry seasons but have played a role in the region's cultural development. South Asia's religious and cultural traditions have influenced—and been shaped by—other parts of Asia. It is the birthplace of three major world religions: Hinduism, Buddhism, and Sikhism. Islam is also important here. With roots in ancient civilizations, the region today has one of the world's largest populations, vibrant software and film industries, rising economic powers, and countries that play an important role in contemporary international **issues**.

East Asia includes China, Japan, and the Koreas. Its many mountainous areas have kept some groups isolated. The region can trace its beginnings to an ancient civilization that arose in China around 2000 B.C. In the centuries that followed, powerful dynasties ruled China. This created a large empire that influenced the cultural development of the entire region. Today, political and economic differences divide the region.

The Southeast Asia culture region is a region of great cultural diversity. The diversity has come about because mountainous terrain and many islands have separated culture groups. It is home to hundreds of ethnic groups, speaking hundreds of different languages. In addition, the region has been affected by the cultures of other regions. Its location at the crossroads of South Asia and East Asia has opened it up to these other influences. Also, parts of Southeast Asia were held as colonies by European powers bringing their own cultural influences.

Australia, New Zealand, and Oceania form another culture region. All of these are islands. The population of Oceania is made up mostly of **indigenous** peoples. The cultures of Oceania's countries reflect these peoples' traditional ways of life. In contrast, the indigenous peoples of Australia and New Zealand are minorities. The majority population in these two countries are peoples descended from European immigrants The cultures of Australia and New Zealand are largely influenced by European colonization and a unique national identity.

✓ **CHECK FOR UNDERSTANDING**

Identifying Which world culture regions were the birthplaces of major world religions?

---

empire an extensive group of countries or places under a single supreme authority

issue an important subject

indigenous the earliest known inhabitants of a place

Identifying Expressions of Culture

GUIDING QUESTION

How do expressions of culture help us learn about a culture?

The study of culture includes examining expressions of culture. Dance, music, visual arts, literature, and food are important expressions of culture. Nearly all cultures have unique art forms that celebrate their history and enrich people's lives. Some art forms, such as singing and dancing, are serious parts of religious ceremonies or other cultural events. Art can also include forms of personal expression or worship and entertainment, as well as ways of retelling and preserving a culture's history.

Traditional Slovak culture is expressed through the dance performance as well as the clothing of these young dancers at a festival in central Slovakia.

Making Connections What other expressions of culture might you find at a festival?

History shapes how culture groups view the world. Expressions of culture often honor the heroes and heroines who brought about successes in a culture's history. Stories about heroes **reveal** the personal characteristics that people think are important. Groups also remember the dark periods of history when they endured disaster or defeat. These experiences, too, influence how groups of people see themselves. Cultural holidays often mark important events or heroes and enable people to remember and celebrate their heritage.

In sports, as in many other aspects of culture, activities are adopted, modified, and shared. Many sports that are played today originated with different culture groups in the past. Athletes in ancient Japan, China, Greece, and Rome played a game similar to soccer. Scholars believe that the Maya of Mexico and Central America developed "ballgame," the first organized team sport. It was played on a 40- to 50-foot-long (12.2- to 15.2-meter-long) recessed court, and the athletes' goal was to kick a rubber ball through a goal.

Customs are also an important outward display of culture. In many traditional cultures, a woman is not permitted to touch a man other than her husband, even for a handshake. In modern European cultures, polite greetings include kissing on the cheeks. People of many cultures bow to others as a sign of greeting, respect, and goodwill. The world's many cultures have countless fascinating customs. Some are used only formally, and others are viewed as good manners and respectful, professional behavior.

✓ **CHECK FOR UNDERSTANDING**

Determining Central Ideas How does history influence how a culture views the world?

reveal to make known

LESSON ACTIVITIES

1. **Narrative Writing** Write a paragraph describing a custom that you and your family practice. Explain how the custom is an expression of your culture.

2. **Collaborating** Work with a partner to learn about a sport with which you are not familiar. Find out where the sport originated, where in the world it is played today, and how it has influenced the culture of the area. Create a poster to share with the class.

04
Analyzing Sources: Culture and Identity

? COMPELLING QUESTION

How does culture impact our daily lives?

Plan Your Inquiry

DEVELOPING QUESTIONS

Think about how culture affects the places we live and the way we eat, dress, and celebrate together. Then read the Compelling Question for this lesson. What questions can you ask to help you answer this Compelling Question? Create a graphic organizer like the one below. Write these Supporting Questions in your graphic organizer.

Supporting Questions	Source	What this source tells me about how culture impacts our daily lives	Questions the source leaves unanswered
	A		
	B		
	C		
	D		
	E		

ANALYZING SOURCES

Next, examine the sources in this lesson. Analyze each source by answering the questions that follow it. How does each source help you answer each Supporting Question you created? What questions do you still have? Write these in your graphic organizer.

After you analyze the sources, you will
- use the evidence from the sources;
- communicate your conclusions;
- take informed action.

Background Information

The term *culture* refers to particular ways of life that are connected to the values and traditions of a group of people. Your own culture helps you make sense of the world and influences how you live.

Culture is passed down through family and community. Your home reflects the cultures of your family members. You might encounter different cultures in the community where you live or when you travel to other places.

A culture region is a geographic area in which people have certain traits in common. Culture regions help us understand where cultural similarities and differences occur. People in a cultural region often live close to one another to share resources, for social reasons, and to keep their cultures and communities strong. Some culture regions are large and others are small. For example, Southwest Asia is a cultural region because it is home to hundreds of millions of people who share the Islamic religion. Along the border between Spain and France, a much smaller culture region is the Basque Country. The Basques have their own unique language, customs, and festivals. Southwest Louisiana is known as Cajun country—a culture region with a unique lifestyle, food, and language.

» Solvang, California, was settled by immigrants from Denmark in the early 1900s. Despite having moved to an area with a very different climate, settlers built structures that reflected the culture of their home country. They built windmills, telephone booths, and houses that are similar to those found in Denmark in this southern California town.

Alexander Reitter/Shutterstock

The Food That Sustains Us

Throughout the world, the types of food people eat often depends on where they live and what foods they have access to. Families might produce their own food, purchase food from markets or stores, or buy prepared foods from restaurants or vendors (sellers).

PRIMARY SOURCE: PHOTO

A family in Kyrgyzstan shares a meal in a yurt, which is a type of circular tent common in Central Asia.

EXAMINE THE SOURCE

1. **Identifying** What foods do you recognize in the photograph? What can you infer from the photograph about the lifestyle of the family?

2. **Making Connections** What factors do you think are most important in determining what this family eats?

The Quilts of Gee's Bend

Around the world, many distinctive cultures have developed. Limited communication and mobility meant that for much of history people lived in relatively isolated groups. Each group created traditions and practices that made sense where they lived. These traditional cultures are often important in establishing a sense of shared identity among people. They can also give places unique character. This article describes an exhibit of homemade quilts that was seen in many American museums.

PRIMARY SOURCE: ARTICLE

Detail of a quilt created in Gee's Bend, Alabama

❝ When enslaved women from the rural, isolated community of Boykin, Alabama—better known as Gee's Bend—began quilting in the 19th century, it arose from a physical need for warmth rather than a quest to reinvent an art form. Yet by piecing together scraps of fabric and clothing, they were creating abstract designs that had never before been expressed on quilts.

These patterns and piecing styles were passed down over generations, surviving slavery, the antebellum South, and Jim Crow. During the Civil Rights movement in 1966, the Freedom Quilting Bee was established as a way for African American women from Gee's Bend and nearby Rehoboth to gain economic independence. The Bee cooperative began to sell quilts throughout the U.S., gaining recognition for the free-form, seemingly **improvisational** designs that had long been the hallmark of local quilt design. As awareness grew, so did acclaim, and the quilts entered the [vocabulary] of homegrown American art.

Since then, quilts from Gee's Bend have been exhibited at the Museum of Fine Arts, Houston, the Philadelphia Museum of Art, the Whitney Museum of American Art, and others. In 2006, the U.S. Postal Service even issued ten commemorative stamps featuring images of Gee's Bend quilts. ❞

— "The Quilts of Gee's Bend: A Slideshow" by Rebecca Gross, *National Endowment for the Arts blog*, October 1, 2015

improvisational done without advanced preparation

EXAMINE THE SOURCE

1. **Explaining** What geographic factors contributed to the uniqueness of the quilts?
2. **Analyzing Perspectives** What do people from outside of Gee's Bend appreciate about the quilts?

Food and Culture

Ethnic food traditions are found in many cities around the world. The special ways of cooking that an immigrant group uses can also become a business opportunity. This photo shows a row of restaurants in London, England that are commonly referred to as *curry houses*. The food served in these restaurants is inspired by traditional Indian and Pakistani cooking, though modified over time to match English tastes. The English began to use the word **curry** to describe this type of food in the 1880s when Queen Victoria made eating Indian cuisine stylish.

PRIMARY SOURCE: PHOTOGRAPH

Indian curry is a popular type of takeout food in London.

curry generic term used to describe many Indian dishes

EXAMINE THE SOURCE

1. **Analyzing** Who do you think is most likely to purchase food from one of these curry houses? Who do you think works in them?

2. **Evaluating** Besides the food they serve, how else might these restaurants reflect their culture?

Culture Regions and Language

When people live close together, they are able to have frequent interactions with each other. They have more shared experiences and can learn from each other. They form a cultural region. On the other hand, physical barriers in the environment, such as mountains and bodies of water, can limit interactions between places and people. **Linguists** look for differences in word usage and pronunciations that are related to cultural regions. This map shows how people around the United States refer to a sweetened, carbonated drink.

PRIMARY SOURCE: MAP

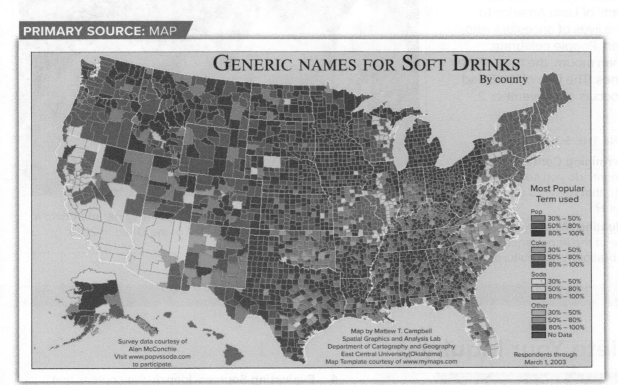

GENERIC NAMES FOR SOFT DRINKS
By county

Most Popular Term used

Pop
30% – 50%
50% – 80%
80% – 100%

Coke
30% – 50%
50% – 80%
80% – 100%

Soda
30% – 50%
50% – 80%
80% – 100%

Other
30% – 50%
50% – 80%
80% – 100%
No Data

Survey data courtesy of
Alan McConchie
Visit www.popvssoda.com
to participate.

Map by Mattew T. Campbell
Spatial Graphics and Analysis Lab
Department of Cartography and Geography
East Central Univerisity(Oklahoma)
Map Template courtesy of www.mymaps.com

Respondents through
March 1, 2003

linguist a person who studies language

EXAMINE THE SOURCE

1. **Inspecting** What spatial patterns can you identify on the map? Which of the terms is least commonly used?

2. **Predicting** Based on this map, in what part of the country would you expect to see the most differences in language? The least?

Cultural Performances

Wearing special types of clothing and participating in traditional activities is part of many holiday celebrations. This photo shows a Day of the Dead parade in Mexico City. The Day of the Dead is a holiday celebrated in Mexico and other parts of Latin America to honor the lives of ancestors who have died. People celebrate, rather than mourn, the lives of their loved ones. The Day of the Dead always occurs on November 2.

PRIMARY SOURCE: PHOTOGRAPH

People with traditional costumes and face painting parade down a major avenue in Mexico City for the Day of the Dead in 2019.

EXAMINE THE SOURCE

1. **Determining Context** What elements do you see in the photo that relate to traditional culture?

2. **Evaluating** What evidence is there of the importance of this celebration to the culture?

Complete Your Inquiry

EVALUATE SOURCES AND USE EVIDENCE

Refer back to the Compelling Question and the Supporting Questions you developed at the beginning of the lesson.

1. **Analyzing Perspectives** Which source(s) give evidence of cultural expressions that are meant to be viewed and appreciated by people of other cultures?

2. **Making Connections** Which source(s) provide examples of how culture is shaped by the natural environment?

3. **Gathering Sources** Which sources helped you answer the Supporting Questions and the Compelling Question? Which sources, if any, challenged what you thought you knew when you first created your Supporting Questions? What information do you still need in order to answer your questions? What other viewpoints would you like to investigate? Where would you find that information?

4. **Evaluating Sources** Identify the sources that helped answer your Supporting Questions. How reliable is each source? How would you verify the reliability of the source?

COMMUNICATE CONCLUSIONS

5. **Collaborating** Share with a partner one of your favorite cultural traditions. Then discuss different ways people participate in cultural activities. How do these examples support your thinking? Use the graphic organizer that you created at the beginning of the lesson to help you. Share your conclusions with the class.

TAKE INFORMED ACTION

Researching a Holiday Identify a cultural holiday celebrated by a group of people in your community. Research the holiday and its meaning. Create a poster that explains the holiday and hang the poster in your school or another location in your community.

Causes and Effects of Cultural Change

READING STRATEGY

Analyzing Key Ideas and Details As you read, use a table like the one below to record key terms and phrases related to the causes and effects of cultural change.

Cultural Change	Causes	Effects
In History		
Migration		
Today		
Globalization		

Cultural Change

GUIDING QUESTION

What causes cultural change?

Cultures are continually changing. New ideas, lifestyles, and inventions create change within cultures. Change can also come from outside of a particular culture. This occurs through spatial interaction such as trade, migration, and war. There can be *cultural convergence,* in which cultures become more alike over time, or *cultural divergence,* in which cultures separate from one another and develop in different ways.

The spread of new ideas from a source is called **cultural diffusion**. Cultural diffusion has been a major factor in cultural development since the dawn of human history. The earliest humans were small groups of hunters and gatherers. They moved from place to place in search of animals to hunt, plants to gather, water, and useful materials. As they moved, they spread culture traits from one group and place to another.

Cultural Change in History

The world's first civilizations arose in **culture hearths**. These were early centers of civilization whose ideas and practices spread to

cultural diffusion the spread of new ideas from a source point

culture hearth a center where cultures developed and from which ideas and traditions spread outward

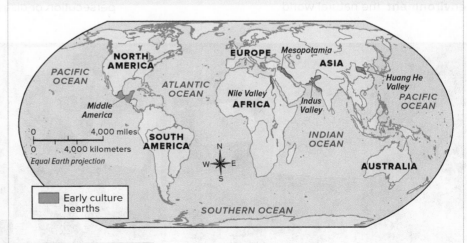

World Culture Hearths

Culture hearths are centers where the world's first civilizations arose.

EXAMINE THE MAP

1. **Exploring Regions** Where in Asia were the first settlements located?
2. **Human-Environment Interaction** What water feature do most of the culture hearths have in common?

surrounding areas. The map shows that some of the most influential culture hearths developed in areas that make up the modern countries of Egypt, Iraq, Pakistan, China, and Mexico.

These five culture hearths had certain geographic features in common. They all began as farming settlements in areas with a mild climate and fertile land. In addition, they were located near a major river or source of water. Making use of favorable **environments**, the people dug canals and ditches to irrigate the land. All of these factors contributed to what is known as the *Agricultural Revolution*. It was a major shift from hunting and gathering food to producing food, and enabled people to grow surplus crops.

Surplus food set the stage for the rise of cities and civilizations. With more food available, there was less need for everyone in a settlement to farm the land. People were able to develop other ways of making a living. They created new technology and carried out specialized economic activities, such as metalworking and shipbuilding. These changes then stimulated the growth of long-distance trade.

The increased wealth from trade reinforced the rise of cities and complex social systems. The ruler of a city or civilization needed a well-organized government to coordinate harvests, plan building projects, and manage an army for defense. Perhaps most importantly, officials and merchants created writing systems that made it possible to record and transmit information.

environment the natural world

Cultural diffusion has increased rapidly during the last 250 years. In the 1700s and 1800s, some countries began to industrialize. They used power-driven machines and factories to mass-produce goods. This period is known as the *Industrial Revolution*. With new production methods, countries produced goods quickly and cheaply, and their economies changed dramatically. These developments also led to social and cultural changes. As people left farms for jobs in factories and mills, cities grew larger.

Migration and Cultural Change

Migration, the movement of people from one place to another, has also promoted cultural diffusion. People migrate for many reasons. Positive factors like better social and economic conditions and religious or political freedoms may draw people from one place to another. These are called *pull factor*s. Many people move from one place to another in search of better economic opportunity. Negative factors such as wars, persecution, and famines also motivate people to migrate away from a place. These are called *push factors*. In most cases, these are voluntary migrations.

Migrations can also be forced. Some people are forced to flee their country because of wars, food shortages, or other problems. They are **refugees**, or people who flee a country to

migration the movement of people from one place to another

refugee a person who flees a country to escape persecution or disaster

People stand together during a rally to welcome refugees escaping a civil war in Syria to a new beginning in Toronto, Canada.

arindambanerjee/Shutterstock

Cargo containers are loaded onto ships in the port of Hong Kong. International trade is the exchange of goods and services between countries. When people trade, they not only trade goods, but they also trade customs and ideas.

escape persecution or disaster. A different instance of forced migration was when enslaved Africans were brought to the Americas. Regardless of the reasons, migrants carry their cultures with them, and their ideas and practices often blend with those of the people already living in the migrants' adopted countries.

Cultural Change Today

At the end of the twentieth century, the world began to experience a new turning point—the *Information Revolution*. Computers now make it possible to store huge amounts of information that include photographs and videos as well as the written word. Combined with advances in communications, this information can be rapidly sent all over the world, thus allowing for the quick spread of ideas and traditions among cultures. The Internet is a central part of these advances in communication. It is responsible for social-networking sites and other sites that allow users to share many types of information and stay connected with others. Consequently, the world feels much smaller than it did previously.

Advances in transportation technologies also began to speed up toward the end of the twentieth century, and they continue to advance today. Air travel makes a particularly important contribution to cultural diffusion. The global average of more than 4 billion passengers a year on commercial airlines means that many people carry their cultures with them to other places.

In addition to people, airplanes also carry freight cargo, particularly for time-sensitive, valuable, or **perishable** products. Such transportation allows us to have food like peaches even in winter because they come from South America. But most international freight transportation occurs by ship, which spreads goods from many places around the world. Increased railroad and over-the-road truck transportation spreads the goods within places and further contributes to cultural diffusion.

✓ **CHECK FOR UNDERSTANDING**

Explaining What factors cause people to migrate?

Global Culture

GUIDING QUESTION

How has global culture impacted the world today?

Today's world is becoming more culturally blended every day. As cultures combine, new cultural elements and traditions are born. The spread of culture and ideas has caused our world to become globalized.

Globalization is the process by which economic, political, and cultural processes expand to reach across the world. Globalization has had the positive effect of people becoming more aware of other cultures and often more

---

perishable likely to decay or spoil quickly

globalization increasing interconnection of economic, political, and cultural processes across the world

understanding and accepting of those other cultures. It also has helped spread ideas and innovations in science and technology. Changes in technology, communications, and travel have resulted in cultural blending on a wider scale than ever before.

There are many **benefits** to globalization. For example, technology developed in one country and shared with others helps increase economic efficiency. International trade has also allowed countries to specialize in the goods they produce well and trade for the goods they do not. This helps economies and gives people access to products and resources they may not otherwise have. Globalization has helped give us new and unique perspectives on the diversity of life.

While globalization has many benefits, it also has drawbacks. Some people do not want their cultures to change, or they want to control the amount and direction of change. Sometimes the changes come too quickly, and cultures can be damaged or destroyed. As ideas, products, and even lifestyles are shared between cultures, traditional cultural heritage can become diluted by outside influences. Language, artistic traditions, clothing styles, and even behaviors can all be altered through interactions with other cultures. For example, globalization promotes a common language across many cultures which can lead to the permanent loss of some of the world's rare languages. Many groups, including the United Nations Educational, Scientific and Cultural Organization (UNESCO), urge the protection of traditional cultures in the face of increasing globalization. Globalization can also result in **pandemics**, or the rapid spread of diseases around the world such as COVID-19.

Just as no one element defines a culture, no one culture can define the world. All cultures have value and add to the human experience. As the world becomes more globalized, it is important for people to respect other ways of life. We have much to learn, and much to gain, from the many cultures that make our world a fascinating place.

The Badaga language is spoken by the Badaga people of southern India. This language is endangered, with fewer than 200,000 speakers.

Identifying Cause and Effect How might globalization make a language endangered?

✓ **CHECK FOR UNDERSTANDING**

Summarizing What is globalization?

LESSON ACTIVITIES

1. **Explanatory Writing** Write a one-page essay summarizing the benefits and drawbacks of globalization.

2. **Analyzing Information** Work with a partner to consider how computers and the Internet have made it easier to spread information and ideas around the world. Think about email, cell-phone apps, and social-media platforms. Research to find examples of how ideas have been spread by each of these means. Share your findings with the class.

benefit a good or helpful result

pandemic the rapid spread of a disease around the world that affects a large number of people over a wide area

06

Global Issues: Continuity and Change in Culture

 COMPELLING QUESTION

Why does culture change over time?

Plan Your Inquiry

DEVELOPING QUESTIONS

Think about how cultures change. Then read the Compelling Question for this lesson. What questions can you ask to help you answer this Compelling Question? Create a graphic organizer like the one below. Write these Supporting Questions in your graphic organizer.

Supporting Questions	Source	What this source tells me about why cultures change over time	Questions the source leaves unanswered
	A		
	B		
	C		
	D		
	E		

ANALYZING SOURCES

Next, examine the sources in this lesson. Analyze each source by answering the questions that follow it. How does each source help you answer each Supporting Question you created? What questions do you still have? Write these in your graphic organizer.

After you analyze the sources, you will
- use the evidence from the sources;
- communicate your conclusions;
- take informed action.

Background Information

Culture is generally passed down from generation to generation. A person often identifies with his or her culture by the types of music listened to, the languages spoken, or the foods eaten. Culture provides a sense of belonging for people. Sometimes, however, cultures face challenges that change them. Cultural elements often cross over from one group to another and blend to form a new culture. Some people believe this leads to a richer cultural identity, while others argue that it destroys cultures.

» Trade often encourages people to interact with others outside of their culture group.

Hula

People maintain traditions to stay connected to their own culture as well as to show respect to their ancestors. Members of a specific culture may choose certain traditions to continue as a way of preserving their identity and values. This article discusses the tradition of hula in Hawaiian culture.

PRIMARY SOURCE: ARTICLE

" [Hula is] an age-old Hawaiian cultural practice enacted through chanting, singing and dancing. Each of hula's movements has a meaning that helps tell a story about gods and goddesses, nature or important events. Rather than simply a performance geared for tourists, the dance is something Hawaiians did for themselves for centuries, at religious ceremonies honoring gods or rites of passage and at social occasions as a means of passing down history.

Young girls practice a hula dance in Hawaii.

After years of Western imperialism—under which hula was first discouraged by Christian missionaries in the early 1800s and later marketed as **kitsch** in the mid-1900s—the dance, in many Hawaiians' eyes, was losing any real sense of history or culture. "Outside influences were making it obsolete," says Rae Fonseca, a *kumu hula*, or hula master, in Hilo on the Big Island. As a result, in the late 1960s and early 1970s, a renewed interest in hula's traditional roots began to sweep across the state.

Today, serious hula is everywhere in Hawaii. The dance can also be found among the mainland **diaspora** and other places such as Japan, Europe and Mexico. *Halaus*, or schools of hula, have cropped up in most Hawaiian towns, and men and women of all ages study the dance diligently.

But while hula historically has involved a merging of different cultural forms, *kumu hulas* of today want blending stopped. Rather than integrate Japanese or, say, Mexican dance traditions with Hawaiian hula in Tokyo or Mexico City, Fonseca says hula must be kept pure, wherever it is performed. "It's up to us teachers to stress that where we come from is important," he says. [Another hula master] strongly agrees: "If the link is not maintained as it should, then we're not passing on something that is hula and we're not being true to our culture. "

---

kitch in poor taste

diaspora the dispersal of a group of people from one location to many others

---

EXAMINE THE SOURCE

1. **Identifying Cause and Effect** What challenges has hula faced in the past? What challenges does it face today?

2. **Making Inferences** What idea is inferred in the article about why Hawaiians should keep the hula tradition pure?

Buildings

Throughout history, towns and cities have been built in ways that reflect the cultures of the people who lived there. Visitors to Cordoba, Spain, can cross this bridge originally built by the Roman conquerors in the 100s B.C. They can also visit a building (center of photo) erected as an Islamic mosque during **Moorish** rule in the 700s A.D. The mosque became a Christian cathedral in 1326 A.D.

PRIMARY SOURCE: PHOTO

Moors northwest African Muslims who controlled territory that is part of present-day Spain from the 700s to 1492 A.D.

EXAMINE THE SOURCE

1. **Inferring** What were the purposes of the bridge and building?
2. **Analyzing Perspectives** Why might Spanish citizens from Cordoba feel a connection to those structures even if they are not part of the cultures that built them?

Blues and the Great Migration

The blues is a unique type of music that originated in African American communities in the southeastern United States. During the Great Migration that began during World War I, African Americans left the South for industrial cities in the North, Midwest, and West in search of work. This map shows how the migration led to the diffusion of three distinct styles of blues music. Each of these styles had evolved in its own way based on regional differences within the South.

SECONDARY SOURCE: MAP

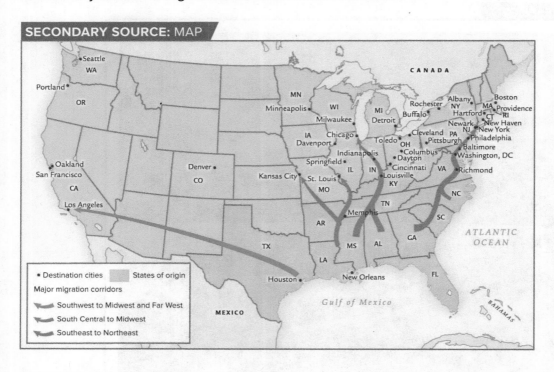

EXAMINE THE SOURCE

1. **Analyzing** Explain why blues music you hear in New York City is likely to sound different than blues music you hear in St. Louis.

2. **Speculating** Why do you think there were three different styles of blues music?

D

Japanese-American Baseball

Leisure-time pursuits and sport are part of every culture. Like food, sport is an element of culture that is easily shared with others through participation. While baseball is often called America's pastime, this article explains how it also became the most popular sport in Japan.

" Japanese immigrants to the United States were deeply familiar with baseball: Japan had embraced it in the late 1800s. In 1871, Hiroshi Hiraoka established the Shimbashi Athletic Club, the first of its kind in the nation, and soon the sport expanded in the Japanese imagination.

During the Meiji Restoration, which lasted roughly from 1880 to 1910, leaders promoted baseball as a way to transform the image of Japan while facilitating [encouraging] international connections. . . .

From 1885 to 1907, as Japan's economy modernized, leaving some Japanese behind, 155,000 traveled east to Hawaii and the American West Coast. Nearly 25,000 of these immigrants had settled in Hawaii by 1896. Unsurprisingly, Japanese-American baseball leagues first emerged in the U.S. territory and often featured multi-ethnic and [multi]-racial competition, as Caucasian, Filipino, Japanese, and Portuguese laborers demonstrated their skills on the diamond. . . .

Part of the game's power lay in its appeal across generations: "Baseball allowed each generation to interpret the meaning of the sport," noted writer Wayne Maeda. First-generation Japanese immigrants, known as *Issei*, saw it as a means to connect their American-born children with Japanese culture, and believed it emphasized Japanese values of loyalty, honor and courage. In contrast, *Nisei* — second-generation immigrants — saw baseball as more modern than Kendo or Judo. Both generations believed the sport testified to their dedication to American ideals. "

—"Japanese-America's Pastime: Baseball," by Wendi Maloney, *Library of Congress Blog*, May 25, 2018

EXAMINE THE SOURCE

1. **Inspecting** When did the game of baseball first become popular in Japan?
2. **Predicting** What different factors helped baseball appeal to two generations of Japanese Americans?

Creating Change

A person's gender, place of origin, or physical characteristics have often been associated with unequal social status in cultures around the world. When people work to win rights for those people who face discrimination, they can cause change within a culture. This article examines the reasons that led two women to organize a movement for equal rights in the workplace in the United States.

66 Originally founded in the early 1970s by political activist Karen Nussbaum and author Ellen Cassedy, 9to5 (National Association of Working Women) started when both Nussbaum and Cassedy met in college in Boston and realized that as **clerical** workers they were not given much opportunity for growth. At that time women took on administrative and clerical roles with no wiggle room. They adhered [followed] to a strict dress code, learned to keep their heads down and perform tasks such as bringing their boss coffee or his dry cleaning as part of their job requirements. And even though their pay was a little more than half of what men were making despite qualifications or experience, they were expected to continue following the outdated employment standards rather than speak up. 99

—"9to5: The Story of a Movement," by Jennifer Vintzileos. Starry Constellation Magazine, July 9, 2020.

clerical general office work

EXAMINE THE SOURCE

1. **Explaining** Why did the women who started the 9to5 Movement want culture to change?
2. **Speculating** What specific changes would best improve the workplace culture described in the source?

Complete Your Inquiry

EVALUATE SOURCES AND USE EVIDENCE

Refer back to the Compelling Question and the Supporting Questions you developed at the beginning of the lesson.

1. **Analyzing Perspectives** Which source(s) refer to forms of entertainment?

2. **Making Connections** Which source(s) show how the movement of people is associated with cultural change?

3. **Gathering Sources** Which sources helped you answer the Supporting Questions and the Compelling Question? Which sources, if any, challenged what you thought you knew when you first created your Supporting Questions? What information do you still need in order to answer your questions? What other viewpoints would you like to investigate? Where would you find that information?

4. **Evaluating Sources** Identify the sources that helped answer your Supporting Questions. How reliable is each source? How would you verify the reliability of the source?

COMMUNICATE CONCLUSIONS

5. **Collaborating** Work with a partner to discuss how new traditions are introduced into a culture. How do these sources support your thinking? Use the graphic organizer that you created at the beginning of the lesson to help you. Share your conclusions with the class.

TAKE INFORMED ACTION

Conducting an Interview About
Culture Interview someone in your family or community to learn more about cultural traditions they experienced during their life. Find out how these traditions might have changed over time and why. Create a class collage with images reflecting the cultures of the people you interviewed, and write a reflection essay on what you learned.

07

Reviewing Cultural Geography

Summary

Elements of Culture	Expressions of Culture	Cultural Change
• Culture is the way of life of a group of people with similar beliefs and customs. • People are not born with culture. They learn it from the people and groups with which they live. • Elements of culture are language, religion, social systems, government, and economic activities.	• A culture region includes different people and groups that have certain culture traits in common. • A culture trait is a central characteristic of the culture that is shared by most of its members.	• Cultural diffusion is a major factor in cultural change. • Culture hearths were early centers of civilization whose ideas and practices spread to surrounding areas. • The Agricultural Revolution and the Industrial Revolution were important to the process of cultural diffusion.
• Language allows people to communicate, sharing beliefs and customs. • Religion influences how people view the world, and helps some people find a sense of identity. • Social systems help people in a culture work together to meet basic needs.	• Broad world culture regions are those of the United States and Canada; Latin America; Europe; Russia; Africa south of the Sahara; North Africa, Southwest Asia, and Central Asia; South Asia; East Asia; Southeast Asia; and Australia and Oceania.	• Migration has promoted cultural diffusion. • Migration can be voluntary or involuntary (forced). • The Information Revolution has increased the speed with which ideas are spread from culture to culture.
• Governments maintain order within a culture, provide protection, and provide services to citizens. • Economic activities influence and shape a culture.	• Dance, music, visual arts, literature, and food are expressions of culture. • History shapes how cultures view the world. • Customs are an outward display of culture.	• Globalization is the process by which economic, political, and cultural processes expand to reach across the world. • Globalization has benefits and drawbacks.

(l) real444/Getty Images, (c) Magdalena Rehova/Alamy Stock Photo, (r) Travel mania/Shutterstock

Checking For Understanding

Answer the questions to see if you understood the topic content.

1. Identify each of these terms as they relate to cultural geography.

   A. culture
   B. custom
   C. culture region
   D. culture trait
   E. cultural diffusion
   F. culture hearth
   G. migration
   H. globalization

2. **Analyzing** How does the language you speak give clues about your culture?

3. **Summarizing** How are the world's languages organized?

4. **Contrasting** What is the difference between a lingua franca and a dialect?

5. **Making Connections** Explain the ways in which religion or a lack of religion is important to a culture.

6. **Evaluating** How are social systems important to culture?

7. **Summarizing** What does a system of government provide to a culture?

8. **Explaining** How do your clothing, the foods you eat, the holidays you celebrate, and the music you listen to give clues about your culture?

9. **Making Connections** What customs do you practice? Make a list of the beliefs, traditions, languages, foods, art, music, clothing, and other elements and expressions of culture that are part of your daily life.

10. **Analyzing** Why are culture regions important for studying and understanding the world?

11. **Identifying** What are the large world culture regions commonly used by geographers?

12. **Identifying** Which world culture region has been described as "a land of immigrants"?

13. **Making Connections** Which world culture region is located at the crossroads of two other culture regions? How does this affect all three culture regions?

14. **Making Generalizations** Describe what is meant by *expressions of culture* and list examples.

15. **Contrasting** How are cultural convergence and cultural divergence different?

16. **Making Connections** What did the five culture hearths have in common?

17. **Identifying Cause and Effect** What was the Agricultural Revolution? How did it lead to the rise of cities?

18. **Analyzing** How is migration related to cultural diffusion?

19. **Identifying** What factors have led to the increased spread of culture today?

20. **Summarizing** Describe one benefit and one drawback of globalization.

NEED EXTRA HELP?

If You've Missed Question	1	2	3	4	5	6	7	8	9	10
Review Lesson	2, 3, 5	2	2	2	2	2	2	3	2, 3	3

If You've Missed Question	11	12	13	14	15	16	17	18	19	20
Review Lesson	3	3	3	3	5	5	5	5	5	5

Apply What You Have Learned

 A Understanding Multiple Perspectives

American restaurant chains have spread across the world. Today, McDonald's is the world's second-largest private employer, with a workforce of nearly 1.5 million people. With such an enormous reach, the company has great influence. McDonald's restaurants in many countries offer food from the American fast-food cuisine that made the company famous as well as local dishes. In countries like China with growing economies, the convenience of fast food has long held appeal. But today, many people are switching from traditional street vendors to American style fast food. However, others criticize the cultural and health effects of switching from traditional meals to fast food.

ACTIVITY **Analyzing Points of View** Work with a partner to explain two points of view about the worldwide influence of fast-food restaurants. Use online sources to support each point of view. Consider these questions: What countries have food cultures that are very different from fast food? What benefits do restaurants like McDonald's bring to the people of those countries? What controversies might those restaurants introduce? Summarize the two points of view in a five-minute report to the class.

 B Writing a Persuasive Argument

A *diaspora* is a group of people from one culture who live somewhere other than their ancestral homeland. Examples include Jewish communities in the United States or Jamaican neighborhoods in London, England. The reasons for such communities being created vary widely, from displacement by war or social conflict to the desire for better jobs and opportunities.

ACTIVITY **Writing an Argument about a Cultural Concept** Identify and research a specific diaspora, either historical or current. Write a paragraph that answers one of the following questions: What makes it important for members of a diaspora to maintain their home culture's cultural identity? What factors lead members of a diaspora to accept or reject elements of their current home's culture?

Geographic Reasoning

When two countries are located near each other, trade with each other, or exchange people through migration, their people often influence each other culturally. Over time, people in one country are likely to acquire ideas, skills, and technology from the other nation. They will also pass along parts of their own culture to that country. For example, the nations of China, Korea, and Japan have distinct and unique cultures, but each has influenced the others in numerous ways over the centuries.

ACTIVITY **Drawing an Illustrated Map of Two Nations** Create a map of two countries with highlights showing examples of structures, objects, or ideas reflecting the unique cultures that exist in both societies. Include at least three structures, objects, or ideas from each culture. Add callouts to the illustrations that explain the influence of one culture on the other. For example, you may illustrate a type of monument that originated in one country and has been imitated in the other country. Label or include names of key physical features, as well as cities and the location of the cultural examples you use to illustrate your map.

 Making Connections to Today

Developments in technology have almost always affected the cultures using that technology. For example, advances in transportation over the centuries made it possible for people to visit, trade with, and otherwise interact with people in more and more places. This led to cultural exchanges and eventually to globalization.

ACTIVITY **Creating a Multimedia Presentation About an Advance in Technology** Conduct online research to learn about an advance that has been made in one type of technology. Examples might include communication, transportation, medicine, or manufacturing. Then create a multimedia presentation, using images or videos, to show how that advance led to changes or developments in the culture responsible for the new technology. Be sure to address these questions in your presentation: What direct cultural changes were the result of the advance? What long-term or indirect effects did that advance have on other cultures?

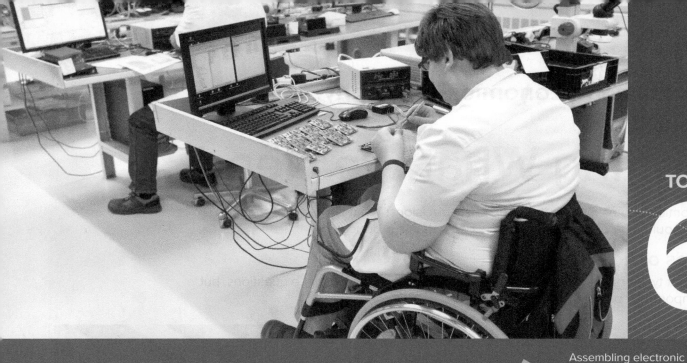

Assembling electronic products is one of many types of economic activities that fuel the global economy.

Economic Geography

INTRODUCTION LESSON

01 Introducing Economic Geography   G186

LEARN THE CONCEPTS LESSONS

02 Economic Systems and Activities   G191

04 Economic Development   G203

05 Economies and World Trade   G209

INQUIRY ACTIVITY LESSONS

03 Global Issues: The Location of Economic Activities   G197

06 Understanding Multiple Perspectives: Participation in Local and Global Economic Systems   G213

REVIEW AND APPLY LESSON

07 Reviewing Economic Geography   G219

industryview/Alamy Stock Photo

Introducing Economic Geography

Getting What We Need

Economic systems must make three basic decisions: What should be produced? How should it be produced? and Who gets what is produced?

Types of Economies

Different types of economic systems have different answers to these questions, but the importance of location is always present.

» Many Inuit communities in northern Canada have traditional economies. Such economies rely on hunting and gathering and are influenced by custom. For example, it is tradition for a hunter to share food from the hunt with other families in the village.

» The type of economy with which you are probably most familiar is a market economy. We, as shoppers, choose what we will buy. Private businesses produce more of what they believe consumers want. There are often many choices available for any one product.

Did You Know?

In some places the government—not the people who need and want the products—decides what is produced and how it is distributed to people. The countries of China and North Korea are examples of countries with this type of command economy.

» the North Korean flag

Economic Activities and Trade

Different types of economic activities provide the things that people need. Countries trade their goods and services with one another, resulting in greater interdependence.

Taking or using natural resources directly from the Earth

Using natural resources to make a new product

Providing services to people and businesses

Processing and managing information

» Economic activities are organized based on what they use or provide. Activities like logging or mining take place near the natural resources that are being taken or used. Other activities create products that are more valuable than the original natural resources. Some activities don't involve gathering or remaking natural resources. Instead, they offer services to other people, or they involve the handling of information.

66 Our urge to trade has profoundly affected the trajectory [path] of the human species. Simply by allowing nations to concentrate on producing those things that their geographic, climatic, and intellectual endowments best enable them to do, and to exchange those goods for what is best produced elsewhere, trade has directly propelled our global prosperity. 99

—William J. Bernstein, American financial theorist

» Some countries enter into trade agreements to set the terms of trade with each other. The United States, Mexico, and Canada have formed such an agreement.

(tl) Kletr/Shutterstock, (tr) Ariel Skelley/DigitalVision/Getty Images, (cl) McGraw-Hill Education, (cr) Morsa Images/DigitalVision/Getty Images, (b) GK Images / Alamy Stock Photo; TEXT: Bernstein, W., A Splendid Exchange: How Trade Shaped the World, May 14, 2009, Grove/Atlantic, Inc

Getting Ready to Learn About . . .
Economic Geography

Economies and Their Activities

Resources are an important factor in economic activities. Some resources like water, soil, plants, and minerals come from the Earth. Other resources come from humans in the form of labor, skills, and ideas. Countries use both kinds of resources to meet their needs.

The way that a society manages its resources determines the type of economy it has. In one type of economy, resources are used to produce items to meet the basic needs of life—food, clothing, and housing. These are produced by people for their own personal use. Another type of economy produces not only the basics but also a variety of additional products and services. The production of items is influenced by the demand people have for products. In this type of economy, private businesses own property and the means of production. In a third type of economy, the government owns the means of production. That government also determines what is produced and how products are distributed. Many economies are mixed, meaning that some parts of the economy are privately owned and other parts are owned by the government.

Within an economy, there are different types of economic activities. One type, which includes farming, fishing, and mining, involves taking resources directly from the Earth. A second category of economic activities makes finished products in factories. This is called manufacturing. A third category involves providing services to people rather than the manufacturing of products. For example, these types of economic activities include banking, education, and health care. Lastly, there are economic activities that are concerned with processing, managing, and distributing information.

Countries work to develop their economies in order to improve the quality of life for the people living there. Economic activities help influence a country's level of economic development. The world's countries can be classified into three different groups based on their level of economic development. The first category is countries that have more technology and manufacturing. The United States is an example of a country in this category. The second category includes countries that have moved from mostly agriculture to more manufacturing activities. Mexico is in this category. The third category is made up of those countries in which agriculture remains the dominant economic activity. Many of the countries in Africa are in this group. The maps in Lesson 4 show the countries that fall into each of these categories.

There are several factors that affect economic activity. Patterns of agriculture and patterns of industrial activity are influenced by such things as climate, topography, soil, nearness to centers of population, and the location of other types of economic activities.

Commercial farming is when crops, like these soybeans, are raised to sell for money. It is an important economic activity for many countries.

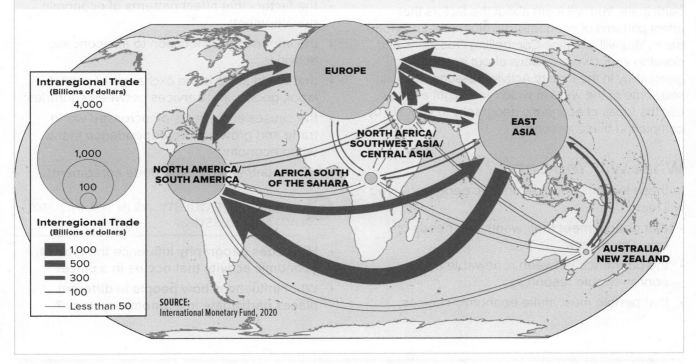

World Trade in Goods

The world's countries trade goods and services to get what they need. The financial impact of the global trade in goods is shown on this map. Notice that the largest money flows are between North and South America, Europe, and East Asia.

Intraregional Trade
(Billions of dollars)

4,000

1,000

100

Interregional Trade
(Billions of dollars)

1,000
500
300
100
Less than 50

EUROPE

NORTH AFRICA/
SOUTHWEST ASIA/
CENTRAL ASIA

NORTH AMERICA/
SOUTH AMERICA

AFRICA SOUTH
OF THE SAHARA

EAST
ASIA

AUSTRALIA/
NEW ZEALAND

SOURCE:
International Monetary Fund, 2020

Connecting Through Trade

Countries around the world interact with one another to get what they need. They trade by buying, selling, or bartering.

This world trade is promoted by the unequal distribution of resources. For example, a country may have a lot of one resource but have a limited supply of another resource. So they engage in trade with another country. For example, the country of Peru in South America trades gold and copper for products its people seek, such as vehicles and televisions. Other things that affect trade are differences in the cost of labor and differences in the level of education and training of the workforce.

Trade occurs within a single country as well as internationally among multiple countries. International trade is usually more expensive than domestic trade due to such things as shipping costs and border delays.

Trade has advantages and disadvantages for the countries involved. It can promote economic growth for a country. However, jobs in one country may be lost if that country gets certain

goods and services from other countries. Both positive and negative consequences result as trade increases and countries around the world become more interconnected, or interdependent.

There are many causes for the increase in relationships between countries. Advances in technology like cell phones and the Internet have made it easier for people to communicate regardless of their location. In addition, improved transportation networks make traveling to other places easier. And lastly, the growth of companies with a presence in many different countries has encouraged trade.

The growing economic interdependence has advantages and disadvantages. The cost of goods is generally lower, and people have access to a wide range of goods. However, a disadvantage is greater environmental damage from the increased use of natural resources.

As countries trade more with each other and become more interconnected, they may join into agreements and form organizations that help regulate trade, monitor economic growth, and provide support for the member countries.

Looking Ahead

You will learn about economic needs and how they are met. You will learn about economic systems and the types of economic activities in which people participate. You will learn about the factors that affect patterns of economic development and world trade. You will examine Compelling Questions and develop your own questions about economic geography in the Inquiry Activity Lessons. You will see some of the ways in which you might already use the ideas of economic geography in your life by completing these lessons.

What Will You Learn?

In these lessons about economic geography, you will learn

- that people meet their wants and needs by using resources;
- the difference between renewable and nonrenewable resources;
- that people must make economic choices;
- the types of economic systems;
- the types of economic activities;
- that the factors of production include land, labor, capital, and entrepreneurship;
- the factors that affect patterns of economic development;
- the importance of location to all economic activities;
- that world trade is the exchange of capital, labor, goods, and services between countries;
- the causes and effects of increased world trade and growing interdependence in the world economy;
- why countries enter into trade agreements.

 COMPELLING QUESTIONS IN THE INQUIRY ACTIVITY LESSONS

- **How does geography influence the type of economic activity that occurs in a place?**
- **What influences how people in different places participate in economic systems?**

Studying Economic Geography

Economic Systems and Activities

- Wants and needs
- Economic choices
- Resources
- Systems: traditional, market, command, mixed
- Activities: primary, secondary, tertiary, quaternary
- Factors of production

Economic Development

- Patterns of economic development
- Factors that affect agriculture
- Factors that affect industry
- Effects of population on land use and economic activities

Economies and World Trade

- Exchange of money, labor, goods, and services between countries
- Specialized products
- Exports and imports
- Causes of increased world trade
- Trade agreements

Economic Systems and Activities

READING STRATEGY

Analyzing Key Ideas and Details As you read, take notes about the key ideas of each economic system.

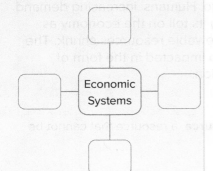

The Economic Question

GUIDING QUESTION

How do people get the things they need and want?

All human beings have wants and needs. To **obtain** them, people use resources. Resources are the supplies that are used to meet our wants and needs. Some types of resources, such as water, soil, plants, and animals, come from the Earth. These are called **natural resources**.

Other resources are supplied by humans. Human resources include the labor, skills, and talents that people contribute. Countries also have wants and needs. Like individuals, countries must use resources to meet their needs.

Wants and Resources

What would happen if 14 students each wanted a glass of lemonade from a pitcher that contained only 12 glasses of lemonade? What if more students wanted a glass of lemonade? No matter how many people want lemonade, the pitcher still contains just 12 glasses. There is not enough for everyone. This is an example of **scarcity**, and it helps us think about how people satisfy unlimited demands with limited supplies. This situation is not uncommon. It happens to individuals and also to countries. You probably can think of many personal examples, as well as current and historical examples, of limited supply and unlimited demand.

---

obtain to gain

natural resource materials or substances such as minerals, forests, water, and fertile land that occur in nature

scarcity the situation in which there are limited resources to satisfy unlimited needs

The mining of iron ore, as shown here in Hibbing, Minnesota, is vital to the iron and steel industries that produce the cars and railroads we use every day.

One type of resource everyone needs is energy. Energy is the power to do work. Energy resources are the supplies that provide the power to do work. Many types of energy resources exist in our world. Energy resources, like all natural resources, can be renewable or nonrenewable. **Renewable resources** are those that can be replenished in a relatively short period of time, usually within a person's lifetime. An important consideration is the time it takes for them to be replenished, because different renewable resources require different amounts of time to recover after they are used. Examples of renewable resources include water for drinking, for agriculture, and for trees that provide lumber. Another type of resource is called a *flow resource*. It is one that is always available naturally. Sometimes this type of resource is considered a renewable resource. An example is water that flows through a hydroelectric dam and produces electricity. Another example is energy from the sun.

In contrast are **nonrenewable resources**. These resources cannot be replaced. Once nonrenewable resources are consumed, they are gone. Examples of nonrenewable resources include minerals such as iron ore and copper. Fossil fuels—oil, coal, and natural gas—are energy resources that are nonrenewable. Fossil fuels received their name because they formed millions of years ago. Humans' increasing demand for energy is taking its toll on the economy as supplies of nonrenewable resources shrink. The environment is also impacted in the form of atmospheric pollution.

renewable resource a resource that can be replenished in a relatively short period of time

nonrenewable resource a resource that cannot be replaced

The Grand Coulee Dam on the Columbia River in Washington state is the largest source of hydroelectric energy in the United States. It provides power to parts of eight western states—Washington, Oregon, Idaho, Montana, California, Nevada, Utah, and Wyoming.

Economic Choices

Every economy must address three basic questions: What to produce? How to produce? and For whom to produce?

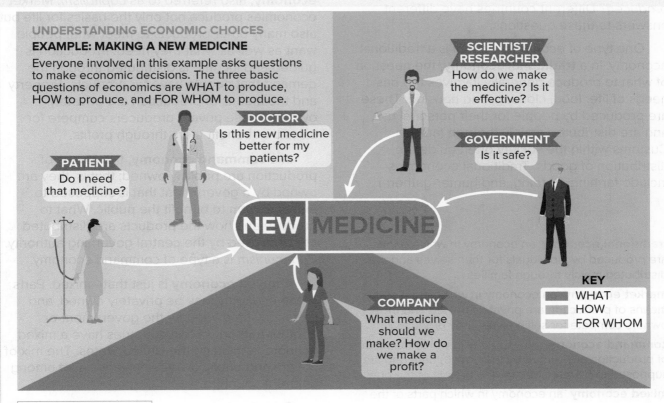

UNDERSTANDING ECONOMIC CHOICES

EXAMPLE: MAKING A NEW MEDICINE

Everyone involved in this example asks questions to make economic decisions. The three basic questions of economics are WHAT to produce, HOW to produce, and FOR WHOM to produce.

DOCTOR
Is this new medicine better for my patients?

PATIENT
Do I need that medicine?

SCIENTIST/ RESEARCHER
How do we make the medicine? Is it effective?

GOVERNMENT
Is it safe?

NEW MEDICINE

COMPANY
What medicine should we make? How do we make a profit?

KEY
| WHAT |
| HOW |
| FOR WHOM |

EXAMINE THE DIAGRAM

Determining Central Ideas What is scarcity? How are the three basic economic questions related to the problem of limited supply?

Making Choices

If the peoples of all countries have unlimited wants but face limited resources, what has to happen? We must make choices. Should we continue to use nonrenewable resources? If so, at what rate should we be using them? Should we switch to renewable resources? Should we switch to flow resources?

To answer such questions, we must weigh the *opportunity cost*, or the value of what we must give up to acquire something else. What opportunity costs exist for the use of renewable, nonrenewable, or flow resources? These costs need to be considered along with many other factors as we make choices now and in the future.

✓ **CHECK FOR UNDERSTANDING**

Contrasting How are renewable and nonrenewable resources different?

Economic Systems

GUIDING QUESTION

What kinds of economic systems are used in the world today?

Economic resources are another important type of resource. They include the goods and services a society provides and how they are produced, distributed, and used. The **economic system** is the way that a society decides who owns the economic resources and how those resources are distributed. There are many spatial aspects to this that geographers consider. Do you ever stop to think about the goods and services you use in a single day? How do these goods and services become available to you? How are they produced? Where do they come from?

economic system the way a society decides on the ownership and distribution of its economic resources

We can break down the discussion on economic systems into three basic economic questions: *What should be produced? How should it be produced? For whom is something produced?* Different countries have different answers to these questions.

One type of economic system is a traditional economy. In a **traditional economy**, the question of what to produce generally involves the basic needs of life: food, clothing, and housing. These are produced by people for their personal use and are distributed mainly through families. Customs within the society often guide the distribution of goods. Traditional economies include farming, herding, and hunter-gatherer

societies. Societies that are mainly agricultural often have traditional economies.

Another type of economic system is a **market economy**, also referred to as *capitalism*. Market economies produce not only the basics for life but also many other products to meet what people want as well as what they need. Production is guided and income is distributed through the demand that people have for products. Property and the means of production are privately owned. These private producers compete for sales that benefit them through profits.

In a **command economy**, the means of production are publicly owned; that is, they are owned by a government that is supposed to manage them to benefit the public. What to produce and how the products are distributed are controlled by the central governing authority. *Communism* is a type of command economy.

A **mixed economy** is just that—mixed. Parts of the economy may be privately owned, and parts may be owned by the government or another authority. Most countries have a mixed economy, including the United States. The mix of private and public ownership varies a lot among different countries.

Socialism is an example of a mixed economy. In socialist societies, property and the distribution of

---

traditional economy an economy in which goods are produced by individuals for themselves and are distributed mainly through families

market economy an economy in which most of the means of production are privately owned and these owners compete for selling what they produce

command economy an economy in which the means of production are owned by the government and are supposed to be managed for the benefit of the public

mixed economy an economy in which parts of the economy are privately owned and parts are owned by the government

---

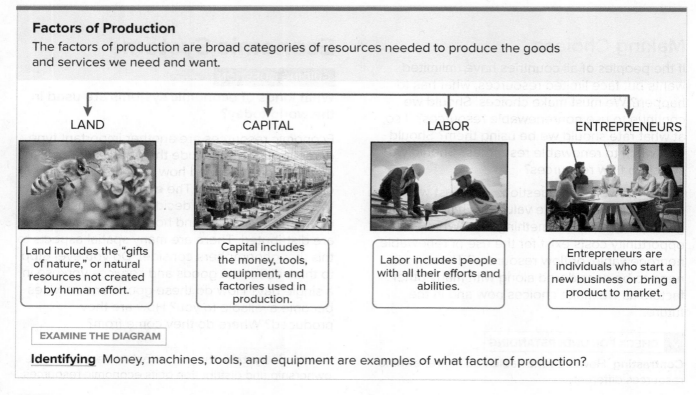

Factors of Production

The factors of production are broad categories of resources needed to produce the goods and services we need and want.

LAND

CAPITAL

LABOR

ENTREPRENEURS

Land includes the "gifts of nature," or natural resources not created by human effort.

Capital includes the money, tools, equipment, and factories used in production.

Labor includes people with all their efforts and abilities.

Entrepreneurs are individuals who start a new business or bring a product to market.

EXAMINE THE DIAGRAM

Identifying Money, machines, tools, and equipment are examples of what factor of production?

goods and income are controlled by the community. They are often managed through democratic processes of cooperation. In some instances, a socialist system turns into a command economy in which the government controls property and the distribution of goods and services.

✓ CHECK FOR UNDERSTANDING

Identifying What is an example of a command economy?

Economic Activities

GUIDING QUESTION

What are the four sectors of economic activities?

Geographers and economists classify the world's economic activities into four sectors, or types. The first type are *primary economic activities* include farming, livestock grazing, fishing, forestry, and mining. These are economic activities that use natural resources directly from the Earth. Such activities may be found near the natural resources. For example, coal mining takes place at the site of a coal deposit.

Secondary economic activities use raw materials to make products that are more valuable than the original raw materials. Such economic activities include manufacturing electric and gas-powered automobiles, assembling cellphones and computers, or producing electric power. These activities occur either close to the resources used or close to the market for the finished goods.

Tertiary economic activities involve providing goods and services to people and businesses. Employees include doctors, teachers, lawyers, bankers, truck drivers, delivery drivers, military service members, restaurant workers, postal workers, store clerks, and many others. The tertiary sector has expended enormously with the rapid growth of the Internet.

Quaternary economic activities are the processing, management, and distribution of information. Cellphones, computers, and other electronic systems have stimulated our modern economy, and created an information **revolution**. Quaternary workers include "white collar"

revolution a sudden or complete change

professionals working in education, government, business, the military, information processing, and research.

The economy can be broken down into parts that are called *factors of production*. In economics, *land* is a factor of production that includes natural resources. A second factor of production is *labor*, which includes all paid workers within a system. The third factor is *capital*, human-made resources necessary to produce other goods. Capital may include money, industrial machinery, vehicles, equipment, and information technology. *Entrepreneurship* is the fourth factor of production and involves human creativity and people's desire to start new projects. These factors of production are involved in all economic activities.

An industry is a vast type of business activity. For example, agriculture is an industry that cultivates land, grows crops, and raises livestock. Manufacturing industries use labor, capital, and technology to produce a variety of finished products for large markets. Service industries provide valuable services rather than goods. Examples include banking, retail sales, food service, transportation, the gaming industry, tourism, air travel, sports competitions, music, and television.

Wholesale and retail distributors buy, sell, and distribute goods and products. Wholesalers buy large quantities of goods from manufacturers, break them down into smaller packages, and then sell them to retail companies. Retail stores such as department stores and shops in malls sell goods directly to consumers.

Location is as much of a factor in economic consumption as it is in production. Location plays a large part in the success of many types of economic activities. For example, businesses that sell cars tend to be grouped together in larger cities. Restaurants are often located side by side along major transportation routes. Shopping centers with dozens of stores offering hundreds of different products also tend to be located near one another or near transportation networks.

Historically, shopping centers have been places that appeal to shoppers of every age group. Shopping centers have something for everyone—music, food, home goods, and clothing. Some may include entertainment such as movie theaters, bowling alleys, and arcades, as well as fitness centers. Consumers today are often looking for more than one product.

Monroe Marketplace in Pennsylvania is an example of agglomeration. This sign at the entrance shows the array of shopping choices available to local consumers.

Combined product locations save consumers a lot of time, travel, and money. This concept of businesses being clustered together is referred to as an *agglomeration*. Not only does this help consumers, but it also helps the businesses themselves through shared parking lots and other common areas.

Another factor to consider with economic consumption is the Internet. It has had a major impact on product availability and delivery for many people around the world. Online shopping increases the types of products that consumers have access to buy. People don't have to live near a shopping center to buy a new pair of shoes, a cellphone, or camping gear. You can order something from a store website, a cellphone app, or an online retailer. It is delivered to your front door by the U.S, Postal Service or another delivery service, usually within a few days.

✓ **CHECK FOR UNDERSTANDING**

Describing Which types of economic activities occur close to the natural resources being used?

LESSON ACTIVITIES

1. **Argumentative Writing** Write a paragraph explaining the benefits of one economic system mentioned in this lesson. Make sure your opinions clear. Conduct research and use facts from the lesson to present a strong argument as to why this economic system is better than other economic systems.

2. **Analyzing Information** Working with a partner, identify and explain the people involved in the creation of a new product. Select a product and then consider who might be involved and the questions they should ask when producing that product. Use the Economic Choices diagram in this lesson as a model.

Global Issues: The Location of Economic Activities

 COMPELLING QUESTION

How does geography influence the type of economic activity that occurs in a place?

Plan Your Inquiry

DEVELOPING QUESTIONS

Think about the different kinds of economic activities that can occur in a place. This could include agriculture, mining, different types of manufacturing, provision of services, retail sales, or knowledge-based activities like education. Reflect on what you know about how economic activities vary from place to place. Then read the Compelling Question for this lesson. What questions can you ask to help you answer this Compelling Question? Create a graphic organizer like the one below. Write these Supporting Questions in your graphic organizer.

Supporting Questions	Source	What this source tells me about how geography influences economic activities	Questions the source leaves unanswered
	A		
	B		
	C		
	D		
	E		

ANALYZING SOURCES

Next, examine the sources in this lesson. Analyze each source by answering the questions that follow it. How does each source help you answer each Supporting Question you created? What questions do you still have? Write these in your graphic organizer.

After you analyze the sources, you will
- use the evidence from the sources;
- communicate your conclusions;
- take informed action.

Background Information

The term *economy* refers to the processes and systems through which goods and services are produced and distributed. Economic systems help determine how resources are distributed and create interdependence among people and places. People participate in economies in order to get the resources they need to support themselves and their families. In the most basic economic system, people need to produce or trade to get what they need. Regional economies allow people and places to specialize in specific types of economic activities. *Primary* economic activities focus on agriculture and the extraction of natural resources. *Secondary* economic activities involve processing and manufacturing. *Tertiary* economic activities provide services. Globalization is increasing the connections among people around the world. This impacts the economic activities that occur in specific places.

» A spice market, or bazaar, in Cairo, Egypt, provides opportunities for buyers and sellers to conduct business.

A

Mining in Australia

Many natural resources are distributed unevenly around the world. When those resources are in demand (needed or desired), it creates economic opportunities for people in the places where they are located. There are many types of minerals that are mined for profit. This article explains why new technologies are creating an economic opportunity in Australia.

PRIMARY SOURCE: ARTICLE

> ❝ Australia leads the world in lithium production and possesses an estimated 6.3m tons of lithium reserves.
>
> The metal is fast becoming a geopolitical bargaining chip, as China, the US and other major powers jostle to secure access to an element expected to surge in demand as the global economy rapidly ramps up production of electric vehicles and renewable energy storage systems, not to mention lithium-ion mobile phone batteries.
>
> Lithium is abundant and even found in seawater, but at such low densities that commercial extraction is not yet feasible. As such, a handful of countries with extractable reserves in hard rock – like Australia – and in the brine of salt lakes are growing in geopolitical importance.
>
> The most common form of extraction in Australia is by crushing a hard rock called spodumene, and from that extracting lithium concentrate using a separation method that Brown says is similar to some coal processing systems. . . .
>
> Miners have long moved to where the resources are, and Queensland and New South Wales coal workers might need to relocate to the **Pilbara** for new lithium mining gigs [jobs]. In the case of **Greenbushes** however, there is a coal mining community right on its doorstep. ❞

A lithium mine at Pilgangoora in Western Australia.

— "How Australia's 'White Gold' Could Power the Global Electric Vehicle Revolution," Max Opray, *The Guardian*, 2020

Pilbara a region in Western Australia
Greenbushes the world's largest hard-rock lithium mine in Western Australia

---

EXAMINE THE SOURCE

1. **Identifying Cause and Effect** Why is the demand for lithium increasing?
2. **Inferring** What other type of mining has also been important in Australia, according to the source?

The Corn Belt of the United States

About 44 percent of the land area of the United States is used for agriculture, according to the World Bank. This includes land used for temporary and permanent crop farming. It also includes pasture and land that is used for keeping livestock. This article provides information about farming in one region of the United States.

PRIMARY SOURCE: ARTICLE

66 Drive down a dirt road, a two-lane country highway, even many Interstates in the Midwest and the view out the window is likely to get monotonous: massive fields filled with acres of corn sprawled in all directions.

The U.S. Department of Agriculture (USDA) expects farmers to harvest about 13.6 billion bushels of corn this season, the third-largest harvest in U.S. history. A fraction of that gigantic crop will sweeten our food and drinks, about a third will be made into ethanol for fuel and, when you figure in exports and byproducts, more than half will go to fattening the livestock that become our chicken filets, pork chops, and burgers. . . .

The Corn Palace in Mitchell, South Dakota has an exterior decorated with corn.

"A lot of that corn, since it's going into livestock feed, ends up in us," said Chad Hart, an agricultural economist at Iowa State University. . . .

While U.S. demand for meat has declined over the past decade, it's rapidly growing around the world, with China biting at the heels of U.S. per capita consumption, and other regions, like North Africa and the Middle East, wanting more animal protein in their diets, as well. . . .

States in the USDA's official Corn Belt, which includes Iowa, Illinois, Indiana, Missouri, and Ohio, grow half of the country's entire corn crop. When you add in other Midwestern states, it jumps to close to 90 percent. 99

— "Massive Corn Belt Crops Form Backbone Of Meat Industry," Abby Wendle, Harvest Public Media, 2015

> ### EXAMINE THE SOURCE
>
> 1. **Identifying** Where is the majority of corn produced in the United States?
> 2. **Making Connections** How does what people eat contribute to making corn a major crop within the Corn Belt?

How Colombia Became an Exporter of Cut Flowers

Cut flowers are a luxury item that is popular in wealthy countries such as the United States. A majority of the flowers sold in the United States are grown in Colombia, where ideal growing conditions exist alongside a low-wage workforce and the aftereffects of political and economic instability. The article explains the origin of this international connection.

PRIMARY SOURCE: ARTICLE

" In 1967 David Cheever, a graduate student in horticulture at Colorado State University, wrote a term paper titled "Bogotá, Colombia as a Cut-Flower Exporter for World Markets." The paper suggested that the savanna near Colombia's capital was an ideal place to grow flowers to sell in the United States. The savanna is a high plain fanning out from the Andean foothills, about 8,700 feet above sea level and 320 miles north of the Equator, and close to both the Pacific Ocean and the Caribbean Sea. Those circumstances, Cheever wrote, create a pleasant climate with little temperature variation and consistent light, about 12 hours per day year-round—ideal for a crop that must always be available. A former lakebed, the savanna also has dense, clay-rich soil and networks of wetlands, tributaries and waterfalls left after the lake receded 100,000 years ago. . . .

After graduating, Cheever put his theories into practice. He and three partners invested $25,000 apiece to start a business in Colombia called Floramérica, which applied assembly-line practices and modern shipping techniques at greenhouses close to Bogotá's El Dorado International Airport. . . .

It's not often that a global industry springs from a school assignment, but Cheever's paper and business efforts started an economic revolution in Colombia. Today, the country is the world's second-largest exporter of cut flowers, after the Netherlands, shipping more than $1 billion in blooms. Colombia now commands about 70 percent of the U.S. market. "

— "The Secrets Behind Your Flowers," John McQuaid, *Smithsonian*, 2011

EXAMINE THE SOURCE

1. **Explaining Cause and Effect** Why is the area around Bogotá ideal for growing flowers for sale in the United States?

2. **Identifying** What other country does more trade in cut flowers than the $1 billion per year that Colombia now does?

McQuaid, John. Smithsonian Magazine. "The Secrets Behind Your Flowers." 2011.

Textile Manufacturing in Bangladesh

Textile and clothing production depends on a large amount of labor and relatively basic technologies. Over the past 25 years, clothing production for U.S. markets has moved to countries where industrialization is in its early development. New factories can be constructed quickly in places where worker wages are relatively low. This helps keep production costs low. This picture shows the printing section of a clothing factory in Bangladesh.

PRIMARY SOURCE: PHOTOGRAPH

EXAMINE THE SOURCE

1. **Identifying** What information does the source provide about the amount of labor required for clothing manufacturing?

2. **Inferring** What evidence indicates that the clothing being produced is for export?

Eco-Tourism in Peru

Tourism is an increasingly important part of the global economy. People all over the world travel for recreation and relaxation. The tourism economy can be especially important for places that have natural or historic features that attract visitors. This article describes one location of *eco-tourism*, which is responsible tourism that aims to protect and sustain natural areas and indigenous populations.

PRIMARY SOURCE: ARTICLE

❝ In the Tambopata Candamo Reserved Zone in southeastern Peru, Rainforest Expeditions, a for-profit eco-tourism company formed by Peruvian conservationists, has entered into a joint eco-tourism venture with the Ese'eja Indian community to attract tourists to a biologically rich site boasting macaws, giant river otters, and harpy eagles. The indigenous community provides labor, lodging, and food for the project, and in return receives 60 percent of the profits from the joint venture. . . .

Both Rainforest Expeditions and the Ese'eja community realize that the success of their tourism venture depends on the protection of local wildlife resources. Accordingly, both sides are actively involved in research, management, and conservation programs to protect the fragile ecosystem. Since its inception, the site has become a highly-rated eco-tourism destination, developed innovative natural and cultural education programs, and played an increasingly important role in the conservation and sustainable development of the region. ❞

— "Eco-Tourism: Encouraging Conservation or Adding to Exploitation?" by Jonathan Nash, Population Reference Bureau website, 2001

EXAMINE THE SOURCE

1. **Explaining** What is the goal of eco-tourism?
2. **Speculating** What are two benefits of eco-tourism for local communities?

Complete Your Inquiry

EVALUATE SOURCES AND USE EVIDENCE

Refer back to the Compelling Question and the Supporting Questions you developed at the beginning of the lesson.

1. **Drawing Conclusions** Which sources provide information about the location of agricultural activities?
2. **Making Connections** Which sources describe economic activities that are not dependent on the climate of a place?
3. **Gathering Sources** Which sources helped you answer the Supporting Questions and the Compelling Question? Which sources, if any, challenged what you thought you knew when you first created your Supporting Questions? What information do you still need in order to answer your questions? What other viewpoints would you like to investigate? Where would you find that information?
4. **Evaluating Sources** Identify the sources that helped answer your Supporting Questions. How reliable is each source? How would you verify the reliability of the source?

COMMUNICATE CONCLUSIONS

5. **Collaborating** Share with a partner which sources you found most useful in answering your Supporting Questions. Discuss how the sources help you understand why specific types of economic activities occur in different locations. Take turns relating how the information you gathered from the sources help explain the economic activities happening in areas near you. How do these sources support your thinking? Use the graphic organizer that you created at the beginning of the lesson to help you. Share your conclusions with the class.

TAKE INFORMED ACTION

Highlight Change in Your Local Economy
Think about economic activities in your local area. Consider which types have been most important in the past and which are most important now. Choose a period in the past when the local economy was different than it is now. Create a visual art piece to show different types of activities that were important to the local economy in the past. Display your art in your school or community.

Nash, J., "Eco-Tourism: Encouraging Conservation or Adding to Exploitation?" April 1, 2001, Population Reference Bureau

READING STRATEGY

Analyzing Key Ideas and Details As you read, take notes about levels of economic development.

Economic Development	
less developed	
newly industrialized	
more developed	

Aspects of Economic Activity

GUIDING QUESTION

How are the world's countries classified in terms of economic development?

Economic development involves improving economic performance in order to improve the quality of life in a society. Economic performance can be a measure of quality of life and how well an economy meets the needs of a society. A common measure of economic performance is gross domestic product. The **gross domestic product (GDP)** is the total dollar value of all final goods and services produced in a country during a single year.

The **standard of living** is the level at which a person, a group, or a country lives. It is measured by how well it meets its needs. These needs include food, shelter, clothing, education, and health care. In many ways, the standard of living is better measured by per capita GDP than simply the total GDP. *Per capita GDP* is the total GDP divided by the number of people in the country.

GDP has many limitations as a measure of how well a society meets its needs. For example, GDP only counts goods and services that are exchanged for money. It fails to count things like volunteer work, housework, raising children, and other things that are important to a society. GDP can also improve when bad things occur that do not improve the standard of living—things like pollution, environmental degradation, natural disasters, and crime. Other measures of meeting the needs of a society have been developed that include more than just economics. These include the Human Development Index (HDI) and the Genuine Progress Indicator (GPI). However, these are not used as often as GDP.

Economic development is different than economic growth. Economic growth just means that the economy has increased in size. It is usually measured by an increase in GDP. Such growth might improve the standard of living, or it might not. Economic growth by itself does not guarantee economic development.

Economic activities help influence a country's level of economic development. Countries that have more technology and manufacturing, such as the United States, Canada, and the countries of Western Europe, are called **more developed countries**. They have highly developed economies and advanced technological infrastructures.

---

gross domestic product (GDP) the dollar value of all final goods and services produced in a country during a single year

standard of living the level at which a person, group, or country lives as measured by the extent to which it meets its needs

more developed country a country with a highly developed economy and advanced technological infrastructure

GO ONLINE Explore the Student Edition eBook and find interactive maps, charts, graphs, and tools.

G203

Because of modern techniques, only a small percentage of workers in more developed countries are needed to grow enough food to feed entire populations. For similar reasons, relatively small percentages of the people are employed in manufacturing industries. People in more developed countries enjoy a higher standard of living, which can be determined not only by the level of economic performance but also by demographic indicators. High life expectancy, a low infant-mortality rate, a low birth rate, and a low population growth rate are all indicators of a high standard of living.

Newly industrialized countries have moved from primarily agricultural activities to more manufacturing and industrial activities. This transition to manufacturing and industry often brings improvements in social development. Examples of newly industrialized countries are Mexico, Malaysia, and Turkey. The standard of living for people in newly industrialized countries is improving but is still lower than for those in more developed countries. Demographic indicators that help determine this include increasing life expectancies, decreasing infant-mortality rates, decreasing birth rates, and decreasing population growth rates.

Those countries that, according to the United Nations, exhibit the lowest indicators of social and economic development are **less developed countries**. Many less developed countries are in Africa, Asia, and Latin America. Agriculture remains dominant in such countries. Even though some **commercial farming** occurs, most farmers

---

newly industrialized country a country transitioning from primarily agricultural to primarily manufacturing and industrial activity

less developed country a country with the lowest indicators of socioeconomic development

commercial farming growing large quantities of crops or livestock to sell for a profit

---

Agriculture and Economic Development

Agriculture makes the largest contributions to GDP in the world's less developed countries.

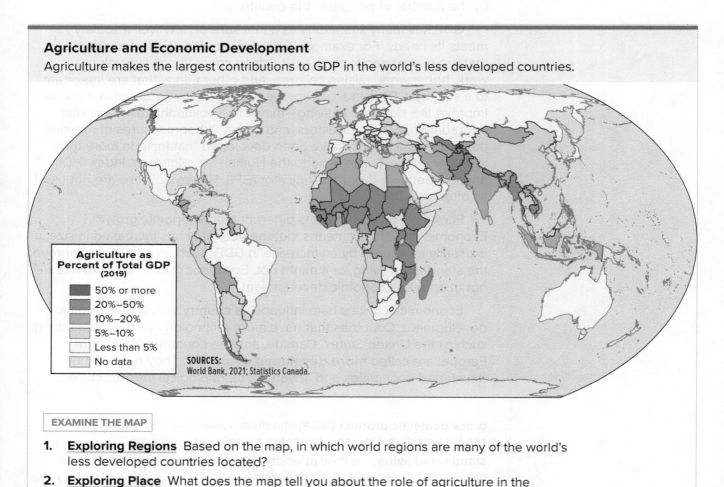

Agriculture as Percent of Total GDP (2019)
- 50% or more
- 20%–50%
- 10%–20%
- 5%–10%
- Less than 5%
- No data

SOURCES:
World Bank, 2021; Statistics Canada.

EXAMINE THE MAP

1. **Exploring Regions** Based on the map, in which world regions are many of the world's less developed countries located?

2. **Exploring Place** What does the map tell you about the role of agriculture in the United States?

Industry and Economic Development

Industrial production and manufacturing contribute significantly to the GDP of newly industrialized countries. Canada is unusual because it is a more developed country that still gets much of its GDP from industry.

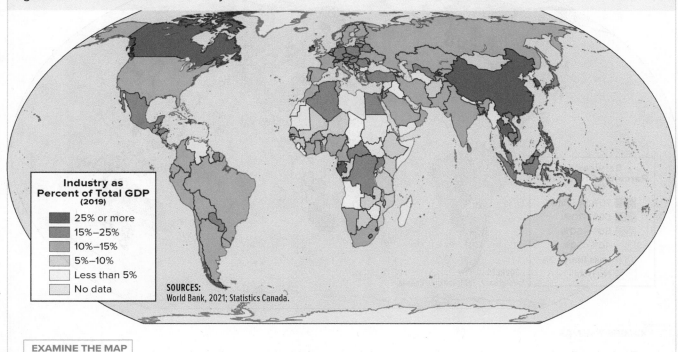

Industry as Percent of Total GDP (2019)
- 25% or more
- 15%–25%
- 10%–15%
- 5%–10%
- Less than 5%
- No data

SOURCES:
World Bank, 2021; Statistics Canada.

EXAMINE THE MAP

1. **Exploring Location** Name at least three newly industrialized countries in which GDP from industry is 15 percent or more?

2. **Exploring Regions** Which world region appears to have the lowest percentage of GDP from industry overall?

---

in these countries engage in subsistence farming. **Subsistence farming** is when all of the livestock and crops grown are only enough to meet the needs of the family. Some countries may have light industry that grows out of a history of *cottage industries*, businesses that employ workers in their homes. Most people in less developed countries have low incomes and a low standard of living. In addition to the economic indicators discussed above, demographic indicators of a low standard of living are low life expectancies, high infant mortality rates, high birth rates, and high population-growth rates.

✓ **CHECK FOR UNDERSTANDING**

Describing How is standard of living a sign of economic development?

---

subsistence farming farming that only provides the basic needs of a family

Patterns of Economic Activity

GUIDING QUESTION

What affects patterns of economic activity?

There are many factors that affect economic activity. Some of these factors are directly related to the economy. *Productivity* is a measurement of what is produced and what is required to produce it. *Sustainable growth* is the growth rate a business can maintain without having to borrow money. The *employment rate* is the percentage of the labor force that is employed. *Exports* are the goods and services that a country sells to other countries.

Elements of physical geography also play important roles. Climate not only helps determine what crops are grown, for example, but also the kinds of goods and services that people want and need. Soybeans are grown where there is

Services and Economic Development

Services as a high percentage of a country's GDP is a characteristic of more developed countries.

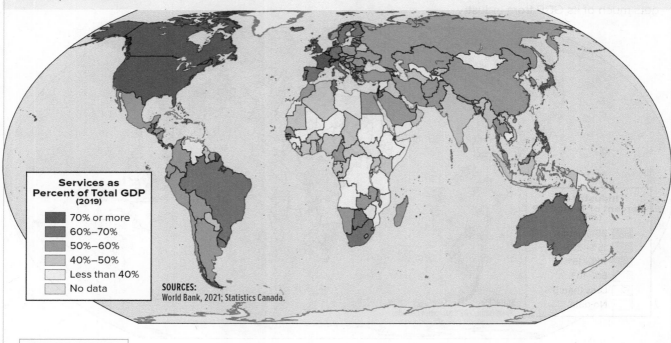

Services as Percent of Total GDP (2019)
- 70% or more
- 60%–70%
- 50%–60%
- 40%–50%
- Less than 40%
- No data

SOURCES: World Bank, 2021; Statistics Canada.

EXAMINE THE MAP

1. **Exploring Regions** According to the map, which parts of the world are heavily involved in tertiary and quaternary economic activities?

2. **Exploring Regions** How does North America compare to Asia when it comes to services as a percentage of GDP?

adequate rainfall and a warm growing season. People who live in cold, snowy climates have more use for boots, car battery warmers, and snow-removal services than those who live in warm, rainy climates.

Topography, or the forms and features on Earth's surface, affects economic activity. For example, mountainous areas make it difficult to build transportation networks like roads and railroads. In addition, rough terrain makes it hard to build and maintain communications networks such as cellphone towers and Internet lines. Without these types of networks to connect people, the distribution of goods and services to people is challenging.

The idea of **complementarity** is important in understanding the location and patterns of economic activities. Complementarity refers to the situation when one place has what another

place wants or needs. The demand for a product in one place and the supply of the same product in another place influences economic decisions. For example, vegetables like broccoli, cabbage, and lettuce are grown in California and eaten by Americans across the country during the winter months. California has the climate and land to grow crops that can't be grown in other parts of the country year-round. An example of complementarity in industry is the movement of manganese—a mineral used in making steel— from Ukraine to steel mills in Western Europe.

Agriculture

As you have read, physical geography influences agricultural activity—what is produced and where it is produced. Agriculture is affected by soil quality, the availability of water, and a mild climate. Corn production in the Midwestern

complementarity the relationship between two places for the demand and supply of a product

United States is favored by all three. Sometimes when one or another of these supportive factors is missing it can be provided by an alternative method. For example, cotton production in the Texas Panhandle relies on water from irrigation.

In addition to physical geography, there are factors related to the financial cost of producing agricultural goods. These also affect the pattern of agricultural activity. Many industries locate near agricultural production in order to keep transportation costs to a minimum. Lumber and wood-processing industries are located near the forests of the U.S. Pacific Northwest.

Commercial agriculture uses the idea of **comparative advantage**. This is the idea that a specific region does a better job of producing a good than another region. In the case of agriculture, a region may have a more advantageous physical environment, better practices of farmers, or better proximity (nearness) to markets. Different crops are grown according to these factors. The choice of how to use the land is dictated by how profitable a particular agricultural activity might be.

In commercial agriculture, the purpose is to make a living by selling products from the farm. The difference between the amount of money farmers receive when they sell their products and the costs they run up to grow the products is their return, or profit. We would expect farmers to use their land as productively as possible to gain the highest return.

One of the major costs for farmers is getting their products to market, usually a city. The distance from the farm to the market influences the farmer's choice of what crops to grow. A dairy farm provides the best example of the importance of proximity to market because milk spoils quickly. Crops that can be shipped long distances without spoiling can be grown farther away from the market.

Generally, as the distance from the market increases, the value of the land decreases. The intensity of agricultural land use corresponds to this. Crops that require more labor and capital and produce perishable goods are produced closest to the market. These may include milk, eggs, and vegetables. As one moves away from the market,

---

comparative advantage a place's ability to produce something more efficiently than another place

emphasis importance

farming becomes less intensive. Areas of forestry produce timber and firewood for building and fuel. Beyond these areas, grains like wheat are grown. Since grains last longer than dairy products, they can be located farther from the market. The zone of agricultural land use farthest away from the market is ranching and livestock.

Industry

Just as with agriculture, there are factors that influence the location and patterns of industrial production. Think about the manufacturing process. It transforms raw materials into finished products and then distributes those products to the people and businesses that need them. Although this seems like a pretty straightforward process, it involves several factors—raw materials, labor, capital, power, technology, transportation, and markets. Each has an element of geography that must be considered.

The factors of distance and transportation costs are especially important. Historically, the best location for manufacturing is somewhere between where the raw materials and markets are located. The site chosen for manufacturing should involve the lowest cost possible of moving raw materials to the factory and finished products to market. However, this has changed in recent times with the rise of centralized fulfillment centers such as those operated by Amazon and other providers. The **emphasis** now is to get the product to market as fast as possible and to deliver it to the consumer as quickly as possible. The Internet and product websites are designed to make it happen.

The cost of labor also affects where a factory is located. High labor costs reduce the profit a factory owner can make. So, a trade-off can be made by locating a factory farther away from the raw materials and markets if cheap labor made up for the increased transportation costs. This is especially the case in today's world, where labor in newly industrialized countries is much cheaper than the costs of transporting goods across the ocean or other long distances.

The concentration of similar businesses in a particular area can affect where industrial activity is located. This is called *agglomeration*, and usually occurs in cities. With agglomeration, industries can share talents, services, and facilities. For example, all manufacturers need office furniture and equipment. The existence of one or more industries that produce these goods satisfies the need for everyone. Therefore, a

Transportation networks like this rail yard in the Ruhr are critical to economic activities. The Ruhr area in western Germany is a major population and industrial center. The existence of large coal supplies contributed to its industrial growth.

location in a city is more appealing **despite** the increased labor costs in the city and the increased transportation costs of the raw materials.

Urban Economies

Understanding the structure of the urban economy is important. Cities in the United States and in many more developed countries display similar patterns of activities. In the center of the city is the **central business district (CBD)**, the commercial and business center of a city. It is characterized by tall office buildings, broad streets, and crowded sidewalks. The CBD includes a variety of economic activities reflected in shopping centers, office buildings, restaurants, sports complexes, and entertainment venues. People travel to and from the CBD daily. It provides employment for a high percentage of the workforce in the cities of more developed countries.

Surrounding the CBD is a residential area of houses, apartments, and condominiums. The people who live here work in the CBD or further out from the city in an industrial zone. Land in the industrial zone is less expensive than land in the CBD. Beyond the industrial zone are the *suburbs*,

residential areas characterized by larger houses situated on larger pieces of land. Transportation networks move people and goods into and out of the city as part of the urban economy.

Many of the huge urban areas that exist today are clusters of many cities, and each city may have its own CBD. All of the CBDs contain a full range of economic activities, but they often specialize in a particular type such as shopping, government, financial services, or corporate headquarters. This leads to complex patterns of service, industrial, residential, and suburban areas. It also means that developing transportation networks is more difficult.

✓ **CHECK FOR UNDERSTANDING**

Summarizing How does physical geography affect patterns of economic activity?

LESSON ACTIVITIES

1. **Informative/Explanatory Writing** Write a paragraph explaining how the major economic activities of a country are related to the country's level of development.

2. **Collaborating** Work with a partner to research agriculture in a more developed, newly industrialized, or less developed country. Create a list of factors that affect the economic agricultural activities in that country today. Use your findings to create a cause-and-effect chart.

despite without being prevented by

central business district (CBD) the center or "downtown" of a city; the shopping, business, and transportation hub of a city

Analyzing Key Ideas and Details As you read, take notes about the factors that stimulate world trade.

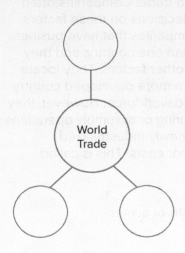

World Trade

Patterns of World Trade

GUIDING QUESTION

What stimulates world trade?

You have read about economic systems and types of economies. These are economic **categories**. All the world's countries can be classified into them. There are many types of systems and economies, but all countries interact with one another to get what they need. Look at the labels on the products you buy. How many different country names can you find? How and why do we get goods from far across the world?

Trade is the business of buying, selling, and/or bartering. When you buy something at a store, you are trading money for a product. On a larger scale, countries trade with each other. *World trade* is the exchange of capital (money), labor, goods, and services across international borders. It involves the **import** and **export** of goods and services.

In most countries, trade represents a significant share of gross domestic product (GDP). Trade among countries has been present throughout history, but its economic, social, and political importance has increased in recent centuries.

category a group of similar things

import to bring a product into one country from another country

export to send a product from one country to another country

Exports and Imports, Leading Countries in 2020

This graph shows the world's top countries by value of imports and exports. Notice that some countries import more than they export. What might this mean for these countries?

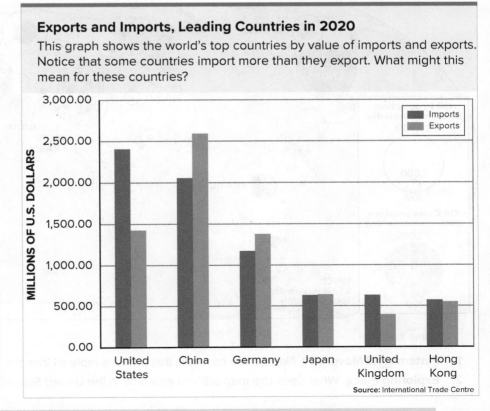

Source: International Trade Centre

The unequal distribution of natural resources is one factor that **promotes** trade among countries. A country might have abundant access to one resource but a limited supply of another resource. Saudi Arabia, for example, is a country rich in petroleum, which is often called crude oil or just oil. But Saudi Arabia has a limited amount of timber. So the country exports petroleum to other countries and imports wood products from other countries.

Trade can benefit countries. Some countries export their specialized products, trading them to other countries that cannot produce those goods. When countries cannot produce as much as they need of a certain good, they import it from another country. That country, in turn, may buy the first country's products, making the two countries trading partners.

In world trade, a country's government will often add extra fees to the cost of importing products. The extra money is a type of tax called a *tariff*. Governments often create tariffs to encourage their people to buy products made in their own country. Sometimes governments set a *quota*, or a limit on the amount of a good that can be imported. By limiting the import of a good, a quota increases the amount of the good produced within the country. A group of countries may decide to set little or no tariffs or quotas when trading among themselves. This is called **free trade**.

In addition to the unequal distribution of natural resources, differences in labor costs and education affect world trade. Companies often base their business decisions on these factors. *Multinationals* are companies that have business operations in more than one country, and they often consider these other factors. They locate their headquarters in a more developed country with a highly educated workforce. However, they locate their manufacturing or assembly operations in less developed or newly industrialized countries with low labor costs. This is called

promote advance
free trade arrangement whereby a group of countries decides to set little or no tariffs or quotas

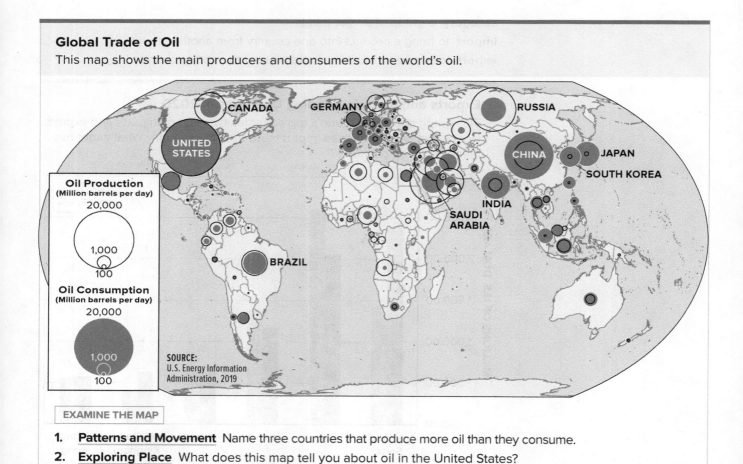

Global Trade of Oil
This map shows the main producers and consumers of the world's oil.

Oil Production
(Million barrels per day)
20,000
1,000
100

Oil Consumption
(Million barrels per day)
20,000
1,000
100

SOURCE:
U.S. Energy Information Administration, 2019

CANADA GERMANY RUSSIA
UNITED STATES CHINA JAPAN
SOUTH KOREA
INDIA
SAUDI ARABIA
BRAZIL

EXAMINE THE MAP

1. **Patterns and Movement** Name three countries that produce more oil than they consume.
2. **Exploring Place** What does this map tell you about oil in the United States?

outsourcing. Economic activities that are outsourced include manufacturing, call-center operations, and information technology (IT). In recent decades, many less developed countries have allowed multinationals to build factories or form partnerships with local companies.

Trade occurs domestically within countries as well as internationally between countries. The two are similar in many ways. The main difference is that international trade is typically more costly because of factors such as tariffs, border delays, shipping costs, and differences between countries such as language, legal systems, or other cultural issues that create barriers.

Another difference between domestic and international trade is that factors of production such as capital and labor are typically more mobile within a country than across countries. Thus, international trade is mostly restricted to trade in goods and services. Monetary capital has become more mobile internationally in recent years, but labor or other factors of production have not.

Trade in goods and services can serve as a substitute for trade in factors of production. Sometimes a country imports goods that make use of a particular factor of production. An example is the import of labor-intensive goods by the United States from China. Instead of importing Chinese labor, the United States imports goods produced by Chinese labor.

Trade has advantages and disadvantages. Trade can help increase economic growth and development as it increases a country's income. On the other hand, jobs might be lost because of importing certain goods and services. Both benefits and drawbacks occur as trade increases the *economic interdependence* that contributes to **globalization**, in which greater interconnections between countries occur in political and cultural affairs as well as economic affairs.

There are multiple causes of increased economic interdependence. Advances in technology have made it easier to communicate and share information with people around the world. Improved transportation networks make global travel easier. The growth of multinational companies with a presence in many different countries encourages trade. All of these factors increase interdependence.

Clothing manufacturing is often outsourced to countries such as Vietnam where labor is cheaper.

Globalization is the result of a modern world economy. The theory of *comparative advantage* has led to a more rigorous form of globalization. Different countries have access to different raw materials, unique labor forces, specialized economies, and specialized technologies. Each country produces what they do best and then trades with other countries for other products. For example, one of the largest computer-chip makers in the world is Taiwan. Lots of clothing comes from the apparel industry in China. Car parts for the U.S. auto industry are manufactured in Mexico.

There are many effects of globalization. In general, economic interdependence decreases the cost of manufacturing. This means that the cost of goods is often lower. Additionally, consumers have access to a greater variety of goods. A more

outsourcing obtaining goods or services from an outside supplier

globalization increasing interconnection of economic, political, and cultural processes to the point that they become global in scale and impact

negative effect of globalization is the environmental damage that results from the increased consumption of goods. This includes rises in the extraction and use of the natural resources used to make goods. In addition, dependence on the economy and production facilities in another country can lead to problems. Countries may have control of strategic products like oil. There may also be supply-chain disruptions. For example, the global COVID-19 pandemic affected the delivery of grocery items, car parts, and computer chips in many countries.

Some effects of interdependence are shaped by the viewpoint a person holds—such as the view from a more developed country or a less developed country. For example, economic globalization often helps more developed countries profit from the resources of less developed countries.

✓ **CHECK FOR UNDERSTANDING**

Analyzing How might globalization affect less developed countries?

Economic Organizations

GUIDING QUESTION

Why are economic organizations important?

As countries become more interdependent, they form economic and political ties. The World Trade Organization (WTO) helps regulate trade among countries. The World Bank provides financing, advice, and research to less developed countries to help them grow their economies. The International Monetary Fund (IMF) monitors economic development. It also lends money to countries and provides training and technical help. One well-known policy and organization that promoted international trade is the North American Free Trade Agreement (NAFTA). It encouraged free trade among the United States, Canada, and Mexico. It was renegotiated in 2018, and the new United States-Mexico-Canada Agreement (USMCA) went into effect in 2020.

The European Union (EU) is a group of European countries that operate under one economic unit and one currency, or type of money—the euro. The EU also facilitates political and cultural interconnections among European countries. The Mercado Común del Sur (formerly called MERCOSUR) is a group of South American countries that promote free trade, economic development, and globalization. The Mercado Común del Sur helps countries make better use of their resources while preserving the environment. The Association of Southeast Asian Nations (ASEAN) is a group of countries in Southeast Asia that promote economic, cultural, and political development. The Dominican Republic-Central America Free Trade Agreement (CAFTA-DR) is an agreement among the United States, five Central American countries, and the Dominican Republic. The agreement promotes free trade.

Whatever balance a country has between producing its own goods and trading, one basic principle exists: sustainability. The principle of **sustainability** is central to the discussion of resources. When a country focuses on sustainability, it works to create conditions where all the natural resources for meeting the needs of society are available. This is the goal not only for present societies, but also for future generations.

What can countries do to ensure sustainability now and into the future? What can you do to plan for your future and the future of your community? Just as every country is part of a global system, you are part of your community. The choices you make affect you and those around you. What can you do now to plan for a bright economic future?

✓ **CHECK FOR UNDERSTANDING**

Explaining Describe the goals of the WTO and the IMF.

LESSON ACTIVITIES

1. **Explanatory Writing** Write an essay explaining the advantages and disadvantages a less developed country might experience by joining a free trade agreement.

2. **Analyzing Main Ideas** Work with a group to discuss the advantages and disadvantages of trade. Consider these questions: *How do countries with large shipping ports have advantages over countries without such ports? What effect does shipping have on the environment? How do quotas and tariffs affect global trade?*

---

sustainability the idea that a country works to create conditions where all natural resources for meeting the needs of society are available

06

Understanding Multiple Perspectives: Participation in Local and Global Economic Systems

? COMPELLING QUESTION

What influences how people in different places participate in economic systems?

Plan Your Inquiry

DEVELOPING QUESTIONS

Think about the different ways in which people can participate in economic systems as producers and consumers. Reflect on what you know about how people make economic decisions. Then read the Compelling Question for this lesson. What questions can you ask to help you answer this Compelling Question? Create a graphic organizer like the one below. Write these Supporting Questions in your graphic organizer.

Supporting Questions	Source	What this source tells me about what influences how people participate in economic systems	Questions the source leaves unanswered
	A		
	B		
	C		
	D		
	E		

ANALYZING SOURCES

Next, examine the sources in this lesson. Analyze each source by answering the questions that follow it. How does each source help you answer each Supporting Question you created? What questions do you still have? Write these in your graphic organizer.

After you analyze the sources, you will
- use the evidence from the sources;
- communicate your conclusions;
- take informed action.

Background Information

The term *economy* refers to the processes and systems through which goods and services are produced and distributed. In a **subsistence** economy, people rely on themselves and direct trade to produce and acquire the things they will consume. As economies develop, governments and private companies organize the production of goods and the provision of services at a larger scale. This means more people participate in paid work rather than working for themselves. Their roles also become more specialized, and this can require increased education and training. People participate in economic systems in different ways. People in different areas might have different opportunities available to them. People can also have individual goals that guide how they participate in economic systems.

subsistence relating to production for one's own use or consumption

» Some people buy fruits and vegetables directly from farmers at small markets like this one in Vimercate, Italy.

A

Sherpas and Tourism in Nepal

The term *Sherpa* originally referred to a group of people who migrated to Nepal from Tibet. This ethnic group settled in high mountain valleys of the Himalaya. This book excerpt explains how Sherpas in the Khumbu Valley of Nepal have found economic opportunities in the world's highest mountain range.

PRIMARY SOURCE: BOOK

66 Khumbu is situated at the edge of Tibet, at the entry to the final, high-altitude **crux** of a major trans-Himalayan trade route linking Nepal and northern India with Tibet. Although the Nangpa La is one of the highest of the many passes that cross the Himalaya to Tibet, it is not a difficult passage for **yak** and for people who are properly equipped for snow and cold and knowledgeable about mountain weather and glacier crossings.[2] Khumbu Sherpas thus have been ideally situated to be middlemen on an important long-distance trade route....

Sherpa porters in Khumbu Valley, Nepal carry heavy loads of gear for a mountaineering expedition.

Sherpas have been associated with tourism since 1907, when several were hired in Darjeeling as **porters** by mountaineers who were making attempts on peaks in Sikkim ...[20] [They] soon developed a reputation for toughness, endurance, courage, and loyalty which made them preferred above all other Himalayan peoples as companions on the great peaks ... Foreign mountaineering was forbidden within Nepal itself, but young Khumbu Sherpas could reach Darjeeling, the British hill station that was the major mountaineering center in the Himalaya until World War II, in only ten days of walking ... Nepal was once one of the world's most remote and most difficult to visit countries ... With the coronation of King Mahendra in 1955 there came the beginning of a new attitude. In that year the first tourist visas were issued, the first hotel was opened, and the first visitors began to arrive.... Since then mass tourism has found Nepal and the Nepal government has responded with increasing interest in tourism development. 99

— "From Tibet Trading to the Tourist Trade," Stanley F. Stevens, *Claiming the High Ground*, 1993

crux the most important or decisive part

yak a type of long-haired domesticated cattle found in the Himalaya region

porter a person hired to carry luggage and other loads

EXAMINE THE SOURCE

1. **Comparing** Why was location important for both of the Sherpas' economic activities described in the source?

2. **Identifying Cause and Effect** What changes increased opportunities for Sherpas living in the Khumbu Valley to participate in the tourist economy?

G214

B

Women in the Workplace

As economies develop, the nature of work changes. In the early 1900s, the vast majority of women in the United States did not participate in employment outside the home. Today, a majority of U.S. women are employed. However, as this article explains, challenges remain in the drive for equal treatment of women in the workplace.

PRIMARY SOURCE: ARTICLE

66 We, as a country, have reaped great benefits from the increasing role that women have played in the economy. But evidence suggests that barriers to women's continued progress remain. The participation rate for prime working-age women peaked in the late 1990s and currently stands at about 76 percent ... a level well below that of prime working-age men, which stands at about 89 percent.

Women working at the U.S. Capitol switchboard, Washington, D.C., 1959 (Library of Congress)

The gap in earnings between men and women has narrowed substantially, but ... women working full time still earn about 17 percent less than men, on average, each week. ...

[T]he difficulty of balancing work and family is a widespread problem. In fact, the recent trend in many occupations is to demand complete scheduling flexibility, which can result in too few hours of work for those with family demands and can make it difficult to schedule childcare. Reforms that encourage companies to provide some predictability in schedules, cross-train workers to perform different tasks, or require a minimum guaranteed number of hours in exchange for flexibility could improve the lives of workers holding such jobs. Another problem is that in most states, childcare is affordable for fewer than half of all families. And just 5 percent of workers with wages in the bottom quarter of the wage distribution have jobs that provide them with paid family leave. This circumstance puts many women in the position of having to choose between caring for a sick family member and keeping their jobs. 99

— "The History of Women's Work and Wages and How It Has Created Success For Us All," Janet L. Yellen, Brookings Institute, 2020

EXAMINE THE SOURCE

1. **Describing** How is the income of women different than that of men?
2. **Making Connections** How are women's roles within families related to employment outside the home?

Consumer Preferences and Behaviors

Being a consumer is one way in which people interact with economic systems. When people can choose how they obtain goods, their consumer behavior and buying preferences shape the economy. The different types of demands consumers create are opportunities for business. To be successful, a business must understand what those consumers want and can afford. This article summarizes the results of a survey about the preferences of consumers in Europe.

PRIMARY SOURCE: ARTICLE

66 A report from shopper insight provider IRI [Information Resources, Inc.] has highlighted the growing importance of geocentric purchasing, with seven out of ten European shoppers identifying "strongly" with ethical purchasing practices and expressing a "clear preference" for buying locally sourced products.

This survey, which gauged the opinion of more than 3,000 European consumers across seven countries, including the UK, Italy, France and Germany, discovered ethical considerations have become closely linked with regional production. . . .

The majority of shoppers included in the survey revealed that they want products made by manufacturers that are local and respect the environment. Just over 70% of millennials, those aged 18–34, said this was important in informing their product purchasing preferences. This rose to 73% of those aged over 35. 99

—"Local Brands are Winning Hearts and Minds," Katy Askew, FoodNavigator website, 2018

EXAMINE THE SOURCE

1. **Inferring** Based on the source, how would you define the phrase "geocentric purchasing"?
2. **Identifying** What are the benefits of geocentric purchasing?

Askew, Katy. Katy Askew. "'Local brands are winning hearts and minds': Rising demand for local food in Europe". 2018

A Border Economy

Countries can have very different levels of economic development and wealth. Sometimes two countries with very different economic circumstances share a border. This can create opportunities for people in that region.

On the border between the United States and Mexico, pedestrians can cross from Mexico into the United States to shop for clothing, household goods, food products, and more. Unlike items bought in Mexico, items bought in the United States are not subject to high import taxes, so they are a good value. Cross-border shopping also offers Mexican shoppers a wider variety of goods than is available in Mexican shops.

People holding Mexican passports can apply to the U.S. State Department for border-crossing cards that allow them to enter and leave the United States for ten years. This program appeals most to people living near the U.S. border; people in large cities further south are less likely to use this system.

The photo shows shoppers returning to Mexico after shopping in Laredo, Texas. Several cities in Texas benefit from the cross-border retail trade, including El Paso, Laredo, and Brownsville.

PRIMARY SOURCE: PHOTOGRAPH

Shoppers cross the Gateway to the Americas International Bridge to return to Mexico after shopping in Laredo, Texas.

EXAMINE THE SOURCE

1. **Evaluating** How do people in both countries benefit from the fact that people from Mexico cross the border into the United States to shop?

2. **Identifying** Why are products more expensive if they are bought in Mexico?

E

Migrating Workers

Many people around the world move to take advantage of economic opportunities, such as education or a job. The willingness of some workers to move long distances, whether temporarily or permanently, to unfamiliar countries creates a global labor market. This article explains how migration by one group of workers is affecting multiple countries.

66 Recent findings revealed that 40% of nursing positions in the Caribbean remain vacant, primarily due to nurse migration [2]. While nurse migration is alleviating [easing] some of the demand for nurses in receiving or destination countries, it is creating a greater deficit in regions from which nurses are migrating such as the Caribbean [4].

Researchers have studied the motivating factors, also referred to as push-pull factors, for nurse migration and found that in general nurses migrate for professional growth, financial benefits, and educational advancement [5, 6]. While the pull factors included improved working conditions and better opportunities, push factors include poor remuneration [pay], lack of professional development, and stressful working conditions [6] . . . . While the detrimental impacts of migration of Caribbean nurses are inarguable, it is maintained that there are benefits, primarily that of **remittances** [7, 10]. 99

—"Caribbean Nurse Migration—a Scoping Review," Shamel Rolle Sands, Kenchera Ingraham, and Bukola Oladunni Salami, *Human Resources for Health*, 2020

remittance money a migrant worker sends back to people in his or her home country

EXAMINE THE SOURCE

1. **Explaining** Why do nurses from Caribbean countries choose to migrate?
2. **Contrasting** What are the positive and negative consequences of nurse migration on the countries of the Caribbean region?

Complete Your Inquiry

Refer back to the Compelling Question and the Supporting Questions you developed at the beginning of the lesson.

1. **Drawing Conclusions** Which sources provide information about how people make choices about consumption?
2. **Making Connections** Which sources describe how people in different places earn money?
3. **Gathering Sources** Which sources helped you answer the Supporting Questions and the Compelling Question? Which sources, if any, challenged what you thought you knew when you first created your Supporting Questions? What information do you still need in order to answer your questions? What other viewpoints would you like to investigate? Where would you find that information?
4. **Evaluating Sources** Identify the sources that helped answer your Supporting Questions. How reliable is each source? How would you verify the reliability of the source?

5. **Collaborating** Share with a partner which sources you found most useful in answering your Supporting Questions. Discuss how the sources help you understand the choices people make regarding participation in economic systems. Take turns relating how the information you gathered might relate to your life or the lives of people in your community. How do these sources support your thinking? Use the graphic organizer that you created at the beginning of the lesson to help you. Share your conclusions with the class.

Supporting Your Local Economy Think about how people contribute to the economy in your area through their career choices or consumer behavior. Consult job listings on employment websites or in your local newspaper to see what kinds of workers are needed in your area and research what products are produced locally. Create a social media post titled "Five Things You Can Do to Support Your Local Economy" based on your research.

Rolle Sands, S., Ingraham, K. & Salami, B.O. Caribbean nurse migration—a scoping review. Human Resources for Health 18, 19 (2020). https://doi.org/10.1186/s12960-020-00466-y

Reviewing Economic Geography

Summary

Economic Systems and Activities	Economic Development	Economies and World Trade

• People use resources to get what they want or need. • Resources include natural resources from Earth as well as human labor. • Humans must weigh the cost of using renewable and nonrenewable resources to get what they want or need.	• Economic performance is measured by GDP. • Countries can be organized into groups based on economic development: more developed, newly industrialized, and less developed. • Agriculture is important to countries at all levels of development.	• World trade involves the import and export of goods and services across international borders. • The unequal distribution of resources stimulates trade among countries. • Trade supports specialization. • Tariffs and quotas are used as ways to control imports.
• Three basic questions help define each type of economic system: What should be produced? How should it be produced? How should it be distributed? • The main types of economic systems are traditional, market, command, and mixed.	• Elements of physical geography affect the location of agricultural and industrial activities. • Complementarity is the relationship between two places—one that has a demand for a product and the other that has a supply of that same product.	• Differences in labor costs and education affect trade. • Outsourcing work to less developed countries with cheap labor is common. • International trade is usually more costly than domestic trade.
• Factors of production are land, labor, capital (money), and entrepreneurship. • The four types of economic activities are primary, secondary, tertiary, and quaternary.	• Comparative advantage is when a country specializes in a certain product for export because they hold an advantage in producing that product. • Farmers use their land as productively as possible to make the largest profit. • Distance to market and transportation costs are important in locating economic activities.	• Economic interdependence leads to globalization. • The causes of globalization include improvements in communications and transportation. The effects include the decreased cost of manufacturing and a greater variety of goods.

Checking For Understanding

Answer the questions to see if you understood the topic content.

IDENTIFY AND EXPLAIN

1. Identify each of these terms as they relate to economic geography.

   A. traditional economy   F. newly industrialized country

   B. market economy   G. less developed country

   C. command economy   H. gross domestic product

   D. mixed economy   I. standard of living

   E. more developed country

REVIEWING KEY FACTS

2. **Contrasting** What is the difference between subsistence farming and commercial farming?

3. **Explaining** What is a nonrenewable resource? How is it different from a renewable resource?

4. **Describing** How do the World Bank and the International Monetary Fund help economic development?

5. **Making Connections** How is outsourcing related to multinational companies?

6. **Explaining** What is scarcity?

7. **Identifying** What are the three basic economic questions that help define different economic systems?

8. **Summarizing** Name and describe the four types of economic activities.

9. **Identifying** What are the factors of production?

CRITICAL THINKING

10. **Making Connections** How are tariffs and quotas related to free trade?

11. **Analyzing** Why is sustainability an important consideration in the discussion of resources and trade?

12. **Identifying Cause and Effect** What are some causes of economic interdependence and globalization? What are some effects?

13. **Contrasting** How are international trade and domestic trade different?

14. **Analyzing** Summarize the differences between a market economy and a command economy.

15. **Determining Central Ideas** How is complementarity important for understanding the patterns of economic activity?

16. **Explaining** What is comparative advantage and how is it related to agricultural activities?

17. **Identifying** What is one of the major costs of doing business for farmers? What influences the farmer's decision of what crops to grow?

18. **Describing** Describe the pattern of the different types of agriculture as one moves farther from the market.

19. **Making Generalizations** What factors affect the location of industrial production?

20. **Summarizing** How can agglomeration affect the location of industrial production?

NEED EXTRA HELP?

If You've Missed Question	1	2	3	4	5	6	7	8	9	10
Review Lesson	2, 4	4	2	5	5	2	2	2	2	5

If You've Missed Question	11	12	13	14	15	16	17	18	19	20
Review Lesson	5	5	5	2	4	4	4	4	4	4

Apply What You Have Learned

Geographic Reasoning

You have learned about gross domestic product (GDP) and about the place of the United States among more developed countries. However, there is a range of development levels among the cities and states within the United States. Different resources, industries, and populations have an impact on the level of economic activity in those areas. Over the past several decades, the GDP of some states has risen along with population growth and the creation of new industries. At the same time, that of other states has fallen due to automation, competition from overseas, and other factors.

ACTIVITY **Comparing and Contrasting GDP** Find online your home state's GDP and compare it to those of the more developed and less developed countries worldwide. The websites of agencies such as the U.S. Bureau of Economic Analysis or the World Bank will be good sources for this data. Create a graphic display (electronically or on a poster) comparing your state's GDP to two countries with a higher GDP and two countries with a lower GDP. Present the display to your class, summarizing the main economic activities in the places you have highlighted.

Understanding Markets

In a market economy, the consumer has a great deal of power over the products and services a business provides. Business owners and employees in such a system must use research and their own experience to decide what they believe consumers want. If they guess correctly, their businesses can be very successful. On the other hand, an incorrect guess can lead to losses for that business.

ACTIVITY **Learning from History** Research online to learn about a consumer product that a business created. This could be an expensive item like a new type of automobile or a technology item, or a smaller product like a food or clothing item. Summarize in a short essay whether the product was successful or (as is often more interesting!) unsuccessful with consumers. Explain the reasons that people did or did not choose to buy the product, and what clues might be learned for future product development.

C Debating Globalization

Globalization is a concept that can be explored in many areas of geography. As people, products, and information move ever more quickly around the world, every region is changed in numerous ways. The economic effects of globalization are possibly the most dramatic, bringing opportunities and challenges alike to the people in more developed and less developed countries.

ACTIVITY Taking a Position Conduct a panel discussion that is focused on the impacts of globalization in your area. What economic changes have come to your community or state that are the result of connections to the economies of other areas? What jobs have been created near you, and what jobs have disappeared? Is there a connection to one particular other region or country that has most changed your community? Have a student act as a moderator to make sure both pros and cons are given equal time.

D Making Connections to Today

Many of the products that we have come to depend on today include parts or ingredients that are found in specific parts of the world. For instance, the batteries that power smartphones and laptop computers include cobalt, which is found in reserves in the Democratic Republic of Congo, Australia, Cuba, and elsewhere. Manufacturers of technology products in China and the United States must import this crucial ingredient.

ACTIVITY Tracking a Resource Choose a single natural resource that exists somewhere other than in your own community, either in another state or another country. Examples include food ingredients, minerals, or chemicals. In a short presentation to the class, explain why the resource is important, why it is not produced or obtained locally, how it is used by your community, and whether it is scarce and/or nonrenewable.

Political geography studies governments and the boundaries and divisions of countries and other levels of government.

Political Geography

INTRODUCTION LESSON

01 Introducing Political Geography   G224

LEARN THE CONCEPTS LESSONS

02 Features of Government   G229

04 Conflict and Cooperation   G241

INQUIRY ACTIVITY LESSONS

03 Global Issues: Political Borders   G235

05 Analyzing Sources: Conflict and
    Cooperation   G247

REVIEW AND APPLY LESSON

06 Reviewing Political Geography   G253

asantosg/Alamy Stock Vector

Introducing Political Geography

What Do Governments Do?

Governments have many different jobs, such as establishing order and providing for the needs of the people within a defined territory. Different types of governments approach these jobs in different ways.

Types of Governments

All countries need some type of formal leadership. What differs among countries is how leaders are chosen, who makes the rules, and how much freedom people have.

» In a democracy, the people hold the power. Their rights and freedoms are protected. In some democracies like the United States, people elect leaders to make and carry out laws on behalf of the people.

» In a monarchy, power and leadership are usually passed down through heredity. The ruler of a monarchy can be a king, a queen, a prince, or a princess. Most monarchs today represent a country's traditions and values, while elected officials run the government. Queen Elizabeth II is the symbolic head of the United Kingdom.

» In a dictatorship, one person has absolute power to rule and to control the government, the people, and the economy. With such power, a dictator can make laws with no concern for how fair the laws are. North Korea is a dictatorship ruled by Kim Jong Un.

Geography and Government

Geography can influence governments through boundaries, resources, and cultural factors. In some instances, conflict can erupt between governments. At other times, cooperation among governments and groups such as the United Nations helps solve problems.

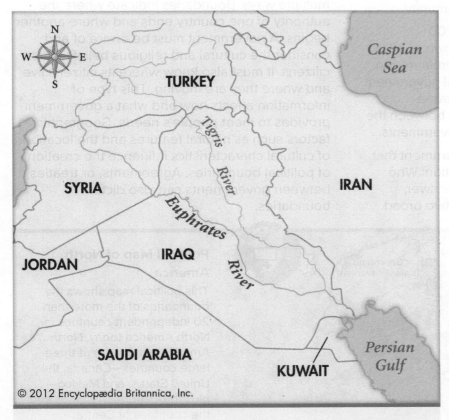

© 2012 Encyclopædia Britannica, Inc.

Conflict between governments can arise over resources and land use. The Tigris and Euphrates Rivers flow from Turkey into Iraq and Syria and have been a source of ongoing conflict among these countries. The construction of dams and irrigation systems in Turkey have resulted in decreased water flow from the rivers in Iraq.

> **66** We the peoples of the United Nations determined to save succeeding generations from the scourge of war, which twice in our lifetime has brought untold sorrow to mankind, and to reaffirm faith in fundamental human rights, in the dignity and worth of the human person, in the equal rights of men and women and of nations large and small, and to establish conditions under which justice and respect for the obligations arising from treaties and other sources of international law can be maintained, and to promote social progress and better standards of life in larger freedom . . . **99**
>
> —Preamble to the United Nations Charter, 1945

Getting Ready to Learn About . . . Political Geography

Governments and Geography

Territory, population, and authority come together to define the government of a country, state, city, town, or village. Governments have many functions. They keep order, provide security, provide services, and guide the community.

The world today contains nearly 200 independent countries, each with its own government. A country's level of government can be one in which the national or central government holds all power. A country may also have a government in which powers are split between the national government and the state governments.

There are different types of government that can be classified by asking the question: Who governs the country? Based on the answer, governments can be organized into two broad categories—governments in which leaders rule with the consent of the citizens and governments in which all power and authority belongs to one person or to a small group.

Geography influences governments in multiple ways. Boundaries indicate where the authority of one country ends and where another begins. A government must be aware of and consider the cultural and religious beliefs of its citizens. It must also know where its citizens live and where they are moving. This type of information affects how and what a government provides to meet people's needs. Geographic factors such as natural features and the location of cultural characteristics influence the creation of political boundaries. Agreements, or treaties, between governments can also dictate boundaries.

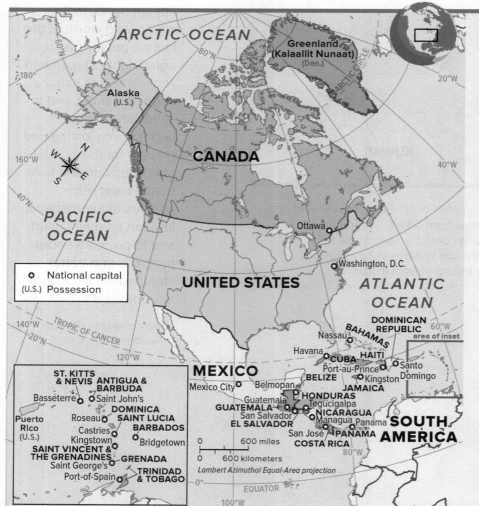

Political Map of North America

This political map shows the boundaries of the more than 20 independent countries in North America today. North America is made up of three large countries—Canada, the United States, and Mexico—plus the Caribbean Islands and the countries of Central America.

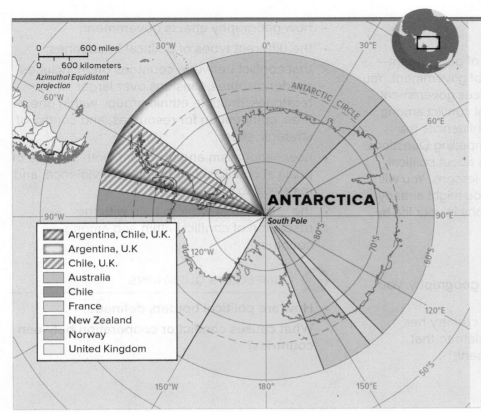

Map legend:
- Argentina, Chile, U.K.
- Argentina, U.K
- Chile, U.K.
- Australia
- Chile
- France
- New Zealand
- Norway
- United Kingdom

ANTARCTICA
South Pole

Conflict and Cooperation

Conflict between and within countries has contributed to and resulted from the political divisions of the world. Some causes of conflict are disputes over borders and disagreements over the possession or control of land. In some instances, conflict arises when multiple ethnic groups with cultural differences live within the same country. Competition for resources like water, oil, and minerals can lead to conflict among countries as they work to expand their economies. Control of strategic sites such as a location on a major shipping route can become a source of conflict when there is disagreement over who controls the site. Conflict can result when people have different levels of devotion and loyalty to their country.

Cooperation between countries has also contributed to and resulted from the political divisions of the world. A shared sense of loyalty to a country can create cooperation. If everyone in the group has the same strong level of loyalty, then they are more likely to work together toward a common goal to build and strengthen the country. Treaties are a reflection of the cooperation the occurs between countries. Treaties help countries establish ways to share and protect resources. For example, the La Paz Agreement between the United States and Mexico was signed in 1983. This treaty was created to protect, conserve, and improve the environment of the border regions of both countries. Treaties also help establish agreements to address economic issues and to strengthen economic ties. Such treaties usually create regional trading groups to support cooperation among countries in a region.

International organizations help countries work together to resolve conflict and foster cooperation. One such organization, the United Nations, works around the world to manage conflict between countries and to provide a means for countries to communicate ideas or information for international discussion. Other organizations create and maintain political alliances among groups of countries. For example the North Atlantic Treaty Organization includes many European countries, the United States, and Canada. Its purpose is to maintain democratic freedom through cooperation. Still other international organizations have the task of monitoring economic cooperation by overseeing trade agreements and settling trading disputes among the world's countries.

Looking Ahead

You will learn about the features of government—levels of government and types of government. You will learn how geography influences government. You will learn about the causes of conflict among countries and how cooperation influences the landscape. You will examine Compelling Questions and develop your own questions about political geography in the Inquiry Activity lessons. You will see some of the ways in which you might already use the ideas of political geography in your life by completing these lessons.

What Will You Learn?

In these lessons about political geography, you will learn

- that the government of each country has unique characteristics that relate to that country's historical development;
- the levels of government;
- the types of governments;
- how geography affects government;
- the different types of political boundaries;
- that conflict between countries often includes border disputes, tensions over larger territories, multiple ethnic groups within one area, competition for resources, and control of strategic sites;
- how nationalism and terrorism, both sources of political conflicts, can breed fear, violence, and may lead to war;
- the groups that work to help with the resolution of conflict within and among countries.

? COMPELLING QUESTIONS

- **How are political borders defined?**
- **What causes conflict or cooperation between countries?**

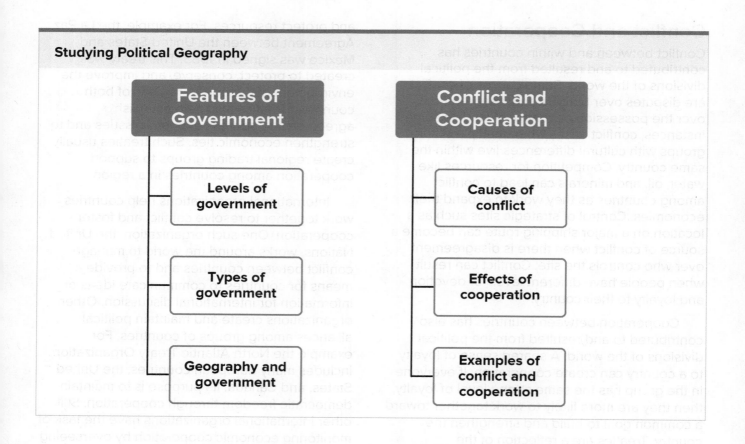

Studying Political Geography

Features of Government
- Levels of government
- Types of government
- Geography and government

Conflict and Cooperation
- Causes of conflict
- Effects of cooperation
- Examples of conflict and cooperation

Analyzing Key Ideas and Details As you read, take notes about the features of government.

Features of Government
Levels of Government
Types of Government
Geography and Government

Levels of Government

What influences each level of a country's government?

The world today includes nearly 200 independent countries. Each of them varies in size, military might, natural resources, and level of influence in the world. A country is defined as having internationally recognized territory over which it has **sovereignty**, or freedom from outside control. All countries have some kind of government that organizes and conducts the actions of the country. A government must make and **enforce** policies and laws that are binding upon all people living within its territory.

sovereignty the supreme and absolute authority within territorial boundaries
enforce to make sure that people do what is required

Functions of Government
Governments serve many purposes. They establish order, provide security, and accomplish common goals.

Keep Order
- Pass and enforce laws to deter crime
- Establish courts

Provide Security
- Establish armed forces
- Protect citizens from foreign attack

Provide Services
- Protect public health
- Protect public safety
- Protect public welfare

Guide the Community
- Develop public policy
- Manage the economy
- Conduct foreign relations

1. **Identifying** What are the four broad functions of government?
2. **Explaining** Why is it important for the government to conduct foreign relations?

GO ONLINE Explore the Student Edition eBook and find interactive maps, charts, graphs, and tools.

G229

The government of each country has unique characteristics that relate to that country's historical development. To carry out their functions, governments are organized in a variety of ways. Most large countries have several different levels of government. These usually include a national or central government, as well as the governments of smaller internal divisions such as provinces, states, counties, cities, towns, and villages.

A **unitary system** of government gives all key powers to the national or central government. This structure does not mean that only one level of government exists. Rather, it means that the central government creates smaller or local governments and gives them limited sovereignty. The United Kingdom and France both created unitary governments as they developed during the late Middle Ages and early modern times.

A **federal system** of government divides the powers of government between the national or central government and state or provincial governments. Each level of government has sovereignty over specified functions in a recognized geographical area. The United States developed a federal system after the thirteen colonies became independent from Great Britain.

Another government structure that is similar to a federal system is a *confederation*, or a loose union of independent territories. The United States at first formed a confederation, but this type of political arrangement failed to provide an effective national government for the new country. As a result, the U.S. Constitution replaced the earlier Articles of Confederation to establish a strong national government while preserving some state government powers. Today, other countries with federal or confederate systems include Canada, Switzerland, Mexico, Brazil, Australia, and India.

✓ CHECK FOR UNDERSTANDING

Explaining How is a unitary system different from a federal system?

unitary system a form of government in which all powers reside with national government

federal system a form of government in which powers are divided between the national government and state or provincial government

Types of Government

GUIDING QUESTION

What influences the type of government a country establishes?

Governments can be classified by asking the question: Who governs the country? Under this classification system, governments can be defined as democratic or authoritarian.

Democratic Governments

A **democracy** is any system of government in which leaders rule with the consent of the citizens. The term *democracy* comes from the Greek words *demos* ("the people") and *kratia* ("rule"). The ancient Greeks used the word *democracy* to mean government by the many rather than government by the few. The key idea of democracy is that people hold sovereign power.

Direct democracy, in which citizens themselves decide on issues, exists in some places at local levels of government. No country today has a national government based on direct democracy. Instead, democratic countries have **representative democracies**, in which the people elect representatives to make laws and conduct government. An assembly of the people's representatives may be called a council, a legislature, a congress, or a parliament.

Many democratic countries, such as the United States and France, are republics. In a republic, voters elect all major officials, who are responsible to the people. The head of state—or head of government—is usually a president elected for a specific term. Not every democracy is a republic.

Some democratic countries today, such as the United Kingdom, Canada, Japan, Jordan, and Thailand, have **constitutional monarchies**. Their monarchs share governmental powers with elected legislatures or parliaments, but the monarchs serve only as ceremonial leaders. The United Kingdom, for example, is a democracy with a monarch as head of state. This monarch's role is

democracy a type of government in which leaders rule with the consent of the citizens

representative democracy a form of democracy in which citizens elect government leaders to represent the people and run the government

constitutional monarchy a form of government in which a monarch is the head of state but elected officials run the government

Governments Around the World

This map shows the different types of government in countries around the world.

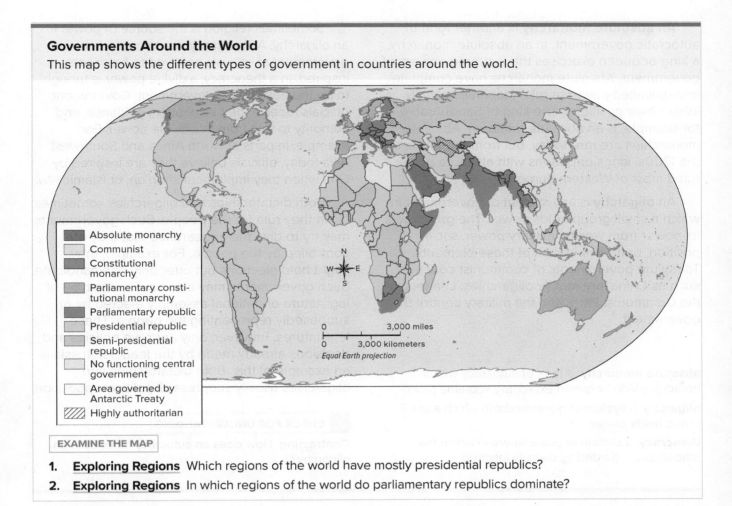

Legend:
- Absolute monarchy
- Communist
- Constitutional monarchy
- Parliamentary constitutional monarchy
- Parliamentary republic
- Presidential republic
- Semi-presidential republic
- No functioning central government
- Area governed by Antarctic Treaty
- Highly authoritarian

0   3,000 miles

0   3,000 kilometers

Equal Earth projection

EXAMINE THE MAP

1. **Exploring Regions** Which regions of the world have mostly presidential republics?
2. **Exploring Regions** In which regions of the world do parliamentary republics dominate?

ceremonial, however. Officials who are elected by the people hold the actual power to rule.

Authoritarian Governments

Any system of government in which the power and **authority** to rule belongs to one person or to a small group is an *authoritarian government*. An **autocracy** is a government in which power is held by a single person. This is the oldest form of government. Most autocrats achieve and maintain their position of authority through inheritance or by the ruthless use of military or police power.

Several forms of autocracy exist. One is an absolute or totalitarian **dictatorship** in which the decisions of a single leader determine government policies. The totalitarian dictator seeks to control all aspects of social and economic life. Examples of totalitarian dictatorships include Adolf Hitler in Nazi Germany, Saddam Hussein in Iraq, and Kim Jong Un of North Korea.

Some dictators abuse their power for personal gain. One negative consequence of abuse of power is lack of personal freedoms and human rights for the general public. **Human rights** are the rights that belong to all individuals simply because we exist as human beings. Those rights are the same for every human in every culture. Some basic human rights are the right to life, liberty, and fair treatment before the law.

authority the power to influence or command thought, opinion, or behavior

autocracy a system of government in which one person rules with unlimited power and authority

dictatorship a form of autocracy in which the one person has absolute power to rule and control the government, the people, and the economy

human rights the rights belonging to all individuals simply because we exist as human beings; these include freedoms and rights, such as freedom of speech, that all people should enjoy

An **absolute monarchy** is another form of autocratic government. In an absolute monarchy, a king or queen exercises the supreme powers of government. Absolute monarchs have complete and unlimited power to rule. Monarchs usually inherit their positions. The king of Saudi Arabia, for example, is an absolute monarch. Absolute monarchies are rare today, but from the 1400s to the 1700s, kings or queens with absolute power ruled most of Western Europe.

An **oligarchy** is any system of government in which a small group holds power. The group gets its power from wealth, military power, social position, or a combination of these elements. Today the governments of communist countries, such as China, are mostly oligarchies. Leaders in the Communist Party and the military control the government.

absolute monarchy a form of autocracy with a hereditary king or queen exercising supreme power

oligarchy a system of government in which a small group holds power

theocracy a system of government in which the officials are regarded as divinely inspired

Sometimes religion is the source of power in an oligarchy. A **theocracy**, for example, is a government of officials believed to be divinely inspired. In a theocracy, a divine power is thought to be the head of the government. Government officials receive their inspiration, guidance, and authority to rule from this divine power. For example, in parts of North Africa and Southwest Asia today, officials believe they are inspired by God when they implement *shari'ah*, or Islamic law.

Both dictatorships and oligarchies sometimes claim they rule for the people. Such governments may try to give the appearance that they are controlled by the people. For example, they might hold elections but offer only one candidate. Such governments may also have some type of legislature or national assembly elected by or supposedly representing the people. These legislatures, however, only approve policies and decisions already made by the leaders. Russia is an example of this. Both dictatorships and oligarchies usually suppress all political opposition.

✓ **CHECK FOR UNDERSTANDING**

Contrasting How does an autocracy differ from an oligarchy?

Comparing Democratic and Authoritarian Governments
This table reveals the key differences between democratic and authoritarian governments.

	Selection of Leaders	Extent of Government Power	Means of Ensuring Obedience	Political Parties
Democracy (Including republic and constitutional monarchy)	Leaders are chosen in free and fair elections.	The government is limited in power by the constitution and laws; citizens' rights and freedoms are protected.	The government relies on the rule of law.	Multiple parties compete for power.
Authoritarian (Including absolute monarchy, dictatorship, and totalitarianism)	Rulers inherit their positions or take power by force.	Rulers have unlimited power; the government may impose an official ideology and control all aspects of political, economic, and civic life.	The government relies on state control of the media, propaganda, military or police power, and terror.	Power lies with a single party.

EXAMINE THE TABLE

1. **Analyzing** What kind of government can a monarchy be?
2. **Inferring** Sometimes authoritarian governments stage elections. Do you think they are usually free and fair? Explain.

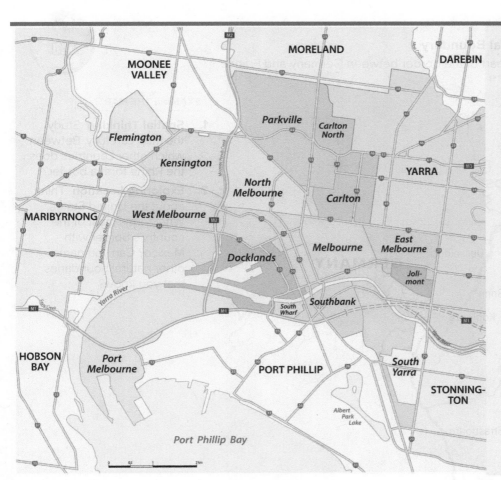

Not only do political boundaries exist between countries, but they also occur at a more local level within cities and neighborhoods. This map of Melbourne, Australia, shows the administrative boundaries of the city's neighborhoods. In some cases, streets are used to mark the boundary between neighborhoods.

Geography and Government

GUIDING QUESTION

How does geography influence a country's government?

Governments can be greatly influenced by geography. Boundaries indicate where the sovereignty of one country ends and another begins. A government must take into account the cultural and religious beliefs of its citizens in order to govern effectively. In autocracies, governments frequently suppress cultural and religious groups to maintain order and power. In democracies, governments usually consider cultural and religious beliefs so as to protect their people's freedoms and ensure their well-being.

Geography influences governments as they develop policy to provide people with goods and services. For example, governments must know where their citizens are moving, why they are moving there, and how that affects their relationship with the environment. Physical infrastructures, such as roads, bridges, and power plants, must be built based on the geographic distribution of people. Governments help organize this activity by using both current demographic data and future projections.

Geographic factors influence the development of boundaries that define the various political units such as countries, states, and provinces. A **natural boundary** follows physical geographic features such as mountains and rivers. For example, the Mississippi River forms the boundaries between several U.S. states. Natural boundaries are relatively easy to identify, and they are often more defensible than other types of boundaries.

Other boundaries develop to separate areas that have cultural differences, such as places with different religions or languages. These **cultural boundaries** geographically divide two identifiable cultures. For example, when Britain partitioned the Indian subcontinent in 1947, it created the countries of India and Pakistan using a cultural boundary based on religion. Muslims were reorganized into Pakistan and Hindus into India.

natural boundary a boundary created by a physical feature, such as a mountain, river, or strait
cultural boundary a geographical boundary between two different cultures

The Rhine River: A Natural Boundary

Europe's Rhine River forms part of the border between Germany and France.

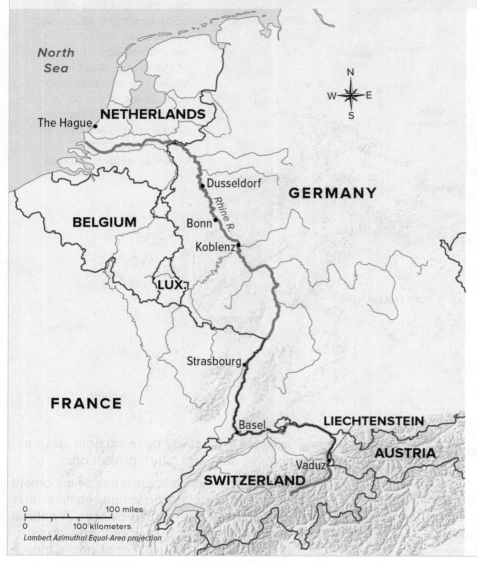

North Sea

NETHERLANDS

The Hague

GERMANY

Dusseldorf

BELGIUM

Bonn

Rhine R.

Koblenz

LUX.

Strasbourg

FRANCE

Basel

LIECHTENSTEIN

AUSTRIA

Vaduz

SWITZERLAND

0   100 miles

0   100 kilometers

Lambert Azimuthal Equal-Area projection

EXAMINE THE MAP

1. **Spatial Thinking** Study the map carefully. Between which other countries does the Rhine form a border?

2. **Exploring Location** Think about the United States. What forms most of the country's border with Mexico? Can you name other natural boundaries that you know?

At other times, culture and natural landforms are not considered when boundaries are drawn. In these cases, treaties might create geometric boundaries to separate countries. For example, European rulers in the 1800s knew little about Africa's political and social systems, and no Africans participated at the Berlin Conference in which boundaries in Africa were established. As a result, Europeans created colonial boundaries that often cut across cultural, religious, or traditional boundaries. **Geometric boundaries** often follow straight lines and do not account for natural and cultural features.

geometric boundary a boundary that follows a geometric pattern

✓ **CHECK FOR UNDERSTANDING**

Summarizing What are the three types of political boundaries?

LESSON ACTIVITIES

1. **Informative/Explanatory Writing** Write an essay explaining the connections between dictatorships and countries in which human rights might not be promoted and protected.

2. **Presenting** Working with a group, research to find additional examples of an autocracy or an oligarchy. Create a presentation explaining how this type of government works in general and how it can appear in the world in different ways.

03

Global Issues: Political Borders

? COMPELLING QUESTION

How are political borders defined?

Plan Your Inquiry

DEVELOPING QUESTIONS

Think about how political borders organize the world into distinct countries and territories, and then further divides those countries and territories into provinces, states, cities, and other types of smaller areas. Reflect on what you know about how political borders are established and how they are represented on the land. Then read the Compelling Question for this lesson. What questions can you ask to help you answer this Compelling Question? Create a graphic organizer like the one below. Write these Supporting Questions in your graphic organizer.

Supporting Questions	Source	What this source tells me about how political borders are defined	Questions the source leaves unanswered
	A		
	B		
	C		
	D		
	E		

ANALYZING SOURCES

Next, examine the sources in this lesson. Analyze each source by answering the questions that follow it. How does each source help you answer each Supporting Question you created? What questions do you still have? Write these in your graphic organizer.

After you analyze the sources, you will
- use the evidence from the sources;
- communicate your conclusions;
- take informed action.

Background Information

A border is a type of boundary that separates political areas. There are many kinds of political borders. Political borders determine the territory where a government has authority. Countries are separated from each other by borders. Maintaining its borders is essential to a country's **sovereignty**. Countries can also have internal political borders that divide the country into separate states, provinces, or other types of territories for the purpose of local governance. For example, the country of Lebanon is divided into nine areas called governorates. The governorates are further divided into 25 districts, and borders are also in place for over 1,000 municipalities (cities and towns) that elect their own local governments. Political borders can impact where people settle and how they interact, or do not interact, with each other.

sovereignty the supreme and absolute authority within territorial boundaries

» Indian soldiers participate in a daily ceremonial closing of the border between India and Pakistan at Wagah Attari.

Jens Benninghofen/Alamy Stock Photo

Borders in Africa

In 1884 a conference led by German Chancellor Otto von Bismarck opened in Berlin. Over the next three months, representatives of 12 European countries, plus the United States and the Ottoman Empire, negotiated the rules for dividing Africa. The European leaders drew borders throughout Africa so that they could establish colonies. These colonies were sources of wealth for the European countries because they provided raw materials for industries. Many of the European borders that were drawn at the Berlin Conference became the borders of today's independent African countries. Africans were not invited to the conference. Ethnic differences between African peoples were also ignored. Many different ethnic groups with no previous association were grouped together within the newly created colonial borders. Other groups were separated by those borders. This 1884 French cartoon illustrates the Berlin Conference, showing European leaders slicing the cake that represents Africa.

PRIMARY SOURCE: CARTOON

EXAMINE THE SOURCE

1. **Comparing and Contrasting** Who was involved in establishing the political borders of colonies in Africa?

2. **Identifying Cause and Effect** How did the boundaries that were created in Africa impact people in Africa?

Water Conflict

Water is the most important natural resource. People depend on reliable supplies of fresh water for their survival. When one country depends on a source of water that has its origins in another country, there is the potential for conflict. One such country is Pakistan, which depends on the Indus River to irrigate over half of its farmland.

PRIMARY SOURCE: ARTICLE

❝ [In 1951,] disagreements over the flow of the Indus River led to the danger of military conflict. . . . India's control of the **headwaters** of the Indus gave it the power to turn fertile Pakistan into a desert . . . [T]he threat of a "water" war became imminent.

. . . [N]egotiations led to the Indus Waters Treaty, signed in 1960. . . . The treaty gives India exclusive use of all of the waters of the Eastern Rivers and their **tributaries** before the points where the rivers enter Pakistan. Similarly, Pakistan has exclusive use of the Western Rivers. . . .

Continuing tensions between India and Pakistan, however, have begun to erode confidence in the treaty.[12] India has greater bargaining power over Pakistan because the upper stream is in its territory, and Pakistan's water rights under the treaty are hard to enforce. . . .

India's plan to build a dam on a tributary of the Indus within its borders has created new tensions. . . . Ongoing political confrontation between India and Pakistan makes the future of the Indus Waters Treaty uncertain.[15] . . . UN Deputy Secretary General Asha-Rose Migiro said, "Projections are that by the year 2050, water accessibility for human consumption will have dropped by 40 percent. Probably the next major conflict will be about water. ❞

— "Pakistan and Water: New Pressures on Global Security and Human Health" by Gregory Pappas, *American Journal of Public Health*, 2011

headwater the source of a stream or river

tributary a stream or river that feeds into a larger stream or river

---

EXAMINE THE SOURCE

1. **Analyzing Perspectives** Why has Pakistan felt threatened by India's control of the headwaters of the Indus River?

2. **Identifying Cause and Effect** What current factors are potential causes of future conflict between India and Pakistan?

Establishment of a Reservation for the Navajo Nation

When the Mexican-American War ended in 1848, the boundary between the two countries shifted. Mexico gave up a large amount of land to the United States. This land included the homeland of the Navajo people. The U.S. government initially removed the Navajo from their lands, as they had done with other groups to the east. Like other Native American groups, the Navajo people have deep connections to their land. Many actively resisted their removal and showed that they were prepared to use force to protect their land. This excerpt is from the treaty negotiated by Navajo leader Barboncito and U.S. General William Tecumseh Sherman.

PRIMARY SOURCE: TREATY DOCUMENT

66 ARTICLE I.

From this day forward all war between the parties to this agreement shall for ever cease. The government of the United States desires peace, and its honor is hereby pledged to keep it. The Indians desire peace, and they now pledge their honor to keep it. . . .

ARTICLE II.

The United States agrees that the following district of country, to wit: bounded on the north by the 37th degree of north latitude, south by an east and west line passing through the site of old Fort Defiance, in Canon Bonito, east by the paralleled of longitude which, if prolonged south, would pass through old Fort Lyon . . . and west by a paralleled of longitude about 109 degree 30' west of Greenwich, provided it embraces the outlet of the Canon-de-Chilly, which canon is to be all included in this reservation, shall be . . . set apart for the use and occupation of the Navajo tribe of Indians, and for such other friendly tribes or individual Indians as from time to time they may be willing, with the consent of the United States, to admit among them; and the United states agrees that no persons except those herein so authorized to do, and except such officers, soldiers, agents, and employees of the government, or of the Indians... shall ever be permitted to pass over, settle upon, or reside in, the territory described in this article. . . .

ARTICLE XII

The tribe herein named, by their representatives, parties to this treaty, agree to make the reservation herein described their permanent Home. 99

— Navajo Nation Treaty of 1868, treaty between the Navajo and the United States

EXAMINE THE SOURCE

1. **Identifying** How did the treaty define the boundaries of the reservation?
2. **Evaluating** How does the treaty limit access to the territory defined as the Navajo Reservation?

A Shifting Border

Physical features, such as rivers and mountain ridges, provide natural barriers that can be used as political borders. These features can change over time, however. For example, while a river can provide a logical basis for a border, rivers can overflow their banks and change course. Italy's border with Austria follows mountain ridges. This article discusses how this border is also facing changes.

PRIMARY SOURCE: ARTICLE

" European borders have proved over history to be wiggly things, and few have wiggled more that Italy's.

Only fully unified as a state in 1870, the new country saw its border with Austria redrawn many times during the First World War. . . . Over their history, Italians have become familiar with negotiating the complexities of frontier geography. Recently, however, they've been faced with an entirely new quandary, one that many countries with mountainous or riverine territory may well soon face: Their border is melting.

Italy's northern land border passes through high alpine landscapes, a land of rock, ice, and seasonal snow. This frontier inhabits a portion of Europe that has been altered by climate change perhaps more than any other. Since 1850, its glaciers have shrunk by 50 percent, and the pace . . . tripled between 1970 and 2000. As ice disappears, the watershed that marks the divide between Italy and Austria has shifted . . . by as much as 100 meters. . . . This leaves both states with a fascinating and unusual problem: how do you fix a border based on a natural barrier that is in constant flux?

You can't. Instead, the Italian government has approved the idea of a movable border, on its icier frontiers at least. Following agreements with Austria in 2008 and Switzerland in 2009, Italy has agreed that its glacial borders can shift depending on the location of the watershed and how it is affected by ice melt. In one year, the nation's territory might expand; in another it might contract slightly. "

— "When Borders Melt," *Bloomberg CityLab*, 2017

EXAMINE THE SOURCE

1. **Summarizing** Why does the location of the border between Austria and Italy need to be clarified?
2. **Inferring** Why is Italy willing to accept the movement of its alpine borders?

Feargus O'Sullivan. Bloomberg. "When Borders Melt". 2017.

Complicated Borders

Typically, we think of a border as a single continuous line that separates two distinct territories. The situation is not always that clear, however. The town of Baarle-Nassau is located on the border between Belgium and the Netherlands. It functions as a normal town in most ways, but it is also unusual. When the two countries established a border in 1843, they did not address the existence of over 5,700 individual parcels of land. Some of those parcels traced their ownership back hundreds of years. Those tiny parcels of land made up a handful of Belgian **enclaves** entirely within Dutch (Netherlands) territory. Today, there are 22 Belgian enclaves in the Netherlands. Within some of them, there are several Dutch enclaves, creating a confusing patchwork of borders that often divides houses and other buildings. The photo shows one such border marked with white paint that leads directly into a doorway.

PRIMARY SOURCE: PHOTO

White crosses painted on the ground are all that marks the location of a portion of the international border between the Netherlands (left) and Belgium (right).

enclave a piece of territory wholly enclosed within another territory

EXAMINE THE SOURCE

1. **Explaining** How did the boundary between the two countries discussed in the source become so complicated?

2. **Inferring** What does this story suggest about relations between the countries of Belgium and the Netherlands since the 1830s?

Complete Your Inquiry

EVALUATE SOURCES AND USE EVIDENCE

Refer back to the Compelling Question and the Supporting Questions you developed at the beginning of the lesson.

1. **Evaluating** Based on the information in the sources, to what extent do modern political borders reflect historical boundaries?

2. **Contrasting** Which sources give examples of how borders are defined on the land, and how are the examples different?

3. **Gathering Sources** Which sources helped you answer the Supporting Questions and the Compelling Question? Which sources, if any, challenged what you thought you knew when you first created your Supporting Questions? What information do you still need in order to answer your questions? What other viewpoints would you like to investigate? Where would you find that information?

4. **Evaluating Sources** Identify the sources that helped answer your Supporting Questions. How reliable is each source? How would you verify the reliability of the source?

COMMUNICATE CONCLUSIONS

5. **Collaborating** Share with a partner which of these sources you found most useful in answering your Supporting Questions. Discuss how the sources help you understand how political borders are defined. Take turns relating how the information you gathered from the sources aligned with or challenged what you previously knew about borders. How do these sources advance your thinking? Use the graphic organizer that you created at the beginning of the lesson to help you. Share your conclusions with the class.

TAKE INFORMED ACTION

Defining Local Borders Think about the political borders that define your school's local area. The school may be located within the borders of a town or city, a county, and a school district. Research online to learn when and how those borders were established. Create a map or poster for your school that shows as many local political borders as possible.

Conflict and Cooperation

Analyzing Key Ideas and Details As you read, take notes about the causes of conflict within and among countries.

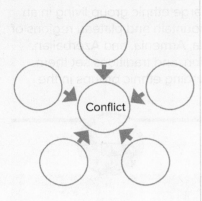

Causes of Conflict

GUIDING QUESTION

How does conflict shape the division of Earth's surface?

Conflict around the world is **induced** by many factors, including border disputes, tensions over territories, multiple ethnic groups within one country, competition for resources, and control of strategic sites.

Political boundaries are referred to as borders, and they are not permanent. Many areas of the world have had changing borders as the result of wars and **territorial** disputes. Border disputes arise from unsettled territorial claims or as a result of one country desiring the resources of another. In February 1848, Mexico and the United States signed a **treaty** ending the war between them. The treaty gave large portions of territory that belonged to Mexico, including present-day California, to the United States. Several days earlier, gold had been discovered near Sacramento, the present-day capital city of California. The treaty assured that the impending Gold Rush occurred within the territory of the United States, which may have also sped up the process of California becoming a state.

induce to cause

territorial of or relating to land belonging to a government

treaty an official agreement, negotiated and signed by each party

HOW WE GROW.

The Treaty of Guadalupe Hidalgo was signed in 1848 to end the war between the United States and Mexico. It gave the United States more than 525,000 square miles (1,360,000 sq. km) of territory—land including what are now the states of California, Nevada, and Utah, as well as most of Arizona and New Mexico and parts of Colorado and Wyoming. Mexico accepted the Rio Grande as its border with Texas.

Other sources of conflict include citizenship practices, public **policy**, and decision making that are influenced by cultural beliefs such as nationalism and patriotism. **Nationalism** is a belief in the right of each people with a unique cultural identity based on common language, religion, and national symbols to be independent. *Patriotism* refers to the love for a country, with an emphasis on values and beliefs. The idea of patriotism carries with it the importance of action in addition to loyalty.

policy an overall plan that establishes goals and determines procedures, decisions, and actions

nationalism belief in the right of each people with a unique cultural identity based on common language, religion, and national symbols to be independent

Pakistanis, Indians, and Kashmiris today all have different views about who should govern Kashmir. For centuries, Kashmir, in the northern areas of Pakistan and India, was part of the Indian kingdoms ruled by *maharajas*, or princes. Conflict began in 1947 when India and Pakistan became independent countries. The maharaja of Kashmir wanted Kashmir to become an independent country that was aligned with India. However, many Pakistanis believed that Kashmir should become part of Pakistan because it included many Muslims. The result has been decades of war and periodic fighting between the two countries.

The Kurds are a large ethnic group living in an area that straddles mountain and plateau regions of Turkey, Iran, Iraq, Syria, Armenia, and Azerbaijan. Their language, religion, and traditions set them apart from the surrounding ethnic groups in the

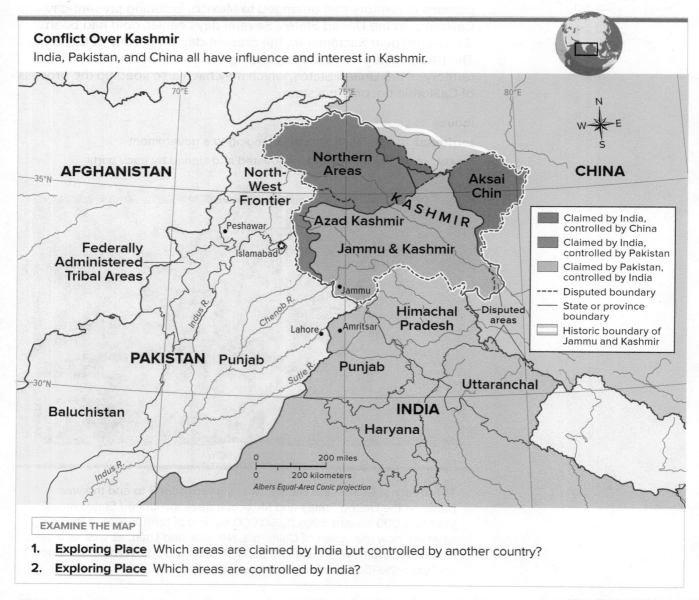

Conflict Over Kashmir

India, Pakistan, and China all have influence and interest in Kashmir.

EXAMINE THE MAP

1. **Exploring Place** Which areas are claimed by India but controlled by another country?
2. **Exploring Place** Which areas are controlled by India?

Refugees wait to get into a water station in a camp in the Darfur region of Sudan. Such camps, watched by armed police, have become home to people forced to flee their villages to escape violence by the Arab militias.

Summarizing What is the cause of the conflict in Darfur?

region. These cultural differences led to a rise in Kurdish nationalism in the early twentieth century. This in turn resulted in many conflicts as the Kurds pursued a self-governing state. In 2005 the Kurdish nationalist movement succeeded in forming a parliament and establishing autonomous civil authority over a portion of northern Iraq. This authority is supported by public policy and upheld by the 2005 Iraqi constitution. Kurdish populations in Syria, Turkey, and Iran continue to push for more political autonomy.

Cultural differences led to conflicts in Sudan in Africa. Arabic-speaking Muslims live mostly in Sudan's northern cities. People in the south, a majority of whom are Christian, live mostly in rural areas and are focused on a subsistence economy. These differences led to a conflict in which nearly 300,000 people died. An estimated 2.7 million people were displaced between 1983 and 2005. A peace agreement was signed in 2005, which provided significant independence for the people of Sudan's southern provinces. South Sudan became a sovereign, independent country in 2011.

Sudan and South Sudan have a long-running conflict over natural resources. In January 2011, South Sudan shut down all of its oil fields after a disagreement about the fees Sudan demanded to transport the oil through its territory. In May 2011, Sudan seized control of Abyei, a disputed oil-rich border region, after three days of clashes with South Sudanese forces. On September 27, 2011, the presidents of Sudan and South Sudan signed an agreement of cooperation. The status of Abyei, however, was not addressed. The future of the region is still uncertain.

Sudan is also experiencing conflict in its western region of Darfur. Beginning in 2003, a civil war broke out as Arab militias attacked African ethnic groups with the support of the Arab-led government. This **genocide** resulted in the burning of entire villages and the killing of more than 400,000 people. The ongoing conflict has created a growing number of displaced people within Sudan as well as **refugees** who have fled to the neighboring country of Chad.

There is also ongoing conflict in western China. The Uyghurs, a Muslim ethnic group that has lived in the region for centuries, has been targeted by the Chinese government with mass imprisonment, efforts to force women to have fewer children, the sending of children to Chinese boarding schools, and attempts to replace Uyghur culture with that of mainstream China. The U.S. government has labeled the conflict a genocide.

In 2014, Victor Yanukovych, the pro-Russian president of Ukraine, was removed from office after months of protests. Putin then claimed that Ukraine had been taken over by extremists.

genocide the intentional destruction of a group of people that can include mass murder, imposing harsh conditions, and forcibly removing children

refugee a person who flees a country because of violence, war, persecution, or disaster

PA Images/Alamy Stock Photo

One of the agencies of the United Nations is the UN High Commission on Refugees (UNHCR). It is dedicated to saving lives and protecting the rights of refugees and displaced people. In this photo, a UNHCR employee is part of a convoy carrying supplies to residents of an area in Ukraine damaged by shelling attacks by Russian forces in 2014.

A separatist rebellion against Ukrainian forces began in the eastern region that killed 14,000 people. At the same time, Russia seized a southern part of Ukraine called Crimea. Since then, Ukraine has attempted to strengthen its bonds with the European Union (EU) and the North Atlantic Treaty Organization (NATO).

In February 2022, Russian president Vladimir Putin launched a full-scale invasion of Ukraine from the north, east, and south. Putin claimed that this invasion was necessary to protect the Russian people from genocide and bullying by Ukraine. He also accused Ukraine of being governed by Nazis. There is no evidence to support Putin's accusations, and Ukraine's president, Volodymr Zelensky, is Jewish. Putin demanded that Ukraine never join NATO, and he insisted that Russia and Ukraine were really one country.

The two countries do have a connection—Ukraine was part of the Soviet Union throughout much of the twentieth century. When the Soviet Union dissolved in 1991, Ukraine declared its independence. It has been recognized as an independent country since then. Today, Russians are the largest ethnic minority in Ukraine and Russian is spoken by about 30 percent of the Ukrainian population.

The Ukrainian response to the 2022 invasion was to call up men aged 18 to 60 years old and to arm its citizens. President Zelensky joined his troops on the front lines. Ukrainian troops and armed civilians provided tough resistance against the Russian military. These events marked the first time that a major power invaded a neighboring

European country since World War II. Millions of refugees crossed the borders into Poland, Hungary, Romania, Moldova, and Slovakia.

Terrorism is also a type of political conflict. Terrorism inspires fear. It includes any violent and destructive act that is done to intimidate a people or a government. Terrorist attacks are usually carried out in such a way as to maximize the severity and length of the psychological impact. Terrorist acts can be supported by the government. They are planned to have an impact on many large audiences.

Terrorists also attack national symbols to show power and to attempt to shake the foundation of the country or society to which they are opposed. For example, there was a series of terrorist attacks on September 11, 2001, at the World Trade Center in New York City, the Pentagon near Washington, D.C., and in the sky over western Pennsylvania. In 2012 there was an attack on the U.S. embassy in Libya. Terrorist acts frequently have a political purpose. The groups who commit such acts desire change so badly that failure to achieve change is seen as a worse outcome than the deaths of civilians.

Terrorism can be influenced by geographic factors, as in the Arab-Israeli conflict in which many lives have been lost. That conflict has resulted from years of differences over cultural, religious, economic, and geographic issues. At the heart of the conflict lie the geographic areas of the West Bank, the Gaza Strip, and East Jerusalem.

terrorism violence committed in order to frighten people or governments into granting demands

Israelis and Palestinians have long disputed control of these lands. These differences have polarized Arabs and Israelis for over 60 years, resulting in ongoing violent conflicts in the region.

✓ CHECK FOR UNDERSTANDING

Identifying Cause and Effect How can nationalism lead to conflict within or between countries?

Alliances and Cooperation

GUIDING QUESTION

How does cooperation shape the division of Earth's surface?

Alliances and cooperation can also be explored from a geographic perspective. Treaties serve many functions. One of these functions is how countries establish ways to share and protect resources. For example, much of the acid rain in Canada comes from air pollution in the United States. As a result, in 1991 the two countries signed

a cooperative agreement. Known as the Canada-United States Air Quality Agreement, this addressed air pollution leading to acid rain.

Another instance of cooperation between countries is the Great Lakes Water Quality Agreement. Originally signed in 1972, the agreement is a commitment between the United States and Canada to restore and protect the waters of the Great Lakes. The agreement has been amended over time to include terms to address habitat degradation and climate change.

In addition to environmental concerns, countries create regional agreements to address economic issues and to strengthen economic ties. In 1992 Canada, the United States, and Mexico signed the North American Free Trade Agreement (NAFTA). This agreement reduced or eliminated tariffs on many goods traded among the three countries. NAFTA was renegotiated in 2020 and is now known as the United States-Mexico-Canada Agreement (USMCA).

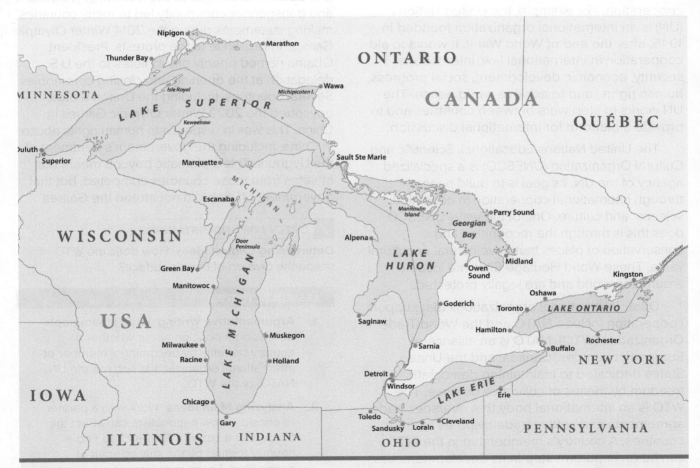

The Great Lakes region is important to both the United States and Canada. As a result, the two countries work together to protect the region's environment.

Identifying What is the purpose of the Great Lakes Water Quality Agreement?

The EU unites much of Europe into one trading community, or bloc. It enjoys one of the largest volumes of trade in the world. The EU has achieved its large trade activity by eliminating tariffs and other trade barriers among member countries. The EU has also formed favorable trade relationships with other countries.

Cooperation among countries also occurs on a larger scale. Since its discovery in 1820, Antarctica has fascinated explorers and scientists. As interest in Antarctica grew, countries began to lay claim to sections of the continent. In 1959, 12 countries negotiated the Antarctic Treaty. This agreement is meant to preserve Antarctica for scientific research and to put all territorial claims on hold. Since then, other countries have established research programs in Antarctica. Today, there are around 70 research stations operated by 29 countries that serve as bases where scientists study physical geography, climate, and wildlife.

International organizations help countries work together to resolve conflict and foster cooperation. For example, the United Nations (UN) is an international organization founded in 1945, after the end of World War II. It works to aid cooperation in international law, international security, economic development, social progress, human rights, and to achieve world peace. The UN works to stop wars between countries and to provide a platform for international discussion.

The United Nations Educational, Scientific and Cultural Organization (UNESCO) is a specialized agency of the UN. Its goal is to build peace through international cooperation in education, science, and culture. One of the ways UNESCO does this is through the recognition and conservation of places that have natural or cultural value. These World Heritage Sites are located around the world and are legally protected.

Other international organizations that support cooperation include NATO and the World Trade Organization (WTO). NATO is an alliance of European countries, Canada, and the United States dedicated to maintaining democratic freedom by means of collective defense. The WTO is an international body that oversees trade agreements and settles trade disputes among countries. A country's membership in the WTO can be an important step in its development. Less developed countries strive to become members. In 2000, hoping that trade might open China to democratic change, the United States granted full trading privileges to China and supported its

entrance into the WTO. The following year China was admitted to the WTO.

Nationalism and patriotism can also result in cooperation. The Olympic Games bring together athletes from hundreds of countries. Participants and spectators alike show their devotion to their countries. The games have also historically highlighted public policy and decision-making by countries working cooperatively. For example, in 1963 the International Olympic Committee (IOC) had broad international support when it banned South Africa from participating in the Games due to its policy of apartheid. The ban remained in effect until 1992. In 1980 the United States led approximately 60 countries in boycotting the Summer Olympic Games in Moscow. This act was in response to the Soviet Union's invasion of Afghanistan. Saudi Arabia, under pressure from the IOC, included women among its athletes for the first time in the 2012 Summer Olympic Games.

Human rights groups protesting Russian laws and discrimination against the lesbian, gay, bisexual, and transgender community led to many countries making statements during the 2014 Winter Olympic Games. In response to the protests, President Obama named openly gay athletes to the U.S. delegation at the opening and closing ceremonies. Several countries, including the United States, boycotted the 2022 Winter Olympic Games in China. This was in response to human rights abuses in China, including the government's treatment of the Uyghurs. This diplomatic boycott meant that athletes from these countries competed, but that government officials did not attend the Games.

✓ CHECK FOR UNDERSTANDING

Determining Central Ideas How does the WTO shape the division of Earth's surface?

LESSON ACTIVITIES

1. **Argumentative Writing** Write a paragraph defending a position about whether a country benefits by becoming a member of international organizations such as the UN, NATO, or the WTO.

2. **Analyzing Main Ideas** Work with a partner to consider how nationalism can affect the borders of a country. Research to find a country formed by the shared spirit of nationalism. Learn about issues relating to the geographical borders of that country. Use your findings to create a report that includes times lines and maps.

05

Analyzing Sources: Conflict and Cooperation

? COMPELLING QUESTION

What causes conflict or cooperation between countries?

Plan Your Inquiry

DEVELOPING QUESTIONS

Think about how countries form the basis of a world community. Consider some of the factors that could motivate countries to enter into conflict or, on the other hand, to cooperate with each other. Then read the Compelling Question for this lesson. What questions can you ask to help you answer this Compelling Question? Create a graphic organizer like the one below. Write these Supporting Questions in your graphic organizer.

Supporting Questions	Source	What this source tells me about what causes conflict or cooperation between countries	Questions the source leaves unanswered
	A		
	B		
	C		
	D		
	E		

ANALYZING SOURCES

Next, examine the sources in this lesson. Analyze each source by answering the questions that follow it. How does each source help you answer each Supporting Question you created? What questions do you still have? Write these in your graphic organizer.

After you analyze the sources, you will
- use the evidence from the sources;
- communicate your conclusions;
- take informed action.

Background Information

Every country in the world has a territory and a government with the authority to represent the interests of its citizens. Those interests include safety, security, and well-being. Maintaining control over a territory and its resources is essential for a country to continue to exist. All of the world's land, except Antarctica, is divided among countries. This means countries often need to interact with each other to obtain resources that they don't have. They also depend on each other to allow for the free movement of their citizens among their territories. Countries can interact in ways that are beneficial to both sides. Countries may also become involved in conflicts with each other for a variety of reasons.

» The United Nations was founded in 1945 as a vehicle to achieve international cooperation, peace, and increased well-being.

The United States and the Soviet Union

The Soviet Union was a country created in the early 1920s after a successful revolution in Russia and its unification with several other countries in Eastern Europe and Central Asia. Different political philosophies meant the United States and the Soviet Union did not have friendly relations from the beginning. The two countries became bitter rivals after the end of World War II. Tensions remained high until the Soviet Union began to break apart in the late 1980s. The website of the U.S. Department of State includes information about a brief period in history when U.S.-Soviet relations were different.

PRIMARY SOURCE: WEBSITE

❝ Although relations between the Soviet Union and the United States had been strained in the years before World War II, the U.S.-Soviet alliance of 1941–1945 was marked by a great degree of cooperation and was essential to securing the defeat of Nazi Germany. . . .

As late as 1939, it seemed highly improbable that the United States and the Soviet Union would forge an alliance. U.S.-Soviet relations had soured significantly following Stalin's decision to sign a non-aggression pact with Nazi Germany in August of 1939. The Soviet occupation of eastern Poland in September and the "Winter War" against Finland in December led President Franklin Roosevelt to condemn the Soviet Union publicly as a "dictatorship as absolute as any other dictatorship in the world," and to impose a "moral embargo" on the export of certain products to the Soviets. Nevertheless . . . Roosevelt never lost sight of the fact that Nazi Germany, not the Soviet Union, posed the greatest threat to world peace. . . .

Following the Nazi defeat of France in June of 1940, Roosevelt grew wary of the increasing aggression of the Germans and made some diplomatic moves to improve relations with the Soviets. . . .

The most important factor in swaying the Soviets eventually to enter into an alliance with the United States was the Nazi decision to launch its invasion of the Soviet Union in June 1941. ❞

— "U.S.-Soviet Alliance, 1941–1945," U.S. Department of State website

This poster created by the U.S. Department of Defense reflected a change in the relationship with the Soviet Union as the United States entered World War II.

This man is your FRIEND

Russian

He fights for FREEDOM

EXAMINE THE SOURCE

1. **Determining Context** Why did the United States have a bad relationship with the Soviet Union before 1940?

2. **Analyzing Perspectives** What caused the United States and the Soviet Union to change their attitudes and enter into an alliance?

Greece, Turkey, and Cyprus

Like many countries in the world, Cyprus is home to people of different ethnicities. The two largest groups are people of Greek and Turkish ancestries. While Cyprus is an independent country, these two groups have strong ties to the countries of Greece and Turkey. The island of Cyprus has a complex history and was conquered and claimed by a series of empires. Cyprus had most recently been a British colony before becoming independent in 1960. This article excerpt provides information about how Cyprus transitioned from British colony to independence.

PRIMARY SOURCE: ARTICLE

66 In 1878 the island of Cyprus in the Eastern Mediterranean Sea, came under British control. Its population is made up of both Greek and Turkish Cypriots. The Greek Cypriot majority desired the removal of British rule and union with Greece, known as Enosis. In 1955, the campaign for Enosis was led by Archbishop Makarios of the Cyprus Orthodox Church and by Colonel George Grivas, Head of [the] National Organisation of Cypriot Fighters (EOKA). They aimed to achieve Enosis by attacking government and military installations and personnel and by mobilising the civilian population to demonstrate against the British presence. . . .

After invading in 1974, Turkey's military remained present in the self-declared Turkish Republic of Northern Cyprus, whose independence is only recognized by Turkey.

EOKA launched its campaign on 1 April 1955 with a series of bombing attacks against government offices in the island's capital Nicosia. . . . The British Government started the process of looking for a political solution. Large numbers of British reinforcements arrived and began a series of operations against EOKA centred on the Troodos Mountains.

In August 1960 Cyprus became a republic but, in the following decades, it was plagued with violence between the Greek and Turkish communities. In 1974 a Greek military coup, which aimed to unite the island with mainland Greece, led to a Turkish invasion and the division of the island between Turkish Northern Cyprus and the Greek Cypriot Republic of Cyprus. Cyprus remains divided to this day. 99

— "What Caused the Division of the Island of Cyprus?" Imperial War Museums (UK) website

EXAMINE THE SOURCE

1. **Summarizing** Which countries have been involved in conflict over the island of Cyprus?
2. **Analyzing Perspectives** Why did conflict occur in Cyprus?

European Integration and the Schengen Agreement

Europe is a densely populated continent and home to some of the world's oldest countries. Centuries of competition and repeated conflict began with the fall of the Roman Empire in 476 A.D. and continued through the 1900s. After World War II, many countries in Europe wanted to become more **integrated** to promote economic prosperity and reduce competition and conflict. This website describes one attempt to move toward integration: the Schengen Area, created in 1995.

PRIMARY SOURCE: WEBSITE

> ❝ Schengen Area signifies a zone where 26 European countries abolished their internal borders, for the free and unrestricted movement of people, in harmony with common rules for controlling external borders and fighting criminality by strengthening the common judicial system and police cooperation.
>
> Schengen Area covers most of the EU countries, except Ireland and the countries that are soon to be part of Romania, Bulgaria, Croatia and Cyprus. Although not members of the EU, countries like Norway, Iceland, Switzerland and Lichtenstein are also part of the Schengen zone. . . .
>
> The external borders of the Schengen Zone reach a distance of 50,000 km long, where 80% of it is comprised of water and 20% of the land. The area counts hundreds of airports and maritime ports, many land crossing points, an area of 4,312,099 km², and a population of 419,392,429 citizens. ❞

— "Schengen Area – The World's Largest Visa Free Zone," SchengenVisaInfo.com, 2021

The Schengen Agreement currently includes 26 countries in Europe, with three more in the process of joining (shown in lighter shade of blue).

---

integrated with functions linked or coordinated

EXAMINE THE SOURCE

1. **Explaining** What is the purpose of the Schengen Agreement?
2. **Speculating** How might participation in the Schengen Agreement affect a country culturally?

China's Growing Role in the Global Economy

As economies around the world have become more connected, the economy of China has experienced rapid growth and increasing influence. The United States enjoyed a position of clear leadership in the global economy for several decades after World War II. However, China has emerged as a competitor for that dominant position. In growing its economy, China has also greatly increased its interactions with other countries all over the world. This article provides information about China's relationships within a particular region.

PRIMARY SOURCE: ARTICLE

> Five years ago, Costa Rica severed diplomatic ties with Taiwan and officially recognized the People's Republic of China. Following the **diplomatic** switch, Beijing negotiated a free trade agreement with the small Central American nation, bought $300 million in Costa Rican bonds, and set up a Confucius Institute in San José. But the most obvious result of the new relationship—Costa Rica National Stadium—opened in March 2011 in the capital, built at an estimated cost of $100 million. Displays of cultural pride from both nations marked the opening ceremonies. The three-tiered, 35,000-seat stadium has two giant television screens, medical facilities, office buildings, a small sports museum, a banquet hall, and an extravagant staircase serving as the southern portal. . . .

Fans fill the new Costa Rican National Stadium in preparation for its opening ceremony in 2011.

> San José Mayor Johnny Araya solicited funding for the project directly from the mayor of Beijing, Guo Jinlong. . . . The mutually beneficial agreement will bring tourism dollars to the site and continue to spread the cultural influence of China throughout the city. . . .
>
> Of the 23 nations that still recognize Taiwan, 11 are found in Latin America. That number continues to dwindle as China makes economic inroads through infrastructure projects and deepening trade relations. Many are catching on to the rewards that come from Chinese friendship. "

— "China's Stadium Diplomacy," Rachel Will, *World Policy Journal*, 2012

diplomatic involving communication among countries to maintain relationships

---

EXAMINE THE SOURCE

1. **Evaluating** How has Costa Rica benefited from its new relationship with China?
2. **Drawing Conclusions** What is the motivation for the Chinese government to spend large amounts of money to build stadiums in other countries?

E

France and the United Kingdom

France and the United Kingdom have had periods of both conflict and cooperation. They fought wars against each other for centuries, but joined to fight against common enemies in World Wars I and II. This article discusses the current state of the relationship between France and the UK regarding their fishing industries.

PRIMARY SOURCE: ARTICLE

❝ PARIS (AP) — France will announce potential sanctions [penalties] over energy prices and trade "by the end of the week" in its fishing dispute with the United Kingdom, the government spokesman said Wednesday.

France **vehemently** protested the decision last month by the U.K. and the Channel Island of Jersey to refuse dozens of French fishing boats a license to operate in their territorial waters. Paris called the move "unacceptable." … "We are obviously in a position to take sanctions if the agreement is not respected," French government spokesperson Gabriel Attal said. "There are several types of sanctions that are possible: energy prices, access to (French) ports, **tariffs** issues."

Jersey, which is only 14 miles (22 kilometers) off the French coast, is a British Crown dependency outside of the U.K. [I]t has its own powers with regard to who is allowed to fish in its territorial waters. ❞

— "France to announce sanctions amid fishing dispute with UK," by Sylvie Corbet, Associated Press News, 2021

vehemently in a forceful or passionate manner
tariff a tax on trade

EXAMINE THE SOURCE

1. **Explaining** What has changed in the relationship between France and the UK?
2. **Predicting** What actions is France considering that could escalate its conflict with the UK?

Complete Your Inquiry

EVALUATE SOURCES AND USE EVIDENCE

Refer back to the Compelling Question and the Supporting Questions you developed at the beginning of the lesson.

1. **Identifying Themes** Based on the information in the sources, what are some of the reasons that countries choose to cooperate with each other?
2. **Making Connections** In what ways do countries engage in conflict?
3. **Gathering Sources** Which sources helped you answer the Supporting Questions and the Compelling Question? Which sources, if any, challenged what you thought you knew when you first created your Supporting Questions? What information do you still need in order to answer your questions? What other viewpoints would you like to investigate? Where would you find that information?
4. **Evaluating Sources** Identify the sources that helped answer your Supporting Questions. How reliable is each source? How would you verify its reliability?

COMMUNICATE CONCLUSIONS

5. **Collaborating** Share with a partner which of these sources you found most useful in answering your Supporting Questions. Discuss how the sources help you understand why countries cooperate with each other and why they have conflicts. Take turns talking about one specific source at a time and try to identify what made the difference between cooperation and conflict. How do these sources support your thinking? Use the graphic organizer that you created at the beginning of the lesson to help you. Share your conclusions with the class.

TAKE INFORMED ACTION

Celebrating Cooperation on the Local Level Conflict and cooperation can happen between countries. They can also happen within a community. Identify an example of cooperation that is helping people in your local community. You might gather information online or by talking with people in your community. Create a social media post promoting cooperation, using the local example that you researched as an example.

Corbet, S., "France to announce sanctions amid fishing dispute with UK" October 20, 2021, AP News

Summary

Political Geography

Political geography is a branch of geography that studies the ways in which geography influences political systems and relationships.

Features of Government

Countries are defined by their territory, their population, and their sovereignty. These elements come together under a government.

- In a unitary system, the national government holds all key powers.
- In a federal system, powers are divided between national and state or provincial governments.
- With democratic governments, leaders rule with the consent of the people. Democratic governments can be representative democracies, republics, and constitutional monarchies.
- With authoritarian governments, the power to rule belongs to a person or small group. Power is inherited or taken by force. Authoritarian governments can be autocracies, absolute monarchies, oligarchies, and theocracies.

- Geography defines the limits of power through boundaries, which designate where sovereignty begins and ends.
- There are three types of political boundaries: natural, cultural, and geometric.
- Geography influences government through the distribution and cultural characteristics of populations, the availability of natural resources, and the strategic locations included in their territories.

Conflict and Cooperation

Conflict and cooperation between countries have contributed to, and resulted from, the political divisions of the world.

- The causes of conflict around the world include border disputes, tensions over territory, multiple ethnic groups with cultural differences within one country, competition for resources, and control of strategic sites.
- Nationalism creates conflict when people have different allegiances to the country. It can also create cooperation when everyone in the group has the feelings and loyalty to the country.
- Terrorism is a type of political conflict.

- Cooperation can take the form of treaties and alliances that outline the ways countries will work together.
- Organizations such as the UN and WTO work to foster cooperation among the world's countries.

Checking For Understanding

Answer the questions to see if you understood the topic content.

1. Identify each of these terms as they relate to political geography.

A. sovereignty   F. autocracy

B. unitary system   G. dictatorship

C. federal system   H. oligarchy

D. democracy   I. theocracy

E. government

REVIEWING KEY FACTS

2. **Explaining** What are the broad functions of government?

3. **Making Connections** What makes the United States a federal system of government?

4. **Identifying** Name two other countries with a federal system of government.

5. **Summarizing** How does a representative democracy work?

6. **Identifying** What are the factors that typically cause conflicts around the world?

7. **Contrasting** How are nationalism and patriotism different?

8. **Making Connections** How is terrorism related to political conflict?

9. **Explaining** What is the main function of the United Nations?

CRITICAL THINKING

10. **Analyzing** How is a republic related to a democracy?

11. **Comparing and Contrasting** How can a monarchy be both a democracy and an authoritarian government?

12. **Making Connections** Why might human rights be an issue for a dictatorship?

13. **Drawing Conclusions** From where do rulers in an oligarchy get their power?

14. **Summarizing** Name and describe the three types of political boundaries.

15. **Determining Central Ideas** How does geography influence a country's government?

16. **Explaining** How is a treaty a form of cooperation between countries?

17. **Identifying** What is the Antarctic Treaty?

18. **Describing** What is the European Union? How is it a form of cooperation?

19. **Summarizing** Describe the cultural differences that led to conflict in Sudan.

20. **Evaluating** How is nationalism important to the Kurds and their desire to be self-governing?

NEED EXTRA HELP?

If You've Missed Question	1	2	3	4	5	6	7	8	9	10
Review Lesson	2	2	2	2	2	4	4	4	4	2

If You've Missed Question	11	12	13	14	15	16	17	18	19	20
Review Lesson	2	2	2	2	2	4	4	4	4	4

Apply What You Have Learned

A Geographic Reasoning

You have learned that there are many reasons for the creation of political borders. Some are tied to geography, such as borders that include a body of water or incorporate desirable natural resources. Some were created in the aftermath of military conflicts, either reflecting conquered territory or from agreements and treaties that ended conflicts. Still others were created by colonizing countries without regard to either geographic or population considerations.

ACTIVITY **Reimagining a Border** Choose a state in the United States and find an online physical map of that state and its surroundings. The map should show features such as mountain ranges, rivers, and forests. On posterboard or in a digital graphic-design program, create a political map of that state with at least one of its borders redrawn to better align with a physical feature of that region. For instance, the new border might follow the path of a river or mark the edge of a desert. Then explain in a short presentation how the redrawn border might affect that state's population, economy, and culture.

B Understanding Multiple Perspectives

The rights and liberties that are enjoyed by people vary widely across the world. Some are granted by constitutions that have been developed over decades or centuries. Others are chosen by a governing body or leader and can change frequently. The reasons for the presence or absence of specific personal rights in a country may be connected to political considerations, religious or cultural traditions, security concerns, or even the whims of a single leader.

ACTIVITY **Taking a Position** Research to learn about a system of government other than democracy in a country today. Then identify one right enjoyed by people in the United States that is not granted to people in the country you researched. Prepare a short speech explaining the point of view of the government or leader in not granting that right to the people in that country. Present that point of view respectfully, considering the reason(s) people might believe that right is not needed or wanted. Have another student counter that speech with an explanation of why people in the United States expect that right to be observed.

Solving Conflicts

Much of the world's history has been shaped by conflict between countries, struggles for power within countries, and other disputes. There are many groups, such as nonprofit organizations and organizations of countries, that continue to try to solve conflicts. Recent history has included examples of successful efforts to end conflicts as well as failed attempts to do the same.

ACTIVITY **Advising a Leader** Research a current conflict that involves a dispute between two or more countries. Create a list of the reasons for the start of that conflict, and another list of reasons why the conflict has not been resolved. Then write a letter of advice to the leader of one of the countries involved in the conflict. Advise the leader on what actions you recommend to end the conflict. Consider what type of action that leader is likely to agree to and note what benefits could be enjoyed by the leader's country if the conflict is resolved.

D Making Connections to Today

The national borders of the United States have remained unchanged for nearly 50 years. The United States gave up territory in the Pacific after World War II. And power over the Panama Canal passed from the United States to Panama in 1979. However, almost every part of the modern United States has at some point been claimed by a different group or country. Some areas have at one time or another been previously controlled by several countries, in addition to control by Native American groups. For example, the current state of Louisiana was at different times claimed as territory by Spain and France. Similarly, both Russia and Great Britain at different times claimed parts of what is now the state of Oregon.

ACTIVITY **Identifying Influences** Choose a location, either in the United States or elsewhere, that has been under the control of more than one country or people in its history. Write a brief essay explaining how that place shows the influence of a previous ruler or occupying group. This may include cultural, economic, architecture and infrastructure, or other characteristics. Include your opinion as to how likely this influence is to endure in future generations in that location.

Scientists monitor the thickness of the ice in the Arctic. Melting of polar ice is an indicator of changes to the global environment.

Human-Environment Interaction

INTRODUCTION LESSON

01 Introducing Human-Environment Interaction   G258

LEARN THE CONCEPTS LESSONS

02 The Environment and Human Settlement   G263

03 Natural Resources and Human Activities   G269

05 Human Impact on the Environment   G279

INQUIRY ACTIVITY LESSONS

04 Analyzing Sources: Human-Environment Interactions Throughout History   G273

06 Global Issues: Adapting to Environmental Change   G285

REVIEW AND APPLY LESSON

07 Reviewing Human-Environment Interaction   G291

01

Introducing Human-Environment Interaction

Our Relationship with the Environment

People and the environment interact—the two are strongly connected. The physical characteristics of a place affect how people live. People impact the environment. These actions can improve life for some, but they can also harm the environment.

How does the environment impact humans?

Physical features, natural resources, and climate affect where people choose to live. Natural disasters can cause destruction of property, injury, and even loss of life.

London, England, began as a settlement established by the Romans. Being on the Thames River meant that ships coming from the Roman Empire could bring goods into London. As a result, the city became an important trading post. Geography—physical features, climate, and natural resources—influence where people choose to settle.

The remains of a house along with power lines, cars, and the rubble of other houses are piled up following an earthquake, which then triggered a devastating tsunami in the coastal city of Ōfunato, Japan. People around the world have to prepare for, and respond to, natural disasters.

PHOTO: (t) Westend61 GmbH/Alamy Stock Photo, (b) MCS 1st Class Matthew M. Bradley/US Navy/DoD

How do humans impact the environment?

Humans take resources from the Earth to feed themselves, to power their cars, and to make a living. Some effects of this are the clearing of forests and the polluting of air, land, and water. These activities have contributed to changes in the global environment, including the rise in temperatures and an increase in the number of severe-weather events like hurricanes.

Natural materials like minerals and water that come from the Earth become resources when people give them value. People use natural resources in their daily lives. This can sometimes result in damage to the natural environment.

> **"** The world's leading scientists and governments have stated flatly, "Warming of the climate system is unequivocal" [clear] and a "settled fact." They have such a high degree of certainty the climate is warming because of the vast and growing amount of evidence pointing to such a conclusion. **"**
>
> —Joseph Romm, *Climate Change: What Everyone Needs to Know*

Getting Ready to Learn About . . . Human-Environment Interaction

Impacting Humans

Humans and the environment each influence the other in different ways. The environment—which includes physical features, climate, and natural resources—affects where people choose to settle and make their homes. People may consider the type of soil, average temperatures, and amount of rainfall if they are involved in agriculture. Other people may decide to settle near a river or sea coast to be close to water for transportation and trade. The location of natural resources like coal, iron ore, and petroleum can determine where some industrial activities are located.

The environment can also present obstacles to which human have to adapt. Since one of the basic human needs is water, people have learned to use tools and technology to get the water they need to live. In some cases, there is too much water and people have found ways to control the water so that they can live in certain areas. Another way in which humans have adapted to the environment is with the farming techniques they use. For example, farmers in mountainous areas may carve out flat areas to plant their crops. Or they may cut down trees and brush to clear areas for farming.

Extreme events in nature, such as earthquakes, volcanic eruptions, wildfires, and hurricanes can also influence human populations, Such events can cause destruction of property and the loss of life. In some instances, humans can plan and monitor for these extreme events with the help of technology.

People today are more aware of the changes occurring in the global environment. As they come to understand the impact that human activities have on these changes, many people work to educate and inform others.

Humans use materials from the Earth such as water, air, soil, minerals, plants, and animals to meet their needs and wants. Once used, some of these materials can be built back up in a relatively short period of time. Water and energy from the sun are examples. Other natural materials from the Earth cannot be replaced once they are used. Copper and coal are examples of these types of materials. An important goal for humans is to use these materials from the Earth responsibly so that they are available to support the needs of people today as well as the needs of people in the future.

Impacting the Environment

Just as the natural environment impacts where and how humans live, human activities also impact the environment. As people settle and build their communities, they change the natural landscape. Towns and cities are sometimes built in areas that are important to the natural systems and wildlife. As a result, the variety of plants and animals is decreased. In some instances, people extract materials from the Earth needed to build their homes, manufacture tools, and fuel their cars. The use of energy resources like coal and oil to power cars and other machinery pollutes the air, land, and water. In some places, such pollution is harmful to humans and their health.

People also change the environment with agricultural activities. For example, they may clear the land of trees and other natural vegetation to plant crops. The overgrazing of livestock can create conditions that make the land vulnerable to erosion. In some instances, people divert rivers and other sources of water to have the water needed to grow their crops. These practices can lead to damage to the environment. The use of agricultural chemicals like fertilizers and pesticides can pollute the land and the water.

Since the basic human need is water, people have learned to use tools and technology to manage water. In some cases, there is too much water and people seek ways to control the water so that they can live in certain areas. But providing freshwater to support human activity can be a challenge. People in some areas of the world face water shortages and groundwater depletion. The problem is with the distribution and quality of Earth's water. Populations are growing in many regions where water supplies are limited. Some regions use technology to process salt water into freshwater.

Large cities like Los Angeles, California, with lots of cars on the road often struggle with air pollution.

Many changes are occurring in the global environment. Some are part of Earth's natural processes, while others are influenced by human activities. Scientists state that some of the most important changes are those in global air temperature and in rising sea levels. The average global air temperature has been increasing. This increased temperature is caused in part by the burning of coal and oil by cars and other machinery. Rising sea levels are a result of the warming air temperatures as ice in the Arctic and Antarctic melts. These rising sea levels can lead to coastal flooding.

Although much human activity produces negative impacts on the environment, many positive outcomes occur as well. Science and technology have helped humans increase food production. The built environment has created safer places in which wild animals don't ravage uncontrolled. It is important to balance meeting human needs and wants while at the same time, ensuring a healthy environment.

Looking Ahead

You will learn how the environment influences human settlement and how people adapt to the environment. You will learn how people prepare for, and react to, natural disasters. You will learn about the relationship between natural resources and human activities. You will learn about human impacts on the environment. You will examine Compelling Questions and develop your own questions about human-environment interaction in the Inquiry Activity Lessons. You will see some of the ways in which you might already use the ideas of human-environment interaction in your life by completing these lessons.

What Will You Learn?

In these lessons about human-environment interaction, you will learn

- how physical geography influences where people live;
- how people adapt to the natural environment;
- how people are affected by natural disasters;

- how people define, use, and acquire natural resources;
- how the value of natural resources is determined;
- how people modify the environment;
- the causes and effects of deforestation, desertification, and pollution;
- the causes and effects of the destruction of wetlands;
- how human activity is related to water availability;
- the causes and effects of global climate change.

 COMPELLING QUESTIONS

- **How have human-environment interactions shaped history?**
- **How do people around the world respond to changes in the environment?**

Studying Human-Environment Interaction

Settlement
- physical geography
- adapting to the environment
- natural disasters

Resources
- renewable
- nonrenewable
- energy resources
- sustainability

Human Impact
- deforestation
- desertification
- pollution
- water availability
- climate change

The Environment and Human Settlement

READING STRATEGY

Analyzing Key Ideas and Details As you read, take notes about how the environment influences humans and human settlement.

The Environment and Human Settlement	
Physical Geography	
Adapting	
Natural Disasters	

Influences of Physical Geography

GUIDING QUESTION

How does physical geography influence human settlement?

The environment affects where people choose to settle and make their homes. The environment includes physical features, climate, and natural resources. Each of these things influence human settlement, depending on how people support themselves and make a living.

During the first Agricultural Revolution, people shifted from hunting and gathering to farming. They settled in areas with environments that could support the growth of crops and the raising of livestock. By about 8000 B.C., people in Southwest Asia had settled near the Tigris and Euphrates Rivers. They began growing wheat and barley. They also domesticated pigs, cows, goats, and sheep. Around the same time, people began growing wheat and barley in the Nile River Valley in Egypt.

Societies today still depend on agriculture. Many people make decisions about where to settle based on the type of natural environment they need for agricultural activities. They consider the type of soil, temperatures, and rainfall. This is true of **subsistence farming** as well as for **commercial farming**. For example, the region of the United States known as the Corn Belt (Indiana, Illinois, Iowa, Missouri, Nebraska, and Kansas) has some of the most fertile soil in the country, and the land is relatively level. The growing season has warm nights, hot days, and

subsistence farming farming that only provides the basic needs of a family
commercial farming growing large quantities of crops or livestock to sell for a profit

Like Bolzono, Italy, many human settlements are in river valleys. Such environments provide fertile soil and access to water.

adequate rainfall. These factors attracted settlers. By the 1850s, the region had begun to dominate corn production in the United States. The Corn Belt also produces soybeans, wheat, and alfalfa as well as livestock.

Similarly, the Northern European Plain is an area of relatively flat and low-lying land. It stretches from southeastern England and western France through central France and across Germany. The plain's fertile soil and wealth of rivers originally drew farmers to the area. It is a region of major agricultural importance.

In some areas of the world, people practice **pastoralism**, or the raising of animals for food.

pastoralism the raising of animals for food

Pastoral societies breed and tend flocks and herds of animals to satisfy the human needs for food, shelter, and clothing. Pastoralism isn't just an economic activity. It is a way of life that often requires people to travel with their herds for water and grazing land. It is usually practiced in cold or dry climates where farming is not possible.

Physical features influence human settlement. Mountains can either encourage settlement or prevent it. For example, the volcanic soil of the Central Highlands in Central America makes these mountains rich agricultural zones. As a result, these are areas of dense settlement. In contrast, the *cordilleras*, or parallel mountain ranges, in the Andes of South America provide

The Fall Line in the United States

The fall line is important to settlement and river traffic in the eastern United States.

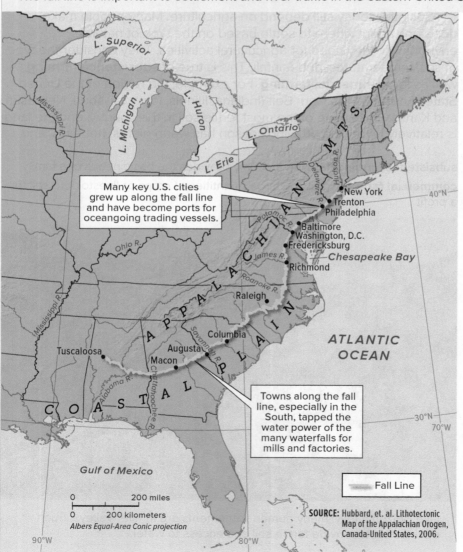

Many key U.S. cities grew up along the fall line and have become ports for oceangoing trading vessels.

Towns along the fall line, especially in the South, tapped the water power of the many waterfalls for mills and factories.

Fall Line

0   200 miles
0   200 kilometers
Albers Equal-Area Conic projection

SOURCE: Hubbard, et. al. Lithotectonic Map of the Appalachian Orogen, Canada-United States, 2006.

EXAMINE THE MAP

1. **Human-Environment Interaction** Think about the physical geography of the fall line. How could people use the rapids and waterfalls here? How did this contribute to the development of cities along the fall line?

2. **Human-Environment Interaction** The land drops from the Piedmont to the Atlantic Coastal Plain in the eastern United States, which prevents ships from traveling further inland. What other physical feature influences people's ability to move westward from the fall line?

natural barriers between surrounding areas. Many indigenous communities developed as isolated groups, and new settlement is difficult.

Rivers provide water for daily living, for agriculture, for power, and for transportation. In the eastern United States, the **fall line** marks the place where the higher land of the Piedmont drops to the lower Atlantic Coastal Plain. Along the fall line, rivers break into rapids and waterfalls. This prevents ships from the Atlantic Ocean from traveling farther inland and the fast-flowing water provides power. Cities such as Philadelphia, Baltimore, and Washington, D.C., were established along the fall line as port cities and industrial centers. Many cities around the world have also developed near rivers as well as along coasts.

Lake Superior, Lake Huron, Lake Erie, Lake Ontario, and Lake Michigan make up the Great Lakes. They formed when glacier basins filled with water. The glaciers also uncovered major deposits of natural resources on the land, including iron ore and coal, that later spurred explosive economic growth.

The Great Lakes serve many economic and recreational purposes, but none are more valuable than the Great Lakes–St. Lawrence Seaway System. This system consists of a **series** of canals, rivers, and waterways linking the Great Lakes with the Atlantic Ocean. The seaway helped make cities along the Great Lakes, such as Chicago, prominent industrial centers that could trade with the world.

Resources such as coal, oil, iron ore, and minerals influence human settlement. Growth of human settlement in cities accompanied the rise of industry in Great Britain during the Industrial Revolution in the 1800s. These developments were fueled by the natural resources and by geography. Great Britain had fine harbors and a large network of rivers that flowed year-round. The country's earliest cotton mills were powered by the flow of river water. Britain also had large supplies of coal and iron. Coal, which replaced wood as a fuel, helped to run machines. Iron was used to build machinery.

Innovations in transportation have influenced human settlement. The invention of the steam engine meant that river boats could travel farther

and faster. It drove the expansion of railroads to connect more distant places. Distance became less of an obstacle to settlement. This occurred to an even greater extent when the automobile was invented. It allowed settlement over a wider area and gave people the convenience of working, shopping, and enjoying entertainment and sports whenever they wanted. With the automobile, they could drive themselves to wherever they wanted, whenever they wanted.

The changes that have occurred in transportation, along with processes of economic growth, increased world trade, and human settlement resulted from access to natural resources. People are also dependent on those resources. As this development of human activity has grown larger over time, it has demanded increasing amounts of resources.

✓ CHECK FOR UNDERSTANDING

Explaining Explain how mountains can both encourage and discourage human settlement.

Adapting to the Environment

GUIDING QUESTION

How do people adapt to the environment?

Since people first began organizing into settlements and communities, they have adapted to the natural environment. They have had to find ways to adjust in order to meet their basic needs while living in a particular place.

One of the basic human needs is water. For centuries, people have built irrigation systems to get water for living and for agriculture. **Irrigation** is the process of bringing in water through human-made means rather than relying only on rainfall. The earliest forms of irrigation involved people carrying water in buckets from wells, streams, and rivers. As better techniques were developed, societies in Egypt and China built dams, canals, and pipes to divert and move water to where they needed it. For example, the ancient Romans built aqueducts to carry water from melted snow in the Alps to cities and towns in the valleys below.

fall line a boundary where a higher, upland area drops to the lower land of a plain

series a set of things arranged in order

irrigation the process of collecting and moving water for use in agriculture

Developments in technology have contributed to today's advanced irrigation systems. They include reservoirs, dams, canals, pipelines, tanks, and wells. Pumps help move water from place to place within the system. Large, commercial farming operations use irrigation systems to keep crops watered. For example, water from the Colorado River is diverted to supply water to communities in several western states for farming, industries, and household use. Subsistence farmers often benefit from small-scale irrigation technologies that use human power to pump water from rivers and canals onto their fields.

Before the 1950s, the people of Egypt depended on the annual flooding of the Nile River. The floodwaters carried sediments that were deposited along the riverbanks. These sediments formed rich **alluvial soil** that made for productive agricultural lands. Beginning in

alluvial soil rich soil made up of sand and mud deposited by running water

Geography and the Industrial Revolution

Rivers and abundant supplies of coal and iron ore were important for the spread of industry in the England and the rest of Western Europe during the Industrial Revolution.

EXAMINE THE MAP

1. **Human-Environment Interaction** Which natural resource influenced the development of Sheffield, Manchester, and Leeds as centers of industry?

2. **Human-Environment Interaction** Describe the relationship between the locations of coal and iron ore fields and the locations of railroads.

the 1950s, dams were built farther upstream along the Nile River. The dams were built to control the river's flow and reduce flooding. Dams were also built to provide year-round irrigation and hydroelectric power in Egypt. The Aswān High Dam, located about 600 miles (966 km) south of Cairo, Egypt, was completed in 1970. The dam controls the Nile's floods. It also created Lake Nasser and provides irrigation for millions of acres of land However, without the annual deposits of new alluvial material, soil fertility diminishes and requires the use of fertilizers.

The world's growing population and increasing urban areas need freshwater for drinking, farming, and manufacturing. In some places, there is not enough freshwater to meet the rising demand. Humans have developed technologies that remove minerals from salt water. **Desalination** is a process that separates most of the dissolved chemical elements to produce freshwater that is safe to drink. People who live in dry regions use desalination to convert seawater into freshwater. This requires a lot of energy and is expensive.

In some places, water can be the enemy. Approximately 25 percent of the Netherlands lies below sea level. There are extensive coastal dunes, but they have not always been helpful in keeping out the North Sea waters. Since the Middle Ages, the Dutch (people who live in the Netherlands) have built **dikes**, or large banks of earth and stone, to hold back the water. With the dikes as protection, they have reclaimed land from the sea. These reclaimed lands, called **polders**, were drained and then kept dry by windmills. Today, other power sources run pumps to remove water. Polders provide hundreds of thousands of acres of land for farming and settlement. Reclaiming land from the sea through the construction of dikes and polders occurs in many places, including Bangladesh, Belgium, China, and the United States.

People have also adapted to their natural environment through the farming techniques they use. For example, terrace farming is used to create flat areas in a hilly or mountainous area.

Land from the Sea

Reclaimed land uses dikes to hold back water from the sea.

Land in 1300
Land reclaimed

Amsterdam

The Hague

NETHERLANDS

Rotterdam

GERMANY

BELGIUM

0   40 miles
0   40 kilometers
Lambert Azimuthal Equal Area projection

EXAMINE THE MAP

1. **Exploring Location** Which two cities are located on reclaimed land?
2. **Exploring Place** Where is most of the reclaimed land in the Netherlands located?

These flat areas are then planted with crops. Terrace farming is practiced in the rice fields of Asia, on the steep slopes of the Andes, and in countries around the Mediterranean Sea.

Another farming technique is **slash-and-burn agriculture**. Trees and underbrush are cut down and the area is burned to create fields for crops or grazing areas for livestock. Slash-and-burn agriculture is used in forest or woodland areas in South and Central America as well as in parts of

desalination the removal of salt from seawater to make it usable for drinking and farming

dike large bank of earth and stone that holds back water

polder low-lying area from which seawater has been drained to create new land

slash-and-burn agriculture a method of farming that involves cutting down trees and underbrush and burning the area to create a field for crops

Southeast Asia. The improved soil fertility that results from the slash-and-burn process only lasts for two or three years. Then the process must be repeated in a new place.

✓ **CHECK FOR UNDERSTANDING**

Analyzing How is irrigation an example of humans adapting to their natural environment?

Natural Disasters

GUIDING QUESTION

How are people affected by natural disasters?

One important way that humans adapt to the environment is by responding to and preparing for natural disasters. A **natural disaster** is an extreme event in nature that usually results in serious damage to human-built infrastructure and often in loss of life. Earthquakes, volcanic eruptions, tsunamis, tornadoes, hurricanes, typhoons, floods, wildfires, droughts, and mudslides are types of natural disasters.

Natural disasters that occur in one community can have effects at the city and state level, or even on an entire country. A natural disaster can cause destruction of property, the loss of economic resources, and personal injury or illness. Such individual losses may lead people to migrate to other places.

Communities that experience natural disasters are left to absorb the impacts of these events. In some places, a community loses so much in economic resources that recovery and rebuilding is very difficult. Population and cultural shifts, often the result of a natural disaster, can affect how a community **recovers**.

People and communities can prepare for some natural disasters. Disaster drills are used to simulate the circumstances of a natural disaster so that people can practice their responses. Such drills are used to help communities prepare for fires, earthquakes, tornadoes, and other types of disasters. You may be familiar with tornado drills at school. Some communities have warning systems that are activated when the threat of a natural disaster is coming. For example, Japan

natural disaster a sudden and extreme event in nature that usually results in serious damage to human-built infrastructure and loss of life

recover to regain

and other countries around the Pacific Ocean have tsunami alert systems.

Science provides an understanding of Earth's natural systems, and modern technology helps us monitor those systems. Together they guide us in predicting when some natural disasters may occur. Scientists monitor air temperatures and atmospheric pressures, wind and ocean currents, and ocean temperatures to predict when a hurricane might hit a particular area. Recording regular rainfall, temperatures, and the condition of the land cover can help scientists predict when an area is more susceptible to forest fires. Computers and other tools help scientists keep a constant watch on the activity of the Earth below the surface to estimate when an earthquake, a volcano, or a tsunami may occur.

Houses and other buildings can be built to make them safer and more able to withstand a natural disaster. For example, houses in typhoon-prone areas like the Philippines are built on stilts so that they are above the floodplain and the storm surge of sea water that accompanies a typhoon. Modern building codes for houses in the southeastern United States that face annual hurricane seasons include concrete walls, hurricane-proof windows, and steel beams that support and protect the structure from high winds that occur. In places that experience earthquakes, buildings have to take up as much of the seismic energy as possible. Many buildings are built with shock absorbers or dampers that can absorb the shock waves produced by the moving Earth.

✓ **CHECK FOR UNDERSTANDING**

Summarizing Describe two ways in which humans live with natural disasters.

LESSON ACTIVITIES

1. **Informative/Explanatory Writing** Select one of the examples from this lesson about how humans adapt to their environment. Conduct research to learn more about it and write an essay explaining the causes and effects of that adaptation.

2. **Analyzing Main Ideas** Working with a group, research to learn more about one of the first river valley civilizations. What factors contributed to its growth? How did the people of that civilization live? How was its location related to its development? Use your answers to create a cause-and-effect chart. Compare your findings with other groups in your class.

Natural Resources and Human Activities

Analyzing Key Ideas and Details As you read, take notes about the types of resources and how they are used.

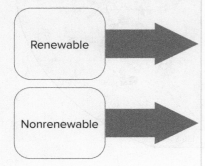

Renewable

Nonrenewable

Natural Resources

GUIDING QUESTION

What are natural resources?

Earth provides all the materials necessary to **sustain** life. These materials occur naturally on Earth. They are not made by people but can be used by people for food, fuel, and other needs and wants. These naturally occurring materials include water, air, soil, minerals, plants, and animals. The become **natural resources** when people learn how to use them to meet their needs and wants.

Natural resources can be classified as either renewable or nonrenewable. **Renewable resources** are those that can be replenished in a relatively short period of time, usually within a person's lifetime. An important consideration is the time it takes for them to be replenished, because different renewable resources require different amounts of time to recover after they are used. One renewable resource is water—for drinking, for agriculture, and for trees that provide lumber.

The Earth's crust contains many **nonrenewable resources** that cannot be replaced once they are consumed. Examples of nonrenewable resources include minerals like iron ore and copper. **Fossil fuels** such as oil, coal, and natural gas are also nonrenewable resources. These fuels received their name because they formed millions of years ago.

sustain to give support to

natural resource materials or substances such as minerals, forests, water, and fertile land that occur in nature and that people use

renewable resource a resource that can be replenished in a relatively short period of time

nonrenewable resource a resource that cannot be replaced

fossil fuel a resource formed in the Earth by plant and animal remains

Classifying Resources
There are important differences to understand between renewable and nonrenewable resources.

Renewable Resources	Nonrenewable Resources
Are renewed or replenished by nature in a short period of time	Are not renewed or replenished in a person's lifetime
Are often available continuously and not likely to be exhausted	Are not available continuously and may be exhausted
Examples: air, wind, water, solar energy	Examples: coal, petroleum, minerals
Not likely to cause pollution	Usually causes pollution when used

Natural Resources Around the World

Natural resources are not evenly distributed throughout the Earth.

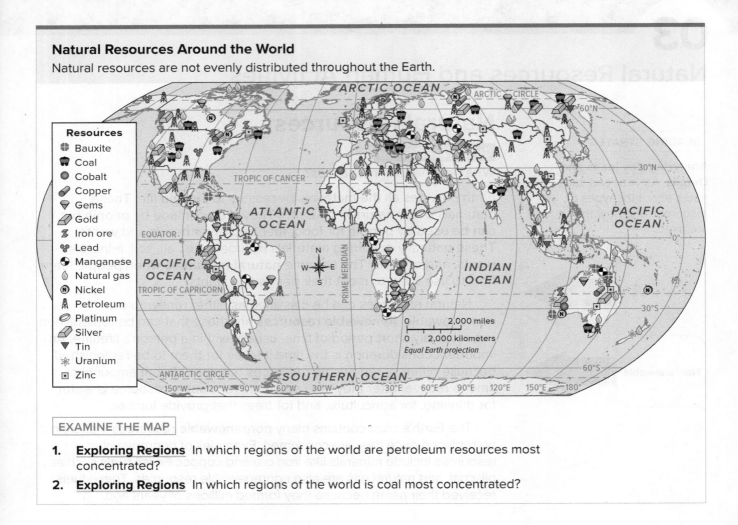

Resources
- ✚ Bauxite
- ⛏ Coal
- ● Cobalt
- ✎ Copper
- ▽ Gems
- ◩ Gold
- Ⅎ Iron ore
- ☙ Lead
- ◐ Manganese
- ◊ Natural gas
- Ⓝ Nickel
- Ⱥ Petroleum
- ⬭ Platinum
- ◿ Silver
- ▼ Tin
- ✳ Uranium
- ⊡ Zinc

EXAMINE THE MAP

1. **Exploring Regions** In which regions of the world are petroleum resources most concentrated?

2. **Exploring Regions** In which regions of the world is coal most concentrated?

Another type of resource is called a *flow resource*. It is one that is always available naturally. Sometimes this type of resource is considered a renewable resource. An example is water that flows through a hydroelectric dam and produces electricity. Energy from the sun is another example.

Many resources are **extracted**, or pulled from the Earth. Resource extraction includes the logging of trees, mining of minerals, drilling for oil and natural gas, and fishing. There are different forms of each type of extraction. Some are more destructive to the environment than others. For example, logging is generally classified into two types. **Clearcutting** is the process of taking out whole forests. *Selective logging* involves only taking trees of similar age and size. Similarly,

there are two general types of mining—mining underground and mining at the Earth's surface. The method used depends on the type of mineral resource and its location at or beneath the surface. The decision to extract any resource usually depends on whether the resource is worth enough money to support extracting it.

One resource that everyone needs is energy. *Energy resources* provide the power to do work. Many types of energy resources exist in our world today. Some are renewable, such as solar, wind, water, or **geothermal energy**. Energy resources can also be nonrenewable, providing power by burning coal, oil, or natural gas. **Nuclear energy** provides electricity through the splitting of atoms.

extract to pull out; to withdraw

clearcutting the removal of all trees in a stand of timber

geothermal energy the electricity produced by natural, underground sources of steam

nuclear energy the energy created by splitting the nucleus of an atom

The uneven distribution of energy resources, as well as mineral resources like copper, gold, and iron ore, influences the economic development of countries around the world. Most of the world's more developed countries have access to energy resources, mineral resources, or both. Countries that do not have access to these types of natural resources typically have a lower *standard of living*. For these less developed countries, the cost of importing resources is a heavy burden.

✓ CHECK FOR UNDERSTANDING

Contrasting How do renewable and nonrenewable resources differ?

Using Natural Resources

GUIDING QUESTION

How do people use natural resources?

Natural materials from the Earth become resources when humans assign value to them. Resources are valuable when they can meet the needs or wants of people. The value of a resource may change over time. Changing technologies can add value to some resources and diminish the value of others. In addition, cultures use and value resources differently. For example, some indigenous cultures believe that the extraction of some natural materials isn't worth the destruction of the land.

Because fossil fuels, such as coal and oil, cannot be replaced, many people believe that they must be conserved. The immediate goal of **conservation** is to manage vital resources carefully so that people's present needs are met. An equally important long-term goal is to ensure that the needs of future generations are met. Together, these goals define the idea of **sustainability**—using natural resources

conservation the act of protecting Earth's natural resources

sustainability the idea of creating conditions where all natural resources for meeting the needs of society are available today and for futire generations

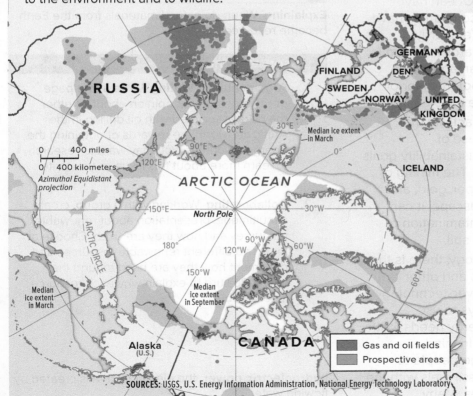

Arctic Resources

The Arctic holds many natural gas and oil resources. Some people support the extraction of these resources. Others are against drilling for gas and oil in the region because of the threat to the environment and to wildlife.

EXAMINE THE MAP

1. **Exploring Location** In which countries are most of the gas and oil fields concentrated?

2. **Global Interconnections** How might the extraction of resources in the Arctic by one country affect other countries in the region?

SOURCES: USGS, U.S. Energy Information Administration, National Energy Technology Laboratory

This photo of coal mining in Campbell County, Wyoming, shows how surface mining involves removing parts or all of mountaintops to expose the coal.

responsibly so they support both present and future generations. These ideas can be applied to discussions on many things including Arctic oil.

Some of Earth's largest oil and natural gas fields lie under the icy waters of the Arctic Ocean. As extraction technology has improved and as demand for these resources has increased, countries that border the Arctic Ocean have staked their claims and energy companies have begun drilling operations. But are energy companies prepared for the hazards involved in Arctic drilling—wildlife habitat damage, noise pollution, air pollution, and the potential for oil spills? Are the rewards worth the risk?

Mining is another extraction activity that requires people to balance the use of nonrenewable resources with sustainability goals and protecting the environment. The mining of coal and other minerals like copper, gold, and iron ore often degrades the landscape. It can result in soil erosion and the contamination of surface water, groundwater, and soil.

As people use more technology, there is the potential to use more fossil fuels and other nonrenewable resources. However, technology also allows us to create new ways to use renewable resources for our energy needs. With the needs of future generations in mind, environmental experts encourage people to replace their dependence on fossil fuels with the use of renewable energy sources. Many

countries, for example, already produce **hydroelectric power**—energy generated by moving water. Another renewable energy resource is solar energy, power produced by the sun's energy. And in many places, giant wind turbines are used to generate electricity.

✓ CHECK FOR UNDERSTANDING

Explaining When do natural materials from the Earth become resources?

LESSON ACTIVITIES

1. **Argumentative Writing** Write a one-page essay arguing your opinion about whether the risks associated with oil drilling in the Arctic are worth the rewards of obtaining the energy resources. You may want to research to learn more about this issue as you write your essay.

2. **Collaborating** Working with a group, create a multimedia presentation about renewable resources and how they are used. Choose at least two different renewable resources and document how they are used around the world. Be sure to explain why each resource is renewable.

hydroelectric power the electricity that is created by flowing water

G272

04

Analyzing Sources: Human-Environment Interactions Throughout History

❓ COMPELLING QUESTION

How have human-environment interactions shaped history?

Plan Your Inquiry

DEVELOPING QUESTIONS

Think about how people in different historical periods interacted with their environments. Reflect on what you know about how people in the past depended on the environment for food, shelter, and clothing, and how they managed to cope with the environmental conditions they experienced. Then read the Compelling Question for this lesson. What questions can you ask to help you answer this Compelling Question? Create a graphic organizer like the one below. Write these Supporting Questions in your graphic organizer.

Supporting Questions	Source	What this source tells me about how human-environment interactions have shaped history	Questions the source leaves unanswered
	A		
	B		
	C		
	D		
	E		

ANALYZING SOURCES

Next, examine the sources in this lesson. Analyze each source by answering the questions that follow it. How does each source help you answer each Supporting Question you created? What questions do you still have? Write these in your graphic organizer.

After you analyze the sources, you will
- use the evidence from the sources;
- communicate your conclusions;
- take informed action.

Background Information

Like all species, humans must adapt to their environment for populations to survive and expand. Humans are one of the most successful species on the planet. Populations have spread across the globe and adapted to a wide range of environmental conditions. Across human history, evidence indicates that skill in adapting to and exploiting environmental conditions is a common characteristic of societies that flourished. On the other hand, environmental hazards and rapid changes have threatened the survival of groups of people as well as entire societies.

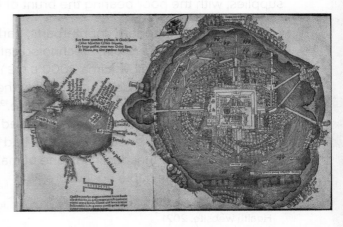

» This map was included in a letter sent to the King of Spain by Spanish explorer Hernán Cortés. It shows the Aztec capital of Tenochtitlán (on the site of present-day Mexico City) as he found it in 1519. The Aztecs built the city on a man-made island located on a lake that no longer exists.

Agriculture in Ancient Times

Ancient civilizations such as Mesopotamia and Egypt were among the first to grow dramatically due to people shifting from hunter-gatherer societies to farming settlements. This website explains how the development of agriculture had both positive and negative impacts on growing populations.

PRIMARY SOURCE: WEBSITE

An aerial view of the Nile Valley in Egypt shows how the irrigated farmlands of the river's floodplain contrast with the dry Sahara.

66 Farming probably involved more work than hunting and gathering, but it is thought to have provided 10 to 100 times more calories per acre.[5] More abundant food supplies could support denser populations, and farming tied people to their land. Small settlements grew into towns, and towns grew into cities.[1] . . .

Agriculture may have made civilizations possible, but it has never been a safeguard against their collapse. Throughout history, increases in agricultural productivity competed against population growth, resource **degradation**, droughts, changing climates, and other forces that periodically crippled food supplies, with the poor bearing the brunt of famine.

Like many of their modern counterparts, early farmers often worked land in ways that depleted its fertility. Technological innovations like irrigation (circa 6000 B.C.E.) and the plow (circa 3000 B.C.E.) brought enormous gains in productivity, but when used irresponsibly they degraded soil—the very foundation that makes agriculture possible.[19, 20] By the beginning of the Common Era, Roman farmers had degraded their soil to the point where they could no longer grow enough food and had to rely on imports from distant Egypt. Rome's eventual decline is one of many cautionary tales about the importance of sustainable agriculture. 99

—"History of Agriculture," Center for a Livable Future, Johns Hopkins Bloomberg School of Public Health website, 2021

degrade to wear out and make lower quality

EXAMINE THE SOURCE

1. **Explaining** How did farming make it possible for the first cities to develop?
2. **Identifying Cause and Effect** In what way did technology both improve and reduce agricultural output?

PHOTO: Mirko Kuzmanovic/Alamy Stock Photo; TEXT:

Dunhuang and the Mogao Caves

Movement and contact between cultures has played an important role in history. Over the centuries, people developed and improved the ability to travel and transport goods over great distances. They used different forms of transportation and identified routes that they could use to travel between places. This made trade and cultural exchange possible. This article explains the role of one place, a city and network of caves founded by Buddhist monks in 366 A.D., along the Silk Road trade routes.

A Buddhist temple sits alongside a spring fed lake surrounded by sand dunes in the desert near Dunhuang, China.

PRIMARY SOURCE: ARTICLE

❝ The Silk Road routes from China to the west passed to the north and south of the Taklamakan Desert, and Dunhuang lay on the junction where these two routes came together. Additionally, the city lies near the western edge of the Gobi Desert … making Dunhuang a vital resting point for merchants and **pilgrims** travelling through the region . . . .

The history of this ancient Silk Road city is reflected in the Mogao Caves . . . 492 caves that were dug into the cliffs just south of the city. Monks and pilgrims often travelled via the Silk Roads, and indeed a number of religions, including Buddhism, spread into areas around the trading routes in this way. . . . . The caves were painted with Buddhist imagery, and their construction would have been an intensely religious process, involving prayers, incense and ritual **fasting**. The earliest wall paintings date back to the 5th century A.D., with the older paintings showing scenes from the Buddha's life, whilst those built after 600 A.D. depict scenes from Buddhist texts . . . . One of the caves . . . contains as many as 40,000 scrolls, a **depositary** of documents that is of enormous value in understanding the cultural diversity of this Silk Road city. ❞

—"Dunhuang," Silk Roads Programme, UNESCO website, 2021

pilgrim a person who travels to a shrine or other holy place
fast to go without eating
depository a place for safe storage

EXAMINE THE SOURCE

1. **Determining Context** Why was Dunhuang an important city in the past?
2. **Speculating** Why do you think visitors come to Dunhuang now?

The Location of Cities in Virginia

Beginning in the early 1500s, Europeans arrived in North America hoping to colonize territory occupied by Native Americans and exploit the resources they found there. Settlers cleared land to establish farms. They also built towns focused on trading agricultural products and raw materials. Some of those towns became manufacturing centers and grew into cities. Much of this development was carried out by enslaved people transported from Africa, who came to make up about one third of Virginia's population by 1860. This website explains how the location of cities was determined during Virginia's early development.

PRIMARY SOURCE: WEBSITE

> 66 As a barrier to travel, it is perhaps easiest to think of the Fall Line from the perspective of the river. For much of Virginia's history, rivers were the main arteries of travel, transportation, and trade. If one were sailing up the James River from Virginia's coast with a boat loaded down with trade goods, the waterfalls of the Fall Line would form a functionally impassable barrier. As a result, these natural junctions [intersections] became the logical place to establish towns to facilitate [assist] the transfer of goods such as agricultural products from inland farms to boats that could transport them to the Atlantic and international markets. Cities such as Fredericksburg and Richmond developed as natural intersections of trade and travel.
>
> Another reason towns formed along the Fall Line rivers was to harness the power of the falls. Industries such as mills that depended on waterpower grew around the valuable rapids. Perhaps Virginia's most famous industrial company, the Tredegar Iron Works made Virginia the Confederacy's largest ironworks and became almost the sole supplier of **munitions**, railroad equipment, plates for iron-clad ships, and other similar products for the Army of the Confederacy. After the Civil War, though damaged by fire, the ironworks transitioned back to peace-time production and **contracted** with the Chesapeake and Ohio railroad to make rails and bridgework. 99

—"Cities and the Fall Line," Virginia Studies website, 2019

munitions military weapons, ammunition, and equipment

contract to create a business arrangement between multiple parties

---

EXAMINE THE SOURCE

1. **Making Connections** Why were cities established along the fall line?
2. **Inferring** Why do you think the fall line became less important as a location for trade and manufacturing over time?

Mining Precious Metals

Most Americans connect the term Gold Rush to California and the "1849'ers" who flooded the state seeking fortunes. In fact, there have been many gold rushes. They have occurred in the United States (Colorado, Alaska, Idaho) and in other parts of the world (Australia, South Africa, Peru). This article describes how the excitement caused by the discovery of the valuable mineral can have lasting impacts on society.

Small-scale gold mining along the Madre de Dios River in Peru in 2018.

PRIMARY SOURCE: ARTICLE

❝ The discovery of the precious metal at Sutter's Mill in January 1848 was a turning point in global history. The rush for gold redirected the technologies of communication and transportation and accelerated and expanded the reach of the American and British Empires.

Telegraph wires, steamships, and railroads followed in their wake; minor ports became major international metropolises [cities] for goods and migrants (such as Melbourne and San Francisco) and interior towns and camps became instant cities (think Johannesburg, Denver and Boise). This development was accompanied by accelerated mobility—of goods, people, credit—and anxieties over the erosion of middle class mores [values] around respectability and **domesticity**.

But gold's new global connections also brought new forms of destruction and exclusion. The human, economic, and cultural waves that swept through the gold regions could be profoundly destructive to Indigenous and other settled communities, and to the natural environment upon which their material, cultural, and social lives depended. Many of the world's environments are gold rush landscapes, violently transformed by excavation [digging], piles of tailings [mine waste], and the reconfiguration of rivers. ❞

—"How gold rushes helped make the modern world," by Benjamin Wilson Mountford and Stephen Tuffnell, *The Conversation*, 2018

domesticity home and family life

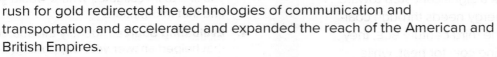

EXAMINE THE SOURCE

1. **Summarizing** How do gold rushes accelerate progress?
2. **Identifying Cause and Effect** What are the negative consequences of gold rushes?

Fueling Advancements in Civilization

The existence of iron ore and coal deposits in Great Britain fueled the Industrial Revolution beginning in the late 1700s. The use of these resources occurred far earlier in a different part of the world, described in this website excerpt.

PRIMARY SOURCE: WEBSITE

❝ Even before civilization began to take root in China 5,000 years ago, ancient people in the region already had been making use of coal [to make carvings] for about 1,000 years.

With the opening of the Fushan mine in northeastern China over 3,000 years ago, the Chinese took a significant step toward meeting their energy needs through coal-fire. By around the third century BCE, they had begun burning coal for heat, while carved coal became a trading commodity [product] in village marketplaces. Around 120 B.C.E., the expansion of their **metallurgy** industry led to mass deforestation [clearing of forests] caused by burning vast amounts of wood-derived charcoal in blast furnaces. Over succeeding centuries, China's deforestation problem developed into a national environmental crisis. The solution to the loss of China's forests was found in coal.

[The Chinese] relocated blast furnaces from deforested areas to coal fields, which allowed the nation's incredible iron industry to continue to thrive while lessening its deforestation problem. ❞

—"Coal," Alberta Culture and Tourism website, 2022

metallurgy the science and technology of purifying and working with metallic elements

EXAMINE THE SOURCE

1. **Identifying** Besides as a fuel source, how did the Chinese first use coal?
2. **Explaining** How did coal help the Chinese adapt to deforestation?

Complete Your Inquiry

EVALUATE SOURCES AND USE EVIDENCE

Refer back to the Compelling Question and the Supporting Questions you developed at the beginning of the lesson.

1. **Making Generalizations** Which sources provide information about how the settlement of specific areas was due to the environmental features of those areas?

2. **Identifying Themes** Which sources provide information about the role of mined materials in history?

3. **Gathering Sources** Which sources helped you answer the Supporting Questions and the Compelling Question? Which sources, if any, challenged what you thought you knew when you first created your Supporting Questions? What information do you still need in order to answer your questions? What other viewpoints would you like to investigate? Where would you find that information?

4. **Evaluating Sources** Identify the sources that helped answer your Supporting Questions. How reliable is each source? How would you verify its reliability?

COMMUNICATE CONCLUSIONS

5. **Collaborating** Share with a partner which of these sources you found most useful in answering your Supporting Questions. Discuss how each helped you understand how human-environment interactions have shaped history. Take turns identifying similarities and differences between the sources. Explain how a source relates to what you previously knew about the topic. Which sources advance your knowledge? Use the graphic organizer that you created at the beginning of the lesson to help you. Share your conclusions with the class.

TAKE INFORMED ACTION

Recalling Past Environments Think about how people have interacted with the environment in the area where you live. Consider its specific resources. Think about how people have changed the environment of the area. Consult local and online sources and create a poster that highlights your area's environmental history and natural resources. Display the poster in a public location to help others learn about your area.

Human Impact on the Environment

READING STRATEGY

Analyzing Key Ideas and Details As you read, take notes about how humans impact the environment.

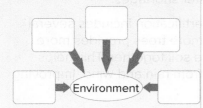

Human Impact

GUIDING QUESTION

How do human activities impact the environment?

As humans expand their communities, they threaten natural **ecosystems**, areas where plants and animals depend on one another and their environment for survival. Because Earth's water, land, and air are all connected, what affects one part of the system affects all other parts—including humans and other living things. As people become more aware of how their actions affect the balance of life, they are managing resources more wisely.

Deforestation and Desertification

About one-third of the Earth is covered in forests ranging from the northern **boreal forests** to tropical rain forests. **Deforestation** and the loss of **biodiversity**, however, are occurring at an alarming rate. As economies grow, so does the demand for timber resources, which are often an important part of a country's export economy. The corporate logging industry is not the only source of deforestation. Growing populations demand more food and energy resources. In response,

ecosystem the complex community of living things and the physical environment that are interdependent in a given place

boreal forest the forest in the far northern regions of Earth

deforestation the loss or destruction of forests, mainly for logging or farming

biodiversity the wide variety of life on Earth

Forests in Malaysia are cleared to make room for palm-tree plantations. Malaysia is one of the world's largest producers of palm oil. Palm oil is used in many processed foods and some lipsticks and eye shadows. It is also found in paints, glues, and soaps.

companies extract oil and other resources. Ranchers and farmers clear forests to create new areas for growing food and grazing cattle. Clear-cutting occurs in many areas today. As a result, forest ecosystems are increasingly becoming less diverse. Wildlife is endangered and land is subject to erosion and flooding.

Given time, forests will regenerate on their own, but with a considerable loss of biodiversity. Laws requiring **reforestation**—the planting of young trees or seeds of trees on the land that has been stripped—can help. Developing new methods of farming, mining, and logging and combining them with conservation practices protects the forests and boosts local economies.

Human activity also negatively affects the arid and semiarid regions of the world. **Desertification** is the process in which arable land becomes desert. It occurs when **shifts** in climate, long periods of drought, and land use destroy vegetation. The land is left dry and barren, unable to support life. Desertification is caused, in part, by animals overgrazing the land, by the cutting of trees for wood to use as fuel, and by the clearing of original vegetation for farming. Once the original vegetation is cleared, it often cannot reestablish itself. The exposed soil is then vulnerable to erosion. Desertification often causes poverty, lack of regular access to enough food, and further water shortages.

Combatting desertification includes several strategies. Planting more trees provides more tree roots to hold the soil together. This helps reduce soil erosion from rain and wind. Improving

reforestation the planting and cultivating of new trees in an effort to restore a forest where the trees have been cut down or destroyed

desertification the process by which an area turns into a desert

shift to change the place, position, or direction of

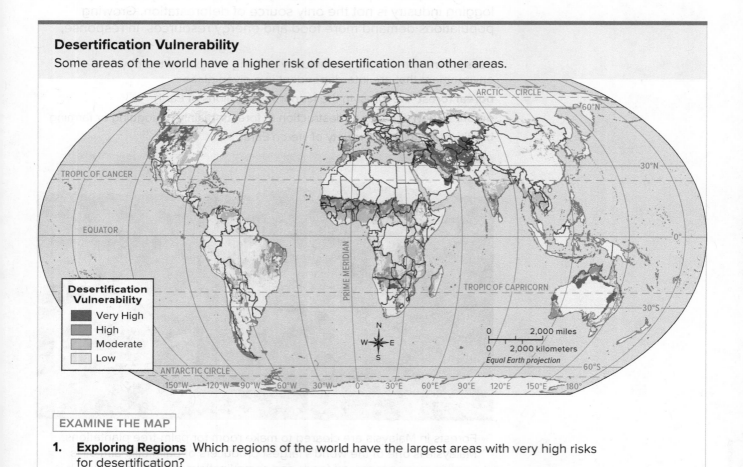

Desertification Vulnerability

Some areas of the world have a higher risk of desertification than other areas.

Desertification Vulnerability
- Very High
- High
- Moderate
- Low

Equal Earth projection

EXAMINE THE MAP

1. **Exploring Regions** Which regions of the world have the largest areas with very high risks for desertification?

2. **Exploring Regions** Which regions of the world have low to moderate risk of desertification?

Many wetland areas are at risk because of human activities. Cardiff Bay Wetlands Reserve in Wales, United Kingdom, was created to protect this transitional area between land and water. It is a habitat for many species of birds, fish, and plants.

the quality of the soil is done by reducing the number of grazing animals and the amount of crops planted. Animal manure is used to fertilize the crops. The roots of the crops hold the soil in place. Managing water use with dams and drip irrigation helps reduce desertification. Water is collected and stored in dams in the wet season and used to irrigate crops in the dry season. Drip irrigation is a method in which water drips slowly into the ground from small holes in a hose placed on top of the soil. This technique minimizes water loss through evaporation. But these technologies are expensive, and many farmers and countries cannot afford them.

Water Availability

Water is a basic need for all humans. Providing freshwater to support human activity is a significant challenge in drought-prone areas around the world. Agriculture, industry, and thriving population centers all depend on a reliable supply of safe, fresh water to support human endeavors.

People in some areas of the world, however, face water shortages and groundwater depletion. The problem is with the distribution and quality of Earth's water. Populations are growing in many regions where water supplies are limited. Increased economic growth and development require more water.

About 97 percent of the planet's water is salt water. Because of its high concentration of minerals, humans and most animals cannot drink salt water. Humans have developed the technology to remove minerals from salt water. People in dry regions of the world use desalination to convert seawater into freshwater. But this process is expensive. Even so, desalination is used around the world because other freshwater sources are scarce. Countries in Southwest Asia and North Africa have the heaviest concentration of desalination plants. Many of these countries also have the energy supplies to run desalination plants.

The environmental concerns surrounding desalination are related to ocean and marine biodiversity. When ocean water is collected, marine life is drawn up in the intake pipe and eventually destroyed during the desalination process. In addition, the waste generated from the process affects coastal water quality. Besides being very salty, this brine may be warmer in temperature and contain chemicals that harm water and marine organisms.

Pollution

In recent decades, economic activities have substantially affected the environment. A major environmental challenge today is pollution—the release of unclean or toxic elements into the water, land, and air.

Earth's bodies of water are normally renewable, purifying themselves over time. However, this natural cycle can be interrupted by human activities. Oil tankers and offshore drilling rigs can cause oil spills. Industrial wastes may be illegally dumped into rivers and streams or may find their way through unnoticed leaks into the groundwater. Industries also cause thermal pollution by releasing heated industrial wastewater into cooler lakes and rivers. Runoff from agricultural chemicals like fertilizers and pesticides can also pollute water.

Water pollution has harmful effects on marine life and the birds and other animals that feed on fish or breed in the wetlands. **Wetlands** include marshes, ponds, and swamps. The toxic chemicals and wastes that pollute the water supply also pose a danger to humans. Water pollution speeds *eutrophication*, the process by which a body of water becomes rich in dissolved nutrients, encouraging the overgrowth of small plants, especially algae. The algae growth can deplete the water's oxygen, suffocating fish. Algae

overgrowth can also turn a lake into a marsh and then, over many years, into dry land.

Wetlands also disappear when they are converted to agricultural or urban land uses. Wetlands are important because they hold valuable water supplies. In many cases they buffer coastal areas from storms and floods. Parts of New Orleans, Louisiana, for example, lie below sea level, making the city susceptible to flooding. In response, humans built *levees*, or raised embankments, around the city to protect it from the torrential rains and storm surges that come from hurricanes in the Gulf of Mexico. However, the building of levees has eliminated many of the wetlands in the area that once naturally protected it from the seasonal flooding of the Mississippi River.

---

wetland land flooded with water permanently or seasonally

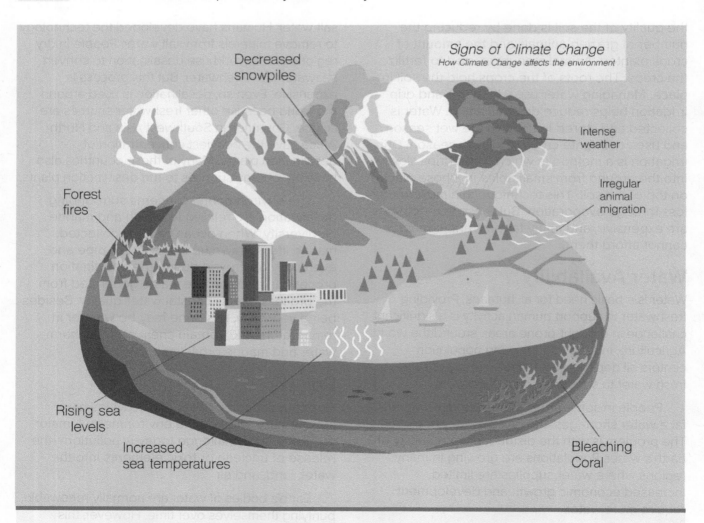

Signs of climate change can be seen in many places in the natural environment.
Identifying Cause and Effect How do increased air temperatures affect the oceans?

Land pollution is caused by human activities such as littering and by waste washed ashore from boats, oil rigs, and sewage treatment plants. Land pollution also occurs when chemical waste poisons fertile topsoil or solid waste is dumped in landfills. Radioactive waste from nuclear power plants and runoff of toxins from chemical processing plants can also leak into the soil.

The main source of air pollution is the burning of fossil fuels by industries and vehicles. Burning fuel gives off poisonous gases that can damage people's health. Acidic chemicals in air pollution combine with precipitation to form **acid rain**. Acid rain can destroy entire forests and it eats away at the surface of buildings. It is especially damaging to bodies of water, as plant life and fish cannot survive in highly acidic waters. Over time, lakes can become unable to support most organisms.

The chemicals that create acid rain also contribute to **smog**. As the sun's rays interact with automobile exhaust gases and industrial emissions, a visible haze forms. This haze can damage or kill plants and harm people's eyes, throats, and lungs. In some larger cities, officials may issue air-quality alerts when emissions and climate conditions together create dangerous levels of smog. These alerts urge children, the elderly, and people with respiratory problems to stay indoors. Authorities may also **prohibit** nonessential driving and the use of other gasoline-powered engines during an air-quality alert.

✓ **CHECK FOR UNDERSTANDING**

1. **Summarizing** What types of human activities drive deforestation?
2. **Describing** Describe the causes of air, land, and water pollution.

Environmental Change

GUIDING QUESTION

What is global climate change?

Many changes are occurring in the environment. One of these is climate change. **Climate change**

refers to the changes in factors used to measure the condition of the atmosphere over long periods of time—especially temperature, precipitation, and weather patterns.

Some changes occur high in the atmosphere. The **ozone layer** that is found high in the atmosphere protects Earth's humans, animals, and plants from ultraviolet (UV) rays that come from the sun. Unfortunately, human activity has caused a thinning of the ozone layer and its protection has decreased. Although steps have been taken to reduce the activities that cause this ozone depletion, it is still a problem.

Changes also occur in local places on Earth. In particular, urban areas are hotter than the surrounding area. These are called *urban heat islands*. The heat in urban areas can be reduced, especially by planting more trees. Some cities are already doing this.

At a global scale near Earth's surface, two clues of climate change are rising global air temperatures and rising sea levels. Other clues are the increase in fierce heat waves, the increase in forest fires, irregular animal migrations in response to changing habitats, and changes in precipitation. In addition, there has been a rise in the strength of severe weather events around the world, such as tsunamis and hurricanes. These events may even be occurring more often.

Earth's air temperatures have been steadily increasing since the Industrial Revolution. Natural processes are the cause of some of this increase. However, evidence shows that human activities play a large role. According to National Aeronautics and Space Administration (NASA) scientists, the average global air temperature has increased by 1.9° F (1.1° C) since 1880. Most of this warming has happened since 1975.

Increased global air temperature is linked to the exhaust gases that are released into the air when cars and factories burn fossil fuels. These are called *greenhouse gases*, and they also cause the acid rain and smog discussed earlier in this lesson. Another effect of increased air temperatures is that many northern villages built

---

acid rain precipitation carrying large amounts of dissolved acids

smog haze caused by chemical fumes from automobile exhausts and other pollution sources

prohibit to forbid somebody from doing something through law or rule

---

climate change changes in the factors used to measure the condition of the atmosphere over time

ozone layer a layer around the Earth's atmosphere that blocks out many of the most harmful rays from the sun

Sandbags are used to prevent the erosion of a an island in the Maldives, a country located southwest of India in the Indian Ocean. According to reports from NASA, nearly 80 percent of the Maldives could become unlivable by 2050 as a result of rising sea levels.

on the hard **permafrost** are now sinking as the ground thaws. Erosion by the sea is also a problem for coastal villages as the ground warms.

The temperature of Earth's ocean is also increasing. Scientists study how ocean temperatures, greenhouse gases, wind patterns, and cloud cover all relate to each other. Through satellite imagery and local measurements, scientists have learned that the Antarctic ice sheet is shrinking as a result of increased ocean and air temperatures. This produces two concerns. First, scientists fear that the impact of climate change will permanently damage Antarctica's environment and ecosystems. Second, Antarctica is seen as a "global barometer," or an indicator of what is happening to the climate of the entire planet. Signs such as shrinking icebergs signal that global temperatures are rising.

As massive chunks of ice *calve*, or break away, from glaciers in the Arctic and Antarctica, more icebergs clutter the surrounding waters. These icebergs also cause global sea levels to rise. Rising sea levels create flooding in coastal areas and make living conditions difficult in many parts of the world. Low-lying islands are losing land to the sea as sea levels rise.

---

permafrost permanently frozen layer of soil beneath the surface of the ground

plankton plants or animals that ride along with water currents

Rising ocean temperatures affect certain types of **plankton** and algae that grow in warm waters. Sometimes this causes overgrowth and the choking out of other life forms. At other times, the algae die off. The breakdown of the relationship between coral and the algae that provide it with nutrients causes *coral bleaching*, or the process when corals become white due to environmental stress. Warmer water temperatures in 1998 and 2002 resulted in massive coral bleaching events in Australia's Great Barrier Reef. Scientists are studying this warming and hope to discover causes, predict consequences, and provide solutions.

✓ **CHECK FOR UNDERSTANDING**

Explaining What are two main clues in the environment that point to climate change?

LESSON ACTIVITIES

1. **Informative/Explanatory Writing** Write an essay discussing the causes and effects of one of the human impacts on the environment described in this lesson. Research to learn more about the spatial distribution of the impact. Where in the world does it occur?

2. **Analyzing Information** One in ten people in the world lacks access to clean drinking water. Working with a partner, select a world region and research to find out what percentage of the population lacks access to clean water. What factors in the region affect access to water? Summarize your findings in a report.

Pix/Alamy Stock Photo

06

Global Issues: Adapting to Environmental Change

? COMPELLING QUESTION

How do people around the world respond to changes in the environment?

Plan Your Inquiry

DEVELOPING QUESTIONS

Think about how changes in the environment can impact people and places. Reflect on what you know about how changes in climate, natural resources, and other environmental characteristics might require people to adapt or change. Then read the Compelling Question for this lesson. What questions can you ask to help you answer this Compelling Question? Create a graphic organizer like the one below. Write these Supporting Questions in your graphic organizer.

Supporting Questions	Source	What this source tells me about how people are responding to environmental changes	Questions the source leaves unanswered
	A		
	B		
	C		
	D		
	E		

ANALYZING SOURCES

Next, examine the sources in this lesson. Analyze each source by answering the questions that follow it. How does each source help you answer each Supporting Question you created? What questions do you still have? Write these in your graphic organizer.

After you analyze the sources, you will
- use the evidence from the sources;
- communicate your conclusions;
- take informed action.

Background Information

The environment is constantly changing. Natural processes like storms, volcanic eruptions, and changes in Earth's orbit can cause changes in the environment and climate. Change can be slow and gradual, such as when the plates that make up the Earth's crust collide and form a mountain range. Change can also occur quickly, such as when lightning sparks a wildfire that sweeps through a forest. Humans depend on the environment. Their activities also change the environment. Clearing of natural vegetation, depletion of natural resources, and introduction of various types of pollution are some of the ways that humans impact the environment. For human societies to be sustainable, they must react to changes in the environment and find ways to adapt and change their activities.

» More than 100,000 people marched to demand that world leaders act on climate change during the United Nations Climate Change Conference of the Parties (COP26) in Glasgow, Scotland, in 2021.

Finlay/Alamy Stock Photo

Relocation of a National Capital

Indonesia is one of the most populous countries in the world. It is also the largest island country in the world by area. Indonesia is made up of more than ten thousand islands located along the Equator. This includes hundreds of tiny, uninhabited islands. It also includes several larger islands, some of which (such as Borneo and New Guinea) are divided between Indonesia and other countries. Jakarta is the largest city in Indonesia and has served as the capital since the country became independent in 1950. This article reports on a big change planned for the country.

Jakarta, Indonesia is one of the largest cities in Southeast Asia.

PRIMARY SOURCE: ARTICLE

❝ Indonesia is moving ahead with the plan to relocate its capital to the island of Borneo in the first half of 2024. . . .

Southeast Asia's biggest economy plans to move the capital from Jakarta to an area of 56,180 hectares in East Kalimantan province. . . . While the move could help secure President Joko Widodo's legacy in the last year of his final term, it has also sparked environmental concern over deforestation.

Widodo, known as Jokowi, said the relocation will help spread economic activities outside of the most-populous island of Java and narrow its income gap with the rest of the country. Java is home to almost 60% of Indonesia's population and contributes more than half to its gross domestic product. Kalimantan, the Indonesian part of Borneo, accounts for 5.8% of the population and makes up 8.2% of the economy.

[Jokowi] deems the move necessary as the current capital of 10 million people is suffering from a traffic gridlock, frequent flooding and its pollution is reaching unhealthy levels. Jakarta is also sinking fast, with two-fifths of the area falling below sea level and some parts are submerging at a rate of 20 centimeters a year.

Not everyone is convinced it was a good call.

Environmental groups have raised concern about the potential damage to Kalimantan's rainforests. Borneo, home to endangered species such as the orangutan, has lost 30% of its forests in a little over four decades, much of it to the paper and pulp industry and palm oil plantations. ❞

—"Indonesia Sets 2024 Deadline to Move Its New Capital to Borneo," by Arys Aditya, *Bloomberg News,* 2021

EXAMINE THE SOURCE

1. **Explaining** What are the specific environmental challenges affecting the current capital city of Jakarta?

2. **Identifying Cause and Effect** What are some of the potential challenges and environmental impacts associated with moving the capital to the proposed site?

PHOTO: Agus D. Laksono/Alamy Stock Photo; TEXT: Aditya, A., "Indonesia Sets 2024 Deadline to Move Its New Capital to Borneo" November 1, 2021, Bloomberg News

Feeding the World

The global population, which is approaching 8 billion, continues to grow. The world's population doubled between 1950 and about 2000! While the pace of population growth is now slowing down, the world's food supply must continue to keep pace in order to feed everyone. Increases in the food supply have been achieved by increasing the productivity of farming methods and by using additional land for agricultural production. This article explains how one country's role in the global food supply is changing.

Large machines called combines are used to harvest wheat in the Canadian province of Manitoba.

PRIMARY SOURCE: ARTICLE

❝ As global warming intensifies droughts and floods, causing crop failures in many parts of the world, Canada may see something different: a farming expansion.

Rising temperatures could open millions of once frigid acres to the plow, officials, farmers and scientists predict.

In the country's three prairie provinces alone—vast swaths of flat land in central Canada covering an area more than twice the size of France—the amount of arable land could rise between 26 and 40 percent by 2040. ...

Farmers hope the country of 35 million will be able to capitalize on the opportunities presented by warmer conditions—including by exporting more food to other regions hard-hit by increasing heat and crop failure. ...

As rising heat and more extreme weather cut harvests in some southern regions, hungry mouths across the developing world may turn to northern nations like Canada for help, experts predict. ❞

—"In Canada, climate change could open new farmland to the plow," by Chris Arsenault, Reuters, 2017

EXAMINE THE SOURCE

1. **Evaluating** How is Canada's role in global food production changing?
2. **Contrasting** How are the impacts of climate change different for Canadians than for people in other regions of the world?

Staying Above Water

One significant challenge associated with climate change is the rising of sea levels around the globe. Glaciers at the poles are melting, which adds more water to the ocean. The warming of the water also causes it to expand and take up more space. Both of these factors contribute to more land being covered up by the rising ocean. Low-lying coastal communities all around the globe are facing an uncertain future. This article discusses actions that are underway in one country that is confronting this challenge.

A road runs atop the 20-mile-long (32 km) barrier known as Afsluitdijk, which holds back the Wadden Sea (right) and keeps large areas of land above water.

PRIMARY SOURCE: ARTICLE

❝ Rising up in a thin line through the waters separating the provinces of North Holland and Friesland, the 87-year-old Afsluitdijk is one of the low-lying Netherlands' key defenses against its ancient enemy, the sea. With climate change bringing more powerful storms and rising sea levels, the **dike** is getting a major makeover.

The Dutch government has embarked on a future-proofing project to beef up the iconic 32-kilometer (20-mile) dam. Work is already underway and is expected to continue until 2023. . . .

The Dutch, whose low-lying country is crisscrossed by rivers and bordered by the sea, have been battling with water for centuries. That challenge will only grow as warmer temperatures cause sea levels to rise. With that in mind, the government this year established a "knowledge program on rising sea levels" that aims to feed expertise into the country's ongoing program of building and maintaining its water defenses.

"The Netherlands is currently the safest delta in the world," the government said, announcing the new program. "We want to keep it that way. . . ."

Engineers are strengthening the Afsluitdijk, including laying thousands of custom-made concrete blocks and raising parts of it.

This kind of innovation and the constant care needed to maintain the Netherlands' thousands of miles of dikes and levees does not come cheap. The government has earmarked nearly 18 billion euros ($20 billion) to fund such projects for the period from 2020–2033. ❞

—"Dutch reinforce major dike as seas rise, climate changes," by Mike Corder and Aleksandar Furtula, AP News, 2019

dike a long wall or earthen bank used to prevent flooding from the sea

EXAMINE THE SOURCE

1. **Comparing** How are the challenges the Netherlands faces in the future similar to challenges the country has faced in the past?
2. **Inferring** How will the need to take action to combat rising sea levels impact the Netherlands?

Water Is Life

All humans depend on access to water to survive. Over time, people have created systems to access, store, and distribute water for farming and to support people living in towns and cities. Reliable water resources have been crucial for populations to expand and thrive. Both the quantity and the quality of water available to people are important. In many locations around the world, water resources are being stretched by ever increasing demand. This article describes the recent experience of one large city.

A public swimming pool in Cape Town, South Africa, sits empty as a result of water shortages.

PRIMARY SOURCE: ARTICLE

66 Cape Town, South Africa has now weathered a full year past its so-called "Day Zero," the day when the municipal water supply for this major city was estimated to run out.

Most of Cape Town relies on dams to supply the city with water. Three years of inadequate rainfall caused dam levels to fall to 25% of capacity by late January 2018, and water was expected to drop to the critical 13.5% of capacity by April 12, 2018. At that critical point [defined as Day Zero], water would only be supplied to critical services such as hospitals, and municipal taps would be shut off. Residents would be forced to line up for daily water rations of 25 liters per person at one of 200 collection points.

Day Zero was pushed back by a full month due to restrictions in allocation of water to surrounding agricultural areas. The city also saved water by implementing a steep tariff [tax] penalizing heavy users of water [and] prohibiting water for pools, lawns, and nonessential uses. . . .

[C]ommunity efforts were equally important. People traded water-saving tips on social media, hotels advised tourists to take short showers and flush toilets only when necessary, and restaurants cut back on making pasta and boiled vegetables.

That rainy season brought average rainfall for the first time in four years, saving the city from its imminent water crisis. . . . However, experts warn that Cape Town is not out of the woods just yet. [F]uture droughts are inevitable. 99

—"Avoiding a water crisis: how Cape Town avoided 'Day Zero,'" *Resilience News*, Global Resilience Institute at Northeastern University, 2019

EXAMINE THE SOURCE

1. **Summarizing** What caused a water-supply crisis in the city of Cape Town?
2. **Identifying Cause and Effect** What enabled the city of Cape Town to avoid reaching Day Zero?

Restoring Natural Systems

Human activities have disrupted natural processes and created negative impacts on the environment. Using land and waterways for agriculture, settlement, and industry have reduced the amount of space available for plants and animals to live. This article describes how people are working to reverse this process with a focus on one particular species.

PRIMARY SOURCE: ARTICLE

❝ Few species manipulate their surroundings enough to make big ecological changes. Humans are one. Beavers are another.

At one point, the rodents numbered in the hundreds of millions in North America, changing the ecological workings of countless streams and rivers. As settlers moved West, they hunted and trapped them to near extinction. Now there are new efforts across the Western U.S. to understand what makes them tick, mimic their engineering skills, boost their numbers, and in turn, get us more comfortable with the way they transform rivers and streams.

Much like us, beavers build dams along streams for their own benefit. They make ponds to protect their lodges and flood areas to increase the vegetation they feed on and use for building materials. While their motivations are selfish, in the process they end up helping their woodland friends, like elk, moose, birds, fish and insects.

Scientists have shown we get lots of benefits, too. Beaver dams improve water quality, trap and store carbon—and in the aggregate could be a significant way of storing groundwater in dry climates. ❞

—"The Bountiful Benefits of Bringing Back the Beavers," Luke Runyon, NPR, 2018

EXAMINE THE SOURCE

1. **Identifying Cause and Effect** What were the reasons that beavers nearly became extinct?
2. **Explaining** What are the benefits of returning beavers to an area?

Complete Your Inquiry

EVALUATE SOURCES AND USE EVIDENCE

Refer back to the Compelling Question and the Supporting Questions you developed at the beginning of the lesson.

1. **Contrasting** Among the responses to environmental change discussed in the sources, which cost a lot of money to implement and which do not?

2. **Identifying Themes** Which sources give examples of how people are responding to environmental change by changing the location of activities?

3. **Gathering Sources** Which sources helped you answer the Supporting Questions and the Compelling Question? Which challenged what you thought you knew when you first created your Supporting Questions? What information do you still need to answer your questions? What other viewpoints can you investigate? Where would you find that information?

4. **Evaluating Sources** Identify the sources that helped answer your Supporting Questions. How reliable is the source? How would you verify the reliability of the source?

COMMUNICATE CONCLUSIONS

5. **Collaborating** Share with a partner which of these sources you found most useful in answering your Supporting Questions. Discuss how the sources help you understand how people are responding and adapting to changes in the environment. Take turns relating how the information you gathered from the sources aligned with or challenged what you previously knew. How do these sources advance your thinking? Use the graphic organizer that you created at the beginning of the lesson to help you. Share your conclusions with the class.

TAKE INFORMED ACTION

Helping Your Local Environment Think about how your community's environment is changing. Research how it is being impacted by human activities. What plants or animals are at risk? Is there a resource being affected by pollution? Determine action you can take to respond to changes in your local environment. Create a social-media post to share your action and the impact you hope to make.

Runyon, L., "The Bountiful Benefits Of Bringing Back The Beavers" 2018, NPR

Reviewing Human-Environment Interaction

Summary

The Environment and Human Settlement	Natural Resources and Human Activities	Human Impact on the Environment

• The natural environment affects where people settle. • Climate, resources, and physical features influence decisions about where people settle. • Innovations in transportation—the steam engine, railroads, and the automobile—have impacted human settlement.	• Natural resources include air, water, soil, minerals, fossil fuels, plants, and animals. • Renewable resources can be replenished in a relatively short period of time. • Nonrenewable resources cannot be replaced once they are used.	• Deforestation is a result of logging, clearing the land for agriculture and other purposes, and the extraction of resources. It leads to a loss of biodiversity. • Desertification is caused by shifts in climate as well as human land use like livestock grazing, cutting trees for fuel, and clearing land for farming.
• Water has shaped how people adapt to the natural environment. • Irrigation and desalination are two ways that people get the water they need. • The use of dams and reclaiming land from the sea allow people to control water in their environment. • Farming techniques help people adapt the environment to suit their needs.	• Resource extraction includes logging, mining, drilling, and fishing. • Energy resources provide the power to do work. They can be renewable or nonrenewable. • The unequal distribution of resources around the world influences economic development.	• People in some parts of the world face water shortages and depletion of groundwater. • Providing freshwater to support human activity is a challenge in drought-prone areas around the world. • The distribution and quality of the world's water varies from region to region. • Water, land, and air pollution are major environmental challenges.
• Natural disasters in one place affect other places. • People adapt to the environment by preparing for and responding to natural disasters. • People prepare for natural disasters through disaster drills, warning systems, building codes, science, and technology.	• Natural materials from the Earth become resources when people assign value to them. • Conservation is necessary to manage Earth's natural resources. • Sustainability is key when using and managing natural resources.	• Climate change refers to the long-term changes in factors used to measure the condition of Earth's atmosphere. • Climate change is caused by natural processes but also by human activities. • Two clues of climate change are rising global air temperatures and rising sea levels.

Checking For Understanding

Answer the questions to see if you understood the topic content.

IDENTIFY AND EXPLAIN

1. Identify each of these terms as they relate to human-environment interaction.

   A. fall line   F. sustainability
   B. alluvial soil   G. deforestation
   C. dike   H. desertification
   D. polder   I. desalination
   E. fossil fuel

REVIEWING KEY FACTS

2. **Explaining** What considerations do people engaged in agriculture make about the natural environment when deciding where to settle?

3. **Making Connections** How is pastoralism dependent on the natural environment?

4. **Identifying** What environmental factor contributed to the growth of Chicago and other cities along the Great Lakes?

5. **Identifying** What are two examples of technology that has allowed people to settle in dry areas?

6. **Describing** What are some examples of the ways in which houses and other buildings can be built to withstand a natural disaster?

7. **Analyzing** How are renewable and nonrenewable resources different?

8. **Making Generalizations** What activities are part of resource extraction?

9. **Explaining** How are wetlands endangered by human activities?

CRITICAL THINKING

10. **Making Connections** What environmental characteristics of the Corn Belt and the Great European Plain make them regions of agricultural importance?

11. **Summarizing** Explain how irrigation helps people adapt to the environment.

12. **Drawing Conclusions** How do natural disasters that occur in one place affect other places?

13. **Analyzing** Why is geothermal energy considered a renewable resource?

14. **Determining Central Ideas** Why is the idea of sustainability especially important when it comes to the use of nonrenewable resources?

15. **Explaining** How does deforestation cause a loss of biodiversity?

16. **Summarizing** What are the environmental concerns associated with desalination?

17. **Making Connections** What types of economic activities contribute to water pollution?

18. **Describing** What are some of the clues of climate change?

19. **Identifying Cause and Effect** What human activities are increased global air temperatures linked to?

20. **Analyzing** How do increased global air temperatures result in increased sea levels?

NEED EXTRA HELP?

If You've Missed Question	1	2	3	4	5	6	7	8	9	10
Review Lesson	2, 3, 5	2	2	2	2	2	3	3	5	2

If You've Missed Question	11	12	13	14	15	16	17	18	19	20
Review Lesson	2	2	3	3	5	5	5	5	5	5

Apply What You Have Learned

 Geographic Reasoning

The places in which people have settled and built communities were in part chosen because of the features of the natural environment. The presence of resources, a tolerable climate, and access to rivers and other bodies of water are some of the factors that have attracted settlement and development. Human development of an area sometimes requires large changes, such as the draining of swampland, clearing of forests, or rerouting of rivers and streams.

ACTIVITY **Describing Settlement** Research online or at a library to learn about how and when the area in which you live—or previously lived—was settled. Determine what features prompted people to build homes and businesses in the area and what challenges the natural environment may have presented to settlement. Then write a brief report explaining what changes to the environment were necessary to make your area livable. Describe what physical or climate obstacles needed to be overcome for your community to be founded and grow.

 Preparing for Disaster

One of the most dramatic ways in which the natural environment impacts the lives of people is the occurrence of natural disasters. Events such as hurricanes, wildfires, and droughts often leave little time for people to react, while other events such as volcanic eruptions, earthquakes, and tsunamis can strike with little or no warning at all. Natural disasters that happen in populated areas of the world can be devastating.

ACTIVITY **Taking a Position** Choose a type of natural disaster that has had an impact on a populated area in recent years. Research what type of preparations or defenses against that disaster were in place. Then write a short speech in which you present your opinion: Were the people of the area as prepared as possible for the disaster? Are there measures that could have been taken to minimize loss of life or property in the affected area? Have the people of the affected area made any new preparations for similar disasters in the future? What challenges limit the degree to which they are able to prepare?

C Overcoming Environmental Challenges

The features of the physical environment have always played a part in determining where humans have been able to live. However, it is more and more possible for technology to help people overcome the restrictions of a given environment. For example, air conditioning makes it possible for people to live comfortably in very hot climates. Advanced transportation methods and housing with controlled climates allow people to travel and work in the extreme cold of Antarctica. It is even possible to visit a snow-skiing facility in the desert of Dubai!

ACTIVITY **Adapting to Extreme Climates**
Research a place where people have used technology to make it possible to live in a specific type of environment. Examples may include regions with extreme temperatures, high altitude, regular flooding, frequent storms, or dangerous wildlife. Then create a slideshow that illustrates the challenges of the environment and the method or methods people have used to make living in that environment possible. Include your opinion as to whether the technology being used sustains or harms that environment.

D Making Connections to Today

You have learned about the importance of sustainability to every society. The need to make sure that there are enough natural resources for the present and the future extends from the local level to the global level. Events such as rising sea levels that affect the world can be impacted by how communities and even individuals live and work.

ACTIVITY **Advising a Leader** Think of a way in which your community can reduce its impact on the environment. For example, you might consider your city's public services, your region's agriculture practices, or your school's energy use. Write a short persuasive letter to a community leader explaining one step that could be taken to reduce the use of nonrenewable resources or otherwise make a change that benefits the environment. Then read your letter to the class, which can then discuss how your suggested change would improve sustainability in your area.

Students stand in line to pay for back-to-school clothes. As consumers, they are part of the U.S. economy.

What Is Economics?

INTRODUCTION LESSON

01 Introducing What Is Economics?   E2

LEARN THE CONCEPTS LESSONS

02 Scarcity and Economic Decisions   E7

03 Economic Systems   E13

05 Demand and Supply in a Market
   Economy   E25

06 Prices in a Market Economy   E31

07 Economic Flow and Economic Growth   E37

INQUIRY ACTIVITY LESSONS

04 Analyzing Sources: Early Economic
   Systems   E19

08 Turning Point: The Industrial
   Revolution   E41

REVIEW AND APPLY LESSON

09 Reviewing What Is Economics?   E47

Royalty-free/Digital Vision/Getty Images

Economics: It's Everywhere!

An *economic system* produces the things people and countries need and want. Economics influences your life in many ways. It determines:

- Why you can't always get what you want

- What you buy—and *when* you buy it

- Which products are available to you—and how much they cost

- Why some jobs pay more than others

- Why some things cost more now than they did 10 years ago (and why other things cost less)

- Why some countries are richer than others

This cartoon shows how expectations about the future may affect your purchasing decisions today. That is economics at work.

Did You Know?

The word *economics* is derived from the Greek word *oikonomia*, meaning "management of the home." Were the Greeks referring to creating a budget and sticking to it? Perhaps. Today, however, *economics* refers to how we can learn to make the best choices to get what we want when we have limited resources, such as money or time.

Scarcity—The Basic Economic Problem

We have unlimited needs and wants, but only limited resources. This situation results in *scarcity*—limits or shortages—in resources or products. Even if everyone in the world were rich, there are not enough resources to satisfy all that we want. Scarcity forces people and societies to make choices.

Prices are a way to reduce the effects of scarcity. Consider what happens during the holiday season when parents are looking to buy the same "hot" toy for their children. Prices go sky-high. Only the people who can afford the high prices will get the toys.

1996

Tickle Me Elmo

The toy first sold for $28.99, but were resold for hundreds of dollars!

2006

Nintendo Wii

The Wii's introductory price was $249.99. When it was discontinued in 2013, it was priced at $99.99!

2016

Hatchimals

Initially priced at $60, but a market shortage resulted in some being sold on eBay for $1,200!

Rationing—Another Way to Deal With Scarcity

What if we're not talking about toys or games, but products necessary to life? How can an economic system distribute necessities fairly? Sometimes it must *ration*, or allow people to have a certain amount of something, using a system other than prices.

» During oil shortages in the 1970s, gas stations often ran out of gas before the end of the day. A system of rationing was introduced: If a license plate ended with an odd number, drivers could get gas only on odd-numbered days. Even-numbered license plates could get gas only on even-numbered days.

» Rationing is not just a thing of the past. Consider the pandemic of 2020 and the shortages that occurred. Stores had to ration toilet paper to keep supplies available.

▶ GO ONLINE — Explore the Student Edition eBook and find interactive maps, charts, graphs, and tools.

E3

Getting Ready to Learn About . . .
What Is Economics?

Economics Includes Goods and Services

Economics affects your life when you earn money or get birthday cash and then decide to spend it on something you need or want. You might buy *goods* such as clothes, food, electronics, or sports equipment. Or you might pay for *services* such as getting your hair styled or attending a concert.

Logs, a natural resource, have been cut, measured, and stacked onto rail cars for distribution and processing.

Economics Includes Resources

Resources make it possible for goods and services to get produced. Resources are at the heart of economics. What resources are shown in these photos? The logs are a *natural* resource. The engineer operating the train is a *human* resource. The machines—referred to as *capital* resources—are tools that create products. Some tools, such as hammers, paintbrushes, and sewing needles, are simple and have been in use for a long time. Other tools—such as robotics and computers—are much more complex machines. Risk-taking individuals who start a new business, introduce a new product, or improve a management technique are also resources—*entrepreneurs*.

Natural resources, human resources, capital resources, and entrepreneurs are called the *factors of production*. You will eventually be part of the human resources or entrepreneur categories—and perhaps both!

At this American auto factory, robotic machines do many of the production tasks. Workers are still needed, however, to run and maintain the machines and the assembly line.

» All of these images show a market.

Economics Includes Markets

When you want to *buy* a good or service, you need to find the people or companies who want to *sell* those goods and services. This happens in markets. A "market" can be a busy shopping mall, a single corner store, or a digital store online. Anywhere that buyers and sellers voluntarily exchange money for a product or a service is a market.

Economics Includes Making Choices

You make choices every day. Should you study another hour to get a better grade on the test? Or should you use that hour to play soccer or practice the violin? Or help get dinner ready? Or play a video game? You can't do everything in that ONE hour, so you have to make choices. The key is to make the BEST choice from all your alternatives. You can do that by considering the benefits and costs of each choice. Understanding the basics of economics will help you make better choices.

You aren't the only one who makes choices. Businesses and consumers do it all the time. For example, businesses have to think about the best locations for their stores and factories—near population centers, near resources, or in a virtual location online. Consumers make choices in how to get the goods and services they need and want.

Economics Includes Prices

Look again at the toys shown earlier, and the ways their prices changed. Prices let you know whether or not to buy something. Prices also let businesses know whether or not to produce something. The government does not set prices in the U.S. economy. Instead, prices vary because of the interaction of you—the consumer—and what you're willing to pay versus the amount that producers are willing to charge.

Demand and supply determine prices in a market economy.

Looking Ahead

Understanding the basics of economics will help you make better decisions. You will learn about scarcity and trade-offs, economic systems, how demand and supply affect prices, the circular flow of the U.S. economy, and how an economy can grow. You will examine Compelling Questions and develop your own questions in the Inquiry Activities.

What Will You Learn?

In these lessons about the basics of economics, you will learn:

- why we must make economic choices.
- the three basic questions societies and their economic systems answer.
- the effects of supply and demand on prices in economic markets.
- how prices aid consumers.
- how the circular flow of economic activity operates in the market system.
- how growth is promoted within an economy.

 COMPELLING QUESTIONS IN THE INQUIRY ACTIVITY LESSONS

- **Why do people need economic systems?**
- **How did changes in production impact the world?**

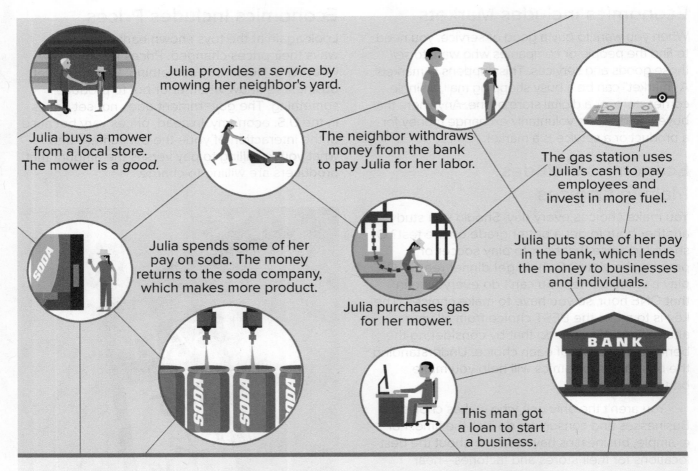

Julia provides a *service* by mowing her neighbor's yard.

Julia buys a mower from a local store. The mower is a *good*.

The neighbor withdraws money from the bank to pay Julia for her labor.

The gas station uses Julia's cash to pay employees and invest in more fuel.

Julia spends some of her pay on soda. The money returns to the soda company, which makes more product.

Julia purchases gas for her mower.

Julia puts some of her pay in the bank, which lends the money to businesses and individuals.

BANK

This man got a loan to start a business.

This diagram shows how everything is connected in an economy. You will learn more about these connections as well as where you fit into the American economy.

02

Scarcity and Economic Decisions

READING STRATEGY

Integrating Knowledge and Ideas As you read, fill out a graphic organizer comparing the four kinds of resources.

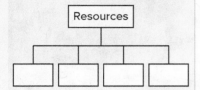

Our Wants and Resources

GUIDING QUESTION

What is scarcity, and how does it affect economic choices?

After an hour of shopping, Jayna found a dress she liked. Then, as she walked toward the cash register, a sweater caught her eye. It was her favorite color, and she liked it immediately. Jayna now had a problem. The sweater cost as much as the dress, and she did not have enough money to buy both. What should she do?

Jayna faced a common problem. She had to decide how to use her limited amount of money to satisfy her competing wants and needs. **Wants** are desires that people have that can be met by getting a product or a service. **Needs** are basic requirements for survival such as food, clothing, and shelter.

Sometimes a good or service can be both a want and a need. For example, the new dress could be both because it is clothing. If Jayna could live without it, however, it would be more of a want than a need.

Unlimited Wants

If Jayna is like the rest of us, her wants are not limited to just two items. If you think about all the things you want, the list is probably so long that we could say your wants are unlimited.

Wants fall into two groups. The first is **goods** and includes things we can touch or hold. The second is **services** and includes work that is done for us. Services include the health care provided by a doctor, the haircut by a hairstylist, or advice about money provided by a banker.

want desire for a good or a service
need basic requirement for survival

goods things we can touch or hold
services work that is done for us

A woman is signing for a package delivered to her home.

Analyzing Visuals What main good and main service do you see in the image?

Societies use many different resources for production. Land (top left) is a natural resource required for the production of crops such as corn. Workers in a lab (top right) are an example of labor, or human resources. Cranes (bottom left) are a capital resource. An entrepreneur (bottom right) designs a new clothing line for her business.

Making Connections Give an example of each type of resource necessary to create a pair of running shoes.

Limited Resources

If resources are limited, and if wants are unlimited, then we have to make choices. **Economics** is the study of how people choose to use their limited resources to satisfy their unlimited wants. **Resources** are all the things that can be used in making products or services that people need and want. Economists talk about four types of resources:

- **Natural resources** include a nation's land and all of the materials nature provides that can be used to make goods or services. Good soil for growing crops, trees for cutting lumber, and iron for making steel are natural resources.
- **Labor** includes workers and their abilities. The more workers a society has, the more it can produce. Workers' knowledge and skills are important, too. The more workers know and the better their skills are, the higher the quality of goods and services they produce.
- **Capital**, which includes buildings and tools, is another type of resource. Businesses build factories to manufacture goods. Equipment such as computers can help work go more quickly. Trucks or trains are used to move goods around. Capital resources make work more productive.
- **Entrepreneurs** are a special resource. They are risk-taking individuals who start a new business, introduce a new product, or improve a method of making or doing something.

Economists refer to these four categories of resources as the **factors of production**. Each factor is necessary to produce goods and services.

---

economics study of how individuals and nations make choices about ways to use scarce resources to fulfil their needs and wants

resources things used to make goods or services

natural resources land and all the materials nature provides

labor workers and their abilities

capital factories, tools, and equipment that manufacture goods

entrepreneurs risk-taking individuals who start a new business, introduce a new product, or improve a method of making something

factors of production four categories of resources used to produce goods and services: natural resources, capital, labor, and entrepreneurs

Scarcity—The Basic Economic Problem

Jayna is not the only one who has the problem of satisfying her competing wants and needs. This is the type of economic problem that everyone—from individuals to businesses to cities, states, and countries—faces every day.

Scarcity occurs whenever we do not have enough resources to produce all of the things we would like to have. In fact, no country has all of the resources it needs, or would like to have. Because of this, *scarcity is the basic economic problem*. Economics is a social science that looks at how we go about dealing with this basic economic problem.

✓ **CHECK FOR UNDERSTANDING**

1. **Contrasting** Explain the difference between a want and a need.
2. **Identifying** What is the basic economic problem faced by people and nations alike?

Economic Decisions

GUIDING QUESTION

Why are trade-offs important in making economic decisions?

Have you ever had to make a choice between two things you really wanted to buy? If so, you have had some practice with economic decision making. Perhaps you had to choose between buying a video game and going to a movie with your friends. To make a good decision, you had to consider the benefits and the costs of each choice. In fact, you already think about many of your choices in the same way that economists do.

Trade-Offs

Making a **trade-off** is giving up one alternative good or service for another. If you choose to buy a pair of running shoes, you are exchanging your money for the opportunity to own the running shoes rather than something else that might cost the same amount.

A trade-off does not apply only to decisions involving money. For example, you might need to decide whether to go to a friend's party or study for an important test. In this case, you would have to make a trade-off with your time. What will you give up—time with friends or studying time?

Businesses also make trade-offs. A company might have to decide whether to invest in research for new products or spend money on advertising to increase sales of existing products. Managers might need to choose whether to give big bonuses to a few workers or small raises to all workers.

Governments face trade-offs, too. If they spend money to build schools, they might not have enough money to build roads or pay for national defense.

When societies and civilizations understand that every decision involves a trade-off, they are better able to use scarce resources wisely. For example, early farmers faced declining crops as the soil became less healthy year after year. Then some farmers changed to a three-field rotating system. One field was left unplanted, and the other two fields alternated types of crops. The next year, a different field was unplanted. The trade-off for fewer crops overall was healthier soil and better harvests in the two planted fields.

Opportunity Costs

When faced with a trade-off, people eventually choose one **option**, or alternative, over all others. For example, you decide to buy a pair of running shoes and give up the chance to buy something else. Or, you choose to study and give up the opportunity to spend time with your friends.

When a city pays to build a new road or bridge like this one, the trade-off is a new school or park or anything else the money could have funded.

---

scarcity situation of not having enough resources to satisfy all of one's wants and needs

trade-off alternative you face when you decide to do one thing rather than another

option alternative, choice

Opportunity cost is the cost of the *next best* use of your money or time when you choose to do one thing rather than another. Economists use the term *opportunity cost* very specifically. The term is reserved only for the next-most-attractive alternative. Other options rejected earlier in the decision-making process are trade-offs but are not considered opportunity costs.

The choices made by businesses and societies also have opportunity costs. Suppose your city has narrowed its choices to spending money on park improvements or fixing city sidewalks. If it decides to spend money on the park, the opportunity cost would be the sidewalks that would not be fixed. If Congress votes to increase spending on preschools rather than on food programs, the opportunity cost is the support not given to the food programs.

Opportunity cost applies to all resources. Choosing to watch a television show one evening will not cost you any money. However, that choice has an opportunity cost. The cost is the time you

---

opportunity cost cost of the next best use of time or money when choosing to do one thing rather than another

---

could have spent doing other things, such as listening to music or visiting your friends.

Businesses also face opportunity costs that do not involve money. For example, some companies may require employees to spend time learning new computer programs. The opportunity cost of that decision is the loss of the employees' work while they are being trained. Why would a company make such a decision? The company might believe the training will help workers be more productive in the long run. Good decisions involve weighing all possible options.

You might think you can avoid opportunity cost, but you really cannot. For example, suppose you want to watch television while you do your homework. You may think you can do both at the same time. However, you risk making careless mistakes while watching TV, or you may do your homework slower. In addition, you may miss some of the story in the TV show. There are costs to both activities when you try to do two things at once.

Opportunity costs will be different for everyone because people are different. If three people are asked to identify the opportunity cost of watching a single TV show, the first person may say it was the homework that could not be completed.

A woman buys a plate at a yard sale. Opportunity costs are everywhere—even at yard sales.
Analyzing Visuals What is a possible opportunity cost for the woman who is buying the plate? Explain your answer.

The second might say the opportunity cost was the time missed talking to friends. The third may lose $20 that could have been earned mowing the lawn. Choosing to do something always has an opportunity cost, and that is why it's important to evaluate your alternatives.

Benefit-Cost Analysis

How can you make the best choices? Individuals, businesses, and governments often perform a benefit-cost analysis of alternative choices. **Benefit-cost analysis** divides the size of the benefit by the cost, and then selects the larger of the two. This type of analysis often helps businesses choose among two or more projects.

For example, suppose a business must choose between investments A and B. If project A is expected to generate $100 at a cost of $80, the benefit-cost ratio is 1.25. We get the number 1.25 by dividing $100 by $80. If project B is expected to generate $150, and if it costs $90, it will have a benefit-cost ratio of 1.67. The business would then choose the one with the higher benefit-cost number, or project B.

	Project A	Project B
Benefits:	$100	$150
Costs:	$80	$90
Benefits/Costs:	1.25	1.67

No decision-making strategy is perfect. But reasonable choices can be made if the short-term and long-term benefits and costs of competing projects are carefully evaluated. For example, suppose a choice must be made to fix an existing two-lane bridge or to build a new four-lane bridge. In the short term, it will cost less to repair the existing bridge, with the benefit of being repaired sooner than a new bridge would take to build. In the long term, building the new, wider bridge will cost more money but will benefit more drivers. People living near the bridge may want construction to be over quickly, so they will vote for repairing the existing bridge. Drivers who must wait in traffic to cross the smaller bridge may vote for funding the new wider bridge.

Finally, the decisions people face cannot always be evaluated in terms of money. Yet even those decisions can be analyzed with

benefit-cost analysis economic decision-making model that divides the total benefits by the total costs

benefit-cost analysis. For example, suppose you are deciding how long a nap to take. The benefit will be the greatest during the first hour of sleep, and then less and less. There would also be a cost: the opportunity cost of other things you could not do. The cost of the first hour of sleep would likely be small. But the longer you slept, the greater the cost would be. Even if you are not aware of it, benefit-cost analysis applies to almost everything we do. This is what it means to "think like an economist."

✓ **CHECK FOR UNDERSTANDING**

1. **Explaining** What is an opportunity cost in an economic decision?

2. **Analyzing** Why is it useful for individuals to do a benefit-cost analysis?

Societies and Economic Choices

GUIDING QUESTION

What determines how societies make economic choices?

Just as individuals and businesses make economic choices, so do entire countries. Scarcity is an economic problem in every nation. Will a society use its limited resources for education or for health care? Will a nation focus on helping businesses grow so they can create more jobs? Or will it spend money on training people for new jobs? Will it spend money on defense or on cleaning up the environment?

Three Basic Economic Questions

Scarcity of resources forces societies to make economic choices. These choices must answer three questions: WHAT goods and services will be produced? HOW will they be produced? FOR WHOM will they be produced?

Each country or society has to decide WHAT goods and services it will produce to meet its people's needs and wants. In making these decisions, societies consider their natural, human, and capital resources. A nation with plenty of land, fertile soil, and a long growing season is likely to use its land to grow crops. A country with large reserves of oil might decide to produce oil.

After deciding what to produce, entrepreneurs and other members of a society must decide HOW to produce these goods and services.

Choices All Societies Face

Societies must deal with the same economic problem of scarcity that individuals do.

```
┌──────────────────┐                    ┌──────────────────┐
│    UNLIMITED      │                    │     LIMITED       │
│     WANTS         │                    │    RESOURCES      │
└──────────────────┘                    └──────────────────┘
            │                                    │
            └──────────────┬─────────────────────┘
                           ▼
                  ┌──────────────────┐
                  │     SCARCITY      │
                  └──────────────────┘
                           │
                           ▼
           ┌──────────────────────────────────┐
           │   CHOICES ALL SOCIETIES FACE      │
           └──────────────────────────────────┘
            │               │                 │
            ▼               ▼                 ▼
   ┌────────────┐   ┌────────────┐   ┌────────────┐
   │   WHAT     │   │    HOW     │   │  FOR WHOM  │
   │ to produce │   │ to produce │   │ to produce │
   └────────────┘   └────────────┘   └────────────┘
```

EXAMINE THE DIAGRAM

1. **Identifying** What are the three basic economic questions that societies face?

2. **Explaining** Why do societies have to make economic choices?

Should they encourage businesses to build factories for large-scale manufacturing of products such as automobiles or shoes? Or should they **promote**, or encourage, small businesses and individual craftwork instead?

After goods and services are produced, a society must decide WHO gets the goods and services. Societies have different ways of distributing goods. The choices they make for distributing goods affect how the goods are consumed. For example, should new housing units be reserved for low-income people, or should they be rented to anyone who can afford them? Should new cars be given to public officials, or should they be sold to the highest bidder? These are not easy questions to answer, but they are ones well-suited to benefit-cost analysis.

Values of a Society

The resources of a nation are not the only reason societies answer the three basic questions differently. What a society *values*, or thinks is most important, also has a big influence. Some societies value individual freedom the most. Others think that economic equality is most important.

promote to support or encourage

Different answers to the three basic questions help a society promote the ideas its people believe are most important. The key thing to remember is that all societies face the same three problems of deciding WHAT to produce, HOW to produce, and FOR WHOM to produce. In a separate lesson, you will learn about the various types of economic systems and how they answer the three basic economic questions.

✓ **CHECK FOR UNDERSTANDING**

1. **Summarizing** How do resources affect WHAT to produce?

2. **Explaining** What role do values play in answering the three basic economic questions?

LESSON ACTIVITIES

1. **Narrative Writing** Write a short story about a student your age who has to make an economic choice. In your story, reveal how plentiful wants conflict with scarce resources.

2. **Collaborating** With a partner, write and create a public service announcement video that explains the benefits of identifying opportunity costs when making decisions. Be sure to use appropriate voice and tone in your presentation.

03
Economic Systems

READING STRATEGY

Integrating Knowledge and Ideas As you read, complete a graphic organizer to identify features of different types of economic systems.

Traditional	Market	Command	Mixed

Traditional Economies

GUIDING QUESTION

What characteristics do traditional economies share?

Societies make economic choices or decisions in different ways. Each country has its own **economic system**, or way of producing and distributing the things people need and want. Economists organize economic systems into four general types: traditional, command, market, and mixed. The way a society answers the three basic economic questions—WHAT, HOW, and FOR WHOM to produce—determines its type of economic system. A country's economic system also relates to the resources it has (land, labor, capital, entrepreneurs) as well as its values and economic goals.

For centuries, most early societies organized their economic systems in what we now call traditional economies. In a **traditional economy**, the economic questions are answered on the basis of traditions and customs—or the way things have always been done. The WHAT to produce question is determined by tradition. If you were born into a family of farmers, for example, you would grow up to be a farmer. If your family always hunted, you would be a hunter. Tradition also determines the HOW to produce question. You would farm or hunt using the same tools your parents and grandparents did. The FOR WHOM to produce question is also answered by tradition. Hunters might keep the best portion of their kill and then share the rest evenly with other families. Farmers might provide crops to the elderly and children first—and then *barter*, or exchange, the rest of their crops for meat or fish from hunters and fishers.

economic system a nation's way of producing and distributing things its people want and need

traditional economy economic system in which the decisions of WHAT, HOW, and FOR WHOM to produce are based on traditions or customs

In the past, this Inuit boy would have learned the tools and techniques of hunting from his father and grandfather.

Analyzing Visuals What evidence identifies the Inuit as a former traditional economy?

Traditional economies had advantages as well as disadvantages. The biggest advantage was less uncertainty about what to do. Everyone knew what was expected of them as they worked to take care of the community. Another advantage was that traditional economies used fewer resources than other types of economic systems. Their biggest disadvantage was they tended to discourage new ideas and were slow to adopt better ways of producing goods. Men and women stayed within their own economic roles.

Most traditional economies slowly changed over time. As they changed, some barter economies invented primitive forms of money. These monies included stones, brightly colored shells or beads, feathers from exotic birds, bundles of tea and tobacco, and eventually early gold and silver coins. The majority of traditional economies **evolved**, or progressed gradually, into ones with well-developed markets. Others with strong political or military leaders became economies based on rigid controls.

Traditional economies are rare today. Certain groups untouched by modern life in the Amazon rain forest or the mountainous regions of Papua New Guinea qualify as having traditional economies. Examples from the last century include the First Nations and Inuit societies of northern Canada and Alaska, and aboriginal peoples of Australia.

✓ **CHECK FOR UNDERSTANDING**

1. **Identifying** How are economic decisions determined in a traditional economy?

2. **Summarizing** What are some advantages and disadvantages of a traditional economy?

Command Economies

GUIDING QUESTION

Who makes the basic economic decisions in a command economy?

Over time, a few traditional economies with strong political or military leaders evolved into command economies. In a **command economy**, the government owns the majority of land, labor, and capital resources. Individuals and businesses do not have much say in how the economy works. Government planners *command* the actions that producers must follow. These powerful planners answer the basic economic questions. They decide WHAT to produce. For example, they decide whether the society will produce military tanks or consumer goods such as shoes.

---

evolve to progress or develop gradually

command economy economic system in which the government owns and directs the majority of a country's land, labor, and capital resources

A North Korean announcer reports on a military parade in Pyongyang, North Korea's capital.

Speculating What kinds of information do you think North Koreans are given on government-controlled programs?

Planners also decide HOW to produce these goods and FOR WHOM the goods and services will be produced. In theory, the government provides all housing, education, health care, and consumer products. In reality, much of the country's wealth goes to the leader or other high government officials.

Command economies are not very efficient. Workers are told where and how to work—and how many products to make. If planners mistake the amounts, people must go without goods and services. In addition, the pay for all workers—from factory workers to doctors—may be roughly similar. People do not get rewarded for working hard, so products are often low quality. And because of the vast number of decisions that central planners must make, command economies tend to grow slowly and inefficiently.

Central planning can provide advantages in rare instances. In times of emergency, planners can direct resources where they are needed most. For example, central planning helped the Soviet Union rapidly rebuild its economy after World War II. Planners shifted resources from farming and consumer goods to industrial factories. As a result, many Soviet people went for decades without decent housing, good food, and everyday products.

Modern examples of pure command economies are limited to a handful of dictatorships. North Korea is perhaps the leading example of a command economy where everything is either owned or controlled by the government. That includes businesses, electricity, transportation, housing, the Internet, and TV programs. Under Fidel Castro and his successors, the Cuban economy was another example of a command economy where consumer choices were limited.

Socialism

Another economy with command elements is socialism. **Socialism** is both an economic and political system in which the government owns some, but not all, of the factors of production. Under socialism, the government's goal is to serve the basic needs of all people, not just its leaders. Therefore, the government directs economic activity to provide transportation, education, jobs, and health care while allowing for some personal property. Sweden experimented briefly with socialism after World War II before it returned to being a market economy. Venezuela, under the direction of Hugo Chavez in 1999, was the most recent country to try socialism.

Communism

Communism is another version of a command economy that involves socialism. The goal of communism is to build a society run entirely by workers without any government involvement. Workers themselves would own all the resources and make all the economic decisions to improve society. Everyone would work to the best of their abilities, and use only the products they need. In this ideal society, no government would be necessary and could be removed.

There has never been a true communist economy. Modern communist leaders say that they must enforce a system of socialism to get ready for the eventual communist state. Meanwhile, they operate their political and economic system in such a way as to accumulate much of society's wealth for themselves. Today the term *communist* generally refers to a country's political system in which a ruling party has all the power. The citizens have limited democratic rights and freedoms.

✓ **CHECK FOR UNDERSTANDING**

1. **Identifying** What is the government's role in a command economy?

2. **Summarizing** What are the major problems with a command economy?

A sign in Cuba states "Young Communist's Union." Cuba has a command economy, but no country has a true communist economy.

---

socialism economy in which government owns some factors of production so it can distribute products and wages more evenly among its citizens; economic system with some command features

communism theoretical state where all property is publicly owned, and everyone works according to their abilities and is paid according to their needs

Dave Moyer

Market Economies

GUIDING QUESTION

What characteristics do market economies share?

Have you ever been to a flea market? You might see a person at one table selling used comic books. He can sell them because they are his to sell. You can choose to buy comic books from that seller, from another seller, or not at all. The seller sets the price of those comic books based on how much he thinks people will pay for them. If no one wants to buy his comic books at that price, he will probably lower the price.

In contrast, if many people are willing to pay that price, he may raise the price even more. Other tables at the flea market will have people selling different goods, which might be things they made by hand or used items they no longer want. The buying and selling in a flea market is an example of a free market economy.

Characteristics of a Market Economy

A **market** is where a buyer and a seller voluntarily exchange money for a good or service. It can be a busy shopping mall, a single corner store, or a digital store online.

In a **market economy**, individuals and businesses have the freedom to use their resources in ways they think best. This freedom helps answer the WHAT, HOW, and FOR WHOM to produce questions. For example, people can spend their money on the products they want most. This is like casting dollar "votes" for products and tells producers WHAT to produce.

Businesses are also free to find the best production methods when deciding HOW to produce. And the income that consumers earn and spend in the market determines the FOR WHOM to produce question.

Prices of goods and services also play an important role in the market economy. Prices are like signals. If the price of a good or service goes up, it is a signal to producers to make more. At the same time, it is a signal to buyers to think about buying less. Or, if the price of a good or service goes down, the opposite happens. It is a signal to producers to produce less, and it is a signal to buyers to buy more. In either case, the market will soon be operating at a new price established by both buyers and sellers.

In a market economy, individuals and businesses act in their own self-interest. No central authority makes their decisions. Instead, the market economy seems to run itself. Private individuals—not the government—own the factors of production. Because individuals own these factors, they have the power to decide how to use them.

The goal is to use limited productive resources to earn **profits**—money over and above the costs of making the product or service. As a result, entrepreneurs and businesses have the **incentive**, or motivation, to come up

---

market place where a buyer and a seller voluntarily exchange money for a good or service

market economy economic system in which individuals and businesses have the freedom to use their resources in ways they think best

profit money earned over and above the costs of making a product or service

incentive motivation

Price Signals in a Market Economy

In a market economy, a change in price signals a change in producer and consumer behavior.

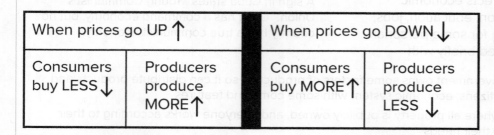

When prices go UP ↑		When prices go DOWN ↓	
Consumers buy LESS ↓	Producers produce MORE ↑	Consumers buy MORE ↑	Producers produce LESS ↓

EXAMINE THE CHART

Identifying What event will signal producers to produce less?

One feature of a market economy is competition. These businesses, located close to one another, must compete for customers. Customers can compare prices, products, and service and make a choice.

Making Connections
Why is competition important to consumers?

with better products and efficient ways of producing them to earn higher profits.

Advantages and Disadvantages of a Market Economy

Market economies give people a lot of freedom. People are free to own property, control their own labor, and make their own economic decisions. Such freedom gives people who live in market economies a high level of satisfaction.

Another advantage of a market economy is the competition that occurs. Sellers compete with each other to attract the most buyers. Buyers compete with each other to find the best prices. Such competition leads to a huge variety of products and services for consumers. If a product can be imagined, it is likely to be produced in hopes that people will buy it. And because sellers must compete against other sellers, products are made with quality in mind.

Market economies do have some disadvantages. Although they enjoy a high degree of success, they usually do not grow at a steady rate. Instead, they go through periods of growth and decline. Although the periods of growth are much longer than the periods of decline, people can be hurt in the down times. For example, many people lose their jobs during down times.

Another problem is that businesses, driven by profit, might not give workers good working conditions or high wages. The profit motive can also result in harmful side effects of business activities, such as pollution.

Finally, a market economy does not provide for everyone. Some people may be too young, too old, or too sick to earn a living or to care for themselves. These people would have difficulty surviving in a pure market economy without help from family, government, or charity groups.

✓ **CHECK FOR UNDERSTANDING**

1. **Identifying** Who owns the factors of production in a market economy?

2. **Analyzing** What are the main advantages of a market economy?

Mixed Economies

GUIDING QUESTION

Why do most countries have a mixed economy?

Today most nations of the world have a **mixed market economy**, or an economic system that has elements of tradition, command, and markets that answer the WHAT, HOW, and FOR WHOM questions. A reason for mixed market economies is that nations tend to evolve, shedding some economic features and adding new ones.

---

mixed market economy economic system in which markets, government, *and* tradition each answer some of the WHAT, HOW, and FOR WHOM questions

What Is Economics? **E17**

Shutterstock / Pix One

Two factory workers wear hardhats, safety vests, and safety goggles. The U.S. government sets safety regulations such as these for businesses.

The United States economy is based on a market system with elements of command and tradition. Individuals and businesses can choose how to use their resources, and prices determine who will receive the goods and services produced. Businesses are usually free to compete for profit with limited interference from the government.

The U.S. economy is not a pure market economy because government has been asked to perform several important functions. It uses taxpayer money to provide some goods and services that competitive markets do not. These include highways, bridges, and airports. The government also provides a system of justice, national defense, and disaster relief.

The government helps markets function smoothly by making sure that sellers are honest and that buyers are protected. Although laws put some restrictions on business owners, people support these laws when they benefit society. The government makes rules for how workers are to be treated. It requires that a minimum wage be paid to most workers, and that workplaces are safe. In addition, the government requires workers to pay taxes on their income and on some of the things they own, such as cars, homes, and other property.

The American economy also has elements of a traditional economy. For example, many people often decide to work in the same job as a parent. Many people also take care of family members who are too old or too sick to care for themselves.

✓ **CHECK FOR UNDERSTANDING**

1. **Analyzing** What are signs that the United States does not have a pure market economy?

2. **Explaining** Why is the U.S. economy called a mixed market economy?

LESSON ACTIVITIES

1. **Argumentative Writing** Choose characteristics from at least two of the main economic systems. Write an essay explaining why these characteristics combined would make a better mixed system than one that was a pure traditional, command, or market system. Use examples and evidence to support your argument.

2. **Presenting** With a partner, think about the disadvantages of a traditional economy. Discuss what actions might be taken to boost economic growth without damaging the social structure on which the society survives. Use your ideas to create a multimedia presentation you might give to an elder member of the society. Show how his or her traditional economy combined with market forces could benefit the society.

04
Analyzing Sources: Early Economic Systems

? COMPELLING QUESTION

Why do people need economic systems?

Plan Your Inquiry

DEVELOPING QUESTIONS

Think ahead a few years when you get your first steady job. In the U.S. free market system, you will work for wages or a salary, and then you'll use those earnings to buy what you need and want—either at a local store or online from anywhere in the world. Have you ever wondered how early people and civilizations obtained what they needed? Read the Compelling Question for this lesson. What questions can you ask to help you answer this Compelling Question? Create a graphic organizer like the one below. Write three Supporting Questions in the first column.

Supporting Questions	Source	What this source tells me about the economic system	Questions the source leaves unanswered
	A		
	B		
	C		
	D		
	E		

ANALYZING SOURCES

Next, read the introductory information for each source in this lesson. Then analyze each source by answering the questions that follow it. How does each source help you answer each Supporting Question you created? What questions do you still have? Write those in your graphic organizer.

After you analyze the sources, you will:
- use the evidence from the sources
- communicate your conclusions
- take informed action

Background Information

Four types of economic systems exist today: traditional, command, market, and mixed. They address the three basic economic questions of WHAT to produce, HOW to produce, and FOR WHOM to produce.

This Inquiry Lesson includes sources about earlier economic systems, including the barter system and early trade routes, manorialism, mercantilism, and the rise of capitalism. As you read and analyze the sources, think about the Compelling Question.

Merchants made long and difficult voyages along the Silk Road, which connected Asia, the Middle East, and Europe. Merchants traveled in caravans of pack animals such as camels to transport valuable goods.

Ratnakorn Piyasirisorost/Moment Open/Getty Images

The Barter System and Early Trade

A barter economy is a moneyless economy that relies on trade. Barter systems have existed since ancient times. The exchange of goods and services is difficult because the products some people have to offer are not always acceptable to others—or easy to divide for payment. For example, how could a farmer with a pail of milk obtain a pair of shoes if the cobbler wanted a basket of fish? It takes time to barter unless two people want exactly what the other has.

In their barter economies, many ancient societies used some form of commodity money—or "money" that has an alternate use as a product, such as clams or tobacco. The image below shows what an Aztec market scene might have looked like in Mexico in the 1400s. In Tenochtitlán and other large cities, Aztec merchants traded goods made by craftspeople. In exchange for the goods, the merchants obtained tropical feathers, cacao beans, animal skins, and gold—items they used as "money" to purchase other goods. When the Spanish arrived in the Americas in the early 1500s, they were astonished to find city markets considerably larger and better stocked than any markets in Spain.

SECONDARY SOURCE: ILLUSTRATION OF A MARKET

An artist has drawn an Aztec market scene. In the illustration, merchants exchange products for local "currency" in Mexico in the 1400s.

EXAMINE THE SOURCE

1. **Analyzing** What products are offered in the Aztec market?
2. **Drawing Conclusions** What appears to be the currency—or "money"—used for payment in this market?

B

The Silk Road

As trading became common, trade routes slowly formed between early societies. Possibly the most famous trade route in history is the Silk Road, which connected China to Europe.

SECONDARY SOURCE: ACADEMIC TEXT

" This framework of roads had its roots in the network of routes that started in Persia and along which **emissaries** with messages galloped throughout the empire in the 4th century B.C. However, in its final **configuration**, the Silk Road was officially opened in 130 B.C., when the Chinese emperor sent his ambassador Zhang Quian on a diplomatic mission in search of new allies. In addition to pacts, the ambassador returned with a new breed of horses and saddles and stirrups used by western warriors. This is the first example of the Silk Road's main function throughout history: the exchange of knowledge and technology.

The current view among historians is that the Silk Road—in service from its opening in 130 B.C. until the 14th century—was used by traders, religious, artists, fugitives and bandits, but above all by refugees and populations of emigrants or displaced persons. It is believed that it was precisely these groups of migrant populations who brought with them knowledge, tools, culture, products or crops (and with them possibly new techniques and irrigation systems). They fostered a cultural and technological "globalization" that was literally going to change the world.

On the **commercial** side, the Silk Road was a small-scale, local trade network, with goods passing from one merchant to another in the markets and exchange centres that lined the route. In both directions, food and animals, spices, materials, ceramics, handicrafts, jewellery and precious stones circulated. And although its name suggests otherwise, silk was not the main **commodity**. What's more, it never received this name during the almost 1,400 years that the Silk Road remained operational. The name was coined centuries later, in 1877, by the German geographer Ferdinand von Richthofen, because silk was the most valued and appreciated product among the nobles and dignitaries of the Roman Empire. "

— From "The Silk Road: The Route for Technological Exchange that Shaped the Modern World" by Miguel Barral, www.bbvaopenmind.com

emissaries agents
configuration arrangement
commercial business
commodity product

EXAMINE THE SOURCE

1. **Identifying** What products went back and forth along the Silk Road?
2. **Analyzing** How did products move from China to the nobles of the Roman Empire?

Manorialism

The manorial system arose in medieval Europe (800–1300 C.E.). A manor was an agricultural estate with a central castle or manor house. In exchange for the manor's protection, unfree serfs worked about three days a week farming the lord's share of land. Serfs paid rent by giving the lord a portion of every product they raised on separate pieces of land the rest of the week. Serfs also paid the lord for the use of the manor's common pasturelands, streams, ponds, and woodlands. Free peasants used the rest of the estate's land to grow food for themselves. Except for a few items such as salt and millstones, the manor provided everything. Most people rarely interacted with those outside of the manor.

Medieval lord and peasants

SECONDARY SOURCE: ILLUSTRATION OF A MANOR

Fields
In the spring, serfs planted crops such as summer wheat, barley, oats, peas, and beans. Crops planted in the fall included winter wheat and rye. Women often helped in the fields.

Church
Village churches often had no benches. Villagers sat on the floor or brought stools from home.

Castle
Castles were built in a variety of forms and were usually designed to fit the landscape.

Serf's Home
Serfs had little furniture. Tables were made from boards stretched across benches, and most peasants slept on straw mattresses on the floor.

EXAMINE THE SOURCE

1. **Analyzing** What obligations did lords and peasants fulfill on a manor?
2. **Summarizing** Summarize the manorial system in one to three words.

D

Mercantilism

From the 1500s to late 1700s, nation-states formed in western Europe. Market towns replaced manors as commercial hubs. Merchants in these towns paid taxes to the government. The taxes supported large armies that protected the nation and fought wars to gain colonies. In return, merchants expected the government to protect them from foreign competition. Thus, the economic system of mercantilism arose. The government increased the nation's wealth by controlling the products that came into and went out of the home country.

SECONDARY SOURCE: ACADEMIC TEXT

66 These [mercantilist] policies took many forms. **Domestically**, governments would provide **capital** to new industries, **exempt** new industries from **guild** rules and taxes, establish monopolies over local and colonial markets, and grant titles and pensions to successful producers. In trade policy the government assisted local industry by imposing **tariffs**, **quotas**, and **prohibitions** on **imports** of goods that competed with local manufacturers. Governments also prohibited the **export** of tools and **capital equipment** and the emigration of skilled labor that would allow foreign countries, and even the colonies of the home country, to compete in the production of manufactured goods. At the same time, diplomats encouraged foreign manufacturers to move to the diplomats' own countries.

Shipping was particularly important during the mercantile period. With the growth of colonies and the shipment of gold from the New World into Spain and Portugal, control of the oceans was considered vital to national power. Because ships could be used for merchant or military purposes, the governments of the era developed strong merchant marines. . . .

During the mercantilist era it was often suggested, if not actually believed, that the principal benefit of foreign trade was the importation of gold and silver. . . . For nations almost constantly on the verge of war, draining one another of valuable gold and silver was thought to be almost as desirable as the direct benefits of trade. 99

— From "Mercantilism" by Laura LaHaye, Library of Economics and Liberty

domestically at home, within the country	**prohibitions** ban, to forbid
capital investment money	**imports** to bring products into a country from a foreign country
exempt free, to excuse	**export** to send or sell to another country
guild union, group of companies	**capital equipment** machines that make products
tariffs taxes on imported products	
quotas limits on imported products	

EXAMINE THE SOURCE

1. **Inferring** Which segments in a country benefited most from mercantilism—producers, consumers, or government?
2. **Drawing Conclusions** How did mercantilism affect trade?

LaHaye, Laura. "Mercantilism." In The Concise Encyclopedia of Economics. Edited by David R. Henderson. Indianapolis: Liberty Fund, 2008.

Capitalism

Adam Smith (1723–1790) is known as the father of economics. In 1776 he published *The Wealth of Nations*, an influential book that promoted free markets rather than markets regulated by governments under mercantilism. Smith also promoted capitalism—an economic system in which individuals privately own the factors of production and use them to make profits. Smith believed that society as a whole benefits when individuals pursue their own economic interests.

PRIMARY SOURCE: BOOK

" It is not from the **benevolence** of the butcher, the brewer, or the baker that we expect our dinner, but from their regard to their own interest. We address ourselves, not to their **humanity**, but to their self-love, and never talk to them of our own necessities, but of their advantages. . . .

. . . [E]very individual . . . neither intends to promote the public interest, nor knows how much he is promoting it. . . . [H]e intends only his own gain, and he is in this, as in many other cases, led by an invisible hand to promote an end which was no part of his intention. "

— From Adam Smith's *An Inquiry Into the Nature and Causes of the Wealth Of Nations*. Edinburgh: Thomas Nelson, 1827.

benevolence generosity
humanity kindness

EXAMINE THE SOURCE

1. **Analyzing** How does Smith describe a successful interaction between a buyer and seller?

2. **Inferring** According to Smith, what motivates people to work hard?

Complete Your Inquiry

EVALUATE SOURCES AND USE EVIDENCE

Refer back to the Compelling Question and the Supporting Questions you developed at the beginning of the lesson.

1. **Evaluating** Which of the sources affected your thinking the most? Why?

2. **Identifying Themes** What theme underlies all of the economic systems?

3. **Gathering Sources** Which sources helped you answer the Supporting Questions and Compelling Question? Which sources, if any, challenged what you thought you knew when you first created your Supporting Questions? What information do you still need in order to answer your questions? Where will you find that information?

4. **Evaluating Sources** Identify the sources that helped answer your Supporting Questions. How reliable is the source? How would you verify the reliability of the source?

COMMUNICATE CONCLUSIONS

5. **Collaborating** Work with a partner to create an illustrated time line of early economic systems and how they answered the basic economic questions. How do these sources provide insight into why economic systems developed and changed over time? Use the graphic organizer that you created at the beginning of the lesson to help you. Share your time line with the class.

TAKE INFORMED ACTION

Discussing How Piracy Affects Entrepreneurs Recall that a free market economy such as that promoted by Adam Smith and capitalism relies on profits as the incentive for entrepreneurs to take risks and create new products and services. Research the amount of money lost from pirated products, such as when individuals illegally download music or when countries illegally copy and sell DVDs. Discuss what actions, if any, you think American companies should take to stop piracy.

Demand and Supply in a Market Economy

READING STRATEGY

Analyzing Key Ideas and Details As you read, complete a chart describing the factors that cause a change in quantity demanded and quantity supplied, as well as factors that cause a change in demand and supply.

Factors that Cause a ...	
Change in Quantity Demanded	Change in Quantity Supplied
Change in Demand	Change in Supply

Demand and Supply

GUIDING QUESTION
What are demand and supply?

The interaction of two forces—demand and supply—is key to the market economy. These forces result from the desires of two groups. **Consumers** are the people who *buy* goods and services. **Producers** are the people or businesses that *provide* goods and services. Let's take a look at the two forces of demand and supply.

Demand

In economics, **demand** is the amount of a good or service that people (consumers) are willing and able to buy at various prices during a given time period. Notice that this definition mentions four parts:

- *Amount*—Demand measures the quantity (or how much) of a good or service consumers are willing to buy over a range of possible prices. So the demand for a certain video game refers to the quantities that consumers would buy at prices such as $20, $40, and $60.
- *Willing to buy*—Consumers must be willing to buy a good or service or there is no demand.
- *Able to buy*—Consumers must have the ability to buy the good or service. Consumers who want a certain good but do not have the money to buy it do not affect the demand for that item.
- *Price*—The quantity that consumers are willing and able to buy is associated with a particular price, be it high or low.

consumer person who buys goods and services

producer person or business that provides goods and services

demand amount of a good or service consumers are willing and able to buy over a range of prices

A woman sells custom-made clothing to a mother and daughter.
Identifying Who is the consumer, and who is the producer?

As the price of a good or service goes up, consumers tend to demand fewer quantities. As the price goes down, they tend to demand larger quantities. Thus, price and quantity demanded have an *inverse*—or opposite—relationship (with price decreasing and quantity increasing—or the other way around). We will see that this inverse relationship can be expressed in a table and graph.

Graphing Demand

The quantity of a particular item that is demanded at each price can be shown in a **schedule**, or table. The information on a schedule can then be drawn as a line on a graph. The line on the graph is referred to as a *demand curve*.

Look at the **Demand Schedule and Graph for Fruit Smoothies**. In the schedule A, every price has a quantity demanded. For example, at a price of $9, the consumer will buy zero fruit smoothies.

schedule table listing items or events

If the price lowers to $6, however, the consumer will demand a quantity of 1 fruit smoothie. If the price is lowered further to $4, he or she will purchase 2. When the price reaches $2, the consumer will buy 4 fruit smoothies. This makes sense. The lower the price, the more quantities will be demanded. The higher the price, fewer amounts will be demanded.

Now look at the demand curve B. Notice how the prices and quantities from the schedule have been put into graph form. Each point on the demand curve shows the amount demanded at a particular price. As the price changes, the points move along the demand curve. At the price of $9, zero quantity is demanded. At $4, a quantity of 2 is demanded. How many quantities will be demanded at a price of $2? Trace the horizontal line that starts at $2. Notice that it connects with the point above the quantity of 4.

Demand Schedule and Graph for Fruit Smoothies

On economics graphs, prices are measured on the left side (called the *y-axis*), starting with $0 at the bottom and moving up to higher prices. Quantities are measured along the bottom of the graph, or *x-axis*. Notice how the quantity increases as you move from left to right on the bottom axis. Both the demand schedule and demand curve show that the quantity demanded changes as the price changes.

A Demand Schedule

Price	Quantity Demanded
$9	0
$6	1
$4	2
$3	3
$2	4

B Demand Curve

EXAMINE THE GRAPH

1. **Analyzing** What quantity of fruit smoothies will consumers demand at a price of $3?
2. **Explaining** What is the relationship between price and quantity demanded?

Supply Schedule and Graph for Fruit Smoothies

The supply schedule and the individual supply curve both show the quantity supplied at every possible price for a certain time period.

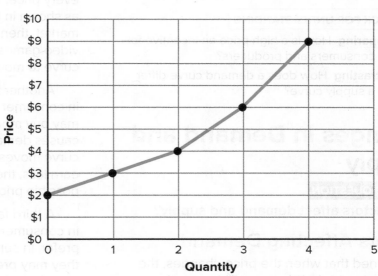

A Supply Schedule

Price	Quantity Supplied
$9	4
$6	3
$4	2
$3	1
$2	0

B Supply Curve

EXAMINE THE GRAPH

1. **Analyzing** What quantity of fruit smoothies are offered for sale at the price of $6?
2. **Explaining** What is the relationship between price and quantity supplied?

The demand curve slopes down to the right. At first glance, this may appear as if demand is decreasing from left to right. That is not the case. In fact, just the opposite is occurring. Notice the quantities along the bottom of the graph. The numbers increase from left to right. Thus, consumers demand more quantities when the price is low, and they tend to demand less (in this case, zero) when the price is high ($9).

Supply

Supply is the amount of a good or service that producers are willing and able to sell at various prices during a given time period. As the price of a good or service goes up, producers tend to supply larger quantities. As the price goes down, they tend to supply less. This happens because a high price is an **incentive**—a motivation or reward—for suppliers to produce more. Your

incentive for studying hard is to earn the reward of a higher grade in class. The reward for producers to supply more quantities at a higher price is to make more **profit**—or the money a business receives for its products over and above what it cost to make the products. A low price is an incentive to produce less.

Graphing Supply

Quantity supplied—like quantity demanded—can also be shown two ways. See the **Supply Schedule and Graph for Fruit Smoothies**. The schedule A shows that at a price of $9, producers will offer 4 fruit smoothies for sale. At a price of $4, suppliers will offer only a quantity of 2. Suppliers will offer a quantity of zero at a price of $2.

The prices and quantities on the supply schedule have been copied over to the supply graph B. Each point on the supply curve shows

---

supply amount of a good or service that producers are willing and able to sell over a range of prices

incentive motivation or reward

profit money a business receives for its products over and above what it cost to make the products

the quantity supplied at a particular price. Notice how the supply curve slopes up when you read the graph from left to right. This shows that if the price goes up, the quantity supplied will go up too. As the price changes, the points move along the supply curve.

✓ **CHECK FOR UNDERSTANDING**

1. **Comparing** How is a high price an incentive for both consumers and producers?
2. **Contrasting** How does a demand curve differ from a supply curve?

Changes in Demand and Supply

GUIDING QUESTION

What factors affect demand and supply?

Factors Affecting Demand

You learned that when the price changes, the *quantity demanded* changes. This was shown along a single demand curve. The curve itself did not change, but the points (quantity demanded) along the curve changed according to the price change.

Sometimes, however, the entire curve moves because of a change in something *other than price*. When this happens, we say there is a change in *demand* (not quantity demanded).

Several factors affect demand, causing the entire demand curve to shift left or right. One factor that affects demand is the number of consumers. Look at the **Change in Demand for Video Games** graphs. If more consumers enter the market, they buy more of the product at each and every price. The demand curve shifts to the right as shown in the top graph. If consumers leave the market, then fewer people are available to buy the video games. This change causes the demand curve to move left, shown in the bottom graph.

Another factor that affects demand is a change in consumer income. If people earn more, they may buy more video games at each price. This causes demand to increase, and the demand curve moves to the right. In contrast, if people earn less, they do not buy as many games at every possible price. Then the demand curve shifts left.

A third factor that affects demand is a change in consumer *preferences*, or what people like or prefer. In summer, people want to be outside, so they may prefer to buy fewer video games at every price. This shifts the demand curve for video games to the left. In winter, consumers prefer indoor activities and would buy more video games at every price. This shifts the product's demand curve to the right.

Factors Affecting Supply

Sometimes the entire supply curve shifts because of a change in something *other than price*. When

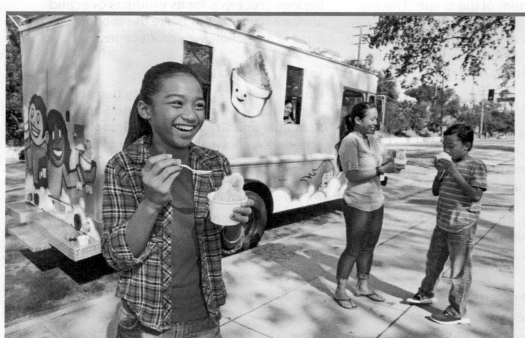

A girl and her family eat shaved ice in front of an ice-cream truck. During warm weather months, demand increases for frozen treats. Because of this change in consumer preferences, people will buy more at every price.

Blend Images / Image Source

Change in Demand for Video Games

Many factors could affect demand for video games. The top graph shows demand expanding, or increasing. The bottom graph shows demand contracting, or decreasing.

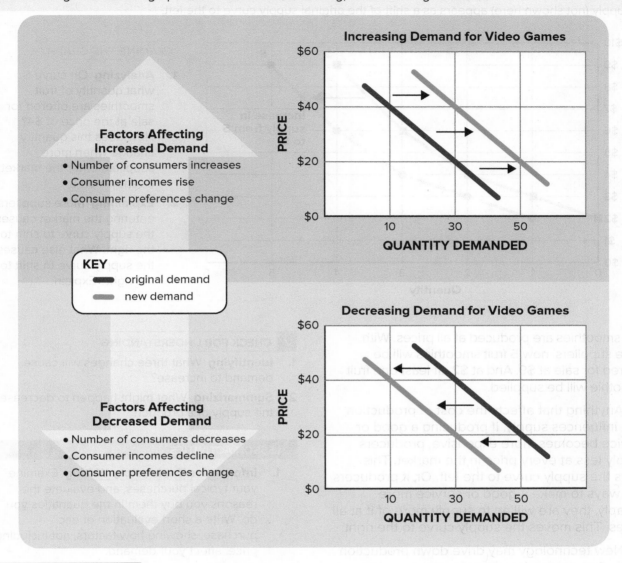

Factors Affecting Increased Demand
- Number of consumers increases
- Consumer incomes rise
- Consumer preferences change

KEY
- original demand
- new demand

Factors Affecting Decreased Demand
- Number of consumers decreases
- Consumer incomes decline
- Consumer preferences change

Increasing Demand for Video Games

PRICE / QUANTITY DEMANDED

Decreasing Demand for Video Games

PRICE / QUANTITY DEMANDED

EXAMINE THE GRAPHS

1. **Identifying Cause and Effect** What economic situation might cause consumers' incomes to go down, and what would happen to the video game demand curve?

2. **Analyzing Visuals** In the top graph, the original demand curve shows that at a price of $40, 10 video games are demanded. Then demand increased because of a change in consumer preferences. How many video games are now demanded at $40 as shown on the new demand curve?

---

this happens, we say there is a change in *supply* (not quantity supplied). Several factors affect supply, which shifts the entire curve left or right. The two key factors are the number of suppliers and the costs of production.

As the number of suppliers increases, more of an item is produced at all prices, and the supply

curve moves to the right. See the **Change in Supply for Fruit Smoothies** graph. The original supply curve **S** shows that at a price of $9, suppliers will offer 4 fruit smoothies for sale. At a price of $2, suppliers will offer a quantity of zero. But notice what happens when more suppliers enter the market. The new supply curve S_1 shows that more

Change in Supply for Fruit Smoothies

A change in supply means that producers will supply different quantities of a product at all prices. An increase in supply appears as a shift of the supply curve to the right. A decrease in supply (not shown here) appears as a shift of the original supply curve to the left.

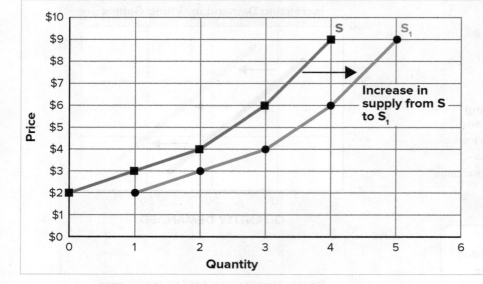

EXAMINE THE GRAPH

1. **Analyzing** On curve S, what quantity of fruit smoothies are offered for sale at the price of $4? How does this quantity change when more suppliers enter the market as shown on S_1?

2. **Explaining** More suppliers entering the market causes the supply curve to shift to the right. What else causes the supply curve to shift to the right? Explain.

fruit smoothies are produced at all prices. With more suppliers, now 5 fruit smoothies will be offered for sale at $9. And at $2, at least one fruit smoothie will be supplied.

Anything that affects the cost of production also influences supply. If producing a good or service becomes more expensive, producers supply less at every price in the market. This shifts the supply curve to the left. Or, if producers find ways to make a good or service more cheaply, they are willing to supply more of it at all prices. This moves the supply curve to the right.

New technology may drive down production costs. Computers, for example, make workers more efficient. A new manufacturing process can reduce waste. This cuts production costs. Lower costs of producing lead to more supply.

Finally, here's a tip that will help you remember the difference between a change in *quantity* demanded (or *quantity* supplied) versus a change in demand (or change in supply). *Only a change in price* can change quantity demanded or quantity supplied. Everything else affects either a change in demand or a change in supply.

✓ **CHECK FOR UNDERSTANDING**

1. **Identifying** What three changes will cause demand to increase?

2. **Summarizing** What might happen to decrease the supply of a product?

LESSON ACTIVITIES

1. **Informative/Explanatory Writing** Examine your typical purchases, and evaluate the reasons you buy them in the quantities you do. Write a short evaluation of each purchase, showing how factors, not including price, affect your demand.

2. **Collaborating** Work with a partner to identify one type of business in which the introduction of a new technology has greatly increased supply. Create a flyer or infographic describing and showing how the new technology lowered production costs and how that had an impact on supply.

READING STRATEGY

Analyzing Key Ideas and Details As you read, complete a diagram like this one to identify the causes and effects of surpluses and shortages.

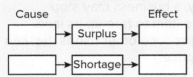

Cause and Effect of
Surplus and Shortage

Cause   Effect

☐ → Surplus → ☐

☐ → Shortage → ☐

The Economic Role of Prices

GUIDING QUESTION

How do prices help consumers and businesses make economic decisions?

Price—the monetary value of a product—does much more than simply tell you how much you have to spend when you make a purchase. Prices help answer the three basic economic questions in a market economy. Prices also measure value. In addition, prices send signals to both consumers and producers.

Prices and the Economic Questions

In a market economy, prices help answer the three basic questions: WHAT to produce, HOW to produce, and FOR WHOM to produce. Prices help determine WHAT to produce by influencing the decisions of producers. If consumer demand for a product is high, demand drives up the price of that product. Businesses react by increasing production of that product to meet the demand and increase their profits. Or, if consumer demand for a product is low, the price of the product falls and businesses produce less of it. For example, why are no large-screen black-and-white televisions produced? The reason is that people will pay a higher price to see TV shows in color. To meet that demand, producers of large-screen TVs focus on making color TVs rather than black-and-white ones. This is how prices help answer the WHAT to produce question.

price the monetary value of a product

A consumer shopping for energy drinks checks prices on his mobile app.

Identifying Cause and Effect How do prices help producers figure out which products consumers are willing and able to buy?

Prices also affect HOW goods and services are produced. For example, cars built by hand would be far too expensive. Instead, automakers use mass production to lower the price. This method lets them produce cars at a price consumers can afford to pay.

Prices also decide FOR WHOM goods and services are produced. Products are made for consumers who can and will buy them at a particular price.

Prices as Measures of Value

Every good and service in a market economy has a price. The price sets the value of each good or service on any particular day. Consumers and producers then use the prices to compare values of goods and services. If a T-shirt costs $10 and a pair of jeans costs $25, then a pair of jeans is worth two and a half T-shirts.

Prices for similar products may be different, however. This often happens when there are small but real differences in quality. It also happens when people *think* the differences are real when in fact they are not. For example, not all jeans have the same value for everyone. You or someone you know may prefer a certain brand or type of jeans. If a jeans company has good advertising, it may convince consumers to value its jeans more highly and pay more for them.

Prices as Signals

Prices send signals to consumers and producers. High prices are a signal for buyers to purchase less, and for producers to produce more. Low prices are a signal for buyers to purchase more, and for producers to produce less.

If consumers think an item is priced too high, some will not buy it. Suppose a bakery charges $5 for a bagel. If nearby bakeries charge less, some consumers will not buy the $5 bagel. This sends a signal to that bakery owner to lower the price.

Prices also serve as incentives to take other actions. If the price of a product goes up at one store, you may decide to shop for it at a different location, or purchase a similar but different item. **Likewise**, or similarly, a business may stop producing a low-price item to free up its land, labor, and capital to produce other things that can be sold for a higher price.

Rationing

What if we did not have prices? Without prices, another system must be used to decide who gets what. One method is **rationing**—a system in

likewise in the same way

rationing system of distributing goods and services without prices

THIS STORE IS PLEDGED TO CONFORM TO THE SUGAR REGULATIONS OF THE U.S. FOOD ADMINISTRATION

Your Sugar Ration
is 2 lbs. per month

SUGAR 2 lbs.

SUGAR 1 lb. 1 oz.

SUGAR 11 oz.

AMERICA'S VOLUNTARY RATION
ENGLAND'S COMPULSORY RATION

FRANCE'S COMPULSORY RATION

ITALY'S COMPULSORY RATION

We must confine our consumption of Sugar to not more than 2 lbs. per person per month in order to provide a restricted ration to England, France and Italy.

This store sign was a reminder for customers to bring their "ration cards" to purchase sugar. During World War I, the government rationed food, gasoline, nylon, tires, and many other items.

Shoppers buy fruit at a farmers market. Sellers at this farmers market—and in every other market—are competing against other sellers, so they keep their prices low enough to attract buyers.

which government decides everyone's "fair" share. Under such a system, people receive a ration coupon or ticket that allows the holder to obtain a limited amount of a product.

Rationing was used widely during World War I and World War II. Since then, rationing has occurred mainly after national disasters. During the 2020-2021 pandemic, for example, stores rationed basic items such as toilet paper and hand sanitizer. Instead of issuing ration coupons, however, stores generally *asked* consumers to personally limit purchases of these items—with mixed success.

✓ CHECK FOR UNDERSTANDING

1. **Explaining** How do prices help us make decisions?

2. **Analyzing** How do high prices serve as incentives for consumers and producers to take "other" actions?

How Prices Are Set

GUIDING QUESTION

How do shortages and surpluses help markets establish equilibrium prices?

As you've learned, markets are vital to the U.S. economy. A **market** is any place where buyers and sellers voluntarily exchange money for a good or service. Markets allow us to choose how we spend our money. In a market economy, buyers and sellers have exactly the opposite goals: buyers want to find good deals at low prices, and sellers hope for high prices and large profits. Neither can get exactly what they want, so some adjustment is necessary to reach a compromise. In this way, everyone who participates—including you!—has a hand in determining prices.

In addition, markets are efficient. To be efficient, markets for identical products must

---

market place or arrangement where a buyer and a seller voluntarily exchange money for a good or service

Market for Fruit Smoothies

Schedule A shows the quantity of fruit smoothies demanded and supplied at each price. When you plot this information on the graph B, you can see that the equilibrium price and equilibrium quantities occur where the two curves intersect—at $4.

A Demand and Supply Schedule

Price	Quantity Demanded	Quantity Supplied
$9	0	4
$6	1	3
$4	2	2
$3	3	1
$2	4	0

B Demand, Supply, and Equilibrium

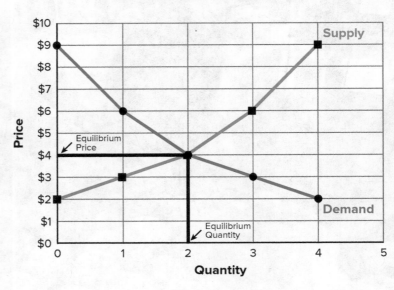

EXAMINE THE GRAPH

1. **Explaining** Why would a seller want to know the equilibrium price of a product?

2. **Analyzing** Why is the quantity demanded for fruit smoothies much lower at a price of $9 than the quantity supplied at that price?

3. **Speculating** What would cause the equilibrium price for fruit smoothies to go up?

---

have many competing buyers and sellers. This **competition**, or struggle among sellers to attract buyers, keeps a product's price at or near a certain level. This level is called the **equilibrium** (EE·kwuh·LIH·bree·uhm) **price**. Here, quantity demanded and quantity supplied are equal. If a market does not have a large number of buyers and sellers, then an exact equilibrium may not be reached. Instead, prices of identical products may be close but not always the same.

When markets operate freely, prices adjust gradually and automatically. Markets also help prevent the production of too many, or too few, goods and services.

Equilibrium

So, how does a market arrive at a compromise price that is "just about right" for both buyers and sellers? To see how this process works, we must put the demand and supply curves together to represent a market—see the **Market for Fruit Smoothies** graph. Note that the curves meet at one point. That point is the price the marketplace sets for the good or service. At this $4 equilibrium price, consumers want to buy the same amount of a good or service that producers are willing to offer. The **equilibrium quantity** at the equilibrium price is two. At the $4 equilibrium price, consumers want to buy a quantity of two fruit

---

competition efforts by different businesses to sell the same good or service

equilibrium price market price where quantity demanded and quantity supplied are equal

equilibrium quantity quantity of output supplied that is equal to the quantity demanded at the equilibrium price

smoothies, and producers want to supply a quantity of two.

Surplus

If the price were higher than the equilibrium price, producers would be willing to produce more. But consumers would not be willing or able to buy more. This would result in a **surplus**, in which the amount supplied by producers is greater than the amount demanded by consumers. A surplus tends to cause prices to fall.

For example, at a price of $6, three smoothies would be supplied but only one would be demanded. This would leave a surplus of two. And at a price of $9, there would be a surplus of four.

Shortage

If the price were lower than the equilibrium price, there would be a **shortage**. This occurs when the quantity demanded is greater than the quantity supplied. For example, at a price of $3, three smoothies would be demanded but only one supplied, leaving a shortage of two. And at a price of $2, there would be a shortage of four. A shortage will cause the price to rise. We often see this happen when a shortage of gasoline drives the price up.

Unless the government steps in to regulate prices, the forces applied by surpluses and shortages work to keep a price at or near its equilibrium level. Surpluses cause prices to go down, whereas shortages cause prices to go up. This process continues until there are no

---

surplus situation in which the amount of a good or service supplied by producers at a certain price is greater than the amount demanded by consumers

shortage situation in which the quantity of a good or service supplied at a certain price is less than the quantity demanded for it

A clearance sale is an example of a surplus.
Drawing Conclusions What must producers do to get rid of their surplus?

Demand and Supply for Oil

The price of crude oil is set by demand and supply. Crude oil is used to make gasoline, heating oil, jet fuel, and other products. Its price affects the whole economy.

Demand Schedule for Crude Oil	
Price Per Barrel	Quantity Demanded
$10	50
$20	40
$30	30
$40	20
$50	10

Supply Schedule for Crude Oil	
Price Per Barrel	Quantity Supplied
$10	10
$20	20
$30	30
$40	40
$50	50

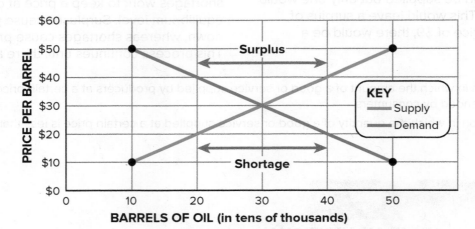

EXAMINE THE GRAPH

1. **Identifying** What is the equilibrium price of oil? At what quantity is this price reached?
2. **Analyzing** What would happen if the producer set the price at $50 per barrel?
3. **Analyzing** What would happen if the government set the price per barrel at $20?

surpluses or shortages, and an equilibrium price is reached.

Thus, a market finds its own price. Look at the graph **Demand and Supply for Oil** to see how surpluses and shortages work together in a large market to get to equilibrium.

✓ CHECK FOR UNDERSTANDING

1. **Explaining** How does competition help consumers?
2. **Identifying Cause and Effect** How do surpluses and shortages help establish an equilibrium price?

LESSON ACTIVITIES

1. **Informative/Explanatory Writing** Suppose you run a company that makes and sells skateboards. If the price of skateboards started to increase, would you choose to make more skateboards or fewer skateboards? In a paragraph, explain your decision.

2. **Collaborating** You and a partner have a new business selling cupcakes. You make 500 cupcakes a day. After a week of selling cupcakes for $4 each, you have a daily surplus of 100 cupcakes. You reduce the price to $3 and sell all the cupcakes daily, but customers would have bought 50 more a day at that price. Working together, discuss how to determine what the equilibrium price is for your product. Write a one-page report, including necessary graphs, explaining why you have settled on a particular price.

Economic Flow and Economic Growth

Analyzing Key Ideas and Details As you read the lesson, complete a diagram like this identifying the four sectors of the economy.

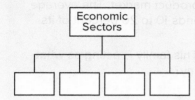

The Circular Flow Model

GUIDING QUESTION

Why do resources, goods and services, and money flow in a circular pattern in a market system?

Economists like to use models to show how a country's economy works. A model is a graph or diagram used to explain something. Demand and supply curves are models. In this lesson we will study another one—the **circular flow model**. This model shows how resources, goods and services, and money flow between businesses and consumers. The model has a circular shape because its flows have no beginning or end. For example, someday you might have a job in a bookstore. Perhaps you use the income you earn to purchase a book. The bookstore uses that money to pay your wages, and so on. The money you earn circles back to the store and then back to you again. This is how the economy works.

circular flow model a model showing how goods, services, resources, and money flow among sectors and markets in the American economy

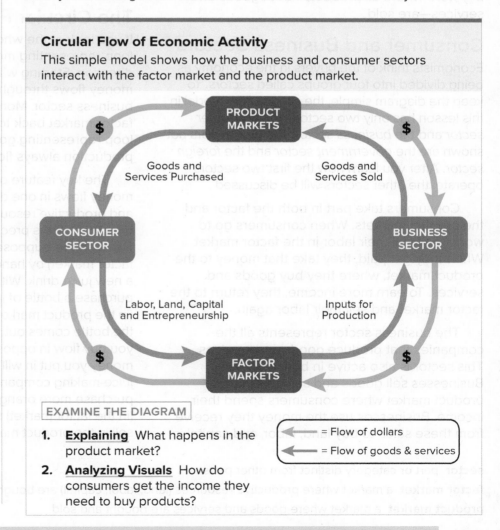

Circular Flow of Economic Activity

This simple model shows how the business and consumer sectors interact with the factor market and the product market.

PRODUCT MARKETS

Goods and Services Purchased

Goods and Services Sold

CONSUMER SECTOR

BUSINESS SECTOR

Labor, Land, Capital and Entrepreneurship

Inputs for Production

FACTOR MARKETS

$

= Flow of dollars

= Flow of goods & services

EXAMINE THE DIAGRAM

1. **Explaining** What happens in the product market?

2. **Analyzing Visuals** How do consumers get the income they need to buy products?

The circular flow model has four parts. Two parts are *markets* where buying and selling take place. Two parts are **sectors** that stand for the two main groups of participants in the markets—people and businesses. We will start by looking at the two markets.

The Factor and Product Markets

The first market is the **factor market**. This is where factors of production are bought and sold. When people go to work, they sell their *labor* in the factor market. *Natural resources* such as oil and timber are also sold in the factor market. *Capital resources* such as machines and tools are bought and sold in the factor market. *Entrepreneurs* organize and use the other three factors in their quest to make profits.

The **product market** is where producers sell their goods and services. Thus, the money individuals receive in the factor market returns to businesses in the product market when people buy products. You can think of this market as one big store where all products—both goods and services—are sold.

Consumer and Business Sectors

Economists think of the buyers in the economy as being divided into four groups called *sectors*. To keep the diagram simple, the circular flow model in this lesson has only two sectors: the *consumer sector* and the *business sector.* The two sectors not shown are the *government sector* and the *foreign sector.* After you learn how the first two sectors operate, the other sectors will be discussed.

Consumers take part in both the factor and the product markets. When consumers go to work, they sell their labor in the factor market. When they get paid, they take that money to the product market, where they buy goods and services. To earn more income, they return to the factor market and sell their labor again.

The business sector represents all the companies that produce goods and services. This sector is also active in both markets. Businesses sell goods and services in the product market where consumers spend their income. Businesses use the money they receive from these sales to buy land, labor, and capital in

When grocery shopping, this family is participating in the product market. The average American family spends 10 to 20 percent of its income on food.
Analyzing Visuals This family is acting as what sector of the economy?

the factor market. They then use these factors to make more things to sell in the product market.

The Circular Flow

If you look at the whole diagram, you see that the loop representing money flows in the clockwise direction. Starting with the consumer sector, money flows through the product market to the business sector. Money then flows through the factor market back to the consumer sector. The loop representing goods, services, and factors of production always flows in the opposite direction.

The key feature of the model is to show that money flows in one direction while the products and productive resources flow in the opposite direction. This is precisely what happens in real life. For example, suppose you earn some cash (in the factor market) by handing out advertising flyers for a new juice drink. With some of your earnings, you purchase a bottle of juice from a vending machine (in the product market). You put the money in, and the bottle comes out. The money and the product you buy flow in opposite directions. In addition, the money you put in will eventually return to the juice-making company. It will use the money to purchase more oranges, containers, and machines (in the factor market) to make more juice drinks to sell in the product market.

---

sector part or category distinct from other parts

factor market a market where productive resources (land, labor, capital) are bought and sold

product market a market where goods and services are bought and sold

The circular flow model also shows that markets link the consumer and business sectors. You probably will never set foot in the factory that makes some of your favorite products. However, you still interact with that factory when you buy its products.

Government and Foreign Sectors

The simplified circular flow model does not show the government and foreign sectors. Both of these sectors are important, but adding them to the model would make it more difficult to understand.

The *government sector* is made up of federal, state, and local government units. These units go to the product market to buy goods and services, just as people in the consumer sector do. Sometimes the government sells goods and services to earn income. For example, state universities charge tuition. If these charges are not enough to fund the universities, governments use taxes and borrowing to get the money needed to operate.

The *foreign sector* is made up of all the people and businesses in other countries. Foreign businesses buy raw materials in U.S. factor markets. They also sell their goods and services to consumers in U.S. product markets. In recent years, about 15 percent of the goods and services bought in the United States have come from foreign countries. Also, about 12 percent of the things we produce are sold outside the United States.

✓ **CHECK FOR UNDERSTANDING**

1. **Explaining** How do businesses and individuals participate in both the product market and the factor market in the U.S. economy?

2. **Making Connections** How does the circular flow diagram reflect the interdependence of the U.S. economy?

This F-16 Fighting Falcon costs roughly $30 million. The government is a big spender.
Making Connections How does the factor market connect the consumer and government sectors?

Promoting Economic Growth

GUIDING QUESTION

How can nations create and promote economic growth?

The United States has experienced mostly steady economic growth. **Economic growth** is the increase in a country's total output of goods and services over time. Government and business leaders work hard to promote economic growth. Why? Because when the economy grows, the nation's wealth increases. This also helps improve the **standard of living**, or quality of life of people living in a country.

A little over 100 years ago, a typical American could expect to live to age 53. Average income, in terms of today's dollars, was under $10,000 per year. Only 55 percent of homes had indoor plumbing, and only 25 percent had phones. No one had a TV, and cars were few. How did Americans, in just one century, manage to improve their standard of living so much? The answer is economic growth.

Two things are needed for economic growth. The first is additional resources. The second is increased productivity.

Additional Productive Resources

As you know, four factors of production are used to make goods and provide services. If a country were to run out of these factors, increasing production would be much more difficult and perhaps impossible. This would cause economic growth to slow or even to stop.

One key resource, land, is in limited supply. Only so much oil is under the ground, for example, and it may run out someday. There is only so much timber to be cut, so it is important to plant new trees regularly. When we make it our goal to save or preserve our trees, streams, and other natural resources, we are helping lay a foundation for future economic growth.

Economic growth also needs either a growing population or one that is becoming more productive. This takes us to the next requirement for growth—productivity.

economic growth the increase in a country's total output of goods and services over time

standard of living the material well-being of an individual, a group, or a nation as measured by how well needs and wants are satisfied

Increasing Productivity

Productivity is a measure of how efficiently businesses use the factors of production to create products. Suppose a factory that has always made 1,000 computers each week begins to make 1,100 a week with the same number of workers. In that case, its productivity has increased by 10 percent. Productivity is increasing when *more* products are made using the original amount of resources in the same amount of time. Productivity is decreasing if *fewer* products are created using the original amount of resources or more time.

Over the years, there have been two key changes in how products are made. These are specialization and the division of labor. Both improve productivity. **Specialization** occurs when people, businesses, or countries focus on the tasks that they can do best. For example, a region with a mild climate and fertile land will specialize in farming. A person who has good mechanical skills might specialize in car repair. By specializing, each becomes more efficient—or productive.

Specialization leads to another development that increases productivity: **division of labor**. This breaks down a job into separate, smaller tasks done by different workers. A worker who performs one task many times a day is likely to be more efficient than a worker who performs many different tasks in the same period. This improves productivity—which increases economic growth.

Businesses always strive to be more productive because their goal is to make more money. They may increase productivity in different ways. One way is to improve existing production methods or invent entirely new processes that do not use costly resources. The invention of the assembly line, for example, sped up production and changed manufacturing forever.

Businesses can also use new and better information technology. Computers originally let one person do the work that was once performed by several people. Now computers connect businesses and workers around the world and control robotics that manufacture products.

Productivity can also improve by using higher-quality factors of production. This is especially true of one factor: labor. When economists talk about the quality of labor, they use the term *human capital*. **Human capital** refers to the knowledge and skills workers can draw on to create products. How can we improve human capital? Three key

The automobile assembly line is a good example of the division of labor. Each worker stays in one place and performs a single task to assemble a car.

Drawing Conclusions How do you think the use of assembly lines in manufacturing affected the country's economic growth?

ways are education, training, and experience. As workers gain more of these, the quality of their work improves and they become more productive. As you know, greater productivity leads to economic growth and a higher standard of living.

✓ CHECK FOR UNDERSTANDING

1. **Explaining** What role does specialization play in the productivity of an economy?
2. **Synthesizing** How do people benefit from economic growth?

productivity the degree to which resources are being used efficiently to produce goods and services

specialization assignment of tasks to workers or factories that can perform them most efficiently

division of labor breaking down of a job into separate, smaller tasks to be performed individually

human capital people's knowledge and skills used to create products

LESSON ACTIVITIES

1. **Informative/Explanatory Writing** Research the ways in which the assembly line changed how Americans made cars. Write a report about your findings. In your report, include details on how these changes affected the price of automobiles.

2. **Presenting** With a partner, present information about a local business that answers these questions: What does the business sell in the product market? What does the business buy in the factor market? Where does the business get the money to make the purchases in the factor market?

08

Turning Point: The Industrial Revolution

? COMPELLING QUESTION

How did changes in production impact the world?

Plan Your Inquiry

DEVELOPING QUESTIONS

Two things are needed for economic growth: additional resources and increased productivity. The Industrial Revolution increased productivity drastically. Think about what life might have been like during the Industrial Revolution. Then read the Compelling Question for this lesson. What questions can you ask to help you answer this Compelling Question? Create a graphic organizer like the one below. Write three Supporting Questions in the first column.

Supporting Questions	Primary or Secondary Source	What the source tells me about the Industrial Revolution's impact on productivity and people's lives	Questions the source leaves unanswered
	A		
	B		
	C		
	D		
	E		

ANALYZING SOURCES

Next, examine the primary and secondary sources in this lesson. Analyze each source by answering the questions that follow it. How does each source help you answer each Supporting Question you created? What questions do you still have? Write these in your graphic organizer.

After you analyze the sources, you will:
• use the evidence from the sources
• communicate your conclusions
• take informed action

Background Information

Most economists use 1776 as the starting date of the Industrial Revolution. That year, James Watt introduced the first commercial steam engine. With the steam engine, mills no longer had to be located near rivers to power their huge water wheels and the creaking machinery inside. Instead, factory owners built their factories closer to their workers or their sources of raw materials. Powerful steam engines propelled trains that carried goods to cities and seaports where they were shipped vast distances. The use of the steam engine also meant that factories could be better organized to mass-produce goods of uniform quality.

Industrialism came at a cost, though. In the years that followed, people no longer lived in villages and made goods by hand in their homes. Instead, people flocked to cities that often became crowded with unhealthy living conditions. Working-class families, including children, worked long hours for low pay. Coal-burning factories belched smoke into the air, and many factory jobs were dangerous.

» Women spin wool into thread and yarn at home in the 1700s. Before the Industrial Revolution, many such "cottage industries" supported rural families.

First Industries

New inventions drove the Industrial Revolution and were often specific to one industry. The *spinning jenny* and *spinning mule* vastly increased the production of thread and yarn. The *power loom* industrialized the production of textiles, or fabrics. Factories were built to house these new machines, and new workers with diverse skills were required to operate the equipment.

The invention of the steam engine powered factory machines as well as steamships and locomotives. By the mid-1800s, trains pulled by steam-powered locomotives were faster and cheaper than any other kind of transportation. Railroads soon connected major cities across Europe and then in other nations, such as the United States. These new industries and faster modes of transportation fueled enormous economic growth.

SECONDARY SOURCE: ENGRAVING

This hand-colored woodcut shows a mill worker tending spinning mules in an industrial textile factory in the 1880s.

PRIMARY SOURCE: BOOK

" [The locomotive] has started into full life within our own time. The locomotive engine had for some years been employed in the **haulage** of coals; but it was not until the opening of the Liverpool and Manchester Railway in 1830, that the importance of the invention came to be acknowledged. The locomotive railway has since been everywhere adopted throughout Europe. In America, Canada, and the Colonies, it has opened up the boundless resources of the soil, bringing the country nearer to the towns, and the towns to the country. It has enhanced the **celerity** of time, and imparted a new series of conditions to every rank of life. "

— From *Men of Invention and Industry* by Samuel Smiles, 1895

haulage transportation of goods
celerity speed

EXAMINE THE SOURCES

1. **Contrasting** Contrast the engraving of the spinning mule with the image on the previous page of the women spinning thread at home. How did the spinning mule increase production of thread and yarn?

2. **Explaining** What does Samuel Smiles think are some benefits of the railway?

New Inventions

The Industrial Revolution also ushered in new inventions that changed people's daily lives. In 1856 Sir Henry Bessemer patented a new process for making high-quality steel efficiently and cheaply, known as the Bessemer process. Steel soon replaced iron to build lighter, smaller, and faster machines and engines. Steel was also used in buildings, railways, ships, and weapons.

Another important innovation was Alexander Graham Bell's telephone, which he invented in 1876. For the first time, telephones allowed people to speak to each other over long distances. Not everyone saw the telephone as an important invention at first, as Bell revealed in the interview excerpted here.

SECONDARY SOURCE: PAINTING

The Bessemer Converter allowed iron to be purified in large amounts so it could be processed into steel.

PRIMARY SOURCE: INTERVIEW

66 I always believed in a practical future for the telephone. But while I was experimenting on it some gentlemen who were paying the expenses of the experiments said to me, 'Mr. Bell, this is a very pretty scientific toy, but of no value practically. We wish you would not waste too much time on it.' 99

— From *Daily Evening Traveller,* September 1, 1880.

EXAMINE THE SOURCES

1. **Drawing Conclusions** Based on the image of the Bessemer Converter, what workplace difficulties do you suppose iron workers faced?

2. **Inferring** Why do you think Bell's investors—the people paying for his experiments—thought the telephone had no practical value? Do you think their response was typical of new inventions? Explain.

3. **Speculating** In what ways do you think people's lives changed when they used a telephone for the first time?

New Methods of Production

Henry Ford became famous for new methods that increased productivity. He pioneered the assembly line for building vehicles in 1913. Between 1916 and 1917, he doubled his production from 500,000 to 1,000,000 cars a year.

PHOTO: Library of Congress, Prints & Photographs Division, Detroit Publishing Company Collection, [LC-DIG-det-4a27966]; TEXT: Ford, Henry and Samuel Crowther. My Life and Work. My Life and Work. Garden City, New York: Doubleday, Page, and Company, 1922.

PRIMARY SOURCE: AUTOBIOGRAPHY

> A Ford car contains about five thousand parts—that is counting screws, nuts, and all. Some of the parts are fairly bulky and others are almost the size of watch parts. . . . The rapid press of production made it necessary to devise plans of production that would avoid having the workers falling over one another. . . .
>
> The first step forward in assembly came when we began taking the work to the men instead of the men to the work. We now have two general principles in all operations—that a man shall never have to take more than one step, if possibly it can be avoided, and that no man need ever stoop over. . . .
>
> In short, the result is this: by the aid of scientific study one man is now able to do somewhat more than four did only a comparatively few years ago. That line established the efficiency of the method and we now use it everywhere. The assembling of the motor, formerly done by one man, is now divided into eighty-four operations—those men do the work that three times their number formerly did.

— From *My Life and Work* by Henry Ford, 1922

PRIMARY SOURCE: PHOTOGRAPH

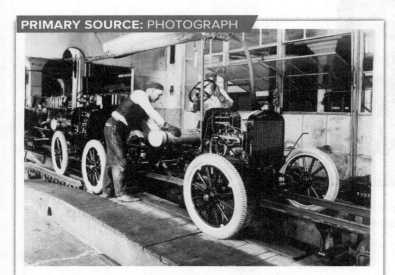

This image shows a motor vehicle on the assembly line at the Ford Motor Company in Detroit, Michigan, in the 1920s.

EXAMINE THE SOURCES

1. **Identifying Cause And Effect** What did Henry Ford do to increase the supply of vehicles?
2. **Determining Context** Ford credits his plant's increased productivity to "scientific study." How can you use context clues to determine the meaning of this term as used in this excerpt?

Benefits of Industrialism

The assembly line and reduced transportation costs let manufacturers mass-produce goods of uniform quality quickly and with greater efficiency. Both supply and demand increased. A middle class emerged, and a consumer culture boomed. In the cities, department stores arose, made possible by steel and electricity. The stores sold a variety of new consumer goods— canned foods, ready-made clothing, clocks, bicycles, electric lights, and typewriters, for example—at affordable prices.

PRIMARY SOURCE: PHOTOGRAPH

Customers shop at Macy's Department Store in New York City around 1907. Electric signs on the building advertise other businesses.

EXAMINE THE SOURCES

1. **Analyzing** What evidence in the photo of Macy's Department Store shows the uses of steel and electricity?

2. **Inferring** What can you infer from the bicycle about daily life among the middle class?

Problems of Industrialization

One cost of industrialization was pollution. Another was that many young children had to work to help support their families.

PRIMARY SOURCE: PHOTOGRAPH

Pollution pours from Carnegie steel factories in Braddock, Pennsylvania, in 1905.

PRIMARY SOURCE: PHOTOGRAPH

Two boys work at a cotton mill in Macon, Georgia, in 1909.

EXAMINE THE SOURCES

1. **Making Connections** What kinds of health issues do you suppose the factories created for people living near them?

2. **Analyzing Points of View** Why do you think the photographer took the image of the young boys?

Complete Your Inquiry

EVALUATE SOURCES AND USE EVIDENCE

Refer back to the Compelling Question and the Supporting Questions you developed at the beginning of the lesson.

1. **Assessing** Which of the sources in this lesson do you find most striking or convincing? Why?

2. **Evaluating** Overall, what impression do these sources give you about the impact of the Industrial Revolution? Was it more positive or more negative? Explain.

3. **Gathering Sources** Which sources helped you answer the Supporting Questions and the Compelling Question? Which sources, if any, challenged what you thought you knew when you first created your Supporting Questions? What information do you still need in order to answer your questions? What other viewpoints would you like to investigate? Where would you find that information?

4. **Evaluating Sources** Identify the sources that helped answer your Supporting Questions. How reliable are the sources? How would you verify the reliability of the sources?

COMMUNICATE CONCLUSIONS

5. **Collaborating** Are the negative and positive impacts you learned about in this lesson just a thing of the past? In a small group, discuss how industrialism still causes problems today. Summarize your discussion and your group's conclusions and share them with the class.

TAKE INFORMED ACTION

Writing a Letter to a State Government Official Think about a problem caused by industrialism that still affects your state. Then write a letter—or work with classmates to write a group letter—to a state official, such as a state legislator or the governor. Let the official know what the problem is, why you think it is a problem, and what negative impacts it causes. Suggest some ways to improve the situation and ask the official to act on the problem. Be sure to include relevant evidence from your sources.

(t) Historicus Inc./Library of Congress, (b) Lewis Wickes Hine/Library of Congress Prints and Photographs Division [LC-DIG-nclc-01581]

Summary

Scarcity and Economic Decisions

Scarcity forces societies to make choices and trade-offs. A benefit-cost analysis can help individuals, businesses, and countries make the best choices.

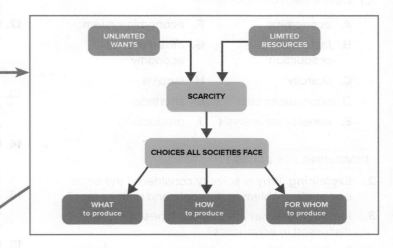

Economic Systems

The way a society answers the three basic economic questions—WHAT, HOW, and FOR WHOM to produce—determines its type of economic system. Economists organize economic systems into four general types: traditional, command, market, and mixed.

Demand, Supply, and Price in a Market Economy

Prices help answer the three basic economic questions in a market economy. The interaction of demand and supply determines price. The forces applied by surpluses and shortages work to keep a price at or near its equilibrium level.

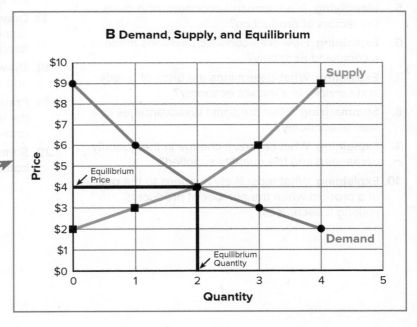

Economic Flow and Economic Growth

The circular flow model shows how resources and products flow in one direction between businesses and consumers, while money flows in the opposite direction. Economic growth relies on additional resources and increased productivity.

Checking For Understanding

Answer the questions to see if you understood the topic content.

REVIEWING KEY TERMS

1. Define each of these terms:

   A. economics
   B. factors of production
   C. scarcity
   D. opportunity cost
   E. benefit-cost analysis
   F. economic system
   G. mixed market economy
   H. surplus
   I. shortage
   J. productivity

REVIEWING KEY FACTS

2. **Explaining** Why is scarcity considered the basic economic problem that people and nations face?

3. **Identifying** What are the four types of resources discussed in economics?

4. **Explaining** What does it mean when a company makes a trade-off?

5. **Identifying** In a command economy, who owns the factors of production?

6. **Explaining** How are economic decisions made in a command economy?

7. **Explaining** What determines the price of goods and services in a market economy?

8. **Summarizing** What are some disadvantages of a market economy?

9. **Explaining** What causes a change in the quantity demanded and the quantity supplied?

10. **Explaining** What most likely happens to the price of a product when the number of producers making it decreases?

11. **Explaining** If consumer demand for a product is high, what impact does that usually have on that product's price?

12. **Identifying** Which four sectors represent the buyers in the economy?

CRITICAL THINKING

13. **Synthesizing** Which type of resource is each of these an example of: delivery van, wind, bus driver, inventor of bar codes?

14. **Identifying Cause and Effect** How does competition affect the quantity and quality of goods and services in a market economy?

15. **Analyzing** What actions of the U.S. government indicate that our economic system is a mixed economy?

16. **Identifying Cause and Effect** A new technology drives down production costs. How will that affect the product's supply?

17. **Drawing Conclusions** There is a surplus of a new brand of cereal on the market. What will likely happen to the price of the cereal?

18. **Drawing Conclusions** How do both specialization and division of labor promote economic growth?

19. **Predicting** How can we ensure that we will have the natural resources needed for future economic growth?

20. **Summarizing** How do consumers take part in both the factor and the product markets?

NEED EXTRA HELP?

If You've Missed Question	1	2	3	4	5	6	7	8	9	10
Review Lesson	2, 3, 5, 6, 7	2	2	2	3	3	3	3	5	5

If You've Missed Question	11	12	13	14	15	16	17	18	19	20
Review Lesson	5	7	2	3	3	5	6	7	7	7

Apply What You Have Learned

Understanding Multiple Perspectives

Consider how economic systems influence societies when they answer the WHAT, HOW, and FOR WHOM to produce questions. Read these excerpts that provide two views of different economic systems.

ACTIVITY **Comparing and Contrasting Perspectives on Economic Systems** Write a brief essay answering these questions: What do the passages have in common? What economic goals does each passage present? What evidence do the authors provide to support their point of view?

❝ The State organizes, directs and controls the national economic activity according to a plan that guarantees the programmed development of the country, with the aim of strengthening the socialist system; satisfying the material and cultural needs of the society and its citizens with constant improvement; and promoting the development of the human being and his dignity, [and] the country's progress and security.

In the preparation and execution of the programs of production and development, an active, conscious role is played by the workers in all branches of the economy, and of those in the other areas of social life. ❞

— Article 16 from *Cuba's Constitution of 1976 with Amendments through 2002*

❝ Ultimately, the best evidence for free market capitalism is its performance compared to other economic systems. Free markets allowed Japan, an island with few natural resources, to recover from war and grow into the world's second-largest economy. Free markets allowed South Korea to make itself into one of the most technologically advanced societies in the world. Free markets turned small areas like Singapore and Hong Kong and Taiwan into global economic players. Today, the success of the world's largest economies comes from their embrace of free markets.

Meanwhile, nations that have pursued other models have experienced devastating results. Soviet communism starved millions, bankrupted an empire, and collapsed as decisively as the Berlin Wall. Cuba, once known for its vast fields of cane, is now forced to ration sugar. And while Iran sits atop giant oil reserves, its people cannot put enough gasoline in their cars.

The record is unmistakable: If you seek economic growth, if you seek opportunity, if you seek social justice and human dignity, the free market system is the way to go. ❞

— President Bush Discusses Financial Markets and World Economy, November 13, 2008

Writing a Persuasive Argument

Consider why and how businesses make trade-offs.

ACTIVITY **Writing a Report About an Economic Choice** Suppose you are the marketing manager of a video game company. The creative team has two ideas for new video games. One game is for younger children and the other game is for teens.

They cost about the same amount to develop. Write a report to the company president explaining why the company can afford to invest in only one of these games, identifying the game you think is a better choice, and explaining why you think so. In your report, use economic ideas you have read about.

C Understanding Economic Concepts

The interaction of demand and supply is key to the market economy. When the price is low, consumers demand more quantities and producers tend to supply less. When the price is high, consumers tend to demand fewer quantities and producers supply more. Recall what affects demand for a product.

ACTIVITY **Applying Economic Concepts to a Graph** The graph shows a demand curve for a package of five trendy stickers. Explain what factors could produce the changes in demand that would move the demand curve to the left or to the right. Do any of the demand curves in the figure show a change in quantity demanded? Explain.

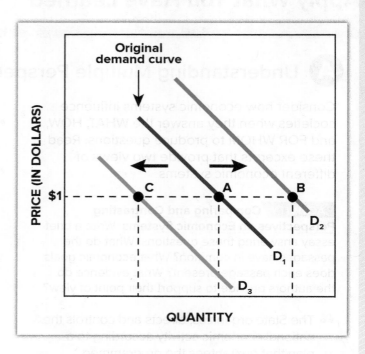

D Making Connections

Prices help answer the three basic economic questions in a market economy. Prices also measure value and send signals to both consumers and producers. Consumers can take advantage of competition to find the best price and save money.

ACTIVITY **Tracking Price Changes on a Chart and Graph** Identify a product or service that people buy regularly and that is widely sold either in your community or via the Internet. Research the price of the good or service at six businesses that offer it. Prepare a chart comparing the prices that different sellers have set for the good or service. Then make a graph showing how much a consumer can save in a year by buying the good or service at the lowest price. Share your findings with friends and family and your class.

Markets, Money, and Businesses

In the American free enterprise economy, people are free to start businesses, employ workers, and spend or save the profits.

INTRODUCTION LESSON

01   Introducing Markets, Money, and Businesses   E52

LEARN THE CONCEPTS LESSONS

02   Capitalism and Free Enterprise   E57

03   Business Organization   E63

04   Roles and Responsibilities of Businesses   E69

05   Employment and Unions   E73

06   Money and Financial Institutions   E79

08   Government and Business   E91

09   Income Inequality   E95

INQUIRY ACTIVITY LESSONS

07   Analyzing Sources: A Cashless Society?   E85

10   Multiple Perspectives: The Minimum Wage   E99

REVIEW AND APPLY LESSON

11   Reviewing Markets, Money, and Businesses   E105

01

Introducing Markets, Money, and Businesses

Based on Freedom

The U.S. economy is called a *FREE enterprise system* for a reason. It allows you the freedom to:

- Spend the money you earn on the products you like best

- Save or invest your money

- Choose your own job

- Start your own business if you don't want to work for someone else

- Produce goods and services that others will pay you for

- Get rewarded with *profits*—monetary rewards—for your own ideas and hard work

FOLKS, SURE THIS POCKETKNIFE IS A BETTER DEAL THAN ANY OTHER BECAUSE IT HAS A SCREWDRIVER, A LEATHER PUNCH, AND A PAIR OF SCISSORS...

...BUT THAT'S NOT ALL! IT ALSO CONTAINS A FLASHLIGHT, A MAGNETIC COMPASS, A CLOCK, A THERMOMETER AND AN ATTENTION-GETTING EMERGENCY WHISTLE!

30 - CALL NOW! 1-800-55

This cartoon shows an entrepreneur trying to convince customers that his product—a pocketknife—is unique and better than any other pocketknife. That is economics at work.

Did You Know?

The word *entrepreneur* is derived from the French word *entreprendre*, meaning "to undertake," and the English word *enterprise*, meaning "difficult or risky project that requires boldness and inventiveness." Today's entrepreneurs are exactly that: risk-taking individuals who start a new business, invent or introduce a new product, or improve a method of making something.

Entrepreneurs in America

Economic freedom allows entrepreneurs to create goods and services that will increase their chances for success. Here are a few famous historical entrepreneurs.

**C. J. Walker,
Beauty Products Entrepreneur**

» C. J. Walker was born Sarah Breedlove in 1867 to parents who were freed African Americans. She was poor and had little formal education but became one of the first female self-made millionaires in this country. With $1.50 in savings, Walker developed a line of beauty products for African American women. Her business earned $500,000 per year.

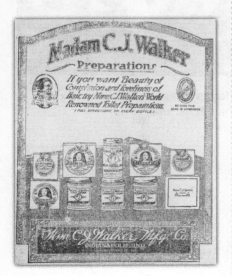

**Walter Elias Disney,
The Walt Disney Company**

» As a boy, Walt Disney loved to draw. In his early 20s, after his first film business failed, he moved to Hollywood to try again. There, Disney made a short cartoon with sound featuring a new character—Mickey Mouse. Disney and his staff then created full-length cartoons. Television provided a new outlet for his shows, and he opened Disneyland in 1954.

**Levi Strauss,
Cofounder Levi's Jeans**

» Levi Strauss migrated to America at age 18 to avoid Jewish discrimination in Germany. After gold was discovered in California, he traveled there to set up a store selling goods to miners. He and a tailor began making pants with rivets out of blue-dyed denim. Workers worldwide wanted the "Levi's," and Strauss & Co. made millions.

Getting Ready to Learn About . . .
Markets, Money, and Businesses

Private Property

How hard would you work if you knew that anything you purchased could be taken away by someone else? Probably not as hard as you would work if you were certain that whatever you purchase is yours—and yours alone—to own.

The right to own private property is central to our democratic principles. The U.S. Constitution guarantees that no citizen will be deprived of "life, liberty, or property" without due process of law. Property ownership is also a vital part of the U.S. economy—also known as *free enterprise capitalism* (or simply referred to as *capitalism*). The ability to gain and keep private property gives citizens an incentive to work hard and save their money.

This young woman has used her hard-earned money to buy a TV. In a capitalist system, she owns her labor as well as anything she purchases from using her labor.

Labor

What job would you like to do when you enter the workforce? Key to a free enterprise economy is the freedom to choose your work. When your grandparents were young, about one-third of Americans worked in factories. They created products such as steel, food, clothing, automobiles, printed materials, and paper.

Today, far more Americans provide services instead of manufacturing goods. A service is work performed using special skills or knowledge that is of value to someone else. A hair stylist provides one kind of service. Other service industries include education, childcare, health care, food preparation, business consulting, beauty care, and social services. Whether in manufacturing goods or providing services, Americans perform a wide range of jobs in a variety of settings. They work for companies large and small, or in their own businesses.

A hairstylist performs a service as he works on his customer's hair at his salon.

Owning a Business

How would you like to wake up before dawn each morning, put in a full day of work making breads, pies, cakes, and cookies, and then work some more after the bakery has closed for the day? Owning a bakery—or any other small business—means hard work and long hours. Yet many Americans enjoy earning a living in this way. Still more dream of becoming business owners themselves.

Owning a business is one way to earn money. Like any other job, it has its risks and its rewards. Many people would rather work as someone else's employee. They may earn less, but they may also work fewer hours with less financial risk. They may also prefer having the freedom to move more easily from one job to another.

Owners of small businesses, such as this bakery owner, have much control over how they run their businesses. They also have many responsibilities.

Money

Money is a *medium of exchange,* which means money is what you give up to purchase a good or service. Money is also a way of assigning a value to a good or service. Suppose you were buying a ticket to a concert by one of your favorite singers. Which would probably get you a seat closer to the stage, the $75 or the $150 ticket? The higher-priced ticket should put you in a better location at the concert. By paying attention to the signals set by prices, we use money to measure value.

Perhaps you have earned some money—but not enough to purchase that large item you want. Where will you put the amount you have for now? You *could* stash it in a box under your bed. But the smart and safe move would be to save it in a financial institution such as a bank. To keep your money safe, the government insures your deposit against loss in case the bank fails.

Government Involvement

Government plays a role in our economy's health, but many people disagree on how big that role should be. Laws and rules regulate businesses to protect consumers. Government rules also help maintain healthy competition among businesses. With the right rules and regulations, our government hopes for a free market in which businesses and consumers feel comfortable and the national economy thrives.

The prices of goods and services, such as a gallon of gas, are given in dollars and cents.

Looking Ahead

In this topic, you will learn about capitalism and free enterprise, business organizations, the roles and responsibilities of businesses, employment and labor unions, money and financial institutions, government involvement in business, and income inequality. You will examine Compelling Questions and develop your own questions in the Inquiry Activities.

What Will You Learn?

In these lessons, you will learn:

- the features of free enterprise and capitalism.
- the characteristics, advantages, and disadvantages of sole proprietorships, partnerships, and corporations.

- how local businesses play a part in people's lives.
- who makes up the U.S. labor force, and why some workers organize into labor unions.
- the types of financial institutions and the services they provide.
- how the government helps maintain competition and regulates the economy.
- what leads to income inequality.

 COMPELLING QUESTIONS IN THE INQUIRY ACTIVITY LESSONS

- **What can be complicated about using money?**
- **Should we increase the minimum wage?**

Characteristics of Free Enterprise Capitalism

This diagram shows the characteristics of the U.S. economy. "Free enterprise capitalism" describes a market economy in which private citizens own the factors of production (land, labor, capital, entrepreneurship), and where businesses compete with little government interference.

- **Freedom** to choose jobs and how to spend money
- **Markets** where goods and services are bought and sold
- **Voluntary exchanges** between buyers and sellers
- Ability to have and control **private property**
- **FREE ENTERPRISE CAPITALISM**
- **Competition** among sellers to get the most customers
- Abillity to make and keep **profits**

02
Capitalism and Free Enterprise

READING STRATEGY

Integrating Knowledge and Ideas As you read the lesson, complete a web diagram by identifying the features of capitalism. Then provide an example of each feature.

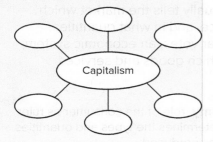

Capitalism in the United States

GUIDING QUESTION

What makes capitalism a successful economic system?

The American market economy is huge. It accounts for about one-seventh of all the economic activity in the world. How did the United States become such an economic powerhouse?

One answer is in the way in which American citizens go about satisfying their basic economic wants. People own the factors of production—the land, labor, capital, and their entrepreneurial efforts—needed to make goods and services. This kind of economic system is called capitalism. In **capitalism**, private citizens own and decide how to use the factors of production in order to satisfy their wants and needs. Our system is also called a free enterprise economy. In a **free enterprise system**, individuals and groups have the freedom to start, own, and manage businesses with little government interference.

Six unique features of the free enterprise system contribute to the economic health of the United States. These features are (1) economic freedom; (2) markets; (3) voluntary exchanges; (4) the profit motive; (5) competition; and (6) private property rights.

capitalism system in which private citizens own most, if not all, of the means of production and decide how to use them within legal limits

free enterprise economic system in which individuals and businesses are allowed to compete for profit with a minimum of government interference

Business partners prepare to open their store. Economic freedom allows entrepreneurs and businesses to choose the skills, markets, and goods and services that will increase their chances for success.

Identifying Cause and Effect How do you think the economy would be affected if people were not allowed to start their own businesses?

Economic Freedom

In the United States, economic freedom is the freedom to make our own economic decisions about the use of our land, labor, capital, and entrepreneurial efforts. As workers, Americans have the freedom to sell their labor. They can decide what jobs they will do and how to save, invest, and spend the money they earn. People are also free to become entrepreneurs and choose what type of goods or services to offer. They may also choose where to locate their business, whom they hire, and how they want to run the business.

These basic economic freedoms give the United States an important advantage over more restricted economies. These freedoms allow the marketplace to adapt quickly to changing economic conditions. As a result, the economy is more efficient and productive.

Markets

Markets are places where buyers and sellers exchange their goods and services. Two forces are at work in these exchanges: demand and supply. Who decides what is supplied and what is demanded? The buyers and sellers themselves—individuals and businesses—make these decisions. The government does not tell producers what to make or consumers what to purchase. Consumers demand products, and businesses supply them.

Because consumers can tell producers what they would like to have produced, consumers are sometimes thought of as the "king" of the market. **Consumer sovereignty** (SAH•vruhn•tee) is the term that describes the role the consumer plays. Think about it. When consumers spend, they are using their dollars to "vote" for the products they want most. This usually tells the market which products to produce, and in what quantities. Without these dollar votes, an economic system would not know which goods and services consumers prefer.

consumer sovereignty role of the consumer as ruler of the market that determines the types and quantities of goods and services produced

---

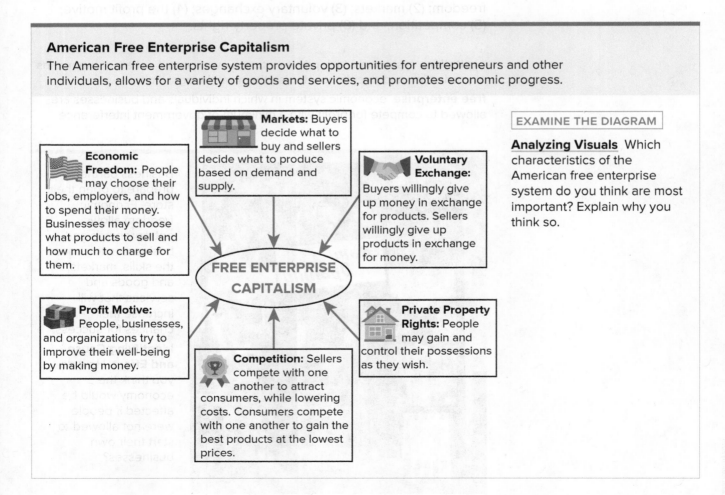

American Free Enterprise Capitalism

The American free enterprise system provides opportunities for entrepreneurs and other individuals, allows for a variety of goods and services, and promotes economic progress.

Economic Freedom: People may choose their jobs, employers, and how to spend their money. Businesses may choose what products to sell and how much to charge for them.

Markets: Buyers decide what to buy and sellers decide what to produce based on demand and supply.

Voluntary Exchange: Buyers willingly give up money in exchange for products. Sellers willingly give up products in exchange for money.

Profit Motive: People, businesses, and organizations try to improve their well-being by making money.

Competition: Sellers compete with one another to attract consumers, while lowering costs. Consumers compete with one another to gain the best products at the lowest prices.

Private Property Rights: People may gain and control their possessions as they wish.

FREE ENTERPRISE CAPITALISM

EXAMINE THE DIAGRAM

Analyzing Visuals Which characteristics of the American free enterprise system do you think are most important? Explain why you think so.

This consumer is shopping at a market on a laptop. A market is wherever a buyer and seller come together, whether in person or on online.

Markets are not perfect, though, because they cannot supply everything we need. Some types of goods, such as a system of justice, public defense, or an efficient highway network, are not easily bought and sold in markets. Still, markets are the best way to answer the WHAT to produce question. Markets establish the prices that help us make economic decisions. This is a major reason why our economic system needs only a limited role for government.

Voluntary Exchange

The activity that takes place in markets is **voluntary exchange**. Voluntary exchange is the act of buyers and sellers freely and willingly choosing to take part in marketplace transactions. These transactions are the buying and selling of goods, services, and the factors of production in exchange for money. In these exchanges, the buyer gives up money in exchange for a product. The seller gives up a product in exchange for money. When these exchanges take place voluntarily, or willingly, both the buyer and seller

are better off. If they did not benefit, the exchange would not have happened in the first place.

The Profit Motive

In a capitalist economy, people risk their savings by investing in new products and businesses. Investing is risky because the effort might not succeed. If it does, successful risk-taking results in making a profit. **Profit** is the amount of money left over from the sale of goods or services after all the costs of production have been paid. The **profit motive** is the driving force that encourages individuals and organizations to improve their material well-being. This is a major reason why capitalism is a successful economic system.

The profit motive pushes entrepreneurs to think of new or improved goods and services. It also leads them to imagine new and more productive ways of making and supplying those goods and services. The new things entrepreneurs do and the businesses they build help the American economy grow and prosper.

---

voluntary exchange the act of buyers and sellers freely and willingly engaging in market transactions
profit the money a business receives for its products or services over and above its costs
profit motive the driving force that encourages individuals and organizations to improve their material well-being

Without entrepreneurs, we would not have so many useful and interesting products such as computers, cell phones, Google maps, social media platforms, medicines and medical cures, and many other things that make life more comfortable or enjoyable. Think of the contributions made by entrepreneurs such as Elon Musk (inventor of PayPal, Tesla, and SpaceX rockets); Jack Ma (creator of Alibaba, the "Chinese Amazon"); Sundar Pichai (Google entrepreneur and Alphabet CEO), Beyoncé (singer-songwriter); and Caine Monroy (9-year-old founder of Caine's Arcade and YouTube star).

Competition

Starting a business does not ensure success. Businesses compete with one another. In fact, capitalism thrives on **competition**—the struggle among businesses with similar products to attract consumers. The most efficient producers sell goods at lower prices. Lower prices attract buyers. If other producers cannot improve their productivity or offer a better-quality product, they might be forced out of business. Competition leads to greater efficiency, higher-quality products, more satisfied customers, and more innovative products.

The Internet is just one innovation that has changed our lives. Consumers use the Internet to search for products, compare prices and features, and stream music or movies to their computers and tablets. Producers can sell their goods directly to buyers without having to put their products in stores. Schools can deliver entire curriculums to students when pandemics prevent them from going to the classroom. Businesses can hire more workers without having to provide for additional offices or more workspaces.

competition the struggle that goes on between buyers and sellers to get the best products at the lowest prices

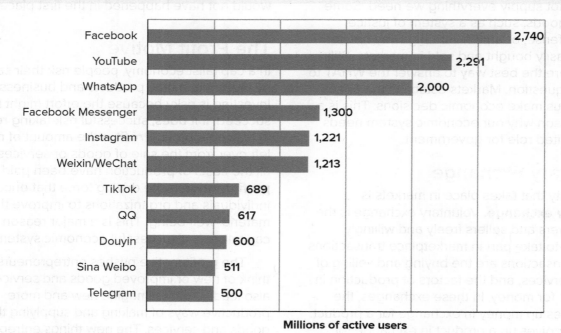

Most Popular Social Networks, 2021

Competition in a free enterprise system leads to a huge variety of choices, including among social media platforms.

Social Network	Millions of active users
Facebook	2,740
YouTube	2,291
WhatsApp	2,000
Facebook Messenger	1,300
Instagram	1,221
Weixin/WeChat	1,213
TikTok	689
QQ	617
Douyin	600
Sina Weibo	511
Telegram	500

Source: www.statista.com

EXAMINE THE GRAPH

1. **Analyzing Visuals** How many people actively used TikTok in 2021?
2. **Making Connections** What role does entrepreneurship have in a free enterprise system?

These shoppers willingly exchange money to purchase products. A voluntary exchange benefits both the buyer and the seller, or the exchange would not occur. In addition, private property rights give this seller the right to use her store property as she wishes.

Private Property Rights

Under capitalism, people and businesses have **private property rights**. This means they have the freedom to own and use their property as they wish. They can even choose to **dispose** of, or get rid of, that property. Property rights give Americans the incentive to work, save, and invest because we can keep any gains we earn. Private property has another important benefit: people tend to take better care of things they own.

✓ **CHECK FOR UNDERSTANDING**

1. **Explaining** Why do people risk their money to start businesses?
2. **Identifying** What are private property rights? How are you affected by them?

Origins of U.S. Capitalism

GUIDING QUESTION

How is the history of capitalism associated with the Founders?

In 1776 Adam Smith, a Scottish philosopher, published a book titled *The Wealth of Nations*. The book opposed the existing economic system of mercantilism. **Mercantilism** was the belief that government should control trade: increase the amount of goods a country exports (in exchange for gold and silver) and decrease the amount of goods imported. Smith argued that, instead, the best way for society to advance is for people to work for their own self-interest, or their own well-being. Because of this, Adam Smith is considered to be the father of economics.

Adam Smith understood that businesses want to earn a profit. That desire will lead them to make products that best meet people's needs and wants. In a similar fashion, people compete to sell their labor, and employers compete to purchase it. The result is an efficient use of resources and a stable society. Smith argued that all of this happens naturally, "as if by an invisible hand." In other words, market exchanges work best with little government interference.

From the writings of Smith and others came the idea of **laissez-faire economics**. *Laissez-faire* (LEH•SAY•FEHR), a French term, means "to let alone." According to this philosophy, government should not interfere in the marketplace. Instead, it should limit itself to those actions needed to **ensure**, or make certain, that competition takes place or in special cases where a market fails.

private property rights the freedom to own and use our property as we choose as long as we do not interfere with the rights of others

dispose get rid of

mercantilism economic system in which government promotes exports and restricts imports to increase the country's stock of gold and silver

laissez-faire economics belief that government should not interfere in the marketplace

ensure to make certain of an outcome

Shoppers purchase from sellers on Mulberry Street in New York City around 1900. Then, as now, the American economy was based on free enterprise. And just as voters in a democracy elect public servants, consumers in a free enterprise system use their dollars to "vote" for their choices of goods and services.

Drawing Conclusions Who or what is "let alone" in a laissez-faire system?

As time went on, economists began to use graphs and math to describe the way markets worked. Still, people never forgot that government was needed to solve some problems that markets could not solve. In fact, government was needed to *protect* markets so they could work efficiently, as Adam Smith thought they could.

Clearly, many of the country's Founders were influenced by *The Wealth of Nations*. James Madison read it, and Alexander Hamilton borrowed from it in his writings. Thomas Jefferson thought Smith's *Wealth of Nations* was the *best* book. The Founders would include Smith's idea of "limited government involvement" when they wrote the U.S. Constitution and Bill of Rights that narrowed the specific powers of government. They reasoned that a democracy, like a market economy, also works best when government power is limited.

✓ **CHECK FOR UNDERSTANDING**

1. **Explaining** What role did Adam Smith believe government should play in the marketplace?

2. **Making Connections** How did the writings of Adam Smith influence the economy of the United States?

LESSON ACTIVITIES

1. **Informative/Explanatory Writing** Adam Smith said that people work for their own self-interest. Write a paragraph describing this idea and whether you agree with it. Use examples of particular jobs to support your opinion.

2. **Collaborating** With a partner, write a public service announcement that explains the key features of capitalism and free enterprise in the U.S. economy.

READING STRATEGY

Analyzing Key Ideas and Details As you read, list the advantages and disadvantages of sole proprietorships, partnerships, and corporations.

Business	Advantages	Disadvantages
Sole Proprietorship		
General Partnership		
Corporation		

Sole Proprietorships

GUIDING QUESTION

What are the advantages and disadvantages of a sole proprietorship?

Have you ever made money by mowing lawns or by babysitting? If you have, you had a small business. Every type of business organization is based on who owns the company and who provides the money to keep it running. Your lawnmowing or babysitting business probably took a simple form. In fact, you were probably the sole proprietor of a sole proprietorship!

A **sole proprietorship** (pruh•PRY•uh•tuhr•SHIP)—also simply called a proprietorship—is a business owned and operated by one person. All businesses in this form are small businesses. Usually they serve the area in which they are located. You see these businesses every day. Dry cleaners, auto repair shops, beauty shops, and local restaurants often take this form.

Sole proprietorships have several advantages. They are easy to organize, which is why they are the most common form of business organization. Sole proprietors are their own bosses. They decide what products or services they will sell. They decide what hours the business will be open. They make decisions without having to consult, or check with, other owners. As the only owner, a sole proprietor receives all the profits from the business.

sole proprietorship a business owned and operated by a single person

A bakery owner offers goods for sale at his store. Among sole proprietorships, the most popular types of businesses are professional, legal, and technical services; construction companies; and retailers—or stores and merchants.

Analyzing Visuals Which business type is shown in the photograph?

Forms of Business Organization

Partnerships and proprietorships together account for more than 80 percent of all businesses but about one-third of total sales.

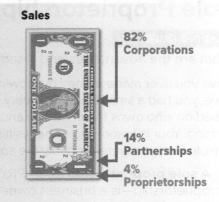

Number of Businesses

- 17% Corporations
- 11% Partnerships
- 72% Proprietorships

Sales

- 82% Corporations
- 14% Partnerships
- 4% Proprietorships

Source: IRS, Statistics of Income Division, 2020

EXAMINE THE DIAGRAM

1. **Analyzing Visuals** What percentage of all businesses are sole proprietorships? What percentage of businesses are partnerships?

2. **Inferring** Why do you think there is such a large difference between the number of proprietorships and their share of all sales?

Sole proprietorships also have some disadvantages. First, the owner may find it difficult to raise **financial capital**. This is the money needed to run or expand a business. Unless the business is run from the owner's home, the proprietor must buy or rent the place from which it operates. The owner might also have to buy equipment and supplies. If the business does not make enough money, the owner will have to use personal money to meet these costs.

Second, sole proprietors have no limits on their **liability** (LY•uh•BIH•luh•tee), or legal responsibility, for the business. This can be a problem if the business cannot pay its debts or loses a lawsuit. Then the owner's personal property—such as a home or car—may have to be sold to pay the business's debts. Third, sole proprietors might have trouble hiring skilled workers. Workers might prefer to take a job with a large company with better benefits, such as health insurance.

✓ **CHECK FOR UNDERSTANDING**

1. **Identifying** In a sole proprietorship, who receives all profits?

2. **Summarizing** What challenges and risks does the owner of a sole proprietorship face?

---

financial capital the money used to run or expand a business

liability the legal responsibility for something, such as an action or a debt

Partnerships

GUIDING QUESTION

What are the advantages and disadvantages of a partnership?

Suppose you are making so much money mowing lawns that you have little time for anything else. You could hire a helper. You have another option, too. You could find someone who could provide the extra help and use their own equipment in return for a share of the business. In this case, your business is no longer a sole proprietorship. It is now a **partnership**—a business that two or more people own and operate together.

As with your lawn-mowing business, partnerships sometimes start as sole proprietorships. In some cases, a single owner cannot raise enough money to expand the business. Or the owner may have enough financial capital but not have all the skills needed to run the business well. In either case, the owner may seek a partner with the money or skills that the business needs to grow.

How Partnerships Are Structured

A partnership is officially organized when two or more people sign a legal agreement called *articles of partnership*. This document states

partnership a business owned by two or more people

what role each partner will play in the business. It tells how much money each will contribute. It **clarifies**, or explains, how each partner will share in the profit or loss of the business. The document also states how each partner can be removed or how new partners can be added. Finally, it describes how the partnership can be ended if the partners decide to do so.

Three kinds of partnerships can be formed. The first is a *general partnership*. In this type, all partners are called general partners. They all own a share of the business, and each partner is responsible for its management and debts.

The second type is a *limited partnership*. In this form, some owners are limited partners, and some are general partners. Limited partners own a share of the business. However, they have no direct involvement in running or managing it. Instead, they usually just provide money the business needs to operate. The general partners run the business and are responsible for everything, including all debts.

The *limited liability partnership* (LLP) is the third and most common kind of partnership. It is popular because it protects all partners, even the general partners, if the business has a lot of debts. This makes the LLP a popular choice of doctors, lawyers, dentists, and most professional people. The initials *LLP* after a business's name show that it is a limited liability partnership.

Partnerships can be any size. They may have two partners, with no other employees. In other cases, a small firm of four or five partners may be just the right size for the market it serves. Other partnerships, such as major law or accounting firms, may have hundreds of partners providing services in many locations across the country.

Advantages of Partnerships

The biggest advantage that partnerships have over sole proprietorships is that they can raise more money to grow and hire more employees. A partnership has more than one owner, so it usually has more capital. Current partners can add new partners to provide additional funds. It is also easier for a partnership to borrow money from a bank.

Another advantage of partnerships is that each partner often brings special talents to the business. For example, one partner in an insurance agency may be good at selling policies

clarify explain

to new customers. The other partner may be better at providing services to people who already have policies. This business will probably be more successful than if just one person owned and operated it.

Disadvantages of Partnerships

The main drawback of a general partnership is the same as that of a sole proprietorship. Each general partner has unlimited liability. He or she is fully responsible for all business debts.

What does this mean? Suppose that you are in a lawnmowing business with two other partners. Each of you had agreed to share one-third of the business profits and one-third of the losses. Now suppose that one of your partners buys an expensive new mower at the end of the season just as your business income drops. Or suppose that one of the business's customers gets hurt by your equipment, or the mower throws a stone through a car windshield. The business could be sued if neither the business nor your partners have enough money to cover the debt or the damages. You, as the remaining general partner, would have to pay 100 percent of the cost out of your personal funds!

✓ **CHECK FOR UNDERSTANDING**

1. **Identifying** What are three types of partnership?
2. **Comparing** How are a sole proprietorship and a general partnership alike regarding liability?

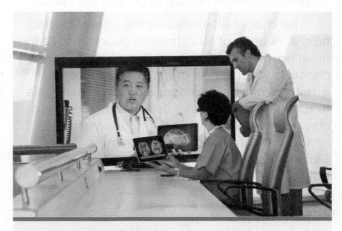

Three doctors consult about a patient's health. Doctors may be sued if patients think they were injured by the doctors' actions. For this reason, many medical practices are limited liability partnerships, or LLPs.

Drawing Conclusions What disadvantage of partnerships does this type of business solve?

Corporations

GUIDING QUESTION

How is a corporation structured, and what are its advantages and disadvantages?

The third major form of business is very different from either a sole proprietorship or a partnership. This third form is the corporation, the most complicated form of business. You can usually tell that a business is a corporation if the abbreviation *Inc.* follows the company's name. *Inc.* stands for "incorporated." Many corporations, such as Amazon and Apple, don't include the abbreviation in their name, however.

A **corporation** is a business owned by many people but is legally treated as though it were a single person. This makes the corporation separate from the people who own it. Under the law, a corporation has many of the rights and responsibilities that an individual has. Like a real person, a corporation can enter into contracts, sue and be sued, own property, and pays taxes.

How a Corporation is Organized

The state where the corporation is formed grants a **charter** to a group of investors. This permits the investors to sell ownership shares in the corporation. The ownership shares are called stock certificates, or **stocks** for short. Those who own the stocks are called stockholders or shareholders. This sale of stocks is how the corporation raises money to go into business.

The charter also requires the corporation to hold a meeting of stockholders every year. At this meeting, the stockholders elect a board of directors to represent them. The **board of directors** hires a president and other managers to run the company on a daily basis. The board also meets several times during the year to review the corporation's performance. Stockholders are not involved in the day-to-day operation of the company.

---

corporation type of business organization owned by many people but treated by law as though it were a person

charter state government document granting permission to organize a corporation

stocks shares of ownership in a corporation; same as stock certificates

board of directors the people elected by the shareholders of a corporation to act on their behalf

The Corporation's Advantages

The corporation has three main advantages. The biggest advantage is the ease of raising financial capital. It can raise huge amounts of money by selling stock. It can also borrow money by selling bonds, which are formal loan agreements between a borrower and a lender. It can then use that money to expand operations, open up businesses in new locations, or buy new equipment. It can also raise money to research new products. This ease of raising money is one reason why the corporation is the most common form of business for large companies.

The ease of raising money and favorable tax rates on corporate profits are why some corporations are so huge. The yearly sales of the biggest corporations are larger than the economies of most of the world's countries. For example, if Walmart were a country, it would be the twenty-third largest country in the world.

A second advantage of the corporation is limited liability. The corporation, not the owners, is responsible for its debts. The owners' property

General Motors (GM) is a major American corporation. Although it is headquartered in Detroit, Michigan, GM does business in more than 120 countries.

Identifying What document spells out how much stock a corporation can sell?

Corporate Structure

Every corporation has shareholders, a board of directors, a management team, and employees. The person who runs the business may be called the president, chief executive officer (CEO), or chief operating officer (COO).

Shareholders (Stockholders)
- Buy stock/ownership shares
- Become owners of the corporation
- Elect board of directors

Board of Directors
- Represents shareholders
- Makes key decisions, sets corporate policies
- Hires president and other key officers

President
- Oversees corporation's daily business
- Hires and manages other officers

Vice President of Sales
- Oversees domestic and international sales
- Hires and manages department heads

Vice President of Production
- Oversees quality control and research and development
- Hires and manages department heads

Vice President of Finance
- Oversees payroll and other financial matters
- Hires and manages department heads

Department Heads
- Hire and manage employees

Employees
- Form the "backbone" of the organization
- Work in all departments contributing to the overall success of the corporation

EXAMINE THE DIAGRAM

1. **Analyzing Visuals** What individual or group chooses the corporation's president?

2. **Comparing** What responsibility do the vice presidents have in common?

3. **Identifying** Who owns a corporation?

cannot be touched to pay those debts. This advantage is important. Suppose some people want to try a risky business like building a nuclear power plant. They would first form a corporation because of its limited liability. If something were to go terribly wrong and the company were sued, the investors would be protected from losing their personal property to settle the case.

The third advantage is that ownership can easily be transferred. Proprietorships and partnerships may end when an owner resigns or dies. When a stockholder no longer wants to own part of a corporation, however, he or she simply sells the stock to someone else. Or, if a stockholder dies, his or her family receives the stock. This means that only the ownership changes, not the company or its name.

The Corporation's Disadvantages

Corporations also have some disadvantages. The government regulates them more than any other form of business. By law, corporations must make their financial records public. This means they have to release reports on expenses and profits on a regular basis. They must also hold a stockholders' meeting at least once a year.

Major corporations have millions of stockholders. If some are unhappy about the way the company is run, it is hard for them to unite and get managers to make changes. They can write a letter to the board or sell their shares and use the money for something else.

Other Business Forms

Corporations have been in the United States long before it was an independent nation. Jamestown, the first English settlement in the Americas, was founded in 1607 by the Virginia Company, a *joint-stock company* similar to today's corporation.

One type of business that has become very common in recent years is the **franchise** (FRAN•CHYZ). A franchise is a business in which the owner is the only seller of a product in a certain area. The owner pays a fee to the supplier for that right. The owner also gives that supplier a share of the profits. Fast-food restaurants and hotels are often franchises. Other popular franchises include Great Clips and Jiffy Lube.

To own a Wendy's franchise, a business owner has to buy the right to operate the restaurant and then pay a percentage of its monthly sales to the Wendy's corporation.

Speculating What might attract a person to buy a franchise like this one?

The franchise owner benefits because there is no competition from another nearby seller. The supplier also helps the franchise owner run the business. The biggest disadvantage is that the franchise owner does not have complete control over the business. The national company often has many rules the franchise owner must follow.

All the businesses you have read about so far have a common goal—to make a profit. However, there is another type of business called a **nonprofit organization**. These organizations provide goods and services without trying to make a profit. Many public hospitals and charitable organizations, such as foodbanks and the Red Cross, are nonprofit organizations.

Cooperatives, or co-ops, are another type of nonprofit organization. This is a business formed by people who want to benefits its members. Different kinds of co-ops exist. A consumer co-op buys goods in large amounts. Members then get those goods at low prices. Service co-ops provide members with services such as insurance or loans. Producer co-ops help members sell their products to large central markets where they can get better prices. Ocean Spray is a producer co-op that promotes the sale of cranberries.

✓ CHECK FOR UNDERSTANDING

1. **Explaining** How does a corporation raise financial capital?

2. **Analyzing** How does a cooperative work?

LESSON ACTIVITIES

1. **Argumentative Writing** If you owned a business, would you rather be a sole proprietor or a partner? Write a paragraph explaining why. Include evidence and examples to support your argument.

2. **Presenting** With a partner, select and research a specific nonprofit organization. Find a description of the organization, who its members are, who benefits from it and how, and the problems it faces. Create a multimedia presentation to present your findings to the class.

franchise company that has permission to sell the supplier's goods or services in a particular area in exchange for payment

nonprofit organization business that does not intend to make a profit for the goods and services it provides

04
Roles and Responsibilities of Businesses

READING STRATEGY

Integrating Knowledge and Ideas As you read this lesson, identify the responsibilities of businesses to their consumers and their employees.

Responsibilities of Businesses	
To Consumers	To Employees

Social Responsibilities of Businesses

GUIDING QUESTION

In what ways do businesses help their communities?

Does your school benefit, or gain, from help given by a business? That might seem like an unusual question, but businesses help schools in many ways. For example, do you play on a sports team? Does a local business sponsor that team? Has your school club ever held a car wash at the parking lot of a local business? Does a store in the area sell school supplies at a discount?

Businesses help their communities in many ways. Businesses play several important roles in society. As producers, they supply the food, clothing, and shelter we use to meet basic needs. They also produce many goods and services that make life more enjoyable and comfortable. Along with being producers, they also have a **social responsibility**. This is the obligation to pursue goals that benefit society as well as themselves.

social responsibility the obligation businesses have to pursue goals that benefit society as well as themselves

The Tide corporation has a truck that carries enough washers and dryers to do 300 loads of laundry a day. It sends the truck to areas hit by natural disasters to help provide clean clothes to people in distress.
Inferring How can this kind of action help a business?

Have you ever eaten a White Castle hamburger? That fast-food chain was founded by a family named the Ingrams. The Ingrams have given more than $29 million to support education. They donate this money through the family's **foundation**, an institution created to promote the public good.

People who have enjoyed success in business, such as the Ingrams, can be very generous. Bill Gates, the founder of Microsoft Corporation, has given away some $50 billion. His foundation aids a wide variety of causes in the United States and around the world. For instance, his foundation **contributed**, or provided, $500 million to public schools. It also gave more than $250 million to fight diseases in Africa, Asia, and South America; and another $250 million to fight the COVID-19 pandemic.

In addition, many corporations have set up their own foundations. These groups give money to support causes they believe are important. The Walmart Foundation has given more than $40 million to help veterans returning to the workforce through job training and education.

Corporations give away about $20 billion each year. Some provide free goods or services.

For example, many drug companies give their products to people who need the medicines but cannot afford to pay for them. Apple and GAP give part of their profits to a fund that fights infectious diseases in 144 countries. GAP also donates millions of pieces of clothing to refugees. American Express has long been involved in helping disaster victims. The company gives money to relief agencies. These agencies use that money to provide food, clothing, and shelter for the victims. Another American Express program helps groups that are trying to preserve important historical sites or natural areas.

Donations do not come from just large American companies. About 75 percent of small companies also give money. Some support groups in their area. Others give to causes they believe in. You have seen examples of how some help schools. Some law firms or accountants provide free services to poor people or nonprofit groups.

✓ **CHECK FOR UNDERSTANDING**

1. **Explaining** What is social responsibility, and how do businesses show it?

2. **Identifying** What are some of the good causes that American companies have donated to?

---

foundation organization established by a company or an individual to provide money for a particular purpose, especially for charity or research

contribute to provide or give

In recent years, the Consumer Product Safety Commission has ordered the recall of many toys painted with dangerous lead paint. It has also recalled millions of riding toys because of a fire hazard.

Identifying Cause and Effect What incentives do businesses have to meet their responsibilities to consumers?

A report provides information about how a corporation is doing financially. Corporate managers must be transparent to stockholders about such information.

Other Business Responsibilities

GUIDING QUESTION

How do businesses carry out their responsibilities to their consumers, owners, and employees?

As they carry out their many activities, businesses have different responsibilities to the groups they interact with. Laws require firms to meet certain obligations. Business owners and managers may face serious problems or even legal action if they do not follow those laws. They also may suffer a loss of reputation, which might result in a loss of business.

Responsibilities to Consumers

Businesses have important responsibilities to their customers, the people who buy their goods and services. First, businesses must sell safe products that work properly. Services must be reliable. A new video game should run without flaws. An auto mechanic should change a car's oil correctly. Many companies guarantee their products and services for a period of time. They replace or redo those that do not work as they should. Second, businesses also have the responsibility to tell the truth in their advertising. Third, businesses should treat all customers fairly.

Responsibilities to Owners

Another responsibility is to the stockholders, who are the owners of the business. This is especially crucial for corporations. In this case, the people who own the company are not the same people who run it. To protect stockholders, corporations have to release financial reports regularly. Making this information public is called **transparency**. The information is published to give investors full and accurate information. People can analyze the facts before they choose to invest money in the company.

Sometimes the managers of a corporation are not honest in these reports. The government can then prosecute them for breaking the law. For example, the scandal involving the company Theranos showed the problems that arise when these reports are not truthful. The company raised hundreds of millions of dollars from investors. Theranos claimed that its miniature testing device could perform hundreds of laboratory tests by analyzing a single drop of blood. But its claims—and its financial statements—were false. When the company went out of business, investors lost all of their money.

Responsibilities to Employees

Finally, businesses have responsibilities to their workers. They are required to maintain a safe workplace. They must also treat all workers fairly and without discrimination. They cannot treat employees differently because of race, religion, gender, age, or disability. Doing so is against the law.

transparency the process of making business deals or conditions more visible to everyone

chormail/123RF

Google, the Internet company, has features in its offices to try to make the workplace enjoyable, comfortable, and healthy for employees. Workers can use company-provided bicycles and scooters to move from one building to another. Game rooms give employees a chance to relax and release tension from work.

Predicting How can a company benefit from taking these steps?

Companies cannot pay different wages to men and women who do the same work, for instance. Nor can they fire workers because they reach an older age. Such decisions must be based on the quality of the work the employees perform.

Many businesses try to help workers by providing benefits or services. For example, many companies help employees with the costs of trade school or college. Some pay for programs to help workers stop smoking, or provide childcare or fitness centers for workers. Others provide breakfast, lunch, and snacks at the office for their workers.

Health insurance is a benefit that many companies have traditionally given to their workers. However, as health insurance costs increased, many businesses grew worried about the cost of this benefit. Some stopped providing it. Others shifted more of the cost of this insurance to their workers. In 2010 Congress passed a law, the Affordable Care Act, requiring businesses to provide health insurance. Some parts of the law are designed to limit increases in the cost of health care and health insurance. The law also gives tax credits to small businesses when they buy health insurance. These limits and credits were meant to make the cost of the benefit easier to meet.

Helping workers in these ways benefits the employer as well as the worker. A worker who is in good health misses less work than one who is not. That worker also has more energy and can be more productive on the job. Benefits also make the worker happier and less likely to leave the job, which helps the employer maintain good workers.

✓ **CHECK FOR UNDERSTANDING**

1. **Explaining** Why is it important for corporations to publish financial information regularly?

2. **Summarizing** How does the government push businesses to act responsibly?

LESSON ACTIVITIES

1. **Argumentative Writing** A business has responsibilities to its customers, its employees, its owners, and its local community. Which of these responsibilities do you think is most important? Explain your point of view in a paragraph.

2. **Collaborating** The Equal Employment Opportunity Commission (EEOC) is a part of the federal government. Its job is to ensure that job applicants and workers are not discriminated against. Go to the EEOC website to learn its powers and how it works to uphold the law. With a partner, summarize what you find. Then write a public service announcement that explains your summary information.

Integrating Knowledge and Ideas As you read this lesson, complete a graphic organizer showing the goals of collective bargaining.

Employment

Who makes up the U.S. labor force, and how has it changed over time?

Every month, the Bureau of Labor Statistics conducts an important **survey**, or examination, that tells us who is working and who is not working in the United States. Specifically, this survey tells us about changes that might be happening to the nearly half of our population that makes up the civilian labor force. By definition, the **civilian labor force (CLF)** includes "all persons 16 years of age or over who are working or not working but are able and willing to work."

Some parts of our population are not part of the civilian labor force. This includes young students like yourselves who are not yet 16 years old, even if you work in a job after school. Nor does it include army, navy, or

survey examination, review

civilian labor force group made up of "all persons 16 years of age or over who are working or not working but are able and willing to work" and who are not in the military

Population and the Civilian Labor Force, 2020

The civilian labor force (CLF) includes both "Employed" and "Unemployed" people over age 16 who are willing and able to work—and are not in the military. In 2020, the total CLF was 160,808,000, and the number of unemployed was 13,025,000.

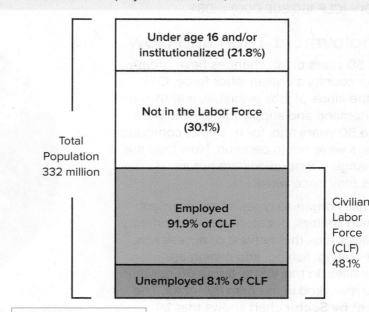

Contrasting What is the difference between those "Not in the Labor Force" and those who are "Unemployed"?

other armed services because the survey includes only "civilian" (nonmilitary) workers. Also, people confined to institutions such as hospitals or prisons are not part of the labor force.

Another major group includes people defined as "not in the labor force" because they choose not to work, even though they could. This group includes stay-at-home parents and retired workers. After all these groups are accounted for, we are left with the civilian labor force.

The **Population and the Civilian Labor Force, 2020** chart shows that 48.1 percent of our total population was in the civilian labor force. The chart also shows that the CLF has two parts. The first includes "employed" people. They made up 91.9 percent of the CLF who were working and had a job. The second category—the "unemployed"—amounted to 8.1 percent of the CLF and included those who did not have a job but were willing to work if they could find one. This is the category we are concerned about.

Unemployment Rate

The term *unemployment rate* is often in the news. The **unemployment rate** is the percentage of people in the CLF who are unemployed at any given time. We get this percentage by dividing the actual number of unemployed people by the size of the CLF. The unemployment rate in the chart is 8.1 percent, or (Number of unemployed 13,025,000) ÷ (CLF 160,808,000) = .081.

U.S. Employment Then and Now

In the past 50 years or so, changes have greatly affected the country's civilian labor force. One change is the kinds of jobs available, and thus a change in demand and supply for specific work skills. Some 50 years ago, for example, computer programmers were not in demand. Now they are. Another change is that unions are not as powerful as they once were.

Economists organize types of employment into four main sectors, or categories. The *primary sector* includes jobs that harvest or access raw materials. Farming, fishing, and mining are primary activities. In the 1850s, about 66 percent of Americans worked in the primary sector. The **Employment by Sector** chart shows that 1.4 percent worked in this sector in 2020.

unemployment rate percentage of people in the civilian labor force who are not working but are looking for jobs

The *secondary sector* includes jobs that change raw materials into finished products. Factory and construction workers perform secondary activities. In the 1970s, about 22 percent were employed in manufacturing jobs. That percentage dropped to 13 percent in 2020.

The *tertiary* [TUHR•shee•air•ee] *sector* is often called the "service sector." People in these jobs perform a service, such as restaurant work or shipping packages, but do not create a finished good. This was the largest sector in 2020.

The fourth, or *quaternary* [kwah•TUHR•nuh•ree] *sector*, is sometimes called the "knowledge sector." It includes specialized workers who provide information-based services, such as educators, doctors, and government officials.

✓ **CHECK FOR UNDERSTANDING**

1. **Identifying** What four groups does the Bureau of Labor Statistics divide the total population into?

2. **Identifying Cause and Effect** What caused the U.S. civilian labor force to change in the past 50 years?

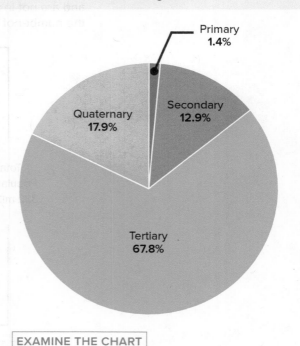

Employment by Sector, 2020

Economists organize types of employment into four main sectors or categories.

Primary 1.4%
Secondary 12.9%
Quaternary 17.9%
Tertiary 67.8%

EXAMINE THE CHART

Making Connections What four sectors do jobs get split into, and what is an example of a job found in each sector?

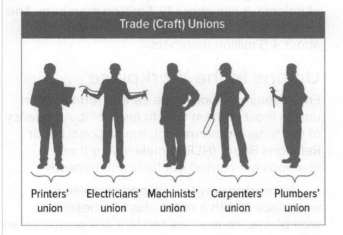
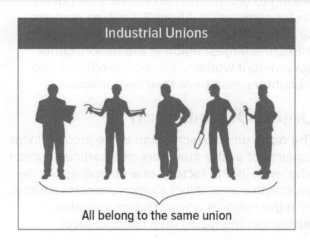
Organized Labor

GUIDING QUESTION

What is the role of organized labor in the U.S. economy?

The list below describes workers' rights that many take for granted in today's workplace:

- Weekends off from work
- The eight-hour work-day
- Safety measures at the workplace
- Paid vacation (or sick days)
- Days off for national holidays
- Extra pay for working overtime
- Minimum wage guarantee

Many of these rights came about because workers "organized" and demanded changes from employers. In other words, organized workers formed labor unions. A **labor union** is an organization of workers that seeks to improve its members' wages and working conditions.

Since 1970, the size of the labor force has nearly doubled. In those years, however, the number of workers belonging to labor unions has fallen. In the early 1970s, about one of every four workers belonged to a union. As of 2020, however, only one worker in ten was a union member. One reason for the decline in union membership is the shift from manufacturing jobs to service jobs. Also, many employers have kept their workplaces union-free.

Unions still play an important role in the United States, however. Workers in many industries belong to unions. Large numbers of coal miners, airline pilots, and truck drivers are union members, for example. Unions have also seen gains in the public sector, where teachers and government employees work.

Types of Unions

There are two types of unions. A union whose members all work at the same craft or trade is called a *trade union*. Examples are the unions formed by bakers or printers. A union that brings together workers from the same industry is called an *industrial union*. An industrial union might have electricians, carpenters, and laborers who work together to manufacture a product. An example is the United Auto Workers (UAW).

Unions have changed over time. In the past, they were formed mostly by industrial workers. Today, however, even actors and professional

labor union association of workers organized to improve wages and working conditions

athletes have unions. Another change is the increase of government workers who are union members. In fact, more government workers belong to unions than do workers for private sector companies. About 2.3 million workers belong to the National Education Association (NEA), the largest union in the nation. Other government workers, like police officers and firefighters, also have their own unions.

Union Organization

The basic unit of each union is the *local*. A local consists of all the members of a particular union who work in one factory, one company, or one geographic area. All of a union's locals together form the national union. This organization represents the locals on a national level.

Many national unions belong to the American Federation of Labor and Congress of Industrial Organizations (AFL-CIO). The AFL-CIO is the nation's largest federation (alliance or partnership of unions). It has about 12.7 million members. The next-largest federation is Change to Win, with about 4.5 million members.

Unions in the Workplace

Employees in a workplace cannot form a union unless most of them vote in favor of it. An agency of the federal government, the National Labor Relations Board (NLRB), makes sure these elections are carried out fairly and honestly.

A common way that unions organize a workplace is with a *union shop*. In these workplaces, companies can hire any person as an

Right-to-Work States

The map shows states with right-to-work laws. Right-to-work laws were made possible by a federal law passed in 1947.

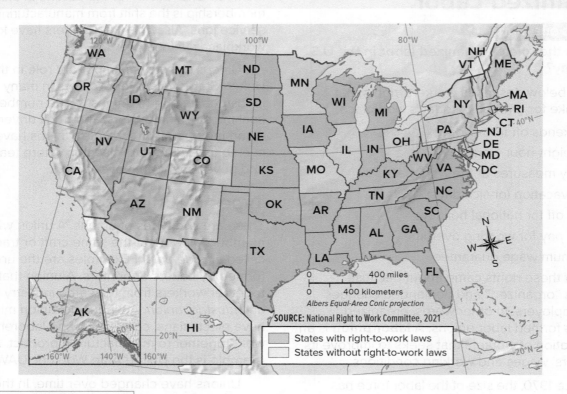

SOURCE: National Right to Work Committee, 2021

States with right-to-work laws
States without right-to-work laws

GEOGRAPHY CONNECTION

1. **Exploring Regions** In what regions of the country are right-to-work states mostly found?
2. **Patterns and Movement** Why do you think states pass right-to-work laws for companies located within their borders?

employee. Once someone is hired, though, they must join the union shortly after starting to work.

Many companies do not like the union shop. In some states, companies have convinced state governments to outlaw these arrangements. Nearly half the states have **right-to-work laws**, which ban union shops.

Other states have what is called a *modified union shop*. In this situation, workers do not have to join a union. But if they do join, they must remain in the union as long as they hold their job. Some workplaces are *agency shops*. In an agency shop, workers who do not join the union still must pay a fee to the union for representing them.

These teachers demand increased funding for public education.

Analyzing Visuals What union tools are the teachers using in this photograph?

✓ CHECK FOR UNDERSTANDING

1. **Contrasting** For what purposes do unions form?
2. **Summarizing** What role do unions play in the United States today?

Labor Negotiations

GUIDING QUESTION

How do labor and management negotiate?

When a company's workers have a union, the union and the company carry out **collective bargaining**. In this process, officials from the union and the company meet to discuss the workers' contract. The contract sets the terms for working at the company. These talks often focus on wages and benefits. Benefits include health insurance, sick days, and holidays. Contracts also cover rules for workers to follow and working conditions, such as breaks for meals during the workday.

With most contracts, the two sides reach an agreement during bargaining. Sometimes, though, negotiations break down. If that happens, unions and employers each have methods to pressure the other side to accept their position.

Union Tools

Labor unions have several tools to try to advance their cause. One method unions use is to call a **strike**. In a strike, all union members refuse to

work. The idea is that without employees, one company—or all companies in the industry—will be forced to meet the union's demands. If they do not, the company or companies lose money every day that they refuse to work.

Striking workers usually stand in public view carrying signs stating that they are on strike. This tactic is called **picketing**. The goals are to embarrass the company and to build public support for the strike. Strikers also hope to discourage other workers from crossing the picket line to work at the company.

Another tool of labor is to put economic pressure on the company. For example, the union may ask people to boycott the company, or refuse to do business with it.

Strikes can cause problems for workers, too. They can drag on for months. If so, strikers may become discouraged. Some might want to go back to work. This can put pressure on the union to give in on some of its demands. Sometimes a strike will end without workers gaining anything they wanted. However, in most cases strikes are settled when the company and the union work out an agreement.

---

right-to-work laws state laws forbidding unions from forcing workers to join

collective bargaining process by which unions and employers negotiate the conditions of employment

strike when workers deliberately stop working in order to force an employer to give in to their demands

picketing a union tactic in which striking workers walk with signs that express their grievances

Employers' Tools

Employers also have ways to try to pressure unions. Their strongest tool is the **lockout**. In a lockout, the employer does not let workers enter the workplace. The employer hopes that the loss of income will force workers to accept company terms. During the lockout, the company often hires replacement workers so it can continue its business. That way, the locked-out workers suffer, but the company does not.

Companies may try to stop union actions by asking for an **injunction** (ihn•JUHNGK•shuhn). An injunction is a legal order from a court to prevent some activity. The company may ask the court to limit picketing or to prevent or stop a strike.

Outside Help

Unions may also seek injunctions. If issued against an employer, the injunction may order the employer not to lock out its workers. In certain industries considered important to the economy or national security, the government can seek an injunction. In 2002, for example, President George W. Bush asked for an injunction to end a lockout of dockworkers on the West Coast. He said keeping ports open was essential to military operations. A court order ended the lockout.

lockout when management closes a workplace to prevent union members from working

injunction a court order to stop some kind of action

A worker and his daughter rally against a company lockout. When unions oppose a wage cut, the company sometimes shuts down the plant or hires other workers until the union agrees to the cut.

Contrasting What is the difference between a lockout and a strike?

When the parties cannot agree on a contract, they have other options. First, they can try **mediation** (mee•dee•AY•shuhn). In this approach, they bring in a third party who tries to help them reach an agreement. They can also choose **arbitration** (ahr•buh•TRAY•shuhn), where a third party listens to both sides and decides how to settle the dispute. Both parties agree in advance to accept the third party's decision.

If a strike threatens the nation's welfare, the government can step in. Federal law allows the president to order a cooling-off period. During this time, the workers must return to work. Meanwhile, the union and the employer must try to reach an agreement. The cooling-off period lasts 80 days. If there is no agreement after that time, the workers have the right to go back on strike.

In an extreme situation, the government can temporarily take over a company or an industry. For example, in 1946 the government seized coal mines when a strike threatened to shut off the nation's coal supply.

✓ CHECK FOR UNDERSTANDING

1. **Identifying** What are the main areas of negotiation between unions and employers in reaching collective bargaining agreements?

2. **Comparing and Contrasting** Compare and contrast the terms *mediation* and *arbitration*.

LESSON ACTIVITIES

1. **Informative/Explanatory Writing** If you were an adult worker, would you join a labor union? Write a paragraph that explains the pros and cons of joining a union.

2. **Collaborating** A local union wants a pay raise and more paid time off. The company argues it cannot give a wage increase. In groups, assign yourselves three roles: union representative, business owner, and mediator. Each should write a paragraph explaining their position. Simulate a collective bargaining session. Write a new contract covering pay, benefits, and time off.

mediation situation in which union and company officials bring in a third party to try to help them reach an agreement

arbitration situation in which union and company officials submit the issues they cannot agree on to a neutral third party for a final decision

Jim West / Alamy Stock Photo

All About Money

GUIDING QUESTION

What gives money value?

Suppose you were selling a bicycle. Would you accept a four-ton stone in payment for it? You might if you lived on the Pacific island of Yap. For centuries, the people of Yap used huge stone disks as money. How could they do that? Money is anything that a group of people accepts as a means of exchange. A wide variety of things—from cheese to shells to dogs' teeth to gold and silver coins—have been used as money. People use money because it makes life easier for everyone.

Functions of Money

Money has three main functions. First, money serves as a *medium of exchange*. A **medium** is a means of doing something. People exchange, or trade, money for goods and services. If we did not have money, we would have to **barter**, or trade for something of equal worth. For example, a person might want to exchange running shoes for a pair of theater tickets. While this might sound like a simple task, the exchange might never take place. The person with the shoes may never find anybody willing to trade theater tickets for them.

Second, money is a *store of value*. This means that we can hold money as a form of wealth until we find something we want to buy with it. The person with the running shoes does not have to wait for someone willing to trade for them. Instead, the person can sell the shoes and hold the money until it is needed later.

medium a means of doing something

barter to trade a good or service for another good or service

Here, $100 bills are printed before being cut into separate bills. U.S. paper money is designed and printed by the Bureau of Engraving and Printing (BEP), which has operations in Washington, D.C., and Fort Worth, Texas. The BEP prints 26 million bills each day—worth about $974 million.

READING STRATEGY

Analyzing Key Ideas and Details As you read, use a chart like the one shown to fill in details about each feature of money.

Money	
Feature	**Details**
Function	
Trait	
Form	

Third, money serves as a *measure of value*. Money is like a measuring stick that can be used to assign value to a good or service. When somebody says that something costs $10, we know exactly what that means.

Characteristics of Money

For something to serve as money, it must have four characteristics:

- *Portable*. Money must be easy to carry around so people have it when they want to buy.
- *Divisible*. Money must be easy to divide into smaller amounts. That way it can be used for large and small purchases.
- *Durable*. Pieces of money should be hardy enough to stay in use for some time.
- *Limited Supply*. To be used as money, an object must be in limited supply. If money were easy to make, everyone would make it. The money would become worthless. That is why making fake money—or counterfeiting—is a crime.

Forms of Money

In the United States, money comes in three forms: coins, paper bills, and electronic money.

Coins are pieces of metal that are used as money. Examples of coins include pennies, nickels, dimes, quarters, and even some dollars. Our "paper" money is made out of high-quality cotton and linen. Coins and paper bills make up the **currency** of the United States.

Electronic money is money in the form of a computer entry at a bank or other financial institution. Electronic money does not exist in any physical form. An example would be the money you have in a checking or savings account. (Do not confuse credit cards with electronic money. A credit card is not money—it just gives the user permission to borrow.)

A new kind of money, according to some, is cryptocurrency. **Cryptocurrency** is electronic money not issued or managed by any country or central bank. Its value can fluctuate widely with

coin metallic form of money, such as a penny

currency money, both coins and paper bills

electronic money money in the form of a computer entry at a bank or other financial institution

cryptocurrency electronic money not issued or managed by any country or central bank

---

Features of U.S. Currency
Paper money has several special features that help prevent criminals from making counterfeit money.

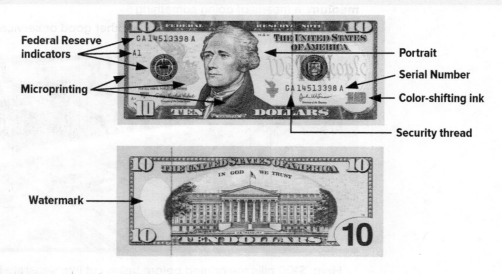

Federal Reserve indicators

Microprinting

Portrait

Serial Number

Color-shifting ink

Security thread

Watermark

EXAMINE THE PHOTOS

1. **Analyzing Visuals** Which features shown will be useful for preventing counterfeiting?
2. **Inferring** Why would a government want to prevent counterfeiting?

These are symbols of Litecoin, Ethereum, and Bitcoin—three popular cryptocurrencies.

Speculating What characteristics must cryptocurrencies have before they are widely accepted as everyday money?

supply and demand. Bitcoin is perhaps the most popular cryptocurrency, and Ethereum is the second-most popular. Although gaining in popularity, none of the cryptocurrencies have achieved widespread adoption. This is due to none having all four required characteristics of money—portability, divisibility, durability, and limited supply.

✓ **CHECK FOR UNDERSTANDING**

1. **Explaining** What does it mean to barter? How does money make bartering unnecessary?

2. **Summarizing** What are three functions of money?

Financial Institutions

GUIDING QUESTION

What do financial institutions do?

The main function of financial institutions is to channel funds from savers to borrowers. When most people receive their pay, they put some money into a financial institution, such as a bank. The money that customers put into a financial institution is called a **deposit**. Businesses, too, deposit the money they receive from selling goods or services. Money is electronic when employers deposit workers' pay directly into workers' bank accounts. The funds in the employer's account go down, and the money in the workers' accounts goes up.

deposit the money that customers put into a financial institution

Taking Deposits

Banks have two main types of accounts—checking and savings. People use money deposited in a **checking account** to pay bills, buy goods and services, and meet other expenses. This is how checking accounts help money serve its function as a medium of exchange. People can withdraw their cash anytime by writing a check, using a debit card, or swiping a cell phone, which is why checking accounts are also called *demand deposit accounts*. Monthly banking fees for using checking accounts are almost always more than the interest paid on the checking deposits.

Deposits in a **savings account** help money serve its function as a store of value or wealth. Banks pay interest on savings accounts to encourage people to keep their money in the bank as long as possible. This allows banks to use some of the deposits to make loans. Savings accounts are usually free, deposits pay interest, and people can make limited but not frequent withdrawals.

A **certificate of deposit (CD)** is a loan you make to a bank, although it seems like a deposit.

checking account an account from which deposited money can be withdrawn at any time by writing a check, using a debit card, or swiping a cell phone; also known as demand deposit accounts

savings account an account that pays interest on deposits but allows only limited withdrawals

certificate of deposit (CD) a timed consumer loan to a bank that states the amount of the loan, maturity, and rate of interest being paid

This person is checking his bank accounts online. Customers often create multiple bank accounts so they can earn higher interest on savings and CDs.

Explaining What do people use checking accounts for?

You must leave the money in the bank for a fixed period, such as a year. Money taken out early faces a penalty in the form of a lower interest rate. The interest paid on CDs is higher than that paid on checking or savings accounts—another way that banks help money serve as a store of wealth.

Making Loans

After people deposit their funds, financial institutions put these deposits to work. Banks might keep some of each deposit in reserve, but the rest can be lent to individuals or businesses. People borrow to purchase a car or home, or pay college tuition. Businesses borrow to expand operations, make new products, or meet a payroll.

Banks charge interest on their loans. That is how they earn money. Banks also *pay* interest on deposits. So banks must be careful to charge a higher rate on loans they lend and pay a lower rate on interest to depositors. Before a loan is final, a lending officer at the bank and the borrower discuss the amount to be borrowed, the interest rate, and when the loan must be repaid.

Types of Financial Institutions

Consumers and businesses have several different financial institutions that can meet their needs. Two main types are commercial banks and credit unions.

Commercial banks offer the most financial services. They accept deposits, provide checking accounts, make loans, and offer other services such as safe deposit boxes where important documents and valuables can be kept. Commercial banks are the largest and most critical part of the financial system. Most businesses deal with commercial banks, and consumers can get loans if they have good credit ratings. Savings and loan associations (S&Ls) also offer many of the same services as commercial banks.

A **credit union** is a nonprofit cooperative that accepts deposits, makes loans, and provides financial services to its members. Credit unions are often formed by people who work in the same industry, work for the same company, or belong to the same labor union. Credit unions are cooperatives that their depositors, or members, own. Sometimes the only place a person can get a loan is from their credit union. Credit unions usually charge lower interest rates on loans, and they lend money only to members.

Regulating Financial institutions

Because financial institutions are an essential part of our economic system, they must be financially sound and secure. To make sure this happens, several different state and federal agencies regulate or oversee the way they do business.

This regulation starts when they go into business. For example, a financial institution must first get a document called a *charter*. Individual states and the federal government grant charters, which are approvals to go into business. The government reviews the finances of the proposed institution to make sure it has enough money to succeed. Officials also examine the people who will run the business. They want to be sure they have the skills to use depositors' money wisely.

After the charter is issued, government officials watch to see how the financial institution is run. They try to make sure it stays in good financial condition and follows all relevant laws. These efforts protect the money depositors entrust to the institution and ensure that business operations are sound.

---

commercial bank a financial institution that offers the most banking services to individuals and businesses

credit union nonprofit service cooperative that accepts deposits, makes loans, and provides other financial services to its members

The money you deposit in a financial institution is safe for another reason—a federal **deposit insurance program** protects it. The Federal Deposit Insurance Corporation (FDIC) is the agency that protects deposits at commercial banks. The National Credit Union Share Insurance Fund (NCUSIF) is the agency that protects credit union deposits. Both programs cover deposits up to $250,000 for one person on all accounts within the same institution. Without these protections, our financial institutions would not be as safe for our deposits.

✓ CHECK FOR UNDERSTANDING

1. **Explaining** What is the role of financial institutions?
2. **Contrasting** How are savings accounts and certificates of deposit different?

How Banking Has Changed

GUIDING QUESTION

How has banking become safer, faster, and more efficient over the years?

Some form of banking has existed since civilization emerged in Mesopotamia around 8000 B.C.E. Temples acted as banks, often storing valuables for the wealthy and loaning seeds to local farmers. The Roman Empire extended loans to finance its empire-building. During medieval times, merchant banks minted coins and loaned money to finance trade on the Silk Road routes. Ever since our nation's founding, banking has also gone through many changes.

deposit insurance program government-backed program that protects bank deposits up to a certain amount if a financial institution fails

The Earliest Banks

States chartered most of the first banks. These banks even printed their own paper currency, which circulated as money. However, many of these banks were also unsound and often went bankrupt, or out of business. Early depositors in these banks usually lost all of their money because there was no form of banking insurance.

The strongest bank in our nation's early history got its charter from the federal government. This was the First Bank of the United States. It was chartered in 1791 to hold the government's money, make its payments, and lend to it during times of need. The bank had a charter for only 20 years. Then, during the War of 1812, the government discovered it had no place to borrow money. As a result, political leaders created the Second Bank of the United States, lasting from 1816 to 1836 when its charter ran out.

From the 1830s to the 1860s, states took over chartering and supervising privately owned banks. This control was inadequate, however, because banks continued to make loans by issuing their own paper currency. Again, private banks printed too many banknotes, leading to a greatly increased money supply and inflation. Demand arose for a uniform currency that was accepted anywhere without risk.

National Banks and the Federal Reserve System

In 1863 the Union government passed the National Bank Act to help bring order to the banking industry. This made it possible for banks to get a national charter, which were better funded than state banks. A government official called the Comptroller of the Currency regulated the national banks, which issued a uniform

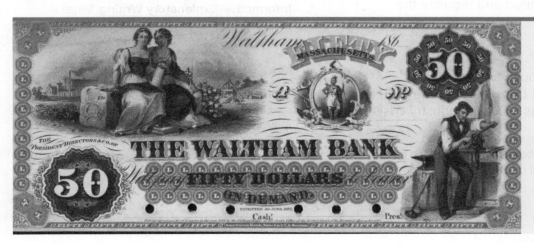

THE WALTHAM BANK
FIFTY DOLLARS
IN DEMAND
$50

By 1860, more than 1,600 state banks issued their own paper currency, such as this one from Massachusetts.

Identifying Cause and Effect Why did the national government want to stop states from issuing their own currency?

currency backed by U.S. government bonds. National banks were safer than state banks but they did not have enough flexibility to deal with an economic downturn.

In 1913 Congress created the Federal Reserve System (Fed) to act as a central bank and oversee a new currency called Federal Reserve Notes. The Fed was a system of 12 regional banks designed to deal with economic problems in each of their districts. But the system was not perfect. Some of the 12 banks did not work together as planned. This problem made the Fed unable to take strong steps to end the Great Depression.

The Great Depression

During the Great Depression (1929–1933), many businesses failed. One out of four workers could not find a job. Panicky depositors all tried to withdraw their money from banks at the same time. This caused a crisis. Banks did not have enough cash on hand because they had loaned most of the money out. The crisis forced many banks to close their doors, which caused more panic, and then even more banks failed.

In response, President Franklin D. Roosevelt ordered all banks closed for four days. He assured people that banks would reopen after they had been checked by a government official and found to be healthy. Congress then passed the Banking Act of 1933. This law created the FDIC program and allowed the Fed to better oversee banks. Another law passed in 1935 gave the Fed even more powers to regulate banks. The Fed's Board of Governors (the main supervisory body of the Fed) also got more authority to regulate the 12 district banks. These reforms established the Fed as an effective **central bank**—a banker's bank that can lend to other banks in time of need and regulate the money supply.

Banks Since the Depression

The economy went through several difficult periods since the Great Depression of the 1930s. The first was the savings and loan industry crisis

---

central bank a banker's bank that can lend to other banks in time of need and can regulate the money supply

of the 1980s. The Great Recession of 2008–2009 was the second. The third was the COVID-19 downturn of 2020–2021. The FDIC, the Fed, and the federal government were responsible for successfully protecting customer deposits in banks during these periods.

The most significant banking regulations took place right after 2008 when banks were required to strengthen their reserves against losses. Banks were also required to pay more for their FDIC insurance, and laws limited risky loans that banks could issue. These changes increased costs and lowered profits for most banks. To reduce costs, free checking accounts disappeared, and many bank mergers took place. Other banks closed branch offices and moved many services online.

New Ways of Banking

New technology has introduced new forms of banking. Banking by cell phone, telephone, or automated teller machines (ATMs) lets people do their banking without setting a foot inside a commercial bank branch. Online banking allows people to use the Internet to check their balances and to see all transactions that have taken place. Many depositors have their bills paid automatically from their accounts. People can make purchases with a swipe of their cell phones rather than using a check or a credit card.

✓ **CHECK FOR UNDERSTANDING**

1. **Explaining** What problem was found with the national banks created under the National Bank Act?

2. **Identifying Cause and Effect** How did the changes after the Great Depression make the banking industry safer?

LESSON ACTIVITIES

1. **Informative/Explanatory Writing** Write a public service announcement (PSA) explaining why financial institutions are safe places for depositors to put their money. Make your PSA lively but informative.

2. **Using Multimedia** With a partner, create a multimedia presentation that explains the characteristics and functions of money in a way an elementary student would understand. Add visuals and music to support your explanations.

07

Analyzing Sources: A Cashless Society?

? COMPELLING QUESTION

What can be complicated about using money?

Plan Your Inquiry

DEVELOPING QUESTIONS

Read the Compelling Question for this lesson. What questions can you ask to help you answer this Compelling Question? Create a graphic organizer like the one below. Write three Supporting Questions in the first column.

Supporting Questions	Source	What this source tells me about the necessity of cash	Questions the source leaves unanswered
	A		
	B		
	C		
	D		
	E		

ANALYZING SOURCES

Next, read the introductory information for each source in this lesson. Then analyze each source by answering the questions that follow it. How does each source help you answer each Supporting Question you created? What questions do you still have? Write those in your graphic organizer.

After you analyze the sources, you will:

- use the evidence from the sources
- communicate your conclusions
- take informed action

Background Information

Money is a medium of exchange. That is, when you make a purchase, money is what you give up in order to get whatever you buy. In the country's early years, coins were the main form of money. Over time, paper money became far more common than coins as a medium of exchange. Today, however, many people carry neither paper money nor coins when they go shopping. Instead, many transactions occur electronically.

This Inquiry Lesson includes sources about the use of cash and noncash payments. As you read and analyze the sources, think about the Compelling Question.

» Do people need to carry cash anymore?

Electronic Transactions

Today consumers can make a purchase simply by swiping a plastic debit or credit card through a card reader, or by tapping or scanning a payment app on a cell phone. These devices are tied electronically to the nation's banking system.

PRIMARY SOURCE: PHOTOGRAPHS

A young woman purchases a book by using her credit card.

Cell phones with near-field communication (NFC) or radio frequency identification (RFID) technology allow customers to make "tap-and-go" purchases if the phone is held within a few inches of the point-of-sale terminal. These "contactless" payments became more popular in 2020 after COVID-19 struck.

EXAMINE THE SOURCE

1. **Analyzing** What must suppliers have available for electronic purchases to occur?
2. **Drawing Conclusions** How do electronic transactions fulfil the characteristic of money as a medium of exchange?

B

Online Shopping

People all over the world shop remotely online, and you can't do that with physical cash and coins. The table identifies the average *revenue*—or sales—created by an online shopper in the countries listed.

Country	Average Revenue Per Online Shopper
United States	$1,804
United Kingdom	$1,629
Sweden	$1,446
France	$1,228
Germany	$1,064
Japan	$968
Spain	$849
China	$626
Russia	$396
Brazil	$350

Source: "Online Shopping Statistics You Need to Know in 2021" by Coral Ouellette, January 2021.

EXAMINE THE SOURCE

1. **Evaluating** Are you surprised by any of the countries listed or the amounts of revenue shown? Why or why not?

2. **Analyzing** What information is missing that might help put this table's data in perspective?

Trends in Noncash Payments

Noncash payments include writing checks and using *debit cards* tied to a person's checking account. The amount a person can spend depends on how much money is in their checking account. Noncash payments also include *prepaid debit cards* not tied to a checking or savings account. An individual can load a certain amount of money onto a prepaid card—and that is the limit they can spend.

Credit cards are another type of noncash payments. Credit card companies offer loans, usually of a limited amount. If you purchase something using a credit card, the store is paid right away—and then you pay back the credit card company over time, with interest.

ACH transfers are yet another type of noncash payment. *ACH* stands for "Automated Clearing House." Computers process payments automatically from one financial institution to another. When you start a job, for example, the company you work for may pay your salary as a "direct deposit" instead of handing you a paper paycheck. In this situation, the ACH network automatically removes your pay from your employer's bank account and transfers it to your bank account.

PRIMARY SOURCE: GRAPH

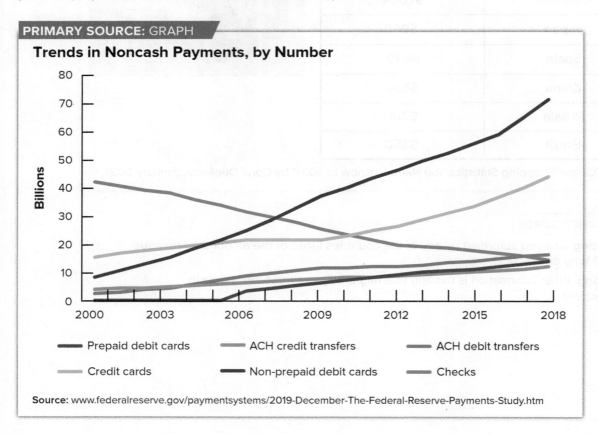

Trends in Noncash Payments, by Number

Legend:
- Prepaid debit cards
- ACH credit transfers
- ACH debit transfers
- Credit cards
- Non-prepaid debit cards
- Checks

Source: www.federalreserve.gov/paymentsystems/2019-December-The-Federal-Reserve-Payments-Study.htm

---

EXAMINE THE SOURCE

1. **Analyzing Visuals** According to the most recent data available in the graph shown, how are most payments made today?

2. **Speculating** The graph ends in 2018. What do you think happened to the trend for noncash payments after the COVID-19 pandemic occurred in 2020?

D

Pros of Using Credit

This article makes the case for using credit cards.

PRIMARY SOURCE: ARTICLE

"Many of us use credit cards irresponsibly and end up in debt. However, contrary to popular belief, if you can use the plastic responsibly, you're actually much better off paying with a credit card than with a debit card and keeping cash transactions to a minimum. . . .

Paying with a credit card makes it easier to avoid losses from **fraud**. When your debit card is used by a thief, the money is missing from your account instantly. . . .

By contrast, when your credit card is used fraudulently, you aren't out any money—you just notify your credit card company of the fraud and don't pay for the **transactions** you didn't make while the credit card company resolves the matter. . . .

Certain purchases are difficult to make with a debit card. When you want to rent a car or stay in a hotel room, you'll almost certainly have an easier time if you have a credit card. . . .

Credit cards are best enjoyed by the disciplined, who can remain **cognizant** of their ability to pay the monthly bill (preferably in full) on or before the due date."

— From "10 Reasons to Use Your Credit Card" by Amy Fontinelle, Investopedia.com, 2021.

fraud illegal action, scam
transactions purchases, trades
cognizant aware

EXAMINE THE SOURCE

1. **Summarizing** In your own words, summarize the author's reasons to use credit cards instead of cash or debit cards.
2. **Drawing Conclusions** Why is it important to be disciplined when using a credit card?

Fontinelle, Amy. "10 Reasons to Use Your Credit Card." Investopedia. May 17, 2021. https://www.investopedia.com/articles/pf/10/credit-card-debit-card.asp.

Cash Is Still Here

In some countries, a large percentage of people still use cash for payments.

EXAMINE THE SOURCE

1. **Identifying** Which five countries shown on the graph have the highest percentage of *noncash* payments?

2. **Drawing Conclusions** Compare the cash payment percentages on this graph to the average revenue per online shopper shown in Source B. What are some conclusions you might draw about shoppers in Spain based on both sources?

PRIMARY SOURCE: GRAPH

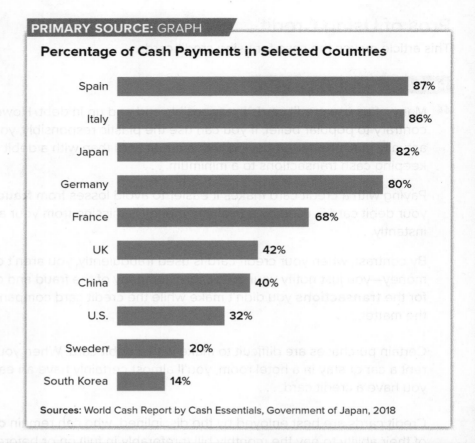

Percentage of Cash Payments in Selected Countries

Country	Percentage
Spain	87%
Italy	86%
Japan	82%
Germany	80%
France	68%
UK	42%
China	40%
U.S.	32%
Sweden	20%
South Korea	14%

Sources: World Cash Report by Cash Essentials, Government of Japan, 2018

Complete Your Inquiry

EVALUATE SOURCES AND USE EVIDENCE

Refer back to the Compelling Question and the Supporting Questions you developed at the beginning of the lesson.

1. **Identifying** Which of the sources affected you the most? Why?

2. **Evaluating** Overall, what impression do these sources give you about the importance of cash? Explain.

3. **Gathering Sources** Which sources helped you answer the Supporting Questions and Compelling Question? Which sources, if any, challenged what you thought you knew when you first created your Supporting Questions? What information do you still need in order to answer your questions? Where would you find that information?

4. **Evaluating Sources** Identify the sources that helped answer your Supporting Questions. How reliable is the source? How would you verify the reliability of the source?

COMMUNICATE CONCLUSIONS

5. **Collaborating** Work with a partner to create an infographic summarizing the various forms of noncash payments and their advantages and disadvantages. How do these sources provide insight into how noncash payments increased over time? Use the graphic organizer you created at the beginning of the lesson to help you. Share your infographic online.

TAKE INFORMED ACTION

Highlighting Smart Money Management
Research negative credit card experiences online. After reading about the situations, create a poster summarizing what NOT to do when students get their first credit card. Display the poster in the school for other students to read.

08
Government and Business

READING STRATEGY

Integrating Knowledge and Ideas As you read, identify ways the government's role in the economy benefits consumers.

Benefits to Consumers

Providing Public Goods

GUIDING QUESTION

What goods does government provide?

What products and services are available in your community without paying a fee? Do you walk on sidewalks or enjoy a local park? Do you have the protection of the police and the use of traffic signs? You do not have to pay for these things each time you use them. They are there for all to use. These products and services are different from those you pay for in stores and use for your own enjoyment.

Private and Public Goods

Businesses produce **private goods**, products that people must buy in order to use or own them. A person who does not pay for a private good is barred from owning or using it. In addition, private goods can be used by only one person. If you eat a meal, no one else can buy and eat it. Clothes, food, and cars are examples of private goods.

Unlike private goods, **public goods** can be consumed by more than one person. For example, your community's sidewalks are public goods. If you walk on a sidewalk, that does not prevent others from walking on it as well. Police protection and national defense are also public goods. A community—not just one person—enjoys the protection of the police.

private good economic good that, when consumed by one person, cannot be used by another

public good economic good that is used collectively, such as a highway and national defense

New York City's Central Park is a public good, meaning it can be "consumed" or used by more than one person.

Public goods are important to a number of people—even an entire community or nation. Yet businesses do not like to produce and sell them. Why is that? The reason is simple. It is difficult to charge everyone who might benefit from using public goods. For instance, how could someone figure out what to charge for your use of a sidewalk? Because it is hard to assign the costs, government takes on the responsibility for providing public goods. It pays for these goods through taxes and other fees it collects.

Externalities

Economic activities of all sorts produce side effects called **externalities** (ehk·stuhr·NAH·luh·teez). These are either positive or negative side effects of an action that impact an uninvolved third person.

Public goods often produce *positive* externalities. That is, there are benefits to everyone who uses those goods. Everyone—not just drivers—benefits from having good roads. Good roads make it faster and cheaper to transport goods. That means those goods can be sold at lower prices. As a result, all consumers benefit. A lower price is one positive externality that comes from having good roads.

Externalities can be *negative*, too. Negative externalities result when an action has harmful side effects. A car provides transportation, but its exhaust pollutes the air. Even people without cars may suffer from air pollution's harmful effects, such as breathing problems.

This airplane's flight path to the airport includes flying over a nearby highway and neighborhoods.

Speculating What positive and negative economic externalities might people living near the airport experience if the airport increases its number of flights?

Government's role is to encourage positive externalities and discourage negative ones. So the government provides public schooling because education leads to positive externalities. A well-educated workforce is more productive. To reduce pollution, the federal government has regulated car exhausts since the 1970s.

✓ **CHECK FOR UNDERSTANDING**

1. **Explaining** How does government pay for public goods?
2. **Making Connections** What is an externality? Provide an example of an externality, as well as its source, and identify it as positive or negative.

Maintaining Competition

GUIDING QUESTION

How does government encourage or increase competition among businesses?

Have you ever played the game Monopoly®? To win, a player tries to control all the properties in the game and bankrupt the other players. In other words, the winner becomes a monopoly. A **monopoly** (muh·NAH·pah·lee) is the exclusive control of a good or service. A *monopolist* is the one who has the monopoly.

Markets work best when large numbers of buyers and sellers participate. If a monopolist gains control of a market, it does not have to compete with other companies for buyers. As a result, it can charge a much higher price. Then consumers suffer because they are forced to pay a high price instead of being able to shop for a better one. To prevent this problem, a goal of the U.S. government has long been to encourage competition so that monopolies do not form.

Antitrust Laws

To protect competition, the government uses antitrust laws. A trust is a combination of businesses that threatens competition. The government's goal in passing **antitrust** (an·tee·TRUHST) **laws** is to control monopoly power and to preserve and promote competition.

---

externality economic side effect that affects an uninvolved third party

monopoly exclusive control of a good or service

antitrust law legislation to prevent monopolies from forming and to preserve and promote competition

In this political cartoon from 1904, artist Udo J. Keppler portrayed the Standard Oil Company as an octopus. The octopus "monopoly" has tentacles wrapped around the steel, copper, and shipping industries and the U.S. Capitol.

Explaining How do antitrust laws prevent monopoly behavior?

In 1890 the government passed its first antitrust law, the Sherman Antitrust Act. This law banned monopolies and other forms of business that prevent competition. In 1911 the government used the law to break up the Standard Oil Company because it had an oil monopoly. In the 1980s the government used the act to break up American Telephone and Telegraph (AT&T). This action ended a monopoly on phone service and created more competition.

In 1914 Congress passed the Clayton Antitrust Act. This law made the Sherman Act stronger and clearer. The Clayton Act banned a number of business practices that hurt competition. For example, a person could no longer be on the board of directors of two competing companies. The government legally took over some mergers.

Mergers

Sometimes two or more companies combine to form a single business. That joining is called a **merger** (MUHR•juhr). Some mergers threaten competition, which might lead to higher prices. In those cases, the government can use the Clayton Act to block the merger. The Federal Trade Commission (FTC) has the power to enforce this law. It looks at any merger that may violate antitrust laws. It also may take actions to **maintain**, or preserve, competition. In 2011, for example, AT&T proposed to merge with T-Mobile, a combination of the second- and fourth-largest wireless carriers in the country. The Justice Department argued that such a merger would lessen competition, so the merger was blocked.

merger a combination of two or more companies to form a single business

maintain to preserve

Natural Monopolies

At times it makes sense to let a single business produce a good or service. For example, it might be better to have one company, instead of two or three, build electric power lines for a city. In these cases, we have a **natural monopoly**. A single business produces and distributes a product better and more cheaply than several companies.

Natural monopolies have great power. They can choose to raise prices whenever they wish. For this reason, a government agency regulates, or closely watches, these companies. The agency usually has to approve any price increases or other changes in business activities.

Sometimes a local government can choose a different approach. It may become the owner of the natural monopoly instead. This is often the case with such services as water and sewers.

In recent years, many governments decided to put an end to certain natural monopolies by restoring competition. About half the states ended the monopoly of electric companies. This policy of ending regulation is called *deregulation*. The new approach has not always led to lower prices, though. Many states are now backing away from deregulation of natural monopolies.

✓ **CHECK FOR UNDERSTANDING**

1. **Summarizing** Why does government promote competition?
2. **Identifying Cause and Effect** Explain what a merger is. How can it lead to a monopoly?

natural monopoly a market situation in which the costs of production are minimized by having a single firm produce the product

Selected U.S. Government Regulatory Agencies

These are just a few of the federal government agencies that regulate businesses.

DEPARTMENT OR AGENCY	PURPOSE
Consumer Product Safety Commission (CPSC)	Protects the public from risks of serious injury or death from consumer products
Environmental Protection Agency (EPA)	Protects human health and the natural environment (air, water, and land)
Federal Trade Commission (FTC)	Promotes and protects consumer interests and competition in the marketplace
Food and Drug Administration (FDA)	Makes sure food, drugs, and cosmetics are truthfully labeled and safe for consumers
Occupational Safety and Health Administration (OSHA)	Makes sure workers have a safe and healthful workplace

EXAMINE THE CHART

Identifying Which regulatory agency has the power to make sure that competition exists in the marketplace?

Protecting Consumers

GUIDING QUESTION

How does government regulate business?

The government also plays a major role in protecting the health and safety of the public. The Food and Drug Administration (FDA) makes sure that foods, drugs, medical equipment, and cosmetics are safe. It also requires companies to tell the truth on product labels and in ads.

The Centers for Disease Control and Prevention (CDC) also tries to improve health. It checks air quality and distributes the flu vaccine, for example. The CDC helped develop and distribute the vaccine for the COVID-19 virus.

The goal of the Consumer Product Safety Commission (CPSC) is to protect consumers from injury. It oversees thousands of products, from toys to tools. The CPSC looks for problems in the design of a product that can create danger. If the CPSC finds a product unsafe, it issues a recall. A **recall** means the unsafe product is removed from stores. For those who bought the product, the manufacturer must change it to make it safe, offer a substitute, or return the customer's money.

The National Highway Traffic Safety Administration (NHTSA) also protects consumers. Between 2016 and 2019, the NHTSA recalled 67 million defective car airbags.

✓ CHECK FOR UNDERSTANDING

1. **Determining Central Ideas** How does government regulation protect the health and safety of consumers?

2. **Identifying** What is the role of the FDA?

LESSON ACTIVITIES

1. **Informative/Explanatory Writing** Identify a public good and a private good that you use regularly. Describe each good, and identify the characteristics that make the public good public and the private good private.

2. **Using Multimedia** Companies that make medicines must test their products before the FDA allows them to be sold. With a partner, research the FDA drug approval process for COVID-19 vaccines and create a slideshow that explains the steps.

recall government action that causes an unsafe product to be removed from consumer contact

READING STRATEGY

Analyzing Key Ideas and Details As you read, identify factors that influence how much income people can earn.

Factors Affecting Income	
Factor	Details

Income Inequality

GUIDING QUESTION

What factors influence income?

The United States is often described as a wealthy country. However, not all Americans are rich. Some people have high incomes, but others are quite poor. Income levels vary for many reasons. Education level, family wealth, and discrimination each play a role.

Education

Education is a key to income. This is why the government wants Americans to graduate from high school and go on to higher learning. Every day, however, more than 7,000 students drop out of high school. The dropout rate hurts the nation's ability to compete economically. Without a high school diploma, people tend to receive lower wages and experience higher unemployment and imprisonment rates than graduates. To fix this problem, the government has been giving money to states and towns for dropout prevention programs since 2010. This federal program, known as High School Graduation Initiative, focuses its efforts on schools with the highest dropout rates.

The level of education a person **attains**, or achieves, has a great influence on their income. Education gives people the skills they need

attain to achieve

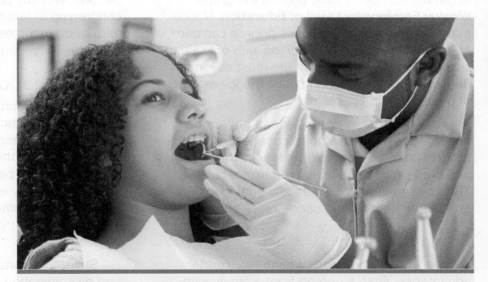

A dentist works on a young patient's teeth. Dentistry is one of the highest-paid occupations in the United States. To become a dentist, a person must first graduate from college and then attend a dental school for at least four years.

Drawing Conclusions What is the relationship between the education and skills of dentists and the amount they earn?

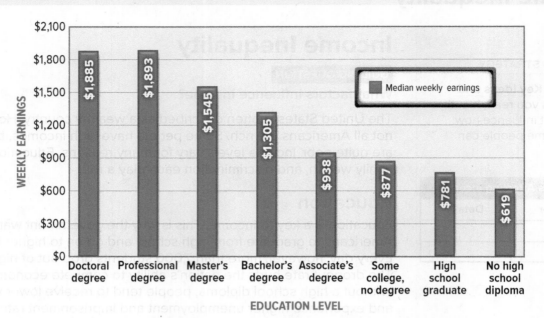

Weekly Earnings by Level of Education, 2020

Education increases human capital and helps people earn more money.

WEEKLY EARNINGS

- Doctoral degree: $1,885
- Professional degree: $1,893
- Master's degree: $1,545
- Bachelor's degree: $1,305
- Associate's degree: $938
- Some college, no degree: $877
- High school graduate: $781
- No high school diploma: $619

Legend: Median weekly earnings

EDUCATION LEVEL

Source: Bureau of Labor Statistics, Current Population Survey.

EXAMINE THE GRAPH

1. **Calculating** What is the difference in earnings between a person with a Bachelor's degree and someone with an Associate's degree?

2. **Evaluating** Do you think a college degree is worth the expense? Why or why not?

to get higher-paying jobs. Look at the graph **Weekly Earnings by Level of Education**. Notice that a person with a bachelor's degree can earn nearly twice as much as a person with only a high school diploma. People with the most advanced degrees earn the most and have the lowest rates of unemployment.

For these reasons, the federal government encourages people to go to college. Some programs help students from low-income families and those with disabilities prepare for college. The government also offers low-cost loans and grants that help make college more affordable.

Family Wealth

People who are born into wealth have certain advantages. First, a person from a family with money has better access to education. As you just read, the more education a person has, the greater his or her **potential**, or possible, income.

Second, wealthy parents can often set their children up in family businesses where they can earn good incomes. Finally, such people usually leave their wealth to their children when they die.

Discrimination

Discrimination limits how much some people can earn. Unfair practices in hiring and promoting people hurt women and members of minority groups. Many of these people are prevented from getting top-paying jobs. For example, women generally earn less than men. In 2020, the American woman working full-time only earned about 81 percent of the American male working full-time.

Several important laws have been passed to end discrimination. These laws have closed the earnings gap between men and women, but some discrimination still exists.

- The *Equal Pay Act of 1963* requires that men and women be given equal pay for equal work. This means that jobs that have the same level of skill and responsibility must pay the same.

potential possible

- The *Civil Rights Act of 1964* bans discrimination based on gender, race, color, religion, and national origin.
- The *Equal Employment Opportunity Act of 1972* gave the government more power to enforce this law.
- The *Americans with Disabilities Act of 1990* gave job protection to people who have physical and mental disabilities.
- The *Lilly Ledbetter Fair Pay Act of 2009* allows workers who suffer unfair treatment because of their gender to sue employers.

The government has also encouraged companies to practice affirmative action. Such a policy is meant to increase the number of minorities and women at work. This effort helps to make up for past actions that held back people in these groups.

✓ **CHECK FOR UNDERSTANDING**

1. **Explaining** How does education affect income?
2. **Identifying** Which laws were the first to give protection to women and to those facing discrimination?

Poverty

GUIDING QUESTION

In what ways does government help those in poverty?

While many Americans are well off, many others are poor. In a recent year, more than 30 million people lived in poverty. This is about 10 percent of the total population in the United States, or about one person in 10. This means they did not earn enough income to pay for basic needs such as food, clothing, and shelter. Because tough economic times can easily add to these numbers, the country has enacted several poverty-fighting programs.

Welfare

To aid struggling families, the federal government provides welfare. **Welfare** is aid given to those in need in the form of money or necessities. The first welfare programs in the United States were established in the 1930s, during the Great Depression. President Franklin D. Roosevelt started these programs to help the millions of Americans who were facing tough economic times.

The government uses poverty guidelines to decide whether a person or family has too little money and therefore needs this help. These guidelines reflect the cost of enough food, clothing, and shelter to survive. The guidelines are updated each year.

Temporary Assistance for Needy Families (TANF) is one welfare program. The federal government provides the money. State governments distribute the funds. TANF began in 1996, replacing a program that was established in the 1930s. This new program set stricter rules for those eligible to take part in welfare programs. It also limited the amount of time during which a person can get benefits. These rules intend to encourage participants to find jobs quickly. In many states, TANF requires those who receive the benefit to work.

welfare aid given to the poor in the form of money or necessities

Temporary Assistance for Needy Families (TANF) welfare program paid for by the federal government and administered by the individual states

Poverty Guidelines for the 48 Contiguous States and Washington, D.C., 2021

The poverty guidelines are adjusted for different sized households. Those with incomes below the official poverty guidelines are eligible for certain federal programs.

Persons in Family/Household	Poverty Guideline
1	$12,880
2	$17,420
3	$21,960
4	$26,500
5	$31,040
6	$35,580
7	$40,120
8	$44,660

For families/households with more than 8 persons, add $4,540 for each additional person.

EXAMINE THE CHART

Calculating What must a family of four earn to stop being eligible for federal aid?

Programs like this are called **workfare** programs. Work activities often take the form of community service. Those getting the aid may be required to attend job training or education programs.

Another important welfare program is **Supplemental Nutrition Assistance Program (SNAP)**. It provides nutritional benefits to supplement the food budgets of needy families. Families wanting to apply for SNAP benefits must apply in the state where they currently live.

Unemployment Insurance

The government also pays some benefits to workers in special cases. One program is unemployment insurance. This program pays compensation to workers who become unemployed through no fault of their own. **Compensation** (KAHM•puhn•SAY•shuhn) is payment to make up for lost wages. If these workers cannot find new jobs, they are usually eligible for unemployment checks for a limited period of time. In addition, workers who are injured on the job may receive workers' compensation benefits, including lost wages and medical care.

Social Insurance Programs

Social Security, a Depression-era program designed to help people provide for their own retirement, is the largest and most powerful anti-poverty program in the United States. In a recent year, Social Security was responsible for keeping more than 27 million people out of poverty. Social Security is funded by a combined tax on workers and employers that amounts to 12.4 percent of a worker's income. This money is then transferred to retired Americans in the form of Social Security payments.

Two other programs are also important. The first is *Medicare*, a program for senior citizens regardless of income. It provides insurance to cover major hospital costs and other medical bills. The second is *Medicaid*, a joint federal-state medical insurance program for low-income persons.

---

workfare programs that require welfare recipients to exchange some of their labor for benefits

Supplemental Nutrition Assistance Program (SNAP) welfare program that provides nutritional benefits to supplement the food budgets of needy families

compensation payment to unemployed or injured workers to make up for lost wages

Social Security anti-poverty program that taxes working people and pays benefits to retirees

Laid-off workers carry their personal belongings out of the workplace. These workers are eligible for compensation.

Making Connections What is compensation? Under what circumstances does someone receive compensation?

Many people who receive payments from Social Security, Medicare, or Medicaid are already in poverty. But these programs are helpful because they ensure that the poorest members of society meet the basic human needs such as food, clothing, and shelter.

✓ **CHECK FOR UNDERSTANDING**

1. **Explaining** What is the purpose of poverty guidelines?

2. **Contrasting** How is workfare different from welfare?

LESSON ACTIVITIES

1. **Argumentative Writing** Suppose a friend of yours is planning to drop out of high school. Write an email using standard writing conventions and proper email etiquette to convince them to stay in school using what you have learned in this lesson.

2. **Collaborating** Which of the following goals do you think would be the most effective in reducing poverty: (1) provide a college education for everyone, (2) ensure living wages for all workers (wages about twice the minimum wage of $7.25 an hour), or (3) provide financial help and job-search assistance to the unemployed? Team up with two classmates, with each selecting one of the goals to research. Present your findings to each other, and decide which goal would be most effective.

10
Multiple Perspectives: The Minimum Wage

 COMPELLING QUESTION

Should we increase the minimum wage?

Plan Your Inquiry

DEVELOPING QUESTIONS

Read the Compelling Question for this lesson. What questions can you ask to help you answer this Compelling Question? Create a graphic organizer like the one below. Write three Supporting Questions in the first column.

Supporting Questions	Source	What this source tells me about the minimum wage	Questions the source leaves unanswered
	A		
	B		
	C		
	D		
	E		

ANALYZING SOURCES

Next, read the introductory information for each source in this lesson. Then analyze each source by answering the questions that follow it. How does each source help you answer each Supporting Question you created? What questions do you still have? Write those in your graphic organizer.

After you analyze the sources, you will:
- use the evidence from the sources
- communicate your conclusions
- take informed action

Background Information

The minimum wage is the lowest hourly rate of pay that employers may pay their workers. This lower limit is set by law. Congress first set the minimum wage in 1938. At the time, the wage was 25 cents. Initially, the law applied only to those employees whose work was part of interstate commerce, or buying and selling products across state lines. Later, Congress changed the law and made it cover most workers. Congress has also raised the rate many times. The current national rate, $7.25, took effect in 2009. The minimum wage has been in place for more than 80 years. Still, the debate over it continues. Critics say the minimum wage discourages businesses from hiring workers. Supporters disagree.

This Inquiry Lesson includes sources about the minimum wage. As you read and analyze the sources, think about the Compelling Question.

Should Congress pass a higher minimum wage law?

Federal Minimum Wage Over Time

The federal minimum wage is the lowest legal wage that can be paid to most workers. As inflation and the cost of living increase, the government reevaluates and increases the minimum wage from time to time. This graph shows the increases in the actual minimum wage from 2000 until it reached its current level—$7.25—in 2009. The second line on the graph shows the minimum wage adjusted for inflation and the cost of living using the value of what the dollar was "worth" in 2009.

PRIMARY SOURCE: GRAPH

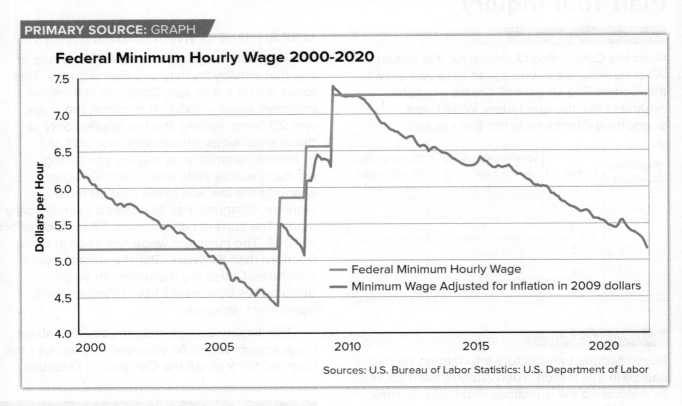

Federal Minimum Hourly Wage 2000-2020

Legend:
— Federal Minimum Hourly Wage
— Minimum Wage Adjusted for Inflation in 2009 dollars

Sources: U.S. Bureau of Labor Statistics; U.S. Department of Labor

---

EXAMINE THE SOURCE

1. **Contrasting** What was the actual federal minimum wage in 2020? What was this wage worth in 2020 after being adjusted for inflation using 2009 prices?

2. **Analyzing** In what years did the inflation-adjusted purchasing power of the minimum wage fall below $7.25? In what years did the inflation-adjusted purchasing power of the minimum wage exceed the actual federal minimum wage?

State Minimum Wages

States can set their own minimum wage. For those states with no minimum wage, they must pay minimum-wage workers the federal minimum wage.

PRIMARY SOURCE: TABLE

State Minimum Wages Per Hour				
AL *None*	**HI** $10.10	**MA** $13.50	**NM** $10.50	**SD** $9.45
AK $10.34	**ID** $7.25	**MI** $9.65	**NY** $12.50	**TN** *None*
AZ $12.15	**IL** $11.00	**MN** $10.08	**NC** $7.25	**TX** $7.25
AR $11.00	**IN** $7.25	**MS** *None*	**ND** $7.25	**UT** $7.25
CA $13.00	**IA** $7.25	**MO** $10.30	**OH** $8.80	**VT** $11.75
CO $12.32	**KS** $7.25	**MT** $8.75	**OK** $7.25	**VA** $7.25
CT $12.00	**KY** $7.25	**NE** $9.00	**OR** $12.00	**WA** $13.69
DE $9.25	**LA** *None*	**NV** $9.00	**PA** $7.25	**WV** $8.75
FL $8.65	**ME** $12.15	**NH** $7.25	**RI** $11.50	**WI** $7.25
GA $7.25	**MD** $11.75	**NJ** $12.00	**SC** *None*	**WY** $7.25

Washington, D.C. $15.00
*As of May 1, 2021
Source: www.dol.gov/agencies/whd/minimum-wage/state

EXAMINE THE SOURCE

1. **Analyzing** Which state paid the highest state minimum wage in 2021?

2. **Drawing Conclusions** What wage do Alabama, Louisiana, Mississippi, South Carolina, and Tennessee pay their minimum-wage workers?

3. **Evaluating** The federal government has one minimum wage for the entire country, but the states' minimum wages differ based on the economic activity and cost of living in each state. Which minimum wage—federal or state—do you think makes more economic sense? Explain your answer.

Argument For a Higher Minimum Wage

Those in favor of increasing the minimum wage argue that it will help low-wage earners make ends meet and bring more people out of poverty.

PRIMARY SOURCE: WEBSITE

The benefits of gradually phasing in a $15 minimum wage by 2025 would be far-reaching, lifting pay for tens of millions of workers and helping reverse decades of growing pay inequality.

The Raise the Wage Act would have the following benefits:

- Gradually raising the federal minimum wage to $15 by 2025 would lift pay for 32 million workers—21% of the U.S. workforce.

- Affected workers who work year round would earn an extra $3,300 a year—enough to make a tremendous difference in the life of a cashier, home health aide, or fast-food worker who today struggles to get by on less than $25,000 a year.

- A majority (59%) of workers whose total family income is below the **poverty line** would receive a pay increase if the minimum wage were raised to $15 by 2025.

- A $15 minimum wage would begin to reverse decades of growing pay inequality between the most underpaid workers and workers receiving close to the **median** wage, particularly along gender and racial lines. For example, minimum wage increases in the late 1960s explained 20% of the decrease in the Black–white **earnings gap** in the years that followed, whereas failures to adequately increase the minimum wage after 1979 account for almost half of the increase in inequality between women at the middle and bottom of the wage distribution.

- A $15 minimum wage by 2025 would generate $107 billion in higher wages for workers and would also benefit communities across the country. Because underpaid workers spend much of their extra earnings, this **injection** of wages will help stimulate the economy and **spur** greater business activity and job growth.

  — From "Why the U.S. needs a $15 minimum wage," Economic Policy Institute Fact Sheet, 2021.

---

poverty line lowest level of income needed to buy necessities in life
median middle
earnings gap difference between the average pay between two different groups of people
injection addition
spur encourage

> **EXAMINE THE SOURCE**

1. **Identifying** What percentage of the U.S. workforce earns a minimum wage?
2. **Explaining** According to the writer, how would an increase in the minimum wage benefit communities?

Economic Policy Institute. "Why the U.S. Needs a $15 Minimum Wage." Economic Policy Institute. February 19, 2021.

D

Argument Against a Higher Minimum Wage

Those against increasing the minimum wage argue that raising the cost of labor hurts employment of low-wage workers as well as family income.

PRIMARY SOURCE: GOVERNMENT WEBSITE

66 The federal minimum wage of $7.25 per hour has not changed since 2009. Increasing it would raise the earnings and family income of most low-wage workers, lifting some families out of poverty—but it would cause other low-wage workers to become jobless, and their family income would fall. . . .

How would increasing the minimum wage affect employment? Raising the minimum wage would increase the cost of employing low-wage workers. As a result, some employers would employ fewer workers than they would have under a lower minimum wage. However, for certain workers or in certain circumstances, employment could increase.

Changes in employment would be seen in the number of jobless, not just unemployed, workers. Jobless workers include those who have dropped out of the labor force (for example, because they believe no jobs are available for them) as well as those who are searching for work. . . .

If workers lost their jobs because of a minimum-wage increase, how long would they stay jobless? At one extreme, an increase in the minimum wage could put a small group of workers out of work indefinitely, so that they never benefited from higher wages. At the other extreme, a large group of workers might shuffle regularly in and out of employment, experiencing joblessness for short spells but receiving higher wages during the weeks they were employed. . . .

How would increasing the minimum wage affect family income? By boosting the income of low-wage workers who had jobs, a higher minimum wage would raise their families' real income, lifting some of those families out of poverty. However, income would fall for some families because other workers would not be employed and because business owners would have to absorb at least some of the higher costs of labor. For those reasons, a minimum-wage increase would cause a net reduction in average family income. 99

— From "How Increasing the Federal Minimum Wage Could Affect Employment and Family Income," Congressional Budget Office, 2021.

EXAMINE THE SOURCE

1. **Analyzing Perspectives** According to the source, why would increasing the minimum wage result in a lower number of jobs available?

2. **Identifying Cause and Effect** How might increasing the minimum wage affect the length of unemployment and average family income, according to the source?

Congressional Budget Office. "How Increasing the Federal Minimum Wage Could Affect Employment and Family Income." Congressional Budget Office. April 5, 2021. https://www.cbo.gov/publication/55681.

Minimum Wage Workers

Many food service workers, especially those working in fast-food restaurants, start out earning the minimum wage.

PRIMARY SOURCE: PHOTOGRAPH

Employees picket to increase wages for fast-food workers.

EXAMINE THE SOURCE

Evaluating Most minimum wage jobs are considered "unskilled labor," meaning the jobs do not require a college education or specialized training. Do you think unskilled labor should be paid at a higher minimum wage? Why or why not?

Complete Your Inquiry

EVALUATE SOURCES AND USE EVIDENCE

Refer back to the Compelling Question and the Supporting Questions you developed at the beginning of the lesson.

1. **Identifying** Which of the sources affected you the most? Why?

2. **Evaluating** Overall, what impression do these sources give you about the importance of raising the minimum wage? Explain.

3. **Gathering Sources** Which sources helped you answer the Supporting Questions and Compelling Question? Which sources, if any, challenged what you thought you knew when you first created your Supporting Questions? What information do you still need in order to answer your questions? Where would you find that information?

4. **Evaluating Sources** Identify the sources that helped answer your Supporting Questions. How reliable is the source? How would you verify the reliability of the source?

COMMUNICATE CONCLUSIONS

5. **Collaborating** Work with a partner to create a storyboard or comic strip showing the advantages and disadvantages of raising the minimum wage. Include at least three frames. After discussing ideas, one of you might create the dialog between characters while the other creates the drawings. Share your storyboards or comic strips with the class.

TAKE INFORMED ACTION

Investigating Minimum Wage Jobs Research the types of jobs in your community that pay minimum wage, or less than minimum wage if tips are included. Identify community problems that might be solved with an increase in the minimum wage. Also identify problems that might arise if the minimum wage were increased. Write a letter to your city mayor or city manager outlining your stance on the minimum wage.

11

Reviewing Markets, Money, and Businesses

Summary

Main Idea

In the U.S. *free enterprise capitalist economy*, private citizens own and decide how to use their land, labor, capital, and entrepreneurial skills—with little government interference. Six features contribute to our economic health: economic freedom, markets, voluntary exchanges, the profit motive, competition, and private property rights. The U.S. economy includes the following elements as well.

Supporting Detail

Businesses are organized in several ways. A *sole proprietorship* is owned and operated by one person. A *partnership* is owned and operated by two or more people. A *corporation* is owned by many people when they purchase stock in the business. LLPs and corporations have limited *liability*. Other business forms are *franchises* and *nonprofit organizations*—charities and cooperatives.

Supporting Detail

Businesses produce goods and services but also have social responsibilities. They also must sell safe products, treat workers fairly, and be *transparent* to stockholders.

Supporting Detail

The U.S. *civilian labor force* includes those who have a job and the unemployed who are looking for a job. Much of our labor force works in the service sector. In the past, *labor unions* fought for workers' rights, such as an eight-hour workday and pay for overtime. Fewer workers today are union members.

Supporting Detail

Income levels in the United States vary for many reasons, including level of education. Family wealth and discrimination also play a role. To aid struggling families, the government provides *welfare*.

Supporting Detail

When you work and earn money, you can spend it, invest it, or *deposit* it into a financial institution. These institutions channel funds from savers to borrowers—paying interest to savers, and charging interest to borrowers. The government regulates banks to protect deposits.

Supporting Detail

Businesses produce *private goods*, and the government provides *public goods,* such as public education. *Antitrust laws* maintain competition among businesses. The government also makes sure foods and drugs are safe.

Checking For Understanding

Answer the questions to see if you understood the topic content.

REVIEWING KEY TERMS

1. Define each of these terms:

   A. capitalism
   B. free enterprise
   C. financial capital
   D. liability
   E. franchise
   F. transparency
   G. collective bargaining
   H. currency
   I. externality
   J. welfare

REVIEWING KEY FACTS

2. **Explaining** How does the limited government interference of capitalism contribute to a healthy economy?

3. **Identifying Cause and Effect** What happens if a sole proprietor cannot pay his or her business debts?

4. **Explaining** Why is it usually easier for a partnership to grow than it is for a sole proprietorship?

5. **Explaining** What is the role of a corporation's board of directors?

6. **Explaining** Why do companies create foundations?

7. **Speculating** How does a company's financial report help to protect potential investors?

8. **Identifying** What options exist for employers and unions when they cannot agree on a contract?

9. **Explaining** How do banks earn money?

10. **Explaining** How does the government pay for the public goods it provides?

11. **Explaining** What are the goals of antitrust laws?

CRITICAL THINKING

12. **Identifying Cause and Effect** How does the profit motive contribute to the economic growth of the United States?

13. **Inferring** In laissez-faire economics, what one government interference would be encouraged?

14. **Contrasting** How does a corporation's liability differ from that of sole proprietorships and partnerships?

15. **Inferring** Why do you think a business would organize as a nonprofit organization?

16. **Identifying Cause and Effect** How does an employer benefit by providing its employees with health insurance?

17. **Identifying Cause and Effect** How does a boycott put economic pressure on a company to meet the demands of the union?

18. **Making Connections** Why must money be limited in supply for it to have value?

19. **Comparing and Contrasting** How are a savings account and a checking account alike and different?

20. **Identifying Cause and Effect** How are consumers affected by a lack of competition in the marketplace?

NEED EXTRA HELP?

If You've Missed Question	1	2	3	4	5	6	7	8	9	10
Review Lesson	2, 3, 4, 5, 6, 8, 9	2	3	3	3	4	4	5	6	8

If You've Missed Question	11	12	13	14	15	16	17	18	19	20
Review Lesson	8	2	2	3	3	4	5	6	6	8

Apply What You Have Learned

A Understanding Multiple Perspectives

Although free enterprise limits government involvement in the economy, the U.S. economic system does include some government intervention. The government tries to ensure that businesses are competitive. The government also helps provide a safety net for those in poverty.

ACTIVITY Comparing and Contrasting Viewpoints on Government Assistance Read these excerpts that provide two views of government assistance to those who need it. Then write a brief essay answering these questions: How did President Johnson view the role of the federal government in combating poverty and aiding society? How did Benjamin Franklin believe poverty could be reduced? How might a speech like President Johnson's, delivered before Congress and televised live to the American people, influence government policy toward social programs?

❝ Let this session of Congress be known as the session which . . . declared all-out war on human poverty and unemployment in these United States; as the session which finally recognized the health needs of our older citizens; . . . and as the session which helped to build more homes, more schools, more libraries, and more hospitals than any single session of Congress in the history of our Republic.❞

— President Lyndon B. Johnson, Annual Message to the Congress on the State of the Union, January 8, 1964

❝ I am for doing good to the poor, but I differ in opinion about the means. I think the best way of doing good to the poor, is not making them easy *in* poverty, but leading or driving them *out* of it. In my youth I travelled much, and I observed in different countries, that the more public provisions were made for the poor, the less they provided for themselves, and of course became poorer. And, on the contrary, the less was done for them, the more they did for themselves, and became richer.❞

— Benjamin Franklin, "On the Price of Corn, and Management of the Poor" Printed in *The London Chronicle,* 1766.

B Writing an Informative Report

Many corporations have set up their own foundations. These groups give money to support causes they believe are important.

ACTIVITY Writing About a Company's Charitable Giving Research online a favorite store or a company whose products you would like to own someday. Go to the company's homepage and use search terms such as "philanthropy" or "foundation." Read about the causes the store donates to, and why. Then write a paragraph informing others about the company's giving practices.

(l) Johnson, Lyndon B. Annual Message to the Congress on the State of the Union, January 8, 1964. Public Papers of the Presidents of the United States: Lyndon B. Johnson, 1963-64. Volume I, entry 91, pp. 112-118. Washington, D.C.: Government Printing Office. 1965., (r) Franklin, Benjamin. "Arator" On the Price of Corn, and Management of the Poor, [29 November 1766]. Founders Online, National Archives, https://founders.archives.gov/documents/Franklin/01-13-02-0194.

C Understanding Economic Concepts

A market is a place where buyers and sellers exchange products and money. Markets are "capitalism in action."

ACTIVITY **Finding Evidence of a Market in a Photograph**
Study the photograph of a farmers' market. Identify in the photograph evidence of each of the following features of capitalism: economic freedom, markets, voluntary exchanges, the profit motive, competition, and private property rights.

D Making Connections

Have you ever heard of "Yankee ingenuity"? How about "American inventiveness"? Americans are famous for creating, inventing, and innovating in the U.S. free enterprise economy.

ACTIVITY **Creating a New Product** The entrepreneurial spirit runs deep in this country. Does it within you? Have you ever wanted to invent something? Here's your chance. A trick to inventing is to think of problems people face and then try to come up with a solution. Make a list of common problems and brainstorm solutions. When you come up with an invention, make a poster that advertises its features.

State and local governments, like the federal government, use taxpayer dollars to pay for many things. Taxes paid for the Main Street Bridge and this city park fountain in Jacksonville, Florida.

Government and the Economy

INTRODUCTION LESSON

01   Introducing Government and the Economy   E110

LEARN THE CONCEPTS LESSONS

02   Gross Domestic Product   E115

03   Measuring the Economy   E119

04   The Federal Reserve System and Monetary Policy   E125

05   Financing the Government   E131

07   Fiscal Policy   E143

INQUIRY ACTIVITY LESSONS

06   Multiple Perspectives: Taxes   E137

08   Multiple Perspectives: The Federal Debt   E149

REVIEW AND APPLY LESSON

09   Reviewing Government and the Economy   E155

01

Introducing Government and the Economy

Signs of the Economy

Economists and government leaders want to know about the health of the economy. They look for signs that will answer questions such as:

- Are more or fewer people spending money at stores and restaurants?

- Is the economy getting better or worse?

- Are prices beginning to rise too fast?

- Is it more expensive or less expensive to get a loan from a bank?

- Are long lines of people waiting at the unemployment office because they've been laid off?

- Are people struggling to pay their bills?

- Is the government going to have a *deficit* (spend more than it receives)?

The government provides many goods and services to the public. Taxpayers must pay for those products. Today those taxpayers are your parents—but someday it will be you!

This cartoon shows—in an exaggerated way—the impact that taxes can have on a worker. Taxes of one kind or another have already taken half of the boy's pay.

Did You Know?

Taxes—and opinions about them—have been around as long as civilization.

- "In this world, nothing can be said to be certain, except death and taxes." –Benjamin Franklin, American Founder, 1789

- "Taxes are what we pay for civilized society." –Oliver Wendell Holmes, Jr., U.S. Supreme Court Justice, 1927

- "When there is an income tax, the just man will pay more and the unjust less on the same amount of income." –Plato, Athenian philosopher, in *The Republic*, 375 B.C.E.

TEXT: (t) Franklin, Benjamin. "To Jean Baptiste Le Roy, November 13, 1789." The Writings of Benjamin Franklin. Volume X: 1789-1790. New York: The Macmillan Company, 1907., (c) Supreme Court Of The United States. U.S. Reports: Compania De Tabacos v. Collector of Internal Revenue, 275 U.S. 87. Washington, DC: Government Printing Office, 1927., (b) Plato. The Republic. Translated by Benjamin Jowett. The Gutenberg Project, August 27, 2008. https://www.gutenberg.org/files/1497/1497-h/1497-h.htm.

What Taxes Pay For

No one likes paying taxes. But our lives would be very different if taxes didn't pay for public goods and services. Education, roads and bridges, national parks, and national defense are just a few things paid for with tax dollars.

» The federal government contributes some tax dollars to pay for public education. However, most funding for public schools comes from state and local taxes.

» Taxes pay for the National Park Service to maintain more than 400 national parks across the country. Two of the most famous include Everglades National Park in Florida and Yellowstone National Park in Wyoming, Idaho, and Montana. Visitors arrive to view Yellowstone's geyser named "Old Faithful."

» A Coast Guard helicopter crew practices a rescue. When people get in trouble at sea, they depend on the Coast Guard to come to their aid. In 2020, the Coast Guard received more than $12 billion from the federal government to fund its operations for the year. With these funds, the Coast Guard buys boats, helicopters, and other equipment. It also pays its highly skilled employees.

Getting Ready to Learn About . . . Government and the Economy

Why the Government Gets Involved

Americans have strong feelings about the role that government ought to play in our lives. Some people feel government does too much; others think it does too little. An important benefit of a democratic political system is that people can freely disagree about what the government should do—and then settle their disagreements peacefully. Yet there are times when people are forcefully reminded of the beneficial side of government involvement.

In 2008–2009, the economy experienced its worst *recession*, or economic downturn, since the Great Depression of the 1930s. Massive spending by the government prevented it from getting even worse.

In 2020, the country entered an even more difficult year. The rapid spread of the COVID-19 virus caused a partial shutdown of the U.S. economy. Millions of Americans and small businesses were hurt. But again, the government passed three spending bills totaling $5 trillion to aid states, families, and small businesses. As difficult as 2020 was, it would have been much worse without the government-provided economic stimulus.

During economic downturns, stores close and people become unemployed.

Measuring the Economy's Health

Everybody wants the economy to grow. Economic growth means businesses earn more profits, people have jobs, and incomes go up. With higher incomes, the government receives more in taxes, which helps the federal budget.

To see if the economy is growing, we like to look at real GDP. *Real Gross Domestic Product (GDP)* is the value of all final products produced in the country during a single year when measured with fixed base-year prices. If GDP increased this year over last year when fixed base prices are used, the economy's growth is real and not an illusion caused by rising prices—hence the term "real GDP."

Economists look at other measures too. They want to know how many people are employed and unemployed. They want to know if prices have changed since last year. They also want to know if stock prices are in a "bull market" or a "bear market." All of these measures help to tell us something about the economy's health.

A "bull market" is strong, like a bull, with stock prices going up. A "bear market" is mean, like a bear, with stock prices going down.

Where the Money Comes From

If you've ever created a budget, you understand that you need to figure out how much money is coming in. This amount tells you how much you can spend.

The federal government prepares a budget in a similar way. The budget has two parts—*revenue*, or money coming in, and *expenditures*, or money being spent. Much of the revenue coming in is from people paying individual income taxes. The more income people earn, the higher their taxes. Other taxes to fund programs such as Social Security are also taken from people's paychecks. In addition, corporations pay taxes on their profits.

The federal government shares some of the revenue it gets with state governments. But state and local governments collect their own taxes too.

Where the Money Goes

The revenue collected is then spent on an astonishing number of goods and services—many of which benefit you. As you learned earlier, taxes pay for education, highways, national defense, and national parks. Social Security payments go to the elderly. Unemployment benefits and welfare payments go to people who have been laid off and are looking for jobs.

Taxes are collected and then used to run the government and its programs.

Fixing the Economy

Have you ever purchased an item—only to find out later that its price had increased? An institution called the Federal Reserve System (the Fed) is in charge of *monetary policy*. Its goal is to manage the money supply as necessary to keep prices stable and keep economic activity strong.

A second policy—*fiscal policy*—is conducted by the government at the federal, state, and even local levels. It refers to how the government uses tax cuts and changes in spending to reach the economic goals of keeping people working, producing, and buying goods and services.

The federal government used fiscal policy when it pumped billions of dollars into the economy during the 2008–2009 Great Recession and the 2020 COVID recession. The government's tax cuts and increased spending led to record-setting *deficits*, or spending more than was collected in revenues. The economy would have been in much worse shape, however, without these policies.

The idea behind fiscal policy is simple. If the economy slows down or heats up too much, the government can help the economy become healthy again by adjusting taxes and spending. Spending is easy, but increasing taxes to pay for the spending is much more difficult.

Looking Ahead

In this topic, you will learn about GDP and other measurements of the economy, the Federal Reserve System and monetary policy, taxes, the federal budget, and fiscal policy. You will examine Compelling Questions and develop your own questions in the Inquiry Activities.

What Will You Learn?

In these lessons, you will learn:

- what GDP is and how it serves as a measure of our economy's health.
- the business cycle and its stages.
- how unemployment and inflation affect the economy.
- what makes up the Federal Reserve System and how the Fed influences the economy through its use of monetary policy.
- how and why governments create budgets.
- the sources of revenue and forms of expenditures for the federal, state, and local governments.
- how and why governments use fiscal policy to maintain growth and a stable economy.
- how the government used fiscal policy during the Great Recession of 2008–2009 and the COVID-19 recession of 2020.
- how deficits happen and why they turn into debt.

 COMPELLING QUESTIONS IN THE INQUIRY ACTIVITY LESSONS

- **Are tax rates fair?**
- **Should the U.S. government be required to balance its budget?**

Government and the Fed's Involvement in the Economy

In this topic, you will learn that Congress (upper left) and the Federal Reserve (upper right) get involved in the U.S. economy to promote economic growth and a higher GDP (bottom).

02

Gross Domestic Product

Gross Domestic Product

READING STRATEGY

Analyzing Key Ideas and Details As you read the lesson, complete a web diagram identifying the three types of GDP a country measures.

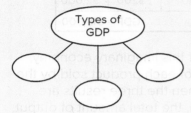

Why GDP Is Important

GUIDING QUESTION

Why is Gross Domestic Product important to a nation?

You can see the busy U.S. economy all around you. Farmers raise crops, and factories produce many kinds of goods. Employees stock goods on store shelves. Shoppers crowd the stores to buy those products. A **product** may be either a good or a service. Goods are something you can touch, such as bicycles, cell phones, books, pens, and clothes. Products also include services, or work done for someone. Vehicle repair, giving a haircut, and babysitting are examples of services. But did you ever wonder *how much* we produce every year?

GDP Measures Total Output

All this economic activity is reflected in **Gross Domestic Product (GDP)**. GDP is the total (gross) market value of all final products produced within a country's borders during a single year.

product anything that is produced; goods and services

Gross Domestic Product (GDP) total market value of all final goods and services produced in a country during a single year

This worker in a bicycle shop adds to the country's GDP. GDP provides two measures of a country's economy. It sums the total market value of all final *output* in a year, including new bicycles made in this country. It also sums the total *income* of the factors of production used to make those products, including this worker's wages. Output equals income in a given year.

Making Connections The word *domestic* comes from the Latin word *domus*, which means "home." How does this information help you better understand the meaning of *gross domestic product*?

What is the GDP of the United States? In 2020, the annual output, or amount produced in the United States, was about $20 trillion. Twenty trillion dollars is a lot of output—$20,000,000,000,000 to be exact. This amount makes the United States the world's second-largest national economy. In fact, U.S. output is about one-seventh of all the goods and services produced in the world. In 2020 only China had a larger economy, with output worth about $24 trillion.

GDP Also Represents Income

Making goods and providing services create income for people in the economy. This is another reason why measuring GDP is important. Although measuring output is GDP's primary function, GDP also represents the nation's income. The workers who make a bicycle, for example, are paid for their labor. But *labor* is not the only factor of production that earns income when a bicycle is produced. So do the other factors of production. *Entrepreneurs*—those who formed the bicycle company and the person who opened a bicycle shop—earn income. Recall that *land* includes forests, soil, and mineral deposits. The natural resources that go into a bicycle—the metal used in the frame and the rubber used in the tires—must be paid for. As a result, this factor of production brings income to the companies that own those resources. The same is true of *capital*—the wrenches, machinery, buildings, and other tools used to make the bicycle. They, too, generate income for the companies and workers that make and sell these capital goods.

✓ CHECK FOR UNDERSTANDING

1. **Describing** What is Gross Domestic Product (GDP), and what does it represent?

2. **Explaining** Why does GDP represent income for all factors of production?

Measuring GDP

GUIDING QUESTION

Why is GDP difficult to measure?

Because so many different goods and services are produced during a year, measuring GDP is difficult. This is true for any economy, not just one the size of the U.S. economy. To calculate this and other measurements, the government uses thousands of highly skilled economists and government workers.

A simple example shows how economists could calculate the GDP of a nation. Suppose a tiny country has an economy that produces only two goods—watches and computers, and one service—concerts. The following table **Estimating Total Annual Output (GDP)** shows how the quantities of each product and their prices are summed to compute GDP.

Estimating Total Annual Output (GDP)			
	Quantity ×	Price =	Value
Watches	10	$100	$1,000
Computers	10	$1,500	$15,000
Concerts	10	$200	$2,000
		GDP =	$18,000

To find the GDP of this imaginary economy, we multiply the price of each product sold by the quantity produced. Then the three results are added. For this nation, the total amount of output, or GDP, would be $18,000 ($1,000 worth of watches, $15,000 worth of computers, and $2,000 worth of concert services).

A modern economy would have many more goods and services, of course, but the process of computing GDP would be about the same. The quantities of every final product times their price would be used to estimate their values, and GDP would be the sum of the values.

Economists do this sort of math to find the GDP of real countries. However, we need to know some additional information about GDP.

GDP Only Includes Final Products

Not all economic activities are included in GDP. GDP reflects only the market value, or prices, of final goods and services produced and sold. A *final* good or service is one that is sold to its final user. A bicycle sold to you is a final good. *Intermediate* goods are ones that go into making a final good. The parts used to make the bicycle—such as the tires, seat, and pedals—are intermediate goods. Intermediate goods are not counted in GDP because the final price of the bike already includes the value of those parts.

Products such as bicycles, clothing, and haircuts are called consumer goods and services. In other words, these are products we consume. Economists use the word *consume* to mean "use as a customer." But what about the final goods businesses use—such as machines or office supplies—to make consumer products? These are called producer goods. They are also known as investment goods or capital goods and are included in calculating GDP.

What GDP Does Not Include

GDP does not include every kind of activity in the economy. It includes only final goods and services produced and sold in the market. It does not include intermediate goods and services. GDP does not count services you do for yourself—such as mowing your lawn. The value of used—or secondhand—goods is also not counted. These products were already counted the year when they were first sold. Transferring them to a new owner creates no new production. Thus, buying a used good is not included a second time in the GDP.

Real GDP

Because of the way it is computed, GDP can appear to increase whenever prices go up. Say the price of concerts shown in the table **Estimating Total Annual Output (GDP)** increases

from $200 to $225. After recomputing, GDP will appear to go up even though the number of concert tickets sold did not change. Therefore, to make accurate comparisons involving different years, GDP must be adjusted for price increases.

To do so, economists use a set of "constant" prices. They choose a year that serves as the basis of comparison for all other years. Suppose we compute GDP for several years in a row using only prices that existed in 2012. Then, any increases in GDP must be due to an increase in quantity and not an increase in prices. This measure is called **real GDP**, or GDP measured with a set of constant base-year prices. Real GDP removes the distortions caused by price changes.

You can think of real GDP as measuring output in an economy where prices do not change. This makes real GDP a better measure of an economy's performance over time because people's welfare ultimately depends on the quantities of products produced, not their prices.

---

real GDP GDP after adjustments for price changes

If a clothing manufacturer buys this colorful fabric to make shirts or dresses for customers to buy, the sale of fabric would not be counted in GDP. If an individual such as you purchases the fabric, however, the sale would be counted as a final good in GDP.

Explaining Why are intermediate goods not included in the measurement of GDP?

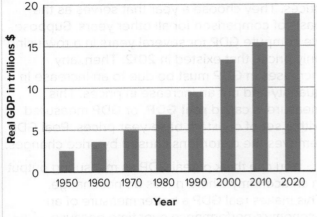

Real GDP 1950–2020 (in constant 2012 prices)

The U.S. economy expanded by 52% from 1960 to 1970, faster than any other decade in the graph.

Real GDP in trillions $

Year

Source: Bureau of Economic Analysis

EXAMINE THE GRAPH

Calculating How would the 2020 GDP in the graph change if it were measured in current dollars instead of constant 2012 dollars?

For example, look at how real GDP for the United States appears in the graph **Real GDP 1950–2020 (in constant 2012 prices)**. The graph shows that the production of real goods and services has actually increased from 1950 to 2020. And none of the increase was due to higher prices. But if GDP is not measured in constant dollars, it is simply called GDP, or current-dollar GDP.

GDP Per Capita

GDP tells how large a country's economy is. But when we compare the output of countries that have different-sized populations, **GDP per capita** is a better measure. *Per capita* means "for each person." GDP per capita is calculated by dividing the country's GDP by its population. The result is the amount of output on a per-person basis. This makes it easier to compare the production of two countries.

The World Bank uses a different way of computing GDP for purposes of comparing two countries. Using its method, China has the world's largest GDP at $24 trillion, and the United States has the second-largest GDP at about $20 trillion. But China's population is much larger. To get a more realistic comparison of the two countries, we would divide China's GDP by its population to get a GDP per capita of about $18,200. If we divide the U.S. GDP by its population, we get a GDP per capita of $62,500. On a per capita basis, the U.S. GDP is more than three times larger than China's GDP; or ($62,500) ÷ ($18,200) = 3.43.

The Standard of Living

The **standard of living** is the quality of life of the people living in a country. GDP is not a measure of the standard of living. This is because GDP is an **aggregate**, or total, number. When we look at GDP, we do not know WHAT was produced, HOW it was produced, or FOR WHOM the production was intended. A country with a GDP that goes to a very few rich people might have a lower standard of living than a country of equal size that produces its goods and services for everyone.

How production takes place is also important. China has a very productive economy but is also a big polluter. If a country does not take steps to reduce its pollution, it could have a lower standard of living than another country.

✓ CHECK FOR UNDERSTANDING

1. **Contrasting** What is the difference between real GDP and GDP per capita?

2. **Analyzing** Why is GDP not necessarily an accurate measure of a nation's standard of living?

LESSON ACTIVITIES

1. **Informative/Explanatory Writing** Research online to find the top five countries based on current GDP, real GDP, and real GDP per capita. Write a short report that displays your findings in a chart and defines the three types of GDP.

2. **Collaborating** With a partner, create a poster illustrating items that are—and are not—counted in GDP. Add captions explaining why you categorized the products the way you did.

GDP per capita GDP on a per-person basis; GDP divided by the population

standard of living material well-being of an individual or a nation as measured by how well needs and wants are satisfied

aggregate total

Expansion	Recession

Economic Performance

GUIDING QUESTION

Why is it important to measure an economy's performance?

When we want to measure the size of the economy, we look at GDP because it is our most comprehensive measure. But when we want to see if GDP is *growing*, we have to use real GDP.

Real GDP

As you recall, *real GDP* is measured with a set of constant base-year prices. Real GDP is useful because it removes the **distortions**, or misleading impressions, caused by price changes. This makes real GDP a better measure of economic growth over time.

Here's why: GDP is the total dollar value of all final goods and services produced by an economy in one year. But if a country has a bigger GDP in one year than it had in the year before, it does not necessarily mean that its economy grew. If the growth was caused only by an increase in prices and nothing else, the same amount of goods and services would only *appear* to be worth more. Because of this, rising prices can make increases in GDP misleading. To avoid this problem, we always use real GDP when comparing GDPs over time.

distortion misleading impression

The Business Cycle

Phases of economic recession and expansion shape the business cycle.

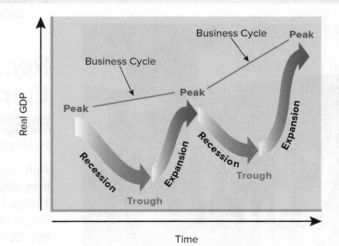

EXAMINE THE DIAGRAM

Analyzing Visuals What is the growing phase of the business cycle called? What happens in this phase?

GO ONLINE Explore the Student Edition eBook and find interactive maps, charts, graphs, and tools.

E119

Business Cycles

The U.S. economy does not grow at a steady rate. Instead, it goes through alternating periods of real economic decline and growth that we call the **business cycle**. Look at the diagram **The Business Cycle**. The diagram shows a simplified version of the economy. It is simplified because business cycles are not as smooth and regular in the real world.

Every business cycle has two distinct parts—a recession and an expansion. A **recession** (rih•SEH•shuhn) begins when real GDP declines after it reaches a *peak*. The recession ends when real GDP hits a *trough* and starts to go up again. Real GDP will then expand for a while, but eventually it will hit another peak and then start to go down.

A business cycle lasts from one peak to the next peak. This means that a business cycle contains one recession and one expansion. Most recessions are fairly short—lasting from about 6 to 18 months. Expansions tend to last longer. Most expansions since the end of World War II have lasted from 5 to 10 years. Economic growth will have taken place if the new peak is higher than the previous peak. Eventually, though, real GDP will start to decline again, marking the start of a new business cycle.

business cycle alternating periods of real economic decline and growth

recession period of declining economic activity lasting about six or more months

Unemployed men in Chicago line up outside a soup kitchen for free food during the Great Depression.

The Great Depression

A recession may turn into a depression if real GDP continues to go down rather than turning back up. A **depression** is a period of severe economic decline with rising unemployment and extreme economic hardships. The United States has had only one major depression. It started in 1929 and reached a trough in 1933. The drop in real GDP was so enormous that it took until 1939—a full ten years—for business activity to get back to the level where it had been in 1929. This is the longest period of no growth in U.S. history.

Most economists think that real GDP fell by almost half from 1929 to 1933. This was a time when one in four workers lost their jobs. About one-fourth of banks went out of business. Many stocks became worthless, and millions of people lost everything they had.

Fortunately, most economists think that something as serious as the Great Depression will not happen again. Laws were passed to prevent the kinds of actions that worsened the situation in the 1930s. We also better understand how the economy works and how to keep it healthy. For example, the government plays a bigger role in the economy today than it did in the 1930s. Because of this, the government is more likely to take steps to fix economic problems before they worsen.

✓ **CHECK FOR UNDERSTANDING**

1. **Identifying** What does a peak on a business cycle mean?
2. **Explaining** What is a recession, and what is a depression?

Recessions and Employment

GUIDING QUESTION

How do recessions affect employment?

Although most modern recessions were much less severe than the Great Depression of the 1930s, two recessions stand out. The first is the Great Recession of 2008–2009. The second is the COVID-19 recession of 2020. These recessions and others can be seen in the graph

depression state of the economy with high unemployment, severely depressed real GDP, and general economic hardship

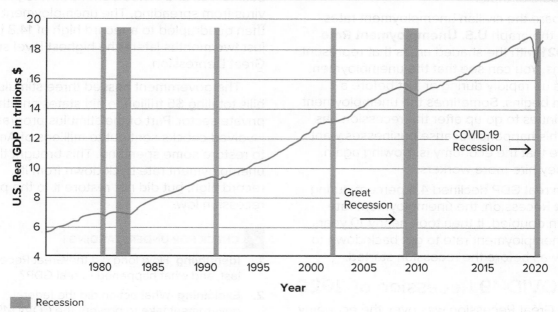

U.S. Real GDP 1975–2021

The United States has experienced mainly economic expansion since 1975.

U.S. Real GDP in trillions $

COVID-19 Recession →

Great Recession →

Recession

Source: Bureau of Economic Analysis

EXAMINE THE GRAPH

1. **Analyzing** What does a downward slope in the real GDP line show? What about an upward slope?
2. **Analyzing Visuals** How many recessions has the United States experienced since 1975?

U.S. Real GDP 1975–2021. The vertical shaded areas in the graph represent the times when the economy was in recession.

The Great Recession of 2008–2009

The Great Recession began in early 2008 and lasted for 18 months. It was the longest and deepest recession in the United States since the Great Depression of the 1930s. Real GDP declined by 4.5 percent during the 18 months of the Great Recession. Yet it took nearly four years for real GDP to get back to the previous high it had reached in 2008.

More than 8 million people lost their jobs during the Great Recession. This is more than the total number of people who live in Washington or Arizona today! Many of these people lost their homes and cars when they couldn't afford to make their monthly payments. Others were forced to use their retirement savings or went into debt just to cover everyday living expenses.

To prevent the economy from declining too much, the federal government passed two major stimulus programs totaling about $1.5 trillion in 2008. Some of the money was used to keep banks from going under. Some was used to help keep General Motors—the maker of Chevrolet, Buick, and Cadillac cars—in business. Other stimulus money went to unemployed people. Some of these expenditures were controversial, but the Great Recession would have been much worse without them.

Recessions and Unemployment

A key measure of how severe any recession can be is the **unemployment rate**. This is the percentage of people in the civilian labor force who are not working but are looking for jobs.

---

unemployment rate the percentage of people in the civilian labor force who are not working but are looking for jobs

In a normally healthy economy, the unemployment rate is low, usually around 4 percent. Because of a recession, however, the unemployment rate can easily double or triple.

Compare the civilian unemployment rates shown in the graph **U.S. Unemployment Rate 1975–2021** with the shaded areas that represent recessions. You can see that the unemployment rate goes up rapidly during or just before a recession begins. Sometimes the unemployment rate continues to go up after the recession has ended. This happens because businesses want to be sure that the economy is growing again before they hire more workers.

When real GDP declined 4.5 percent during the Great Recession, the unemployment rate more than doubled. It then took nearly 10 years for the unemployment rate to get back down to where it was before the recession started.

The COVID-19 Recession of 2020

After the Great Recession was over, the economy began an unprecedented expansion that lasted almost 11 years. Then, when the unemployment rate was at a record low of 3.5 percent in early February 2020, the COVID-19 virus struck. Many businesses and schools shut down to prevent the virus from spreading. The unemployment rate then quadrupled to a record high of 14.8 percent just two months later—the highest level since the Great Depression.

The government passed three stimulus bills totaling $5 trillion to aid states and the private sector. Part of the stimulus programs involved checks sent out to millions of Americans to restore some spending. This brought the unemployment rate back down from its record high, but did not restore it to the pre-recession low.

☑ **CHECK FOR UNDERSTANDING**

1. **Identifying** How long did the Great Recession last, and what happened to real GDP?

2. **Explaining** What action did the federal government take to prevent the COVID-19 recession of 2020 from getting worse?

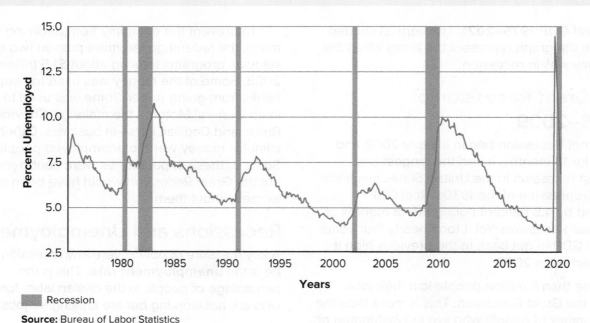

U.S. Unemployment Rate 1975–2021

The graph shows changes in the unemployment rate from 1975 to 2021. Keep in mind that an upward movement of the unemployment rate line is bad because it means there are more unemployed workers.

Recession

Source: Bureau of Labor Statistics

EXAMINE THE GRAPH

1. **Analyzing Visuals** How would you describe the unemployment rate from 2010 to 2019?

2. **Identifying Cause and Effect** Why does unemployment go up in a recession?

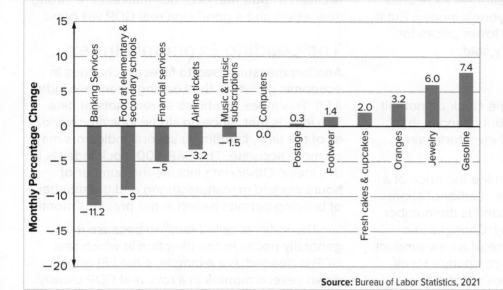

Changes in Prices, January 2021

The consumer price index (CPI) tracks how the prices of different goods most people buy can change over time. Ten of the 400 categories in the CPI are shown in the chart. The CPI is used to track inflation, or changes in the general level of prices in the economy.

Monthly Percentage Change

Banking Services: −11.2
Food at elementary & secondary schools: −9
Financial services: −5
Airline tickets: −3.2
Music & music subscriptions: −1.5
Computers: 0.0
Postage: 0.3
Footwear: 1.4
Fresh cakes & cupcakes: 2.0
Oranges: 3.2
Jewelry: 6.0
Gasoline: 7.4

Source: Bureau of Labor Statistics, 2021

EXAMINE THE GRAPH

1. **Analyzing Visuals** What does each bar in the graph show?

2. **Calculating** Would consumers be more or less likely to buy jewelry in January 2021? Why?

Other Measures of Performance

GUIDING QUESTION

What are other signs of an economy's health?

Suppose someone asked you to describe your economic situation. Before answering, you might think about whether you get an allowance, have money to buy the clothes or music you want, or have money saved for a class trip. Adults think about their economic health in a similar way. Nations have economic health, too. To judge that health, economists look at specific *economic indicators*. These are signs that "indicate" how the economy is doing, such as the unemployment rate just discussed. Other key indicators are price stability, stock indexes, and a measurement called the Leading Economic Index®.

Price Stability

One sign of an economy's health is the general level of prices. If prices remain stable or steady, consumers and businesses can better plan for the future. This is especially important for people who are retired and live on a **fixed income**. A fixed income remains the same each month and does not have the potential to increase when prices go up.

When prices remain stable, money has the same purchasing power, or value. When prices go up, money loses some of its purchasing power. For example, suppose an ice-cream cone that costs a dollar doubles in price to two dollars. The higher price means that your dollar buys less. You need to spend twice as many dollars to buy the same ice-cream cone. As the chart **Changes in Prices, January 2021** shows, monthly percentage price changes are not always stable.

An increase in the price of one good does not affect purchasing power very much. If most prices rise, though, the situation is different. **Inflation** (ihn•FLAY•shuhn) is the name for a long-term increase in the general level of prices. Inflation hurts consumers and people on fixed incomes because it reduces everyone's purchasing power.

Every month the government tracks the prices of about 400 products consumers usually buy. Prices of these 400 products make up a measure called the *consumer price index* (CPI). Typically, the prices of some items in the CPI go up every month, and other prices go down. If the overall level of the CPI goes up, inflation is taking place.

fixed income an income that remains the same each month and does not have the potential to go up when prices are going up

inflation a long-term increase in the general level of prices

Of course, prices can go down as well as up. **Deflation** is the name for a prolonged decrease in the general level of prices. Deflation doesn't occur very often, but it did happen for 11 months during the Great Recession of 2008–2009. This had the opposite effect of inflation—it *increased* the purchasing power of everyone's money. But it also hurt sellers who received lower prices for the same products they usually sold.

Stock Indexes

Changes in the value of a single stock do not tell us much about the economy. But changes in *all* stock prices do. That tells us if investors have confidence in the economy.

Supply and demand determine the price of a company's stock. Supply is the number of shares people are willing to sell. Demand is the number of shares investors want to buy. Changes in a company's profits or the release of a new product can change the demand for a company's stock, and its price. A stock index consists of many stocks, so the change in the value of an index shows how the prices of all stocks are changing.

Two popular stock indexes are the Dow Jones Industrial Average (DJIA) and the Standard and Poor's (S&P) 500. The DJIA tracks the prices of 30 stocks. These include companies such as Coca-Cola, McDonald's, Walt Disney, and Walmart. The S&P 500 index tracks the total market value of 500 stocks.

When stock indexes go down, the stock market is called a "**bear market**." A bear market is a mean or "nasty" market (like a bear). Bear markets cause a drop in investors' wealth and can be a sign of an unhealthy economy.

This bull statue in New York City's financial district symbolizes hope for a strong stock market.

Identifying What do stock indexes measure?

If investors feel good about the economy and expect it to grow, they are more likely to buy stocks. These purchases will drive stock prices up. Rising prices mean rising stock indexes. A rising stock market fueled by confident investors is called a "**bull market**." Bull markets are strong (like a bull) and a good sign real GDP will grow.

The Leading Economic Index®

Another measure used to forecast changes in economic growth is the Leading Economic Index (LEI). This index combines several sets of data. The idea is that, since no single indicator works all of the time, combining several indicators may be more accurate. The S&P 500 stock index is in this index. Other data include the number of hours worked in manufacturing and the number of building permits issued in the previous month.

The index is called *leading* because it generally points to the direction in which real GDP is headed. For example, if the LEI goes down several months in a row, real GDP usually goes down a few months later. If the LEI goes up, real GDP usually goes up several months later. The Leading Economic Index, then, is a good tool for predicting the future of the economy.

✓ **CHECK FOR UNDERSTANDING**

1. **Explaining** What does inflation do to purchasing power? Why?
2. **Describing** How does the government keep track of inflation? Why does it do so?

LESSON ACTIVITIES

1. **Narrative Writing** Suppose you lived through a swing in the business cycle from recession to peak economic growth. Write a story contrasting life during a recession with life during a period of peak economic growth.
2. **Using Multimedia** With a partner, research images and music of the Great Depression. Prepare a slide show with audio that reveals how difficult life was during much of the 1930s. End the slide show with images of economic recovery in the United States.

deflation a prolonged decrease in the general level of prices

bear market period during which stock prices decline for a substantial period

bull market period during which stock prices steadily increase

The Federal Reserve System and Monetary Policy

Analyzing Key Ideas and Details As you read the lesson, identify the functions of the Federal Reserve System.

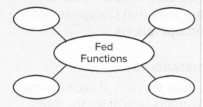

Fed Functions

The Fed's Structure

GUIDING QUESTION

What is the structure of the Federal Reserve System?

In the early 1900s, the United States suffered several severe recessions. During these hard times, banks were unable to make new loans. This crisis hurt even the largest and strongest banks. As a result, most people thought conditions would be better if the nation had a central bank. A **central bank** is a bankers' bank that lends to other banks when times are difficult.

Although the country needed a central bank, the government did not have enough money to finance one. As a result, it decided to require all banks with a national charter to contribute funds to build the new central bank. In return, they would receive some stock in that central bank. The result was the 1913 creation of the Federal Reserve System, or "the Fed" as it is often called. This was the first true central bank for the United States.

Today, the Fed has several very important responsibilities. It manages our currency, regulates more than 5,000 commercial banks and savings institutions, serves as the government's bank, and conducts specific policies to keep the economy healthy and strong. To understand how it does all these things, we first need to see how the Fed is organized.

central bank a bankers' bank that lends money to other banks in difficult times

The Federal Reserve Building is located in Washington, D.C. The "Fed" is our nation's central bank.

Explaining What is the role of a central bank?

The Board of Governors

At the top of the Fed is a seven-member Board of Governors. Each member is nominated by the president and must be **confirmed**, or approved, by the Senate. The members serve staggered 14-year terms, so they are fairly free of influence from elected officials. This enables them to make decisions that are in the best interest of the economy. The Board typically meets every other week.

District Banks

The Federal Reserve System has 12 districts, each with a district bank. These banks are also called Federal Reserve Banks. Each bank carries out the Fed's policies and oversees banking within its district. Nine directors run each district bank. Any profits earned by these banks are paid to the U.S. Treasury.

The Federal Open Market Committee

The **Federal Open Market Committee (FOMC)** influences the whole economy by managing the money supply and thus affecting interest rates. The next section explains why and how the Fed makes changes to the money supply.

confirmed approved

Federal Open Market Committee (FOMC) powerful committee of the Fed that makes decisions affecting the economy by managing the money supply in order to affect interest rates

The Federal Reserve Bank of Chicago carries out Fed policy in District 7.

Advisory Councils

The Board of Governors receives advice from three advisory councils. One advises on consumer borrowing. The second advises on matters relating to the banking system. The third works with the Board of Governors on issues related to savings and loan institutions.

Member Institutions

The Federal Reserve System is more than just a bank. It is a true "system" that supervises more than 5,000 member institutions in all U.S. states and territories. The Fed also ensures that its members carry out the laws that Congress has passed regarding Fed operations.

✓ **CHECK FOR UNDERSTANDING**

1. **Speculating** How does the term of each member of the Board of Governors help make the Board independent of political influence?
2. **Making Connections** How are the parts of the Federal Reserve System related to one another?

What the Fed Does

GUIDING QUESTION

What are the functions of the Federal Reserve System?

The Fed has several important functions. One of these is conducting monetary policy. The Fed also regulates and supervises banks, maintains the currency, promotes consumer protection, and acts as the government's bank.

Conducting Monetary Policy

The Fed is in charge of conducting monetary policy. **Monetary policy** means managing the monetary base to affect the cost and availability of credit. The *monetary base* is the currency in circulation and the deposits that banks and other depository institutions have at the Fed. While this may sound complicated, think of it as a case of supply and demand that determines the interest rate—which is the price of borrowing money.

The two **Monetary Policy and Interest Rates** graphs show how this works. In panel A, the Fed expands the monetary base, shifting the money supply to the right.

monetary policy Fed's management of the money supply to affect the cost and availability of credit

The Federal Reserve System

The Federal Reserve System has a complex structure. Each part carries out specific functions.

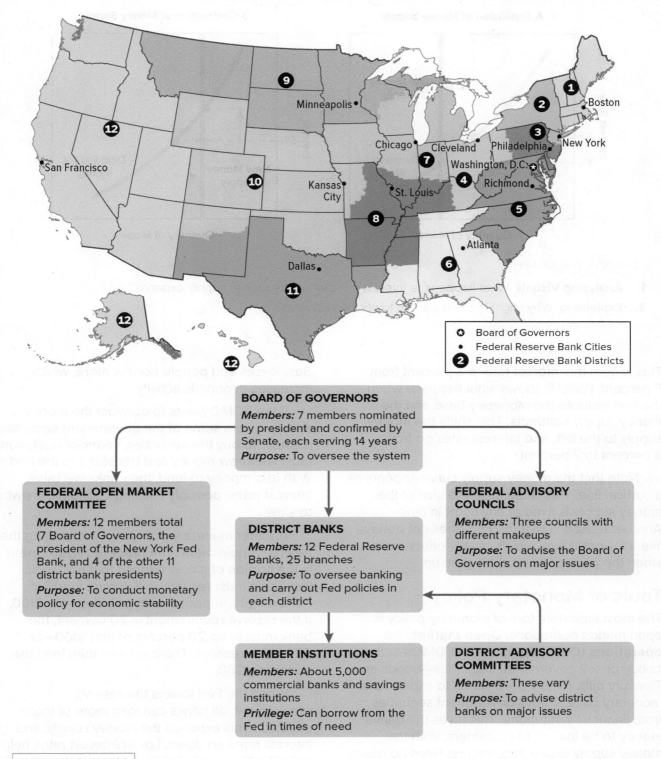

Legend:
- ✪ Board of Governors
- • Federal Reserve Bank Cities
- ❷ Federal Reserve Bank Districts

Map labels: Minneapolis, Boston, New York, Chicago, Cleveland, Philadelphia, Washington, D.C., San Francisco, Kansas City, St. Louis, Richmond, Dallas, Atlanta

BOARD OF GOVERNORS

Members: 7 members nominated by president and confirmed by Senate, each serving 14 years
Purpose: To oversee the system

FEDERAL OPEN MARKET COMMITTEE

Members: 12 members total (7 Board of Governors, the president of the New York Fed Bank, and 4 of the other 11 district bank presidents)
Purpose: To conduct monetary policy for economic stability

DISTRICT BANKS

Members: 12 Federal Reserve Banks, 25 branches
Purpose: To oversee banking and carry out Fed policies in each district

FEDERAL ADVISORY COUNCILS

Members: Three councils with different makeups
Purpose: To advise the Board of Governors on major issues

MEMBER INSTITUTIONS

Members: About 5,000 commercial banks and savings institutions
Privilege: Can borrow from the Fed in times of need

DISTRICT ADVISORY COMMITTEES

Members: These vary
Purpose: To advise district banks on major issues

EXAMINE THE MAP

1. **Exploring Regions** How many district banks are in the Fed? What are the roles of the district banks?
2. **Analyzing Visuals** Which component oversees the Federal Reserve System?

Monetary Policy and Interest Rates

The two graphs show the effects of the Fed's efforts to expand or contract the money supply.

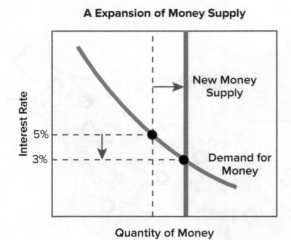

A Expansion of Money Supply

Interest Rate / Quantity of Money

5%
3%

New Money Supply

Demand for Money

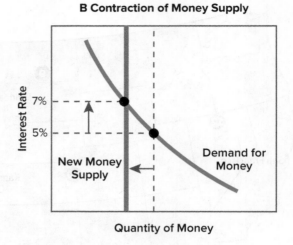

B Contraction of Money Supply

Interest Rate / Quantity of Money

7%
5%

New Money Supply

Demand for Money

EXAMINE THE GRAPHS

1. **Analyzing Visuals** What happens to interest rates when the money supply expands?
2. **Explaining** Why might the Fed want to lower interest rates?

This lowers the interest rate to 3 percent from 5 percent. Panel B shows what happens when the Fed reduces the monetary base and the money supply contracts. This shifts the money supply to the left, and interest rates go from 5 percent to 7 percent.

Note that the money supply curve appears as a vertical line. This is because the size of the money supply is fixed at any point in time. Another reason is that the Fed does not behave like a for-profit firm, which usually offers more when the price of its product goes up.

Tools of Monetary Policy

The most important tool of monetary policy is open market operations. **Open market operations (OMO)** refers to the FOMC's action to buy or sell *government securities*—bonds and Treasury bills. If the FOMC wants to expand the monetary base, it buys government securities from financial institutions. This gives the banks money to be loaned to customers. With the money supply expanding, interest rates go down.

Businesses and people borrow more, which increases economic activity.

If the FOMC wants to *contract* the money supply, it sells some of the government securities it holds. To buy the securities, financial institutions must withdraw money and transfer it to the Fed. With less money to lend, the banks will raise interest rates, possibly causing economic growth to slow.

Another monetary policy tool is changing the reserve requirement. The **reserve requirement** is the portion of a new deposit that a financial institution must reserve and *cannot* lend out. For example, suppose someone deposits $100. If the reserve requirement is 20 percent, the bank must keep 20 percent of that $100—or $20—as a reserve. The bank can then lend the remaining $80.

When the Fed lowers the reserve requirement, all banks can loan more of their deposits. This *expands* the money supply, and interest rates go down. Lower interest rates help people who want to take vacations or buy things

open market operations (OMO) Fed's purchase or sale of U.S. government securities—bond notes and Treasury bills

reserve requirement percentage of a deposit that banks have to set aside as cash in their vaults or as deposits in their Federal Reserve district bank

such as cars and houses. In early 2020 during the COVID-19 recession, the Fed cut the reserve requirement to zero. This meant that banks could lend all the deposits customers made.

In contrast, raising the reserve requirement makes fewer reserves available to be loaned out, which *contracts* the money supply. This happens if the Fed thinks the money supply is getting too big, or that interest rates are too low.

A third way the Fed controls the money supply is by changing the discount rate. The **discount rate** is the interest rate the Fed charges on loans to member institutions when they borrow from the Fed. If the Fed wants to *expand* the money supply, it could lower the discount rate. The reduced rate might encourage banks to borrow from the Fed. Banks could then loan this borrowed money, and the money supply grows. If the Fed wants to *contract* the money supply, it raises the discount rate. A higher rate discourages borrowing.

A bank might also need to borrow money from the Fed's "discount" window if it faced a "liquidity crisis"—such as customers suddenly wanting to withdraw some or all of their deposits. If a bank had already loaned out all of its deposits, it would not have enough money on hand to give customers their withdrawals. But the bank could borrow from the Fed, the discount rate would be charged on the borrowed money, and depositors would get their withdrawals.

Regulating and Supervising Banks

The Fed writes other rules member banks must follow. These involve everything from ways that banks must report the reserve requirements to how they make loans to bank officers. The Fed also has the power to approve mergers between member banks. Many of the world's largest banks located in other countries have branch offices in the United States. The Fed is responsible for ensuring that all foreign bank branch offices that operate on U.S. soil follow all of our banking laws and regulations.

Maintaining the Currency

The Bureau of Engraving and Printing prints our "paper" money, which is really made of linen and cotton. After the new money is printed, it is sent to the Fed for safekeeping and distribution. It will not be counted as part of our money supply until it is released to commercial banks and other lending institutions.

The Fed is also responsible for pulling old money out of use. Whenever paper money becomes tattered and worn, it is sent to the Fed. The Fed exchanges the old bills for newer currency and destroys the worn-out currency.

---

discount rate interest rate the Fed charges on its loans to financial institutions

A newspaper headline announces that the Fed will lower interest rates. This is front page news because people, businesses, and governments react to information from the Fed.

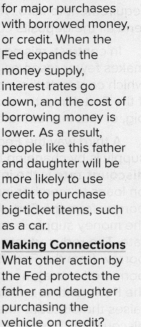

Consumers often pay for major purchases with borrowed money, or credit. When the Fed expands the money supply, interest rates go down, and the cost of borrowing money is lower. As a result, people like this father and daughter will be more likely to use credit to purchase big-ticket items, such as a car.

Making Connections What other action by the Fed protects the father and daughter purchasing the vehicle on credit?

Thus, the Fed plays a major role in keeping our currency in good condition.

Protecting Consumers

Based on laws passed by Congress, the Fed creates rules all lenders must follow. For example, the Fed designed most of the forms people have to sign when they take out a car loan. These forms require lenders to spell out the terms of the loan clearly. The Fed also writes rules that prevent lenders from using deceptive practices when making loans.

Acting as the Government's Bank

Finally, the Fed acts as the government's bank. Whenever people write a check to the U.S. Treasury to pay their taxes, those checks are sent to the Fed for deposit. The Fed also holds other money the government receives and keeps it until it is needed.

The government can write checks on these deposits whenever it needs to make a purchase or other payment. Any federal agency check, such as a monthly Social Security check, is taken out of an account at the Fed.

✓ **CHECK FOR UNDERSTANDING**

1. **Identifying** What is monetary policy?
2. **Explaining** What are two things the Fed can do to expand the money supply?

LESSON ACTIVITIES

1. **Informative/Explanatory Writing** Write an article for an online student resource that explains the functions of the Federal Reserve System and its role in the nation's economy.

2. **Collaborating** With a partner, make flashcards that give details about parts of the Fed. For example, one flashcard might state on the front: "There are 12 of these." The answer on the back of the card would be "District Banks, also called Federal Reserve Banks." After preparing 15 to 20 flashcards, challenge another pair of students to a quiz-off.

Financing the Government

Integrating Knowledge and Ideas As you read the lesson, create a concept web linking related concepts about the federal budget, such as *revenues* and *expenditures*.

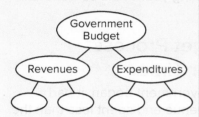

The Federal Budget Process

How does the federal government prepare a budget?

Do you know how to make a personal budget? First, you figure out what your income is for a period, such as a month. Then, you estimate your savings and expenses for that month. Tally your regular expenses, such as weekly food costs. Account for occasional costs too—such as a birthday gift for a parent. If expenses are greater than income, you need to cut spending or find a way to earn more income. Another strategy is to borrow to cover overspending. Borrowing will lead to debt, however, and getting out of debt is often difficult.

The federal government also has a budget. It has two main parts—revenues and expenditures. **Revenue** is the money a government collects to fund its spending. The federal budget covers a period called a fiscal (FIHS•kuhl) year. A fiscal year is any 12 months chosen for keeping accounts. The **fiscal year** of the federal government begins October 1 and ends September 30 of the next year. For example, fiscal year 2022 began October 1, 2021, and ended September 30, 2022.

revenue money a government collects to fund its spending

fiscal year any 12-month period chosen for keeping accounts

The Federal Budget Process

Congress faces deadlines for acting on the budget. It is supposed to approve the budget resolution by April 15. It does not always meet this date, however.

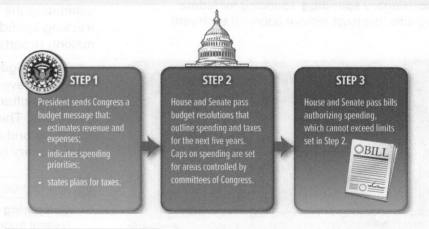

STEP 1
President sends Congress a budget message that:
- estimates revenue and expenses;
- indicates spending priorities;
- states plans for taxes.

STEP 2
House and Senate pass budget resolutions that outline spending and taxes for the next five years. Caps on spending are set for areas controlled by committees of Congress.

STEP 3
House and Senate pass bills authorizing spending, which cannot exceed limits set in Step 2.

BILL

1. **Making Connections** Why are tax revenue estimates included in the budget?
2. **Inferring** What relationship does the budget resolution have to the president's budget?

The Federal Budget Process diagram shows just the general steps in making a federal budget. This process is complex because of the size of the budget and the number of parties involved.

Steps in the Budget Process

The process starts when the president **transmits**, or sends, a budget message to Congress. This message states how much the president wants to spend on each federal program. This message must be sent no later than the first Monday in February.

Next, key members of Congress agree on a *budget resolution*. This is Congress's plan for revenue and spending on broad categories such as health. The budget has two different kinds of spending: mandatory (MAN•duh•TOHR•ee) and discretionary (dis•KREH•shuh•NEHR•ee). **Mandatory spending** is set by laws outside the budget process. One example is Social Security, which makes payments to retirees. Mandatory spending is generally fixed from year to year. **Discretionary spending** involves spending choices made and approved each year. It includes items such as national defense and highways. The amount of discretionary spending can differ from year to year.

Next, Congress must set spending on each program for the coming year. That process starts

---

transmit send

mandatory spending federal spending required by law that continues without the need for congressional approval each year

discretionary spending spending for federal programs that must receive approval each year

---

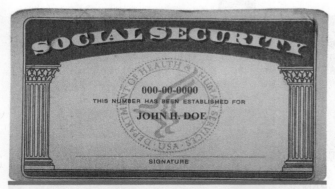

Americans obtain a Social Security number when they are born. Social Security falls under the "mandatory spending" category of the federal budget.

Explaining What is the difference between mandatory and discretionary spending?

---

when committees in the House write appropriations (uh•proh•pree•AY•shuhnz) bills. An **appropriations bill** gives official approval for the government to spend money. All appropriations bills start in the House of Representatives, but they must be approved by both the House and Senate. After the Senate and House pass each bill, it is sent to the president. The president can either sign it into law or veto it. If the bill is vetoed, Congress can rewrite the bill or override the veto.

Sometimes Congress does not pass the budget in time. When this happens, Congress approves a *continuing resolution*. This law sets spending for the coming year at the same level as the year before.

How the Budget Process Changed Over Time

When the federal government began, it had fewer sources of revenue and spent less than the government does today. The budget process was very informal, with little overall planning.

Over time, federal spending increased. As a result, the budget process had to be improved. Congress passed a law in 1921 that made the process more formal. For the first time, the president was required to send a budget to Congress each year. In 1974 Congress passed another law to improve the budget process. It required Congress to set up committees to focus on the budget. It also set up the Congressional Budget Office (CBO). That office has the job of estimating the cost of proposed legislation, tracking spending and revenue measures, and making reports to Congress.

The budget process still faces difficulties today, however. For example, members of Congress often add spending for pet projects to major bills. These add-ons can increase overall spending and often increase the category of discretionary or even mandatory expenditures.

✓ **CHECK FOR UNDERSTANDING**

1. **Comparing and Contrasting** How is making the federal budget similar to and different from making a personal budget?

2. **Analyzing** What kinds of choices are involved in making the federal budget?

---

appropriations bill legislation that sets spending on particular programs for the coming year

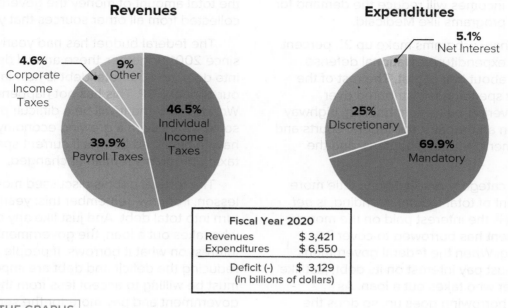

The Federal Budget, Fiscal Year 2020

Federal spending for Fiscal Year (FY) 2020 was much higher than revenues. This happened because the economy was in recession that year and there had been massive tax cuts.

Revenues

- 4.6% Corporate Income Taxes
- 9% Other
- 46.5% Individual Income Taxes
- 39.9% Payroll Taxes

Expenditures

- 5.1% Net Interest
- 25% Discretionary
- 69.9% Mandatory

Fiscal Year 2020	
Revenues	$ 3,421
Expenditures	$ 6,550
Deficit (-)	$ 3,129
(in billions of dollars)	

EXAMINE THE GRAPHS

1. **Analyzing Visuals** What are the two largest sources of revenue?
2. **Predicting** What happens to federal revenue when many people lose their jobs? Why?

Understanding the Federal Budget

GUIDING QUESTION

How does the federal budget reflect choices?

Federal Budget Revenues

The circle graph of **Federal Revenues** shows the individual income tax as the biggest source of federal revenue. This tax is paid by all people who earn income above a certain amount. The second-largest share of revenue is from payroll taxes. These are taken from workers' paychecks to fund social insurance programs such as Social Security and Medicare. Medicare provides some health care coverage for people age 65 and older. Finally, the third-largest source of revenue comes from the tax that corporations pay on their profits.

Taxes fall into three categories: progressive, proportional, and regressive. With a *progressive tax*, the tax rate goes up as income goes up. The federal income tax is a progressive tax. A *proportional tax* has a constant tax rate, regardless

of income. The tax for Medicare is proportional because it is the same rate for all wage earners. A *regressive tax* takes a smaller percentage of your income as the amount you earn goes up. The sales tax is an example of a regressive tax.

Federal Budget Expenditures

The circle graph of **Federal Expenditures** shows mandatory and discretionary federal spending for Fiscal Year 2020. The mandatory part of federal spending is the largest category, accounting for almost 70 percent of all federal spending. Social Security (not shown) is the largest component of mandatory expenditures. Medicare, a federal health insurance program for senior citizens regardless of income, is the second-largest category. Expenditures for the Small Business Administration was third-largest. The COVID-19 recession pushed unemployment compensation expenditures into fourth place. Medicaid, a joint federal-state medical insurance program for low-income people, was the fifth-largest mandatory expenditure.

Collectively, mandatory expenditures are expected to increase as our population gets older and medical expenses continue to rise. Unemployment compensation expenditures will get smaller as the economy moves out of recession. And programs designed to help low-income Americans find better jobs and earn more secure incomes will reduce the demand for welfare-type programs like Medicaid.

Discretionary programs make up 25 percent of all federal expenditures. National defense accounts for about half of that. The rest of the discretionary spending is distributed over programs covering education, training, highway transportation and repairs, the federal courts and law enforcement, natural resources, and the environment.

The final category, amounting to little more than 5 percent of total federal spending, is *net interest*. This is the interest paid on the money the government has borrowed to cover its overspending. When the federal government borrows, it must pay interest on its debt, just like any consumer who takes out a loan. As government borrowing goes up, so does the interest owed. This category is expected to go up significantly if government borrowing continues and if interest rates rise.

Federal Budget Deficits

The **Federal Budget, Fiscal Year 2020** graph shows the federal budget situation in 2020.

An Amtrak train pulls out of a station in New Orleans. Funding for Amtrak, along with other public transportation systems, is a discretionary federal expenditure.

Identifying What other programs are included in discretionary expenditures?

With total revenues of $3,421 billion and total expenses of $6,550 billion, a *deficit*—or shortfall—of $3,129 billion remained. This was a staggering amount by any historical standard! In fact, the deficit for 2020 was almost as large as the total amount of money the government collected from all other sources that year.

The federal budget has had yearly deficits since 2001. Together, these annual deficits turned into debt, and our total debt is now larger than our annual GDP. This has not happened since World War II, and it will be a difficult problem to solve. Why? Even a growing economy will not have a balanced budget if current spending and taxing programs remain unchanged.

The federal debt is discussed more in a later lesson. For now, remember this: yearly deficits turn into total debt. And just like any consumer who takes out a loan, the government must pay interest on what it borrows. If people believe that reducing the deficit and debt are important, they must be willing to accept less from the federal government and pay more for the services they receive. Enacting necessary changes will require cooperation and courage by our elected leaders—soon. Solutions will be harder to achieve the longer they are delayed.

✓ CHECK FOR UNDERSTANDING

1. **Describing** What is a progressive tax? What is an example of a progressive tax?

2. **Explaining** Why does the government owe interest?

Budgeting for State and Local Governments

GUIDING QUESTION

How do state and local revenues and expenditures differ from those of the federal government?

State and local governments also prepare budgets. The governments of all states except Vermont cannot, by law, spend more than they receive in revenue. Local governments also have to limit spending so they do not exceed revenues.

State Governments

The **State and Local Government Revenue and Expenses** graphs are for all states combined. As you can see, the largest source of state income is

Robert Kaufmann/FEMA

State and Local Government Revenue and Expenses

The federal government provides money to state governments. State governments, in turn, provide funds for local governments.

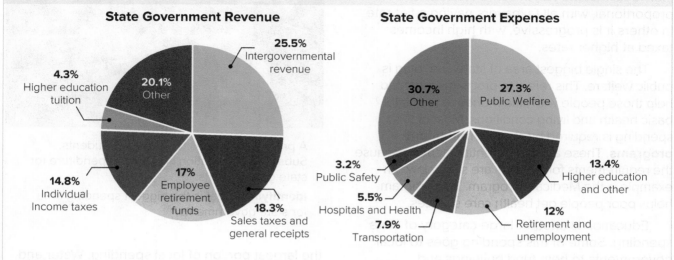

State Government Revenue
- 25.5% Intergovernmental revenue
- 20.1% Other
- 4.3% Higher education tuition
- 14.8% Individual Income taxes
- 17% Employee retirement funds
- 18.3% Sales taxes and general receipts

State Government Expenses
- 30.7% Other
- 27.3% Public Welfare
- 3.2% Public Safety
- 5.5% Hospitals and Health
- 7.9% Transportation
- 13.4% Higher education and other
- 12% Retirement and unemployment

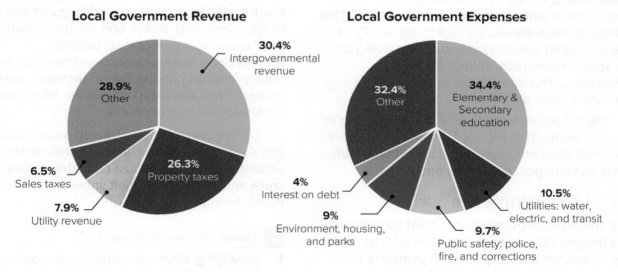

Local Government Revenue
- 30.4% Intergovernmental revenue
- 28.9% Other
- 6.5% Sales taxes
- 7.9% Utility revenue
- 26.3% Property taxes

Local Government Expenses
- 32.4% Other
- 34.4% Elementary & Secondary education
- 4% Interest on debt
- 9% Environment, housing, and parks
- 9.7% Public safety: police, fire, and corrections
- 10.5% Utilities: water, electric, and transit

EXAMINE THE GRAPHS

1. **Analyzing Visuals** What is one specific tax that local governments collect and state governments do not?

2. **Inferring** Which level of government—state or local—would be hurt more by an end to intergovernmental revenue? Why?

Intergovernmental revenue. These are funds that one level of government receives from another level of government. States receive this money from the federal government.

Sales taxes are the states' second-most-important source of revenue. A **sales tax** is paid when someone buys a good or service. All but five states have sales taxes. A 5 percent sales tax on clothing means that a person spending $100 on clothes pays another $5 in taxes. Sales taxes can be regressive, meaning that they are a bigger burden on low-income than on high-income earners. As a result, many states do not tax essential goods such as food and medicine. Some states also declare sales tax holidays.

intergovernmental revenue funds that one level of government receives from another level of government

sales tax tax paid by consumers at the time they buy goods or services

During these periods, sales taxes on most products, including school supplies, are not taxed.

All but nine states have a personal or individual income tax. In some states this tax is proportional, with all taxpayers paying a flat rate. In others it is progressive, with high incomes taxed at higher rates.

The single biggest area of state spending is public welfare. This refers to programs meant to help those people with little money to maintain basic health and living conditions. Most of this spending is required by states' **entitlement programs**. These are called "entitlements" because the requirements for benefits are set by law. An example is the Medicaid program. This program helps poor people get health care services.

Education is another large category of state spending. Some of this spending goes to local governments to help fund buildings and operations at state colleges and universities. This spending helps **subsidize** (SUHB•suh•DYZ), or pay for, higher education for students going to the states' community colleges and state universities. This subsidy helps make higher education more affordable.

"Other" refers to additional state spending in all other areas. This includes insurance payments to retired state employees and spending in such areas as state police, prisons, and parks.

Local Governments

Like states, local governments must raise money. The biggest difference is that local governments rely heavily on property taxes. A **property tax** is a tax based on the value of land and property that people own. Generally, the higher the value of the property, the higher the property tax that is paid. Many states allow their local governments to charge sales taxes and collect income taxes as well. Fines for traffic and other violations, along with fees for permits and special services, also provide income for local governments.

Local governments provide many of the basic services on which citizens depend. Education is

entitlement program a government program that makes payments to people who meet certain requirements in order to help them meet minimum health, nutrition, and income needs

subsidize to aid or support a person, business, institution, or undertaking with money or tax breaks

property tax tax on the value of land and property that people own

A professor lectures to university students. Subsidizing education is a large expenditure for state governments.

Identifying What is the biggest spending area for state governments?

the largest portion of local spending. Water and electric utilities are the second-largest category of local spending. Police and fire protection are the third-largest part of local budgets. Professionals always provide police services. Volunteers can provide fire protection in small communities, although larger cities often have professionals as firefighters.

Local governments build and maintain city and county streets, too. In areas with harsh winter weather, city workers must clear the streets of snow and apply salt to melt snow and ice to make driving safe.

✓ CHECK FOR UNDERSTANDING

1. **Identifying** What is the largest local government expenditure?

2. **Explaining** What is intergovernmental revenue? How does it work?

LESSON ACTIVITIES

1. **Informative/Explanatory Writing** Write a paragraph explaining how the federal government prepares a budget and makes spending decisions. Include all steps through appropriations.

2. **Presenting** With a partner, use the Internet to find out how your state ranks among other states in terms of state sales taxes and individual income taxes. Then create a poster highlighting economic advantages of living in your state. Use the tax information—along with other economic statistics—to convince people to live in your state.

Shutterstock

06

Multiple Perspectives: Taxes

? COMPELLING QUESTION

Are tax rates fair?

Plan Your Inquiry

DEVELOPING QUESTIONS

Read the Compelling Question for this lesson. What questions can you ask to help you answer this Compelling Question? Create a graphic organizer like the one below. Write three Supporting Questions in the first column.

Supporting Questions	Source	What this source tells me about tax rates	Questions the source leaves unanswered
	A		
	B		
	C		
	D		
	E		

ANALYZING SOURCES

Next, read the introductory information for each source in this lesson. Then analyze each source by answering the questions that follow it. How does each source help you answer each Supporting Question you created? What questions do you still have? Write those in your graphic organizer.

After you analyze the sources, you will:

- use the evidence from the sources
- communicate your conclusions
- take informed action

Background Information

Taxes fall into three categories: progressive, proportional, and regressive. With a *progressive tax*, the tax rate goes up as income goes up. The federal income tax is a progressive tax. After you get a job and start earning wages or a salary, you will be required to give some of your paycheck to the federal and state and local governments. The more you earn, the more taxes you'll pay.

This Inquiry Lesson includes sources about tax rates. As you read and analyze the sources, think about the Compelling Question.

Is it fair to ask people to pay more in taxes than they already do?

Time Line of Top Federal Income Tax Rates

The percentage of income tax paid at the top bracket, or level, of wealth has changed over time.

PRIMARY SOURCE: ONLINE ARTICLE

❝ The tax law, like almost all laws, grows as lawmakers use it for **pork**, try to make it fairer, use it to stimulate a sector of the economy, or just want to raise revenue.

In 1913, the top tax bracket was 7 percent on all income over $500,000 ($11 million in today's dollars); and the lowest tax bracket was 1 percent.

World War I In order to finance U.S. participation in World War One, Congress passed the 1916 Revenue Act, and then the War Revenue Act of 1917. The highest income tax rate jumped from 15 percent in 1916 to 67 percent in 1917 to 77 percent in 1918. War is expensive. After the war, federal income tax rates took on the steam of the roaring 1920s, dropping to 25 percent from 1925 through 1931.

The Depression Congress raised taxes again in 1932 during the Great Depression from 25 percent to 63 percent on the top earners.

World War II As we mentioned earlier, war is expensive. In 1944, the top rate peaked at 94 percent on taxable income over $200,000 ($2.5 million in today's dollars). That's a high tax rate.

The 1950s, 1960s, and 1970s Over the next three decades, the top federal income tax rate remained high, never dipping below 70 percent.

The 1980s The Economic Recovery Tax Act of 1981 slashed the highest rate from 70 to 50 percent, and indexed the brackets for inflation. Then, the Tax Reform Act of 1986, claiming that it was a two-tiered flat tax, expanded the tax base and dropped the top rate to 28 percent for tax years beginning in 1988. . . .

The 1990s–2012 During the 1990s, the top rate jumped to 39.6 percent. However, the Economic Growth and Tax Relief and Reconciliation Act of 2001 dropped the highest income tax rate to 35 percent from 2003 to 2010. The Tax Relief, Unemployment Insurance Reauthorization, and Job Creation Act of 2010 maintained the 35 percent tax rate through 2012.

2013–2017 The American Taxpayer Relief Act of 2012 increased the highest income tax rate to 39.6 percent. The Patient Protection and Affordable Care Act added an additional 3.8 percent on to this making the maximum federal income tax rate 43.4 percent.

2018–2021 The highest income tax rate was lowered to 37 percent for tax years beginning in 2018. The additional 3.8 percent is still applicable, making the maximum federal income tax rate 40.8 percent. ❞

— From "History of Federal Income Tax Rates: 1913–2021," bradfordtaxinstitute.com

pork projects paid for with taxes that benefit only a local area or a pet (personal) project

| EXAMINE THE SOURCE |

1. **Identifying** When was the tax rate highest for the wealthy?
2. **Evaluating** What rate do you think is a fair rate for the wealthy to be taxed? Explain.

TEXT: Bradford Tax Institute. "History of Federal Income Tax Rates: 1913-2021." Bradford Tax Institute. Accessed July 26, 2021. https://bradfordtaxinstitute.com/Free_Resources/Federal-Income-Tax-Rates.aspx.

Income Before and After Taxes

The graph below includes four separate bar graphs—each dividing the U.S. population into *quintiles*, or fifths. The first graph on the left shows that the highest one-fifth—or 20% of Americans—earned, on average, more than $300,000 in 2017. The lowest quintile, or 20% of Americans, earned about $20,000.

The second set of bars shows the amount of "transfers" each quintile received. *Transfers* are cash payments and similar welfare benefits from federal, state, and local governments. They are designed to help individuals and families who have low income.

The third set of bars shows the amount of federal income tax each quintile paid in 2017.

The fourth set of bars shows the amount of income each quintile earned after *adding* in any transfer payments and *subtracting* the amount paid in federal income taxes.

PRIMARY SOURCE: GRAPH

Income Before and After Transfers and Taxes

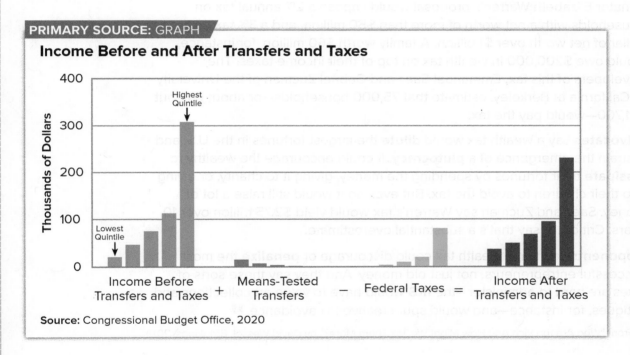

Source: Congressional Budget Office, 2020

EXAMINE THE SOURCE

1. **Speculating** Why are there essentially only four bars shown for "Means-Tested Transfers" and for "Federal Taxes"?

2. **Analyzing** About how much in federal taxes did the wealthy, or highest quintile, pay? What was their income after taxes were paid?

Wealth Tax

Some government leaders and economists have suggested taxing not only the *income* of the richest Americans, but also their *wealth*.

66 Income is different from wealth. Income is what you earn from your labor each year as well as interest, dividends, capital gains, and rents (if you're lucky enough to have any). Wealth is the value of the things you own, such as stocks, bonds, houses, etc. The federal government taxes income, but generally doesn't tax wealth except when someone makes a profit on the sale of assets, such as a share of stock or a piece of property. The Federal Reserve estimates that the top 1% holds slightly more wealth (31.1%) than the entire bottom 90% of the population (29.9%), and their share has been rising over time.

Senator Elizabeth Warren's proposal would impose a 2% annual tax on households with a net worth of more than $50 million, and a 3% tax on every dollar of net worth over $1 billion. A family worth $60 million, for instance, would owe $200,000 in wealth tax on top of their income taxes. The developers of this tax, Emmanuel Saez and Gabriel Zucman of the University of California at Berkeley, estimate that 75,000 households—or about one out of 1,700—would pay the tax.

Advocates say a wealth tax would **dilute** the largest fortunes in the U.S. and restrain the emergence of a **plutocracy**. It could encourage the wealthy to **dissipate** their fortunes by spending the money, giving it to charity, or giving it to their children to avoid the tax. But even so it would still raise a lot of money. Saez and Zucman say Warren's tax would yield $2.75 trillion over 10 years. Critics . . . say that's a substantial overestimate.

Opponents say that a wealth tax could discourage or **penalize** the most successful entrepreneurs, not just old money. And they say these sorts of taxes are hard to administer—the **IRS** would have to value art collections and antiques, for instance—and would spur creative tax avoidance. **99**

— From "Who Are the Rich And How Might We Tax Them More?" by David Wessel, Brookings, 2019.

---

advocates those who are in favor of something

dilute reduce or weaken

plutocracy country or society governed by the wealthy

dissipate break up

opponents those who are against something

penalize punish

IRS Internal Revenue Service; institution that collects federal taxes

> **EXAMINE THE SOURCE**

1. **Identifying** According to this source, what percentage of wealth does the top 1 percent of people hold in the United States?

2. **Summarizing** What reasons are given for advocating the wealth tax? What reasons are given for opposing the wealth tax?

TEXT: Wessel, David. "Who Are the Rich and How Might We Tax Them More?" Policy 2020. The Brookings Institute. October 19, 2020. https://www.brookings.edu/policy2020/votervital/who-are-the-rich-and-how-might-we-tax-them-more/.

D

The Rich Pay Enough

The article below makes the case that the rich already pay their fair share of taxes.

PRIMARY SOURCE: ONLINE ARTICLE AND CHART

> High-income Americans already pay the large majority of taxes, and the U.S. tax system is highly progressive when compared to those of other countries around the world.
>
> The latest government data show that in 2018, the top 1% of income earners—those who earned more than $540,000—earned 21% of all U.S. income while paying 40% of all federal income taxes. The top 10% earned 48% of the income and paid 71% of federal income taxes.

— From "In 1 Chart, How Much the Rich Pay in Taxes" by Adam Michel, www.heritage.org, 2021.

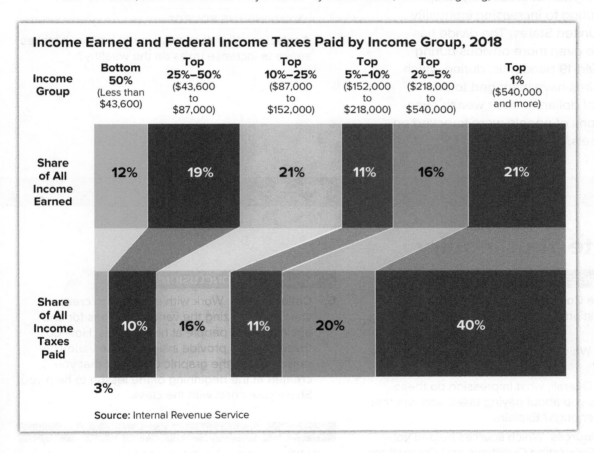

Income Earned and Federal Income Taxes Paid by Income Group, 2018

Income Group	Bottom 50% (Less than $43,600)	Top 25%–50% ($43,600 to $87,000)	Top 10%–25% ($87,000 to $152,000)	Top 5%–10% ($152,000 to $218,000)	Top 2%–5% ($218,000 to $540,000)	Top 1% ($540,000 and more)
Share of All Income Earned	12%	19%	21%	11%	16%	21%
Share of All Income Taxes Paid	10%	16%	11%	20%	40%	

3%

Source: Internal Revenue Service

EXAMINE THE SOURCES

1. **Analyzing Visuals** According to the chart, what share of total taxes were paid by those with earnings of $87,000–$152,000?

2. **Evaluating** Do you agree with this source, that the wealthy already pay their fair share of taxes? Explain.

The Rich Should Pay More

Some people think raising tax rates for the wealthy is fair.

66 Sometimes, the haggling and hemming and hawing over what to do about the debt overshadow a point that many Americans find obvious: It's simply a good, fair idea to tax the wealthy. They have **disproportionately** reaped the benefits of economic growth and the stock market in recent years, contributing to increasing inequality in the United States. The divide has become even more obvious during the Covid-19 pandemic, during which billionaires have managed to add heaps of dollars to their wealth even as millions of people were knocked on their heels. . . .

The chips of the economy are stacked in rich people's favor, and they're getting handed more chips constantly. So why not take a few chips away? 99

— From "Seriously, Just Tax the Rich" by Emily Stewart, vox.com, 2021.

disproportionately excessively

EXAMINE THE SOURCE

Analyzing Why does this source believe it makes sense to increase taxes on the wealthy?

Complete Your Inquiry

EVALUATE SOURCES AND USE EVIDENCE

Refer back to the Compelling Question and the Supporting Questions you developed at the beginning of the lesson.

1. **Identifying** Which of the sources affected you the most? Why?

2. **Evaluating** Overall, what impression do these sources give you about paying taxes, and whether people pay enough? Explain.

3. **Gathering Sources** Which sources helped you answer the Supporting Questions and Compelling Question? Which sources, if any, challenged what you thought you knew when you first created your Supporting Questions? What information do you still need in order to answer your questions? Where would you find that information?

4. **Evaluating Sources** Identify the sources that helped answer your Supporting Questions. How reliable is the source? How would you verify the reliability of the source?

COMMUNICATE CONCLUSIONS

5. **Collaborating** Work with a partner to create a chart summarizing the various reasons for and against taxing people at higher rates. How do these sources provide insight on the various reasons? Use the graphic organizer that you created at the beginning of the lesson to help you. Share your chart with the class.

TAKE INFORMED ACTION

Writing an Editorial Take a side on the issue of whether people should pay higher taxes. Write an opinion paragraph—also known as an editorial—stating your position. Support your opinion logically by using evidence and details from the sources. Remember to analyze the validity of the sources for bias, and to cite—or give credit to—the sources you quote from. Work with your teacher to get your opinion published in the school paper or website or local newspaper.

TEXT: Stewart, Emily. "Seriously, Just Tax the Rich." Vox. May 18, 2021. https://www.vox.com/22432338/joe-biden-tax-plan.

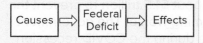

READING STRATEGY

Integrating Knowledge and Ideas As you read the lesson, complete a chart to identify the causes and effects of a federal deficit.

Causes → Federal Deficit → Effects

Managing the Economy

GUIDING QUESTION

How does the government try to influence the economy?

For much of U.S. history, the government had a limited economic role. It left most matters to consumers and businesses. In the 1930s, however, the government's economic role changed.

Franklin D. Roosevelt became president in 1933. The Great Depression had been going on for nearly four years, and millions were suffering. Roosevelt decided the nation could not wait any longer for

During the Great Depression, government programs hired workers to build roads, buildings, and dams (top left). Other workers were hired to record the histories of formerly enslaved women and men (top right), and to photograph the struggles of farm and migrant workers (bottom). Such programs were designed to stimulate the economy.

Analyzing Visuals What were two ways these government programs helped the country during the Depression?

the economy to recover. Believing the government had to act, Roosevelt started new government programs that put people back to work. These programs paid people to build schools, post offices, bridges, parks, and more. These new programs helped the economy recover. Later, when America entered World War II, a stronger economy helped people provide supplies for the military.

Fiscal Policy

Government spending programs like those during the Great Depression are known as fiscal policy. **Fiscal policy** is the government's use of taxes and spending to stimulate the economy. The idea behind fiscal policy is simple: If the economy slows or enters a recession, the government can increase spending, reduce taxes, or do a combination of the two. Using fiscal policy to boost growth is called *stimulating* the economy.

By boosting spending, the government creates demand for goods and services. The extra spending helps convince producers to hire back workers they had laid off. Cutting taxes puts more money in people's pockets, which can also lead to increased demand. The government could also cut taxes that businesses pay. Lower taxes might encourage businesses to pay workers more or increase the production of goods and services.

Advantages of Fiscal Policy

Fiscal policy was used successfully several times since 1900. The first was during the Great Depression when the government spent billions to

fiscal policy government use of taxes and spending to stimulate the economy and reach economic goals

The U.S. Treasury sent stimulus checks to Americans during the COVID-19 recession. Stimulating the economy is a goal of fiscal policy.

reduce unemployment and stimulate production. This was followed by massive government spending on the military during World War II—spending that also stimulated economic growth. But when World War II ended, the country began to worry about falling back into depression. As a result, Congress passed the Employment Act of 1946, which set the three official economic goals. The three goals were to keep people working, to keep producing goods, and to keep consumers buying goods and services.

When the Great Recession of 2008–2009 struck, Congress passed two massive stimulus programs totaling about $1.5 trillion to prevent the situation from becoming worse. The government played a similar role during the COVID recession of 2020. By dramatically increasing spending, and by cutting taxes, the government helped restore demand for goods and services.

Without the massive government involvement during these periods, peoples' economic situation would have been much worse. Consequently, many people today feel that some government involvement is needed to improve their situation in life.

Problems with Fiscal Policy

Fiscal policy comes with problems, however. Even when politicians agree to pursue a stimulus program, they may disagree on how much to spend or which programs should be funded. For example, President Biden and the Democrats wanted to spend $1.9 trillion to stimulate the economy because of the COVID recession. But Republicans objected because they thought the proposed spending was too high. Disagreements like these are to be expected in a democracy, but not all sides can get what they want.

Fiscal policy can also be slow to take effect after a policy is approved. It may take months for the U.S. Treasury to print and mail stimulus checks to the right addresses. Other taxpayers may not receive the benefits of a tax break until they file their annual tax returns. By the time the money is spent, the state of the economy may have already improved.

Another problem is the difficulty of judging whether a stimulus bill is too big or too small. As mentioned above, about $1.5 trillion in stimulus spending was passed during the Great Recession. Yet the monthly unemployment rate took 10 years to fall back to its pre-recession low.

Unemployment insurance will help this laid-off worker pay bills while she looks for a new job. Unemployment insurance is an automatic stabilizer.

Economists agree that the recession would have been much worse without the stimulus bills, but they still did not know exactly how many jobs were saved.

Automatic Stabilizers

Because of problems that can occur with fiscal policy, economists like programs called **automatic stabilizers**. These are stabilization programs that do not need legislative approval to activate because of worsening economic conditions. In other words, programs already in place can automatically provide benefits without getting caught up in partisan congressional gridlock.

The most important automatic stabilizer is *unemployment insurance*. Think about how unemployment insurance works. In a recession, millions of people lose their jobs. After they are unemployed long enough, they are eligible to receive unemployment insurance payments. Using these insurance payments, unemployed workers can then pay their bills until they are rehired or find a new job.

Another automatic stabilizer is the progressive *income tax* system. When people work less or lose their jobs, they are taxed at a

lower rate. This means they can keep a larger percentage of their income. Thus, lower tax rates partially **offset**, or counterbalance, the loss of income. When the economy recovers and they go back to work, they begin to make more money. Then, when they can better afford it, they are taxed at a higher rate. That, in turn, helps lower the deficit made worse by recessions.

Some people would like the individual income tax to be proportional instead of progressive. This may sound fair because a proportional tax requires everyone to pay the same percentage rate. But the individual income tax cannot work as an automatic stabilizer if it is proportional. This is a strong argument for maintaining a progressive income tax.

Automatic stabilizers prevent incomes from falling too far in hard times. They thus help offset periods of high unemployment that severely threaten economic growth. And when times are good, government spending falls and taxes rise.

✓ CHECK FOR UNDERSTANDING

1. **Identifying Cause and Effect** Under what conditions does the government use fiscal policy? Why?

2. **Explaining** What problems sometimes accompany fiscal policy?

---

automatic stabilizer program that works to preserve income without additional government action during economic downturns

offset to counterbalance

Budgets—Balanced or Unbalanced?

GUIDING QUESTION

What do governments do when the budget does not balance?

A **balanced budget** occurs when a government's annual revenues and spending equal each other. Most states *must* have a balanced budget. States can borrow money to invest in long-term projects but they cannot borrow to fund normal operating expenses or regular programs. States can also save money during the years in which they have a budget surplus. A **budget surplus** occurs when government collects more money than it spends. If revenues fall in another year, states can use the reserves they saved to balance the budget. Of course, they also have the option to cut spending.

The federal government is different. It is allowed to have a **budget deficit**, or spend more than it collects in revenues in a fiscal year. To make up the difference, the federal government borrows money by selling government securities: bonds, Treasury notes, and Treasury bills. A government bond is a contract in which the government promises to repay borrowed money with interest at a specific time in the future. Most government bonds are repaid in 10 to 30 years.

balanced budget annual budget in which expenditures equal revenues

budget surplus situation that occurs when a government collects more revenues than it spends

budget deficit situation that occurs when a government spends more than it collects in revenue

Savings bonds like these are a government security, or money borrowed from the purchaser of the bond that will be repaid with interest at some future date.

Treasury notes have shorter maturities and must be paid back in one to 10 years. Treasury bills, or T-bills, are short term obligations that must be repaid in one year or less.

Deficits Can Be Good or Bad

Deficit spending can be good if not abused. If the economy is in a recession, for example, then deficit spending can help stimulate economic growth again. In addition, deficit spending can be useful if it funds expensive infrastructure improvements. **Infrastructure** includes the highways, levees, bridges, power, water, sewage, and other public goods needed to support a population. Deficit spending for these projects might be justified because they have benefits for several generations. Deficit spending can also be justified if the deficits are repaid at some future time. But it is easier to spend money than to repay it. That is why deficit spending can be a problem.

Deficit spending by the federal government has several negative effects. First, deficits can turn into debt. The **federal debt** is money the government has borrowed and not yet paid back. The yearly interest payments on the federal debt strains the federal budget. More than 5 percent of the 2020 budget went to paying interest on the federal debt. If the government continues to borrow, or if interest rates go up, payments on the debt will go up, and the federal government will have less to spend on other programs.

More borrowing has another effect. People think of government securities as safe ways to invest money, so they are popular. The more money people invest in bonds, Treasury notes, or Treasury bills, the less they have to invest in private businesses. Businesses will have less money to increase productivity, which could cause slower economic growth.

Supply and demand could also drive up interest rates. If the federal government needs to borrow more funds, less money is left for other investors, so they must pay more to borrow. For example, if your parents take out a loan to buy a car, more government borrowing could result in the loan costing them more than it would otherwise.

infrastructure highways, levees, bridges, power, water, sewage, and other public goods needed to support a growing economy

federal debt money the government has borrowed and not yet paid back

Government Deficits Become Debt

A massive federal deficit during the COVID-19 recession of 2020 dramatically increased the federal deficit and the federal debt as a percentage of GDP.

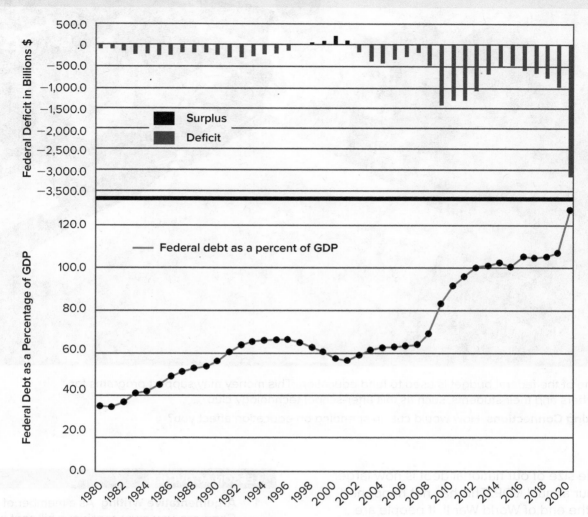

Source: White House, Office of Management and Budget

EXAMINE THE GRAPHS

1. **Analyzing Visuals** In which years did the federal government have a budget surplus?

2. **Inferring** How did the federal government respond, in terms of spending, to the recession that began in 2007 and worsened in 2008? How can you tell?

3. **Summarizing** For the most part, has the federal government had a balanced budget, a surplus, or a deficit over the last two decades?

Deficits Create Debt

Budget deficits create debt when the government borrows money. Each year that the federal government runs a deficit, its total debt goes up. If the federal government runs a surplus one year, it can use the extra money to pay down the debt. A budget surplus, then, can cause the federal debt to become smaller.

The government seldom runs a surplus that reduces the debt, however. The bar graph in **Government Deficits Become Debt** shows that the federal government has had only four budget surpluses in the last 40 years. The line graph in the same figure shows how those deficits have added to the total debt.

Some of the federal budget is used to fund education. This money may support programs for teachers and their students, such as this after-school technology club.

Making Connections How would cuts in spending on education affect you?

The size of our national debt is now larger than our annual GDP. This has not happened since the end of World War II. If people are concerned about the size of the debt, they must be willing to live with fewer government goods and services. Or they must be willing to pay higher taxes, which will help reduce the deficit. Economic growth alone will not be enough to enable the federal government to have a balanced budget. This is because the gap between annual spending and annual revenue collections has become too large.

✓ **CHECK FOR UNDERSTANDING**

1. **Analyzing** How are budget deficits related to debt?

2. **Contrasting** How does the federal government differ from state governments in its reaction to an unbalanced budget?

LESSON ACTIVITIES

1. **Argumentative Writing** As a member of Congress, you must consider a bill sent by the president cutting income taxes in the hope of stimulating the economy during a tough recession. Write a short speech for or against the president's plan. Give reasons for your position.

2. **Collaborating** When Congress works on a stimulus package, your local representative often attempts to get federal funding for programs in your area. With a partner, think of a local program or project that would benefit your community. Then create a presentation outlining your project idea and explaining why you think your community should get federal funding for it.

08

Multiple Perspectives: The Federal Debt

 COMPELLING QUESTION

Should the U.S. government be required to balance its budget?

Plan Your Inquiry

Read the Compelling Question for this lesson. What questions can you ask to help you answer this Compelling Question? Create a graphic organizer like the one below. Write three Supporting Questions in the first column.

Supporting Questions	Source	What this source tells me about the federal debt	Questions the source leaves unanswered
	A		
	B		
	C		
	D		
	E		

Next, read the introductory information for each source in this lesson. Then analyze each source by answering the questions that follow it. How does each source help you answer each Supporting Question you created? What questions do you still have? Write those in your graphic organizer.

After you analyze the sources, you will:
- use the evidence from the sources
- communicate your conclusions
- take informed action

Background Information

The federal government is not required to have a balanced budget. Instead, it's allowed to have a *budget deficit*, or spend more than it collects in revenues in a fiscal year. To make up the difference, the federal government borrows money by selling bonds, Treasury notes, or Treasury bills. The government then has to pay back the borrowed amount with interest. Each year that the federal government runs a deficit, its total debt goes up.

This Inquiry Lesson includes sources about the federal debt and whether the country should have a balanced budget. As you read and analyze the sources, think about the Compelling Question.

State governments are required to balance their budgets. Should this be a rule for the federal government as well?

The Federal Deficit and Debt

The bar graph shows that the federal government has had budget deficits for the past twenty years. The line graph in the same figure shows how those deficits have added to the total debt. The size of our national debt is now larger than our annual GDP.

PRIMARY SOURCE: GRAPH

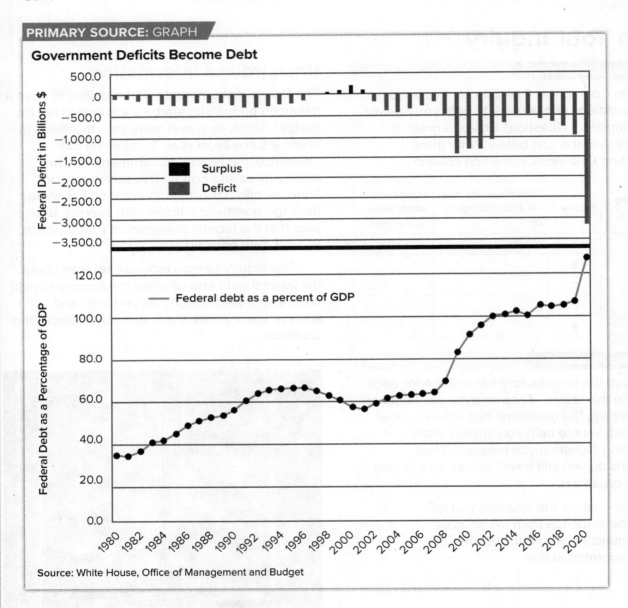

Government Deficits Become Debt

Source: White House, Office of Management and Budget

EXAMINE THE SOURCE

1. **Analyzing** What happened to the federal debt when the government had a budget surplus instead of a deficit?

2. **Drawing Conclusions** What can government leaders do if they are worried about the size of the federal debt?

National Debt—The Clock's Ticking

The national debt clock is located in New York City. It shows the accumulation of government deficits over the decades. Seymour Durst, an investor, created the first debt clock in 1989. President Franklin D. Roosevelt explains why the national debt is not something to worry about.

PRIMARY SOURCE: PHOTOGRAPH

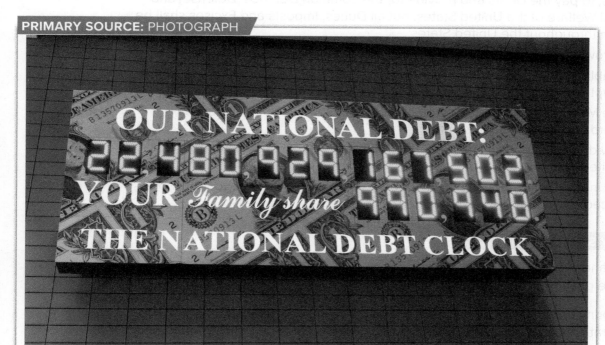

OUR NATIONAL DEBT:
22,480,929,167,502
YOUR *Family share* 990,948
THE NATIONAL DEBT CLOCK

The clock is constantly changing. During years when the deficit is low or there is a surplus, the clock runs backward. The debt was $22.48 trillion in 2019. It rose to $28 trillion by April 2021.

PRIMARY SOURCE: GOVERNMENT SPEECH

❝ National income will be greater tomorrow than it is today because government has had the courage to borrow idle capital and put it and idle labor to work. . . . Our national debt after all is an internal debt owed not only by the Nation but to the Nation. If our children have to pay interest on it, they will pay that interest to themselves. A reasonable internal debt will not impoverish our children or put the Nation into bankruptcy. ❞

— President Franklin D. Roosevelt, "Address Before the American Retail Federation, Washington, D.C." 1939.

EXAMINE THE SOURCES

1. **Speculating** Why do you suppose Durst created the clock?

2. **Calculating** According to the photo, what is the debt if divided by households in the United States?

3. **Analyzing Perspectives** Why was President Roosevelt not worried about deficit spending?

4. **Speculating** Do you think President Roosevelt would consider the current federal debt "reasonable"? Why or why not?

Balanced Budget Amendment

Regarding taxes, money, borrowing, and bankruptcy, the U.S. Constitution states in Article I, Section 8:

- The Congress shall have Power To lay and collect Taxes, Duties, Imposts and Excises, to pay the Debts and provide for the common Defence [Defense] and general Welfare of the United States; but all Duties, Imposts and Excises shall be uniform throughout the United States;

- To borrow Money on the credit of the United States; . . .

- To establish an uniform Rule of Naturalization, and uniform Laws on the subject of Bankruptcies throughout the United States;

- To coin Money, regulate the Value thereof, and of foreign Coin, and fix the Standard of Weights and Measures;

- To provide for the Punishment of counterfeiting the Securities and current Coin of the United States; . . .

Nowhere in the Constitution, however, does it state that the federal budget should be balanced. Some people would like to change that.

PRIMARY SOURCE: ARTICLE

❝ A balanced budget amendment would make it Constitutionally **mandatory** for the government to operate without a deficit in each fiscal year.

The primary benefit of such an amendment is that it would protect future generations against accumulated debt. In 1979, the national debt of the United States was $827 billion. In 2017, the national debt was $20.2 trillion. With the amendment, this debt could begin to be reined in to prevent fiscal irresponsibility.

The primary issue with a balanced budget amendment is that it would limit the tools available to the government during times of economic difficulty. Countering recessions or responding to a national emergency would require the costs be offset on other budget lines, which would likely limit the help people may need to simply survive. ❞

— From "12 Key Balanced Budget Amendment Pros and Cons" by Louise Gaille, vittana.org, 2018.

mandatory legally binding

EXAMINE THE SOURCE

1. **Analyzing Perspectives** What reasons does the source give for having a balanced budget amendment?

2. **Explaining** What is the primary issue against a balanced budget amendment, according to the source?

Does the Debt Matter?

For decades, many people believed the huge federal debt would cause a collapse of the U.S. economy. That has not happened . . . yet. Even so, others suggest we might want to tackle the debt sooner rather than later.

PRIMARY SOURCE: ARTICLE

❝ As soon as it's reasonable to do so, we need to begin to repair the immense fiscal imbalance we have wrought. Post-pandemic, this suggests the need to make a commitment to fiscal responsibility—a pledge to take unpleasant but essential steps that politicians on all sides have paid lip service to but have rarely put into practice. Everything needs to be on the table—entitlement reforms, a restructuring of the tax code (including tax increases), changes to the federal budget process, etc.—to ensure that the United States can gradually bring the public debt down to more sustainable and safer levels. . . .

The debt threatens our country's ability to prosper in the decades to come, but it probably will not bring on a sharp and sudden catastrophe. That means it poses a particularly difficult challenge for our politics. If the only way to motivate politicians and voters to take the debt seriously is to insist that we are on the edge of an **abyss**, we will almost certainly fail to address the challenge. Rather than scare ourselves senseless, we need to take responsibility for the future—and to treat Americans like serious adults who can understand what responsibility might mean.

For decades, those warning that rising deficits and debt will lead to a sharp and catastrophic economic calamity were crying wolf. According to Aesop's fable, the villagers, tired of being tricked by the bored boy, started to ignore him. But that is not where the story ends. Eventually, the wolf did come. And things did not turn out well for the villagers when they ignored the boy then. ❞

— From "Does the Debt Matter?" by Peter Wehner and Ian Tufts, National Affairs, 2020.

---

abyss bottomless pit

EXAMINE THE SOURCE

1. **Summarizing** What actions does this source recommend taking to reduce the federal debt?
2. **Making Connections** How do the authors of this piece use a fable to get their point across?

Money In, Money Out

The graph shows the ratio of each dollar of spending relative to each dollar of income by fiscal year.

EXAMINE THE SOURCE

1. **Analyzing Visuals** Of the years shown, which ones spent less than what was coming in?

2. **Speculating** What two years had the highest ratio of spending to income? What occurred during those years?

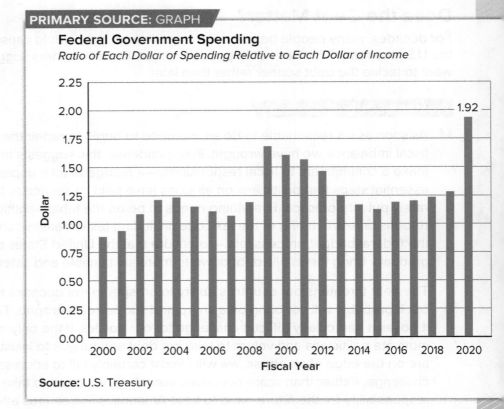

PRIMARY SOURCE: GRAPH

Federal Government Spending
Ratio of Each Dollar of Spending Relative to Each Dollar of Income

Source: U.S. Treasury

Complete Your Inquiry

EVALUATE SOURCES AND USE EVIDENCE

Refer back to the Compelling Question and the Supporting Questions you developed at the beginning of the lesson.

1. **Identifying** Which of the sources affected you the most? Why?

2. **Evaluating** Overall, what impression do these sources give you about the federal debt? Explain.

3. **Gathering Sources** Which sources helped you answer the Supporting Questions and Compelling Question? Which sources, if any, challenged what you thought you knew when you first created your Supporting Questions? What information do you still need in order to answer your questions? Where would you find that information?

4. **Evaluating Sources** Identify the sources that helped answer your Supporting Questions. How reliable is the source? How would you verify the reliability of the source?

COMMUNICATE CONCLUSIONS

5. **Collaborating** Work with a partner to create a short skit of a conversation between someone who wants a balanced budget and someone who does not think it is necessary. Include evidence from the sources in your dialog. Use the graphic organizer you created at the beginning of the lesson to help you. Share your skit with the class.

TAKE INFORMED ACTION

Writing a Letter to a Member of Congress Take a poll, or survey, of 10 adults, asking whether they believe the federal government should strive for a balanced budget, or whether it should consider passing a balanced-budget amendment. Write down their reasons for their opinion. Then write a letter to your representative or senator in Congress outlining your findings. Ask the lawmaker to reply with his or her opinion on a balanced budget.

Reviewing Government and the Economy

Summary

GDP

A major U.S. goal is to keep the economy strong. A strong economy provides a higher standard of living for Americans. The best way to measure the strength of the economy is by looking at the nation's *real Gross Domestic Product*—the total value of all final products produced in a country during a single year.

The Fed and Monetary Policy

The Federal Reserve System was set up to act as the country's *central bank*. The Fed uses *monetary policy* to keep the economy healthy and strong, and to maintain stable prices. The Fed also maintains the currency, regulates and supervises banks, promotes consumer protection, and acts as the government's bank.

Measuring the Economy

The U.S. economy does not grow at a steady rate. Instead, it goes through alternating periods of growth and decline known as the *business cycle*. Every business cycle has two distinct parts—a *recession* and an expansion.

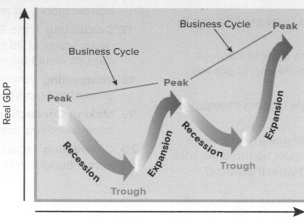

Financing the Government

Federal, state, and local governments prepare a budget every *fiscal year*. A budget is made up of two parts—*revenues* and *expenditures*. State and local budgets must balance, but the federal budget does not. The biggest source of federal revenue is the individual income tax. The largest federal expenditure is Social Security. If the government spends more than it receives in revenues, it runs a deficit—and *deficits* turn into *debt*.

Fiscal Policy

The government's goals are to keep people working, to keep producing goods, and to keep consumers buying goods and services. *Fiscal policy*—the government's use of taxes and spending—is used to achieve these goals. If the economy enters a recession, the government can increase spending, cut taxes, or a combination of the two.

Checking For Understanding

Answer the questions to see if you understood the topic content.

REVIEWING KEY TERMS

1. Define each of these terms:

   A. recession
   B. inflation
   C. bear market
   D. central bank
   E. monetary policy
   F. open market operations
   G. federal debt
   H. fiscal policy
   I. automatic stabilizer
   J. subsidize

REVIEWING KEY FACTS

2. **Explaining** How does real GDP differ from current GDP?

3. **Identifying** What economic indicators are used to judge an economy's health?

4. **Identifying** What name is given to a rising stock market fueled by investors optimistic about the economy's future?

5. **Identifying** By what name is the central bank of the United States known?

6. **Explaining** What three tools does the Fed use to control the money supply?

7. **Identifying** Which part of the budget involves spending choices that are made and approved each year?

8. **Identifying** What is the single biggest category of state spending?

9. **Explaining** When the federal government has a budget deficit, how does it make up the difference between what it spends and what it collects?

10. **Identifying** Which two automatic stabilizers are in place to preserve income when the economy slows?

11. **Identifying Cause and Effect** How does a budget deficit create debt for the federal government?

12. **Drawing Conclusions** What is the goal of a stimulus program?

CRITICAL THINKING

13. **Explaining** Why is real GDP per capita a better measure of comparing countries' economies than current GDP or real GDP?

14. **Predicting** If the Leading Economic Index goes up, what usually happens to the real GDP several months later?

15. **Identifying Cause and Effect** How does the Federal Open Market Committee (FOMC) affect the money supply when it buys and sells government bonds and Treasury bills?

16. **Drawing Conclusions** How does a lowered discount rate encourage financial institutions to borrow money from the Fed?

17. **Calculating** If the Fed has set the reserve requirement at 35 percent, how much of a $100 deposit would be available to lend out?

18. **Interpreting** Why is the federal income tax considered a progressive tax?

19. **Making Connections** Why are sales taxes considered regressive taxes?

20. **Contrasting** How does a budget surplus differ from a budget deficit?

NEED EXTRA HELP?

If You've Missed Question	1	2	3	4	5	6	7	8	9	10
Review Lesson	2, 3, 4, 5, 7	2	3	3	4	4	5	5	7	7

If You've Missed Question	11	12	13	14	15	16	17	18	19	20
Review Lesson	7	7	2	3	4	4	4	5	5	7

Apply What You Have Learned

A Understanding Multiple Perspectives

How involved should the federal government be in our economy?

ACTIVITY **Using Evidence in an Essay About Fiscal Policy** Read these excerpts that provide two opinions of government fiscal policy. Then use information in the excerpts to write a brief essay answering these questions: What are some benefits of fiscal policy? What are some drawbacks? What evidence from the lessons can you provide to support the two views?

❝ The great thing about fiscal policy is that it has a direct impact and doesn't require you to bind the hands of future policymakers. ❞

— Paul Krugman, "An Interview with Paul Krugman" in *The Washington Post*, 2012.

❝ Popular as Keynesian fiscal policy may be, many economists are skeptical that it works. They argue that fine-tuning the economy is a virtually impossible task, and that fiscal-stimulus programs are usually too small, and arrive too late, to make a difference. ❞

— James Surowiecki, "The Stimulus Strategy" in *The New Yorker*, 2008.

B Writing an Informative Report

Can a country's economy ever just shut down? Not really, but when government leaders cannot agree upon a national budget, selected shutdowns and closings can occur. In December 2018, parts of the federal government shut down until January. Many federal offices and departments closed. Museums and monuments run by the government were closed to visitors. Only federal workers who were essential to the country's safety, such as members of the armed forces, stayed on the job. At the root of this crisis was a disagreement between the president and members of Congress about how to spend the U.S. revenues. The budget crisis made Americans more aware of how complicated financing the government can be.

ACTIVITY **Writing About the Federal Budget**
As the U.S. economy has expanded, both its revenues and expenditures have grown to amounts totaling trillions of dollars. As a result, the process for creating a federal budget has become more complex. Write a description of how the process for making the federal budget has changed since the early 1900s. Describe how the process works today.

Then prepare a family or personal budget. Assume a sum for monthly income. Include these "mandatory" expenses: housing, transportation, communication (phone and Internet). Also include "discretionary" expenses, such as leisure and recreation, or miscellaneous. Compare your income and spending categories to federal government budget allocations.

(t) Klein, Ezra, and Paul Krugman. "An Interview with Paul Krugman." The Washington Post. May 4, 2012. https://www.washingtonpost.com/blogs/ezra-klein/post/an-interview-with-paul-krugman/2012/05/04/gIQAR9xnIT_blog.html., (b) Surowiecki, James. "The Stimulus Strategy." The New Yorker. Condé Nast. February 17, 2008. https://www.newyorker.com/magazine/2008/02/25/the-stimulus-strategy.

C Understanding Economic Concepts

The federal government is not required to have a balanced budget. Many people would like to see that change, however.

ACTIVITY **Apply Evidence in a Photograph to the Federal Budget** What do you see in this photograph? What natural occurrence might have caused this situation? Write a paragraph to explain how such an event might turn a national budget surplus into a budget deficit and, eventually, a debt. Discuss how the federal government responds to emergencies and disasters that strike states unexpectedly.

Houston, Texas 2017

D Making Connections

Budgets are created at the national, state, and local levels. Budgets are also created at district levels, such as your school district.

ACTIVITY **Being an Active Citizen** Find out when your local school board will meet to discuss the school system's budget for the upcoming year. If possible, attend the meeting, listen to the discussion, and contribute your own ideas. Perhaps prepare a position on a school health-related issue, such as bullying prevention, Internet safety, or nutritional choices. Be sure to support your position with accurate health information. Then write a summary of the meeting. If you cannot attend, read a report of the meeting, and write a summary of it.

Currencies from all over the world are pictured here. These currencies are used to finance international trade.

The Global Economy

INTRODUCTION LESSON

01 Introducing The Global Economy   E160

LEARN THE CONCEPTS LESSONS

02 Why Nations Trade   E165

03 Exchange Rates and Trade Balances   E169

04 Global Trade Alliances and Issues   E173

05 The Wealth of Nations   E179

INQUIRY ACTIVITY LESSON

06 Analyzing Sources: Environmental Balance   E185

REVIEW AND APPLY LESSON

07 Reviewing The Global Economy   E191

Ralf Siemieniec/Shutterstock.com

Introducing The Global Economy

Worldwide Connections

You might not realize it, but you are connected to the world. How so? You're connected by:

- The clothes you wear
- The foods you eat
- The cellphone you're using
- The games you play
- The Internet
- The sheets or blankets you're sleeping on
- And much more!

International trade gives you choices from around the world. Although some people would like to limit world trade, it's safe to say that it's here to stay.

This cartoon portrays the United States in the form of "Uncle Sam." He has built a wall to protect American industries from foreign competition. As you can see, however, there is also a drawback of blocking free trade: it prevents *American* firms from selling their products in international markets.

Did You Know?

Trading began in 6000 B.C.E. with the Mesopotamians in the earliest cultural hearths of human society. Ancient peoples used trade, or barter, to get such items as food, weapons, and spices.

Global Trade

Economies around the world rely on international trade. In the United States, businesses of all sizes influence trade with other countries. When countries rely heavily on one another, good trade rules and relations become vital. For countries that share borders, such as the United States with Canada and Mexico, the effects of trade are even more important.

» A small grocery store in New York City specializes in selling gourmet foods from France.

» People in London, England, buy DVDs of American movies filmed in Toronto, Canada.

» Halfway around the world, workers in an office in India answer service calls for the products of a company in Dallas, Texas.

» A farmer in Brazil harvests coffee beans to sell to a coffee store in Seattle, Washington.

Getting Ready to Learn About . . .
The Global Economy

Why Nations Trade

Individual nations do not always have the necessary resources to make the products their people need and want. To solve this problem of scarcity, nations trade with one another. They trade food, manufactured goods, services, and raw materials.

Nations *import*, or bring into the country, goods produced in other nations. They *export*, or sell to other nations, goods they produce. Trade between nations today is more extensive and important than ever.

When countries trade, they specialize in the things they can produce relatively better than other countries can. In other words, trade allows each country to focus on things it can make at a lower opportunity cost. This is called *comparative advantage*, and is the basis for international trade.

But while international trade is beneficial to almost everyone, it can be painful to some.

Companies that lose sales to lower-priced foreign-made goods often want to limit foreign trade. So do workers in those companies who may lose their jobs. Governments can help protect home industries with *tariffs*, or taxes on imports, which makes them more expensive than home products. Governments sometimes place *quotas*, or limits on the number of products that can come into their country. A third option is to give *subsidies*, or payments, to the home company so it doesn't raise its prices for home consumers.

Paying for Trade Goods

Most countries use their own currency, or money, to pay for the products they import. Currencies are bought and sold in markets, just like goods and services. As a result, the value of a currency, like the value of a good or service, can go up or down. The value of one currency in terms of another is called its *exchange rate*.

Evidence of international trade is hard to miss in the port of Barcelona, Spain. In this bird's-eye view, you can see colorful cargo containers stacked high as they wait to be loaded with massive cranes onto ships.

Global economic interconnections can be seen by this McDonald's restaurant in Bangkok, Thailand.

Economic Interdependence

Countries around the world rely upon one another for resources, goods, and services. Many of the products we use are made by *multinationals*, or corporations that have manufacturing operations in a number of different countries. This results in *economic interdependence*, and is the reason countries must cooperate when they trade. Over the decades, some countries have formed *trade blocs*, or alliances, to increase the benefits of international trade.

The European Union (EU) is the most famous trade bloc. Today the EU creates a free-trade zone that covers 27 European countries. Within this regional zone, goods, services, and workers can travel freely across national borders. The United States, Canada, and Mexico also merged to create the United States-Mexico-Canada Agreement (USMCA). This is also one of the world's largest free-trade areas.

The economic interdependence resulting from trade alliances provides benefits to almost everyone, but there are still problems. Jobs are often lost when multinational companies *outsource*, or move their factories to other countries with lower labor costs.

As mentioned earlier, protectionist policies such as tariffs, quotas, and subsidies are often used to help domestic industries that are threatened by free trade. These policies raise the prices of international goods to domestic consumers and encourage them to buy domestic products instead. These policies also weaken the potential gains from a country's comparative advantage.

Rich vs. Poor Countries

In *developed countries*, citizens experience a relatively high *standard of living*, or quality of life. In *developing countries*, average citizens go without plentiful food, medicine, and good housing. Developing countries face many challenges to economic growth. These include high population growth rates, few resources, little industry, war, disease, and corruption. Many countries are transitioning to a market-oriented economy to improve their standard of living. Others have stalled and are having difficulty making the transition.

Looking Ahead

In this topic, you will discover why nations trade. You will also learn information about trade balances, trade alliances, and issues with global trade. You will understand why some countries are wealthier than others, and how some countries are transitioning to market-based economies. You will examine a Compelling Question and develop your own questions in the Inquiry Activity.

What Will You Learn?

In these lessons, you will learn:

- why nations trade, and how this helps reduce the problem of scarcity.
- why some countries set up trade barriers to protect domestic jobs and businesses.
- how international trade relies on the ability to exchange foreign currencies.
- how a positive or negative balance of trade affects a nation's currency.
- how economies are globally interdependent.
- why trade agreements and international organizations are formed to facilitate trade among countries.
- about issues that have arisen because of economic interdependence.
- the difference between developed and developing countries.
- the obstacles developing countries must overcome to move toward a market economy.

 COMPELLING QUESTION IN THE INQUIRY ACTIVITY LESSON

- **How do we balance the needs of consumers and the needs of a sustainable planet?**

In this topic, you will learn how the world is interconnected through trade.

02

Why Nations Trade

READING STRATEGY

Integrating Knowledge and Ideas As you read the lesson, complete a graphic organizer listing effects of trade barriers.

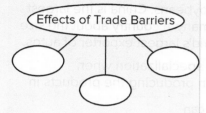

Effects of Trade Barriers

Trade Between Nations

GUIDING QUESTION

Why do nations trade with one another?

How is it that you can enjoy fresh summer fruit year-round? Why do Americans import cars from Japan or Korea when we can produce them here? Why is the United States the world's largest corn producer?

Individual nations do not always have the necessary resources to make the products their people need and want. To solve this problem of scarcity, nations trade with one another. Nations **import,** or bring into the country, goods produced in other nations. They **export,** or sell to other nations, goods they produce. The graph **Leading Exporters and Importers** shows that trillions of dollars' worth of products are exchanged by countries every year.

import to buy goods from another country

export to sell goods to other countries

Leading Exporters and Importers

The five leading export countries are also the five leading importers.

Five Leading Exporters

Exports in $billions: China $2,490; United States $2,377; Germany $2,003; Japan $1,032; France $969

Five Leading Importers

Imports in $billions: United States $3,214; China $2,140; Germany $1,804; Japan $1,084; France $1,022

EXAMINE THE GRAPHS

Analyzing Visuals
Which nation exports more than the United States?

📌 **GO ONLINE** Explore the Student Edition eBook and find interactive maps, charts, graphs, and tools.

E165

Trade between nations today is more extensive than ever. The total value of exports globally is about $20 trillion per year. As the map **Export Partners: Selected Nations** shows, countries in widely different locations are closely tied to one another through international trade.

Comparative Advantage

As Adam Smith observed in 1776, nations trade for the same reasons people trade. It simply makes sense for people to focus on the things they can produce best and then exchange those products for the things other people can produce best.

However, even if we could produce something cheaper than anyone else could—a concept known as **absolute advantage**—we still would not have the time to produce all the things we need. The best use of our time is to make more of the things we produce best—and then trade them for other things we still need.

Thus, countries focus on their *comparative* advantage instead. **Comparative advantage** is the ability to produce something *relatively* more efficiently or cheaper—or at a relatively lower opportunity cost—than anyone else can. (Recall that opportunity cost is the value of the *next* best thing that is given up when choosing to do or produce one thing instead of another.)

The goods and services countries produce and sell in abundance are due to their comparative advantages. Brazil, for example, is the world's largest exporter of soybeans. China is the largest exporter of tea. Germany, a country about the size of Indiana, is the world's largest exporter of autos.

Trade increases specialization when everyone focuses on producing the products in

absolute advantage the ability to produce something cheaper than anyone else can

comparative advantage a country's ability to produce a good relatively more efficiently than other countries can (or, a country's ability to produce a good at a lower opportunity cost than another country)

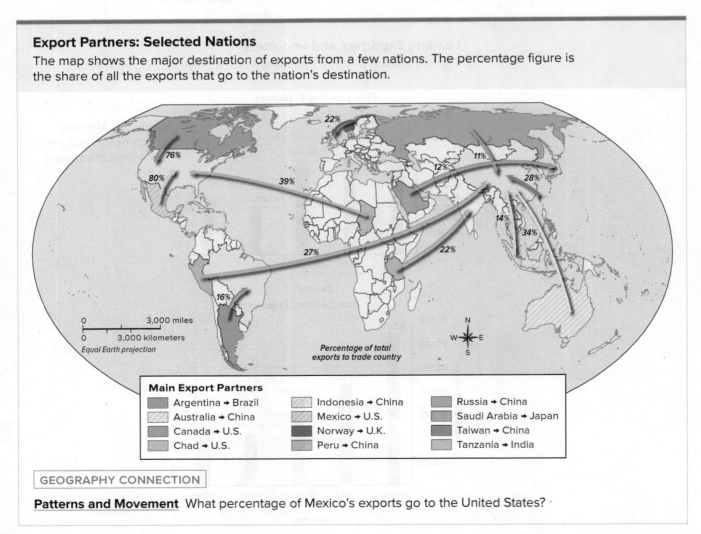

Export Partners: Selected Nations

The map shows the major destination of exports from a few nations. The percentage figure is the share of all the exports that go to the nation's destination.

Percentage of total exports to trade country

Main Export Partners

Argentina → Brazil	Indonesia → China	Russia → China
Australia → China	Mexico → U.S.	Saudi Arabia → Japan
Canada → U.S.	Norway → U.K.	Taiwan → China
Chad → U.S.	Peru → China	Tanzania → India

GEOGRAPHY CONNECTION

Patterns and Movement What percentage of Mexico's exports go to the United States?

Farmers harvest bananas in Ecuador. Ecuador's natural resources give it a comparative advantage in banana production.

which they have a comparative advantage. When people specialize, they produce the things they do best and exchange those products for what other people do best. States also specialize. For example, New York is a financial center for stocks and bonds, while vehicles are a major industry in Michigan. Texas is known for oil and cattle, while Florida and California are famous for citrus fruit.

Available Resources

A country's factors of production—its natural resources, labor, capital, and entrepreneurs—often determine its comparative advantage. For example, the climate and natural resources in Ecuador are excellent for growing bananas. It has a comparative advantage in producing bananas. It is also South America's largest exporter of bananas.

The United States is a major producer of gasoline. It has large petroleum reserves and specialized capital equipment needed to turn petroleum into gasoline. The United States has a comparative advantage in producing gasoline.

Because of these comparative advantages, Ecuador sells bananas to the United States, and the United States sells gasoline to Ecuador.

An economy with fewer resources can still specialize in a single export if it has a comparative advantage. These economies are known as single-resource economies. But to **rely**, or depend, on a single export makes a nation vulnerable to price changes in the marketplace. Diversified economies, which export a variety of products, are better able to respond to market changes.

✓ **CHECK FOR UNDERSTANDING**

1. **Contrasting** What is the difference between an import and an export?

2. **Explaining** What determines whether a country has a comparative advantage in producing a specific product?

Barriers to Trade

GUIDING QUESTION

How do trade barriers affect producers and consumers?

Home companies that lose sales to lower-priced foreign-made goods argue for **protectionism**—the use of tactics that make imported goods more expensive than domestic goods. Governments try to protect home industries and their jobs in three different ways: tariffs, import quotas, or subsidies. These policies can impact trade dramatically.

A **tariff** is a tax on imports. The goal is to make the price of imported goods higher than goods produced at home. Tariffs can give domestic industries some protection from foreign competition, but they raise prices for consumers. Tariffs should not be raised too high or last too long—or problems occur. When tariffs were widely used during the Great Depression, other countries struck back by putting high tariffs on U.S. goods. Everyone suffered. More recently, President Trump placed high tariffs on goods from major trading partners. These tariffs sparked a global *trade war* that caused other countries to put high tariffs on American goods. The result was that prices increased for consumers in all countries, and international trade was impacted.

rely to depend on

protectionism use of tactics that make imported goods more expensive than domestic goods

tariff tax on an imported good

Mehmet Hilmi Barcin/iStock/Getty Images

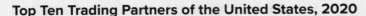

Top Ten Trading Partners of the United States, 2020

The map shows the United States at the center of a global trading system with its ten major partners.

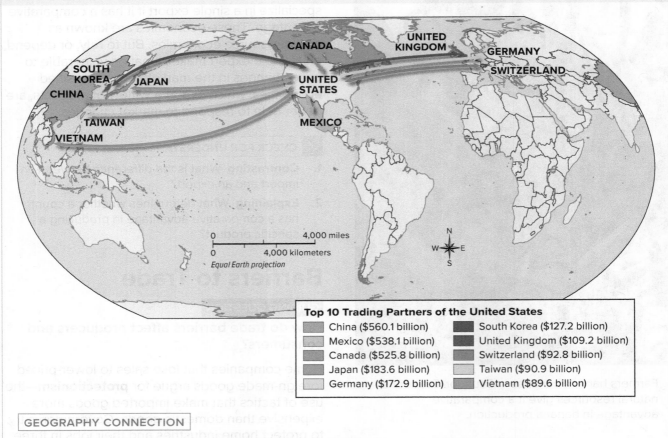

Top 10 Trading Partners of the United States

China ($560.1 billion)	South Korea ($127.2 billion)
Mexico ($538.1 billion)	United Kingdom ($109.2 billion)
Canada ($525.8 billion)	Switzerland ($92.8 billion)
Japan ($183.6 billion)	Taiwan ($90.9 billion)
Germany ($172.9 billion)	Vietnam ($89.6 billion)

GEOGRAPHY CONNECTION

1. **Global Interconnections** Which country is the top trading partner of the United States?

2. **Spatial Thinking** What might be a reason the United States trades so much with Mexico and Canada?

Tariffs aren't the only barriers to trade. People may want a product so badly that higher prices do not stop them from buying it. In this case, a home country can block trade by using a quota. An import **quota** limits the amount of a particular good that enters the country.

Domestic firms can also ask for a subsidy. A **subsidy** is a payment or other benefit given by the government to help a domestic producer. A subsidy helps a producer offset some of its cost of production so it can choose to not raise its prices for the home market. Thus, the home product becomes more competitive with cheaper imported products.

quota limit on the amount of foreign goods imported into a country

subsidy payment or benefit given by the government to help a domestic producer keep prices low

✓ **CHECK FOR UNDERSTANDING**

1. **Explaining** Why do nations sometimes impose tariffs?

2. **Identifying Cause and Effect** How does protectionism harm consumers?

LESSON ACTIVITIES

1. **Argumentative Writing** Suppose you are an adviser to the president. Write a letter to persuade the president to lift or impose trade barriers. Cite specific reasons for your recommendation.

2. **Presenting** You must explain to a small country why it should engage in international trade. Prepare an outline for your speech that answers these questions: What should the country consider when deciding what to trade? How might trade benefit the country?

Exchange Rates and Trade Balances

READING STRATEGY

Analyzing Key Ideas and Details As you read the lesson, complete a graphic organizer like this one to describe the effects of trade deficits.

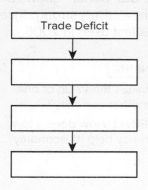

Exchange Rates

How do exchange rates affect international trade?

Nations use their own **currency**—or system of money—to carry out trade with other countries. For example, when American companies trade with firms in Japan, they use the dollar ($). When Japanese companies trade with U.S. companies, they use their currency called the yen (¥). The domestic currencies used to finance foreign trade are called **foreign exchange**, whether they be dollars or yen. Supply and demand set the value of these currencies in relation to each other.

Currencies are bought and sold in foreign exchange markets, just like goods and services in other markets. As a result, the value of a currency, like the value of a good or service, can go up or down. The value of one currency in terms of another is called its **exchange rate**. For example, in the United States, the value of $1 might be ¥100. In Japan, the value of one ¥ is one U.S. penny, or one one-hundredth of a dollar.

currency system of money in general use in a country

foreign exchange the domestic currencies that are used to finance foreign trade

exchange rate the value of a nation's currency in relation to another nation's currency

Currency exchange rates are shown on a digital display board.

Suppose businesses in the United States want to import goods from Japan. Because Japanese firms want to be paid in their own currency, U.S. firms must sell dollars to buy yen. They can then use this yen to buy Japanese products. This action has two results. First, more dollars become available in foreign exchange markets where currencies are bought and sold. Second, the purchase of yen reduces the number of yen available for other countries to buy. The combination of the two pushes the value of the U.S. dollar down and the yen's value up. If this continues for too long, excessive imports can lower the value of the dollar.

The lower value of the dollar might not be completely bad. It will help companies in the United States increase their exports. The reason for this is simple. When a country's currency is less expensive, the prices of its goods are lower. That makes the goods in the United States more attractive to other countries. When companies in other countries buy these relatively cheaper American goods, they will have to buy U.S. dollars. This will push the value of the dollar back up again. And if a country continues to export more than it imports for too long—the value of its currency will rise.

If we want to know the value of the dollar in international markets, we have to look at the exchange rate of the dollar against several countries together. To do this, we look at a statistic called the *International Value of the U.S. Dollar.* This measure shows the dollar's value compared to a broad group of U.S. trading partner currencies. The broad dollar index graph—**International Value of the U.S. Dollar**—shows that the dollar's value has been relatively strong since the statistical measure was first used in 2006.

✓ CHECK FOR UNDERSTANDING

1. **Explaining** What determines the value of one currency in relation to another currency?

2. **Identifying Cause and Effect** How does a low value of a country's currency help its economy?

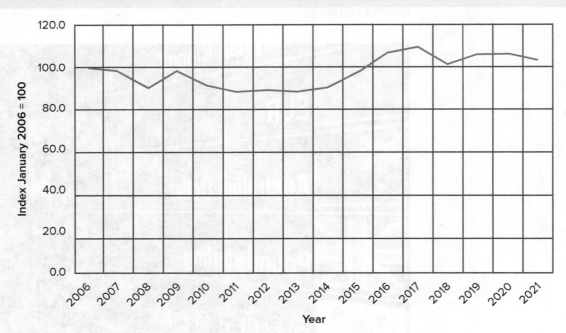

International Value of the U.S. Dollar

The forces of supply and demand help set the international value of the dollar.

EXAMINE THE GRAPH

1. **Analyzing Visuals** According to the graph, in what year was the value of the dollar highest?

2. **Inferring** In which years were U.S. exports least expensive for other countries to buy? What influence did this have on U.S. exports during those years?

Two cargo ships—one representing the United States and the other representing China—symbolize the desire of each country to increase exports to the other.

Balance of Trade

GUIDING QUESTION

How does a nation's trade balance affect its economy?

A country may sell a different amount of goods and services to another country than it buys from that country. The difference between the value of a nation's exports and the value of its imports is the **balance of trade**. That balance can be positive or negative.

Positive Balance of Trade

When the value of a nation's exports is greater than its imports, it has a positive balance of trade. For example, if a country's exports are worth $100 billion and its imports are worth $70 billion, the country has a positive trade balance of $30 billion.

A positive balance of trade is also called a *trade surplus*. A country with a trade surplus for a long time will find that the value of its currency goes up in international currency markets. This will eventually decrease the demand for the country's products because they have become more expensive.

Negative Balance of Trade

When a nation imports more than it exports, it has a negative balance of trade. Suppose a country exports $70 billion in goods and imports $100 billion. If so, it has a negative trade balance of $30 billion. This is called a *trade deficit*. If the country has a trade deficit for too long, it will find that the value of its currency will start to decline.

A large trade deficit with a single country—like the trade deficit the United States has with China—can also cause other problems. For example, it may cause job losses in American industries hurt by foreign competition. This situation usually gains the attention of politicians who often want to do something about it. That is why President Trump launched a trade war against China shortly after taking office. However, a deficit with one country is usually offset by surpluses with other countries. What counts is the overall trade balance, not the trade balance with a single country. This is why the dollar's international value, as shown in the graph **International Value of the U.S. Dollar**, has remained relatively strong despite significant trade deficits with China from 2006 to 2020.

---

balance of trade the difference between the value of a nation's exports and its imports

A Strong Dollar vs. a Weak Dollar

The diagram shows how trade imbalances tend to be self-correcting.

Strong Dollar
Exports are low, and imports are high.

This results in a **Trade Deficit.**

The value of the dollar decreases.

Weak Dollar
Exports increase, and imports decrease.

This results in a **Trade Surplus.**

The value of the dollar increases.

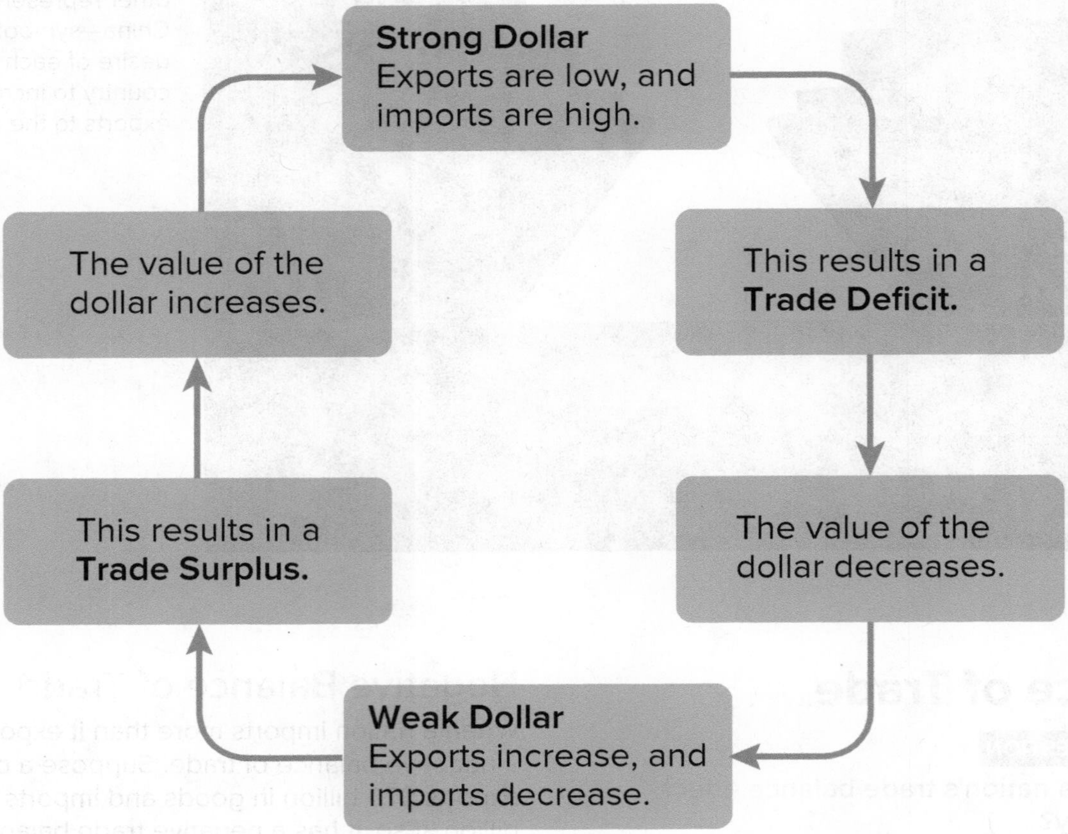

EXAMINE THE DIAGRAM

Analyzing Visuals Which is best—a strong dollar or a weak dollar? Explain.

There was a time when countries worried about their trade balances. The trade balance of the U.S. economy is a modest percentage of its GDP. But the imbalance is not a major concern because trade imbalances tend to be self-correcting. Trade imbalances slowly change the value of a country's currency in international markets, and these changes will usually work to correct the imbalance.

✓ **CHECK FOR UNDERSTANDING**

1. **Explaining** What is the balance of trade, and how does it go from negative to positive?
2. **Identifying Cause and Effect** How does a trade surplus affect the value of a nation's currency?

LESSON ACTIVITIES

1. **Argumentative Writing** You are listening to a political debate. One candidate complains about her opponent's support of a trade agreement that contributed to the weakening of the dollar. Write one or two paragraphs explaining why a weak dollar is or is not a problem.

2. **Using Multimedia** A group of entrepreneurs have asked you to explain how the international value of the dollar affects their businesses. Working with a partner, create a short multimedia presentation that explains the causes and effects of changes to the international value of the dollar.

Global Trade Alliances and Issues

Analyzing Key Ideas and Details As you read the lesson, complete a graphic organizer to identify the functions or benefits of the trade agreements listed.

Trade Agreement	Functions/Benefits
GATT	
WTO	
EU	
NAFTA/USMCA	

Global and Regional Cooperation

GUIDING QUESTION

Why do nations depend on one another?

Look at the tags on the clothes you wear. Examine the labels on your headphones, laptop, stuffed animals, staplers, and any other products on your desk or in your room. You will find that workers in other countries made most of your possessions. Worldwide transportation, communication networks, and technology have supported international trade, which has made it possible for us to have an incredible variety of goods and services. International trade has also made for more economic interdependence than ever before. **Economic interdependence** means that we rely on others—and others rely on us—to provide for many of our wants and needs.

Economic interdependence is not an economic theory. It is a fact of life. Economic interdependence is due to the extensive global and regional cooperation that characterizes international trade. How did this come about? What are the implications? And is it possible that we have gone too far? These are important questions, but it helps if we first understand how we got to where we are today. It also helps to remember that nations trade for the same reason people trade: the value of the things we get in trade is more valuable to us than what we give up to get them.

economic interdependence the reliance of people and countries around the world on one another for goods and services

Bar codes and clothing tags specify the country where a product was made.

Making Connections What goods does your family regularly use that were produced in other countries? When you purchase one of these products, who profits from it?

GATT

Nations have a long history of trading with each other. But two world wars and the Great Depression of the 1930s severely damaged international trade. After World War II was over, countries faced two critical trade problems. First, the war had significantly disrupted trade between nations. Second, many tariffs that nearly doubled the prices of goods during the Great Depression were still in place.

Along with a small number of nations, the United States created the General Agreement on Tariffs and Trade (GATT). Founded in 1947, the GATT was a multilateral trade agreement designed to reduce tariffs—or taxes on traded goods—among its member nations. In a **multilateral agreement,** all parties agree to do the same thing at about the same time. For example, all nations might agree to reduce their tariffs on steel by 15 percent by a specific date.

The reduced tariffs worked, and more international trade took place. This encouraged the GATT countries to make more tariff reductions, which in turn led to more trade. Other countries also joined the GATT. As of the early 1990s, 120 countries had become members, with thousands of multilateral tariff reduction agreements between them.

Despite this success, the GATT had no way to ensure that all countries honored the tariff agreements. This led to the founding of the World Trade Organization (WTO) in 1995.

World Trade Organization (WTO)

The WTO functions like a world court where member countries can take complaints about tariff violations. The WTO brings countries together in front of a panel of nations to present their arguments. After countries have had a chance to register their complaints, they usually agree to settle their trade disputes. If they don't, the WTO will recommend a solution. If a country fails to follow final WTO recommendations after a dispute, its trading partners can seek compensation or impose their own trade sanctions. The WTO has no power to enforce its solution, however.

Today, with a membership of 164 countries, the WTO is the only international organization that tries to administer the global rules of trade. It oversees trade agreements, tries to settle trade disputes among its member nations, and it also helps developing countries negotiate successful trade agreements.

European Union (EU)

While GATT was working to reduce tariffs, some nations tried to reduce their tariff barriers even more by forming trade blocs. A **trade bloc** is a group of countries, usually located in a particular region, that cooperate to expand free trade.

---

multilateral agreement treaty in which multiple nations agree to give each other the same benefits

trade bloc group of countries, usually located in a particular region, that cooperate to expand free trade among one another

The World Trade Organization (WTO) headquarters are located in Geneva, Switzerland. The WTO deals with the global rules of trade.

Martin Good/Shutterstock

European Union

Since 1993, the EU has expanded its membership to include several Eastern European nations that were transitioning from a command economy structure to a market economy. The map shows other countries that want to become part of the EU.

Major Trade Bloc Comparisons	USMCA	EU*	U.S.
GDP (PPP), in billions	$24,893	$19,886	$20,525
GDP per capita	$43,786	$44,436	$62,530
Population, in millions	503.1	450.1	335.0
Labor Force, in millions	215.2	238.9	146.1

*2019 Data, Includes the UK

Source: CIA World Factbook, 2021

GEOGRAPHY CONNECTION

1. **Global Interconnections** Why do you think additional countries want to join the EU?
2. **Exploring Regions** How do the EU and USMCA compare in GDP and GDP per capita?

The European Union (EU) is the most famous trade bloc. The purpose of the EU was to **integrate**, or combine, the economies of its members. Politically, doing so would also help ensure that future wars on European soil would not happen. The effort was a great achievement, and the EU became a major success story. In 1993 the EU became the largest single unified market in the world in terms of population and GDP. As the map **European Union** shows, it continues to add members as certain requirements are met.

Today the EU creates a free-trade zone that covers 27 European countries. Within this regional zone, goods, services, and workers can travel freely across national borders. In addition, more than two-thirds of the EU nations share a common currency called the euro, making trade easier.

integrate combine

The United Kingdom was once a member but left the EU in 2020. Still, as the chart **Major Trade Bloc Comparisons** shows, the European Union rivals the United States-Mexico-Canada Agreement (USMCA), a trading bloc of which the United States is a member.

NAFTA and the USMCA

In 1994 the United States, Canada, and Mexico merged to create the North American Free Trade Agreement (NAFTA). The goal was to remove trade barriers and encourage economic growth among the three countries. Since then, trade among the three nations has more than tripled. This brought lower prices and a greater variety of goods to consumers and producers in all three countries. NAFTA was renamed the United States-Mexico-Canada Agreement (USMCA) in 2020 after President Trump secured changes to NAFTA's dairy, automobile, labor, and copyright laws.

Other nations have also created their own trading blocs. For example, the Andean Community is a trading bloc in South America between Colombia, Ecuador, Peru, and Bolivia. The Association of Southeast Asian Nations (ASEAN) was formed in Southeast Asia by ten nations in 1967 and has expanded recently to include other Asian countries. Despite these efforts, none have been as successful as the EU or the USMCA.

✓ **CHECK FOR UNDERSTANDING**

1. **Summarizing** What is economic interdependence?
2. **Explaining** How do trade blocs help member nations?

Global Issues in Trade

GUIDING QUESTION

What are some consequences of economic interdependence?

Increased specialization and trade make countries wealthier yet more interdependent. International trade also lowers costs for producers and consumers. Consider that many of the products we use are made by **multinationals**.

multinationals corporations that have manufacturing or service operations in several different countries

These are corporations that have manufacturing or service operations in several different countries. Apple, British Petroleum, Dell Technologies, General Motors, Nabisco, Mitsubishi, and Sony are examples of multinational corporations with worldwide economic importance.

Multinationals are important because they can move resources, goods, services, and financial capital across national borders. Multinationals are usually welcome because they bring new technologies and create new jobs in areas where jobs are needed. Multinationals also produce tax revenues for the host country, which helps that nation's economy.

Interdependence brings change, however, and change can be unwelcome or even harmful to certain groups. So who feels the pain, and should countries be concerned if international trade threatens their independence?

Factory workers assemble computers at a Dell factory in India. Dell Technologies is a multinational corporation with operations in 180 countries.

REUTERS/Alamy Stock Photo

Workers in Detroit, Michigan, protest the outsourcing of auto jobs.

Free Trade Can Be Painful

Free trade provides benefits to almost everyone, but there are still problems. Workers—and sometimes even entire industries—are often threatened by free trade. In the United States, free trade has enabled companies to **outsource**, or move factories or business operations to other countries with lower labor costs. Many Americans consider outsourcing a problem because they fear losing their jobs to overseas workers. In the United States, for example, many automotive and steel factories have shut down, increasing unemployment in those industries. This happened because other nations were producing automobiles and steel relatively cheaper than U.S. factories.

Because the overall benefits of trade generally outweigh its costs, government policy can be used to limit the harms done by international trade without necessarily resorting to protectionism. For example, to offset the problems caused at home by outsourcing, the government can pay to retrain workers for other jobs. If a region is hard-hit by unemployment, the government can help people move to more prosperous locations. Retraining programs are expensive, however, and workers are often reluctant to leave the places where they have put down their roots.

There may never be a satisfactory way to resolve all the unemployment caused by international trade. As long as countries continue to specialize according to their comparative advantages, some workers will be displaced. Compensating these workers with other jobs might help heal the pain. Still, the overall positives of free trade—especially lower prices that consumers pay for goods and services—outweigh the negatives.

Loss of Independence

Free trade agreements can also threaten a country's independence. This is the major reason the United Kingdom left the European Union. The United Kingdom was worried that too many immigrant workers would be allowed to come in from other countries. It also worried that it could not control its own economic and trade policies because it had to follow EU laws and policies.

When President Trump was elected, he followed an "America First" policy for similar reasons. He thought too many immigrant workers would enter the country and work for lower wages, potentially displacing some of the American work force. He also thought international trade agreements under NAFTA

outsource to move factories to other countries with lower labor costs

Workers sew jeans at a garment factory in China. Many companies outsource their production to countries where labor costs are low, such as China.

Speculating Why do countries welcome multinationals into their borders?

threatened American independence, so he revised the treaty to make it the USMCA.

Some people believe multinationals also limit the power of governments. For example, some multinationals have been known to abuse their position and pay very low wages to workers or demand lower tax rates from the host country. They may also interfere with traditional ways of life and business customs in the countries where they do business. If government leaders voice disapproval, the multinational may threaten to move its operations to another country.

Yet close economic cooperation between nations also has a major political advantage—it reduces the threat of future military conflicts. Nations economically dependent upon one another will find it less appealing to start a war. For example, a war between any of the current European Union countries—such as France and Germany—is hard to imagine. Likewise, a military conflict between the United States and Canada or the United States and Japan is highly unlikely. The additional economic stability helps offset the threat to a nation's independence.

✓ **CHECK FOR UNDERSTANDING**

1. **Explaining** What are some reasons people and groups object to economic interdependence?

2. **Describing** What are the advantages and disadvantages of multinational corporations?

LESSON ACTIVITIES

1. **Informative/Explanatory Writing** Consider the products and services you use in your daily life. Identify six items that are the result of trade with other nations. Then write a paragraph explaining how you benefit from economic interdependence and why international trade is important to you.

2. **Using Multimedia** With a partner, research one of the trade blocs presented in this lesson, such as the EU, Andean Community, or ASEAN. Or you might research a trade bloc not mentioned, such as the Common Market for Eastern and Southern Africa (COMESA) or the Organization of Petroleum Exporting Countries (OPEC). Then create a multimedia presentation that displays a short history of the trade bloc; the countries involved; what the bloc does for its members; agencies or other institutions that make decisions; and the output the bloc produces.

The Wealth of Nations

Analyzing Key Ideas and Details As you read the lesson, complete a graphic organizer identifying some of the obstacles developing countries face as they try to improve their economies.

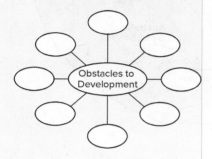

High-Income, Low-Income Countries

GUIDING QUESTION

How do we know that some countries are wealthier than others?

Are some countries wealthier than others? Does the type of economic system matter? These types of questions assess a country's standard of living. **Standard of living** is the quality of life measured by such things as having plentiful goods and services, high per capita incomes, and good health care. The countries with these features are often called **developed countries**. Their characteristics include:

- Strong industry due to investment in machines and factories
- Large service sector that includes shops and restaurants
- Strong *infrastructure*, such as highways, power, and sewers
- Public goods such as libraries, parks, and museums
- Private ownership of the factors of production
- Good institutions, such as courts and hospitals
- Low population growth rate

Countries without these characteristics are called **developing countries**. Their standard of living is lower than developed countries.

Income levels are often used to classify countries as "developed" or "developing." As the chart **Country Classification Levels** shows,

standard of living the material well-being of an individual or nation
developed country industrialized country with a high standard of living
developing country nonindustrial country with a low standard of living

Country Classification Levels

The chart classifies countries into two "Developed" categories and two "Developing" categories.

Country Classification	2021 Per Capita Income Levels	% of Global Population	% of Global GDP
Developed: High Income	$12,536 or more	17%	49%
Developed: Upper-middle Income	$4,046–$12,535	36%	34%
Developing: Lower-middle Income	$1,036–$4,045	40%	16%
Developing: Low Income	$1,035 or less	8%	Less than 1%

Source: World Bank, 2021

Analyzing Visuals Which category makes up the largest percentage of the global population? What percentage of GDP does it produce?

GDP Per Capita—Selected Countries

The map shows GDP per capita for a few nations in the world.

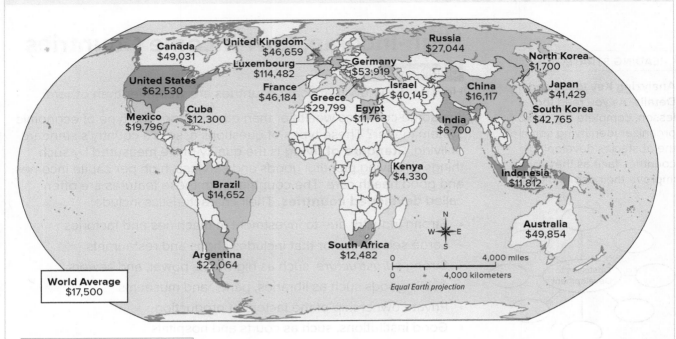

GEOGRAPHY CONNECTION

1. **Global Interconnections** Of the countries shown, where does the United States rank in GDP per capita?

2. **Patterns and Movement** Which country—North Korea or South Korea—do you think has a market economy? Why?

high-income countries account for about 17 percent of the global population. But they produce nearly half of the total world GDP.

Gross Domestic Product (GDP) per capita is one of the best indicators of living standards. Recall that GDP is the total market value of all final products produced within a country's borders during a single year. And *per capita* means "for each person." *GDP per capita* is calculated by dividing the country's GDP by its population. The result is the amount of output on a per-person basis. As the map **GDP Per Capita—Selected Countries** shows, per capita incomes range from $114,482 per person in Luxembourg, a tiny country with rich royalty, to North Korea, with an estimated $1,700 per person income. Other countries' per capita incomes may be higher or lower than the 23 countries shown.

So the answer to the earlier question is "yes," some countries *are* wealthier than others. And the gap between the wealthier and the poorer countries is getting wider every year. Why is this

happening, and can anything be done about it? Although many factors are involved, a country's economic system is one of the most important characteristics of high-income countries.

✓ **CHECK FOR UNDERSTANDING**

1. **Contrasting** What distinguishes developed countries from developing countries?

2. **Identifying** Which economic measurement is the best indicator of a country's standard of living?

Transitioning to a Market Economy

GUIDING QUESTION

What evidence shows that a market economy leads to the highest per capita incomes?

One way to see if market economies lead to the highest per capita incomes is to look at the alternatives—command economies. Command economies are not very efficient. People often face shortages of goods and services and products of low quality. When shortages occur, leaders usually provide for themselves first. The result is that government leaders have plentiful food, nice cars, and good houses, while average citizens often go without. In the past several decades, several command economies have begun transitioning to a market-based economy.

Transitioning Economies

Command economies have never performed as well as market/capitalist economies. As the map **GDP Per Capita—Selected Countries** shows, Cuba and North Korea—both of which have command economies—have some of the world's lowest GDPs per capita. Russia's GDP per capita was low when the Soviet Union collapsed around 1990. Russia's progress since then is partially due to its efforts to adopt some features of a market economy.

Russia's attempt to change its command economy to a market economy proved to be especially difficult. The biggest hurdle was privatization. **Privatization** is the process of changing state-owned businesses, factories, and farms into ones owned by private citizens. To do this, government-owned factories were converted into corporations with millions of shares of common stock. These shares were given directly to people or used as payment for factory wages. If done correctly, factory ownership would have been **transferred** to private citizens. But many of these ownership shares ended up in bureaucrats' pockets. This left the ownership of many important industries in the hands of a few high government officials.

Another difficulty was that people had to learn how to make decisions in an economy that had markets and prices. In the former Soviet command economy, people were told when and where to work. Wages were minimal but about equal, regardless of the work performed. The government provided some things for free, such as transportation, education, medicines, and some foods. People were able to have some of their needs satisfied even if they did not work, or work very hard. When converting to a market economy, however, people had to learn to work hard if they wanted to provide for their food, shelter, and clothing. This transition was difficult for many people to make.

Hungary and other former Soviet countries had some experience with black markets, which helped acquaint them with market economies. A **black market** is a market where illegal goods are bought and sold. This doesn't mean the goods are necessarily bad. Blue jeans were

---

privatization process of changing state-owned businesses, factories, and farms to ones owned by private citizens
transfer to shift or reassign
black market market where illegal goods are bought and sold

Russian women purchase shoes at a black market in the Soviet Union in the late 1980s.

Comparing Economies

The United States has long enjoyed the benefits of having a market-oriented economy. Russia and China are making the transition. North Korea lags far behind with its command economy.

	North Korea	China	Russia	United States
GDP per capita	$1,700	$16,117	$27,044	$62,530
Real GDP	$40 billion	$22.5 trillion	$3.9 trillion	$20.5 trillion
Population	25,831,360	1,397,897,720	142,320,720	334,998,398
Labor Force (total)	14,000,000	774,710,000	69,923,000	146,128,000
Labor Force % in Agriculture % in Industry % in Services	37.0% 14.0% NA	27.7% 28.8% 43.5%	9.4% 27.6% 63.0%	0.7% 20.3% 79.0%
Exports	$222 million	$2.5 trillion	$551 billion	$2.3 trillion
Imports	$2.3 billion	$2.1 trillion	$367 billion	$3.2 trillion

Source: CIA World Factbook, 2021

EXAMINE THE CHART

1. **Comparing** In what area does China still lag well behind the United States?
2. **Analyzing Visuals** In what area of economic performance has China surpassed the success of the United States?

popular black-market goods in the Soviet Union because the government thought they were unnecessary and did not produce them.

Few of the former Soviet countries ever changed completely to market economies. Many kept some of their previous command features. This left them with a mixture of command, socialism, and market economies—known as a mixed economy. Today, Russia has a mixed economy with strong elements of a command economy, such as friends of high government officials owning significant properties and industries. At best, Russia and some of its former satellite countries *tried* to transition to capitalism, but stalled along the way.

China has been more successful at transitioning to a market economy. Starting in the late 1970s, China began adopting some features of a market economy. The result has been a dramatic rise in GDP and GDP per capita. China now has the world's largest economy when measured in terms of total real GDP produced. It also helps that it is one of the largest countries, which gives it one of the largest potential markets.

Yet some industries and markets are controlled by the government, and only one party—the Communist Party—controls the government. The government has been successful in propelling economic growth, but that success is not shared by most of the people. Whether entrepreneurs can flourish in a centrally controlled economy is yet to be seen.

Cultural Obstacles to Development

Command economies with few market structures are not the only obstacle to development. Cultural obstacles also hinder development. For example, some cultures place high value in having large families. This can lead to a high rate of population growth. When the population grows faster than GDP, GDP per capita declines. The result is that each person has a smaller share of what the economy produces. Therefore, countries with the highest rates of population growth often tend to have the lowest GDP per capita.

Countries with high rates of population growth face another difficulty. As more people are added to the population, more jobs need to be created for workers. A growing population will also need expanded services, such as transportation, health care, and education. Additionally, a high rate of population growth itself can result in poverty, or a rise in poverty.

Some cultures also hinder the ability of women to contribute to the economy. In Saudi Arabia and much of the Arab world, women are restricted to few occupations. Female education is limited. These women have few to no chances to help their country's economy to develop.

Political Obstacles to Development

War is a huge obstacle in some developing countries. Fighting claims lives and forces civilians to move away to safer areas. It also damages a nation's resources. Productivity slows and people face food, health care, and education shortages. As a result, warring countries have more difficulty investing in their economies.

Political corruption slows growth in many developing countries as well. Some leaders steal money that was meant to pay for economic development or other projects to help their people. Corrupt leaders base their economic decisions not on what is best for their country but on what is best for themselves. Some corruption often extends down to minor government officials who demand bribes for the smallest of favors.

Environmental Obstacles to Development

Diseases have crippled some developing countries. Malaria and dengue fever are mosquito-borne diseases found in tropical climates. Other diseases such as AIDS and COVID-19 have also been major problems in developing countries. These diseases affect the working population and can reduce the size of

the labor force. All too frequently, young children and elderly parents are left behind to fend for themselves.

Even geography can be a problem for economic growth. Land-locked countries with no access to the sea may have difficulty getting goods to and from other countries. Many developing nations often lack the means to extract, use, and sell their resources.

Economic Obstacles to Development

Many developing countries have barriers to international trade. They do this because they want to protect domestic jobs and young industries. Trade barriers usually protect industries that are not efficient, however. Most economists think these barriers actually hold back the countries' ability to build their economies.

Developed countries have diversified economies, whereas less-developed countries may have single-resource economies. In a single-resource economy, a nation depends on a single export product for its economic growth. A failure of a crop, a decrease in the price of oil, or a similar problem with the one resource a country exports makes it difficult for a developing country to experience progress. In addition, focusing on only a single resource may lead to environmental problems, such as over-using the soil or polluting the water sources near mines.

Finally, many developing countries also face the problem of severe debt to developed nations.

A street in Syria lies in ruins after being bombed. In addition to the human toll, war destroys infrastructure necessary for development.

These students in India have benefited from the country's higher standard of living. India has overcome many obstacles to development by lowering trade barriers and investing in technology and infrastructure. As the industrial and service sectors have grown, the standard of living has improved as well.

Explaining How do trade barriers hurt development?

Many once borrowed large sums of money from wealthy nations to encourage economic growth. Much of that money was siphoned off by corruption, which slowed investment in capital resources and other development projects. The result was that economic growth was not fast enough to pay off those debts. Now the countries must use too much of their national income to pay off their debt.

Any one of these problems would be difficult for a nation to solve. Unfortunately, many developing nations have to deal with two or three of these problems at the same time. As a result, satisfactory economic progress has been difficult for them to achieve and the gap between the wealthy and the poorer countries gets wider every year.

The Wealth of Nations

In 1776, Adam Smith wrote that the wealth of a nation was its *people*, not its stock of gold and silver. The same thing is true today—but people also need an economic system that lets them grow and maximize their abilities. Economies with free markets generally do this successfully. The proof is that they have some of the highest per capita incomes in the world. These income levels are possible because free markets give people the opportunity to make decisions that are best for them, and ultimately best for the economy. Market economies are not perfect, but they are the best way to organize the billions of minor economic decisions that people make daily—decisions that ultimately determine our standard of living.

✓ **CHECK FOR UNDERSTANDING**

1. **Determining Central Ideas** What is the first step that a command economy must do before it can successfully become a market economy?

2. **Identifying Cause and Effect** In what two ways do high population growth rates hurt developing countries?

LESSON ACTIVITIES

1. **Informative/Explanatory Writing** Suppose you have friends who live in a country that is transitioning from a command economy to a market economy. Write a letter explaining to your friends how they need to adjust their thinking and work ethic in the new market economy. Include information about learning to make decisions on their own, taking initiative, interpreting prices, and fending for themselves in free markets.

2. **Presenting** With a partner, choose two developed countries and two developing countries. Find data on their GDP per capita, rate of population growth, type of government, type of economic system, and level of debt. Identify whether they are single-resource or diversified economies. You can find the data at the website for the CIA World Factbook, or you can consult another equally reliable source, such as the World Bank. Make a table to display your data. For the poorest-performing country you research, identify the obstacles to its development. Share your table and findings in a presentation to the class.

06

Analyzing Sources: Environmental Balance

? COMPELLING QUESTION

How do we balance the needs of consumers and the needs of a sustainable planet?

Plan Your Inquiry

DEVELOPING QUESTIONS

Read the Compelling Question for this lesson. What questions can you ask to help you answer this Compelling Question? Create a graphic organizer like the one below. Write three Supporting Questions in the first column.

Supporting Questions	Source	What this source tells me about environmental balance	Questions the source leaves unanswered
	A		
	B		
	C		
	D		
	E		

ANALYZING SOURCES

Next, read the introductory information for each source in this lesson. Then analyze each source by answering the questions that follow it. How does each source help you answer each Supporting Question you created? What questions do you still have? Write those in your graphic organizer.

After you analyze the sources, you will:
- use the evidence from the sources
- communicate your conclusions
- take informed action

Background Information

In recent years, people have become aware of dangers to the world's environment. Chemicals released by factories and vehicles pollute the air and water. *Deforestation*—the mass removal of trees—causes flooding, leads to mud slides, and lessens the amount of carbon dioxide that trees absorb. Plastic waste overflows landfills and finds its way into the oceans.

Some people have turned to conservation to reduce environmental damage. Conservation means carefully using resources and limiting the harmful effects of human activity. Points of view about conservation differ. Some people think conserving natural resources is less important than economic growth. They argue that limiting the ways that businesses operate drives up costs. Others claim that not conserving resources today will lead to greater future economic and environmental costs.

This Inquiry Lesson includes sources about finding that environmental balance. As you read and analyze the sources, think about the Compelling Question.

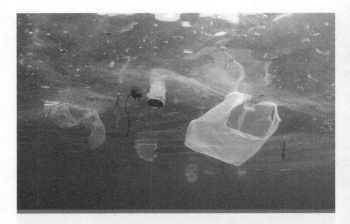

Trash litters the ocean. Modern life offers many comforts, but they can come at a cost to the environment.

Priority: Economic Growth?

Solving environmental problems is not easy. The process of switching to cleaner sources of energy can be costly. Protecting forests may deprive farmers of land they need to grow crops. Many developing nations fear that taking steps to curb pollution will slow their economic growth—growth they need in order to improve the lives of their people.

PRIMARY SOURCE: ONLINE ARTICLE

❝ Developing countries currently cannot sustain themselves, let alone grow, without relying heavily on **fossil fuels**. Global warming typically takes a back seat to feeding, housing, and employing these countries' citizens. . . .

Fossil fuels are still the cheapest, most reliable energy resources available. When a developing country wants to build a functional economic system and end **rampant** poverty, it turns to fossil fuels.

India hopes to transition to renewable energy as its economy grows, but the investment needed to meet its renewable energy goals "is equivalent to . . . over ten times the country's annual spending on health and education."

Unless something changes, developing countries like India cannot fight climate change *and* provide for their citizens. ❞

— From "Developing Countries Can't Afford Climate Change" by Tucker Davey, futureoflife.org, 2016.

---

fossil fuels fuels such as coal, gas, or petroleum that were formed long ago by the remains of living organisms

rampant widespread

EXAMINE THE SOURCE

1. **Identifying** Why do many developing countries rely on fossil fuels?
2. **Explaining** Why has India not transitioned to renewable energy sources?

B

Gaming and Energy

Conserving energy is one way to balance the needs of consumers and the planet. Here, an energy company offers tips on how gamers can help conserve energy.

Two teens in Texas play video games. One person is having more fun than the other, but both can help conserve energy.

PRIMARY SOURCE: ONLINE ARTICLE

❝ **Turn controllers off when not in use** Automatic shutdown functions can save power, but not all gaming consoles have them. Be proactive and turn off your console and its controllers when you're not using them to cut down on energy waste.

Unplug when you can Like many advanced electronics, gaming consoles can draw power even when turned off, which adds up over time. Unplugging them when not in use cuts game console energy consumption that's pure waste.

Avoid streaming on consoles The energy costs of streaming on game console systems can be considerable. Game consoles are not optimized for streaming like a streaming device such as an Apple TV or Roku. ❞

— From "How Much Energy Do Game Consoles Really Use?" by blog.constellation.com.

EXAMINE THE SOURCE

1. **Making Connections** Are you a gamer? Did you find these suggestions helpful? Explain.
2. **Speculating** In addition to gaming, what other home functions do you think require large amounts of energy?

Pollution Tax

Economic incentives can help solve the global problem of pollution. Most economists argue that the best way to attack the problem is to attack the incentives that caused pollution in the first place.

SECONDARY SOURCE: TEXTBOOK

66 Pollution does not occur on its own: it occurs because people and firms have an incentive to pollute. If that incentive can be removed, pollution will be reduced. For example, factories historically located along the banks of rivers so they could discharge their **refuse** into the moving waters. Factories that generated smoke and other air pollutants often were located farther from the water with tall smokestacks to send the pollutants long distances. Others tried to avoid the problem by digging pits on their property to bury their toxic wastes. In all three situations, factory owners were trying to lower production costs by using the environment as a giant waste-disposal system. . . .

A market-based approach [to reduce pollution] is to tax or charge firms in proportion to the amount of pollutants they release. Depending on the industry, the size of the tax would depend on the severity of the pollution and the quantity of toxic substances being released. A firm can then either pay the fees or take steps to reduce the pollution. . . . As long as it is cheaper to clean up the pollution than to pay the tax, individual firms will have the incentive to clean up and stop polluting.

An expanded version of pollution fees is the **EPA's** use of pollution permits—federal permits allowing public utilities to release specific amounts of **emissions** into the air—to reduce sulfur dioxide emissions at coal-burning electric utilities that contribute to the problem of **acid rain**.

Under this program, the EPA awards a limited number of permits to all utilities. . . . If the level of pollutants is still too high, the EPA can distribute fewer permits. A smaller number of permits will make each one worth more than before, which will again cause firms to redouble their efforts to reduce pollution. In the end, the market forces of supply and demand will provide the encouragement to reduce pollution. 99

— From *Principles of Economics* by Gary Clayton, McGraw Hill, 2024.

---

refuse waste

EPA Environmental Protection Agency

emissions polluting gases

acid rain pollution in the form of rainwater mixed with sulfur dioxide to form a mild form of sulfuric acid

---

EXAMINE THE SOURCE

1. **Identifying Cause and Effect** Why do companies have an incentive to pollute? How can governments counteract this incentive?
2. **Making Connections** Why did the author call pollution fees a "market-based approach"?

Clayton, Gary E. Principles of Economics. Bothell, WA: McGraw Hill LLC, 2019.

Changing Views

The perception of energy use and economic progress has changed over time, as shown in this cartoon.

PRIMARY SOURCE: POLITICAL CARTOON

EXAMINE THE SOURCE

1. **Analyzing Visuals** The cartoonist is trying to convey a shift in the national conversation about the use of coal. In the first panel, why does the factory owner encourage the farmer to use coal?

2. **Speculating** Why have their positions reversed in the second panel?

Consuming Less

Another way to balance the needs of the planet is to use less.

PRIMARY SOURCE: PHOTOGRAPH

Some 140 million tons of waste are emptied into U.S. landfills like this one every year.

EXAMINE THE SOURCE

Predicting What environmental problems might arise from too much trash?

Complete Your Inquiry

EVALUATE SOURCES AND USE EVIDENCE

Refer back to the Compelling Question and the Supporting Questions you developed at the beginning of the lesson.

1. **Identifying** Which of the sources affected you the most? Why?

2. **Evaluating** Overall, what impression do these sources give you about the importance of balancing environmental and human needs? Explain.

3. **Gathering Sources** Which sources helped you answer the Supporting Questions and Compelling Question? Which sources, if any, challenged what you thought you knew when you first created your Supporting Questions? What information do you still need in order to answer your questions? Where would you find that information?

4. **Evaluating Sources** Identify the sources that helped answer your Supporting Questions. How reliable is the source? How would you verify the reliability of the source?

COMMUNICATE CONCLUSIONS

5. **Collaborating** With a partner, conduct research about a renewable form of energy, including its advantages and disadvantages. Then debate whether you think this energy form will ever be economically viable. Each of you should give specific reasons for your opinions.

TAKE INFORMED ACTION

Writing an Editorial About Conservation Governments, businesses, and people have different ideas on conservation, reuse, and/or recycling. Think about the views of each of these groups. Write a letter to the editor focused on one audience in which you take a position on one issue. Explain what conservation, reuse, and/or recycling efforts you think are important or are unnecessary, and why. Explain why you recommend those steps.

Design Pics/Leah Warkentin

07

Reviewing The Global Economy

Summary

Why Nations Trade

Nations trade because of comparative advantage. Nations *import* goods produced in other countries. They *export* goods they produce. To protect home industries that lose sales to lower-priced imported goods, governments use trade barriers such as *tariffs*, *quotas*, or *subsidies*.

Trade Balances

The difference between the value of a nation's exports and the value of its imports is the *balance of trade*. That balance can be positive (trade surplus) or negative (trade deficit). Trade imbalances tend to be self-correcting over time. Currencies used to finance world trade are called *foreign exchange*.

Economic Interdependence

International trade has made for more *economic interdependence* than ever before. We rely on others—and others rely on us—to provide for our wants and needs.

Trade Alliances and Issues

International organizations such as the WTO oversee trade agreements and help settle trade disputes among member nations. Many countries form regional *trade blocs* to lower trade barriers and increase trade. Free trade can be painful for those who lose their jobs due to *outsourcing*, but overall, the net benefits of trade are positive even if countries lose some political independence.

The Wealth of Nations

Standard of living is the quality of life measured by plentiful goods and services and high per capita incomes. Countries with these features are *developed countries*. Countries without these features are *developing countries*. Many face obstacles such as high population growth rates, single-resource economies, low education and poor health care of citizens, corruption, and trade barriers.

Travel mania/Shutterstock

Checking For Understanding

Answer the questions to see if you understood the topic content.

REVIEWING KEY TERMS

1. Define each of these terms:

   A. import
   B. export
   C. comparative advantage
   D. protectionism
   E. tariff
   F. exchange rate
   G. balance of trade
   H. trade bloc
   I. outsource
   J. developing country

REVIEWING KEY FACTS

2. **Identifying Cause and Effect** Why do nations trade with one another?

3. **Explaining** What policies do governments use to restrict international trade and protect home industries?

4. **Identifying** When U.S. citizens travel to Japan, what type of currency will they receive in exchange for their U.S. dollars?

5. **Identifying** What does the exchange rate measure?

6. **Interpreting** What does it mean when a country has a positive balance of trade?

7. **Explaining** How do ongoing trade deficits hurt a country's economy?

8. **Identifying** What is the European Union?

9. **Contrasting** What is the difference between a developed country and a developing country?

10. **Explaining** What are black markets, and are they bad?

11. **Explaining** What happens when a developing country's population grows faster than the GDP?

12. **Contrasting** What challenges does a single-resource economy face that a diversified economy does not?

CRITICAL THINKING

13. **Inferring** How are comparative advantage and specialization related?

14. **Identifying Cause and Effect** What effect do trade barriers have on global interdependence?

15. **Drawing Conclusions** How does a country's ongoing balance of trade affect the value of its currency?

16. **Evaluating** Is a weak exchange rate for the U.S. dollar good or bad for the country?

17. **Summarizing** What roles do international trade organizations play in trade?

18. **Drawing Conclusions** Why do multinationals increase economic interdependence?

19. **Explaining** What is privatization and why have some governments used it?

20. **Summarizing** What are some obstacles to development that developing countries face?

NEED EXTRA HELP?

If You've Missed Question	1	2	3	4	5	6	7	8	9	10
Review Lesson	2, 3, 4, 5	2	2	3	3	3	3	4	5	5

If You've Missed Question	11	12	13	14	15	16	17	18	19	20
Review Lesson	5	5	2	2	3	3	4	4	5	5

Apply What You Have Learned

A Understanding Multiple Perspectives

Should companies outsource their work force?

ACTIVITY Using Sources to Write About Outsourcing Read these excerpts that provide two opinions on outsourcing American jobs. Then write a brief essay answering these questions: What are some benefits of outsourcing? What are some drawbacks? What evidence from the lessons can you provide to support the two views?

66 The phenomenon of foreign outsourcing creates tangible benefits for the U.S. economy and American workers. Whatever negative impact it has had on specific firms and workers has been limited and is far outweighed by the benefits. 99

— Daniel Griswold, director of the Center for Trade Policy Studies at the Cato Institute.

66 One: How many more jobs must we lose before they become concerned about our middle class and our strength as a consumer market? Two: When will the U.S. have to quit borrowing foreign capital to buy foreign goods that support European and Asian economies while driving us deeper into debt? Three: What jobs will our currently 15 million unemployed workers fill, where and when? 99

— Lou Dobbs, former anchor and managing editor of Lou Dobbs Tonight, CNN.

B Writing an Argument

The gains from trade help an economy to grow. The growth comes from three sources: more overseas customers for a country's exported goods and services, more resources to import that allow for more home production, and lower prices of goods and services that can increase savings for consumers. Yet while free markets and international trade can bring benefits, some people still object to trade when it harms home industries and their workers.

ACTIVITY Writing About a Trade Policy for Athletic Shoes Suppose you are in charge of trade policy for the United States. Would you recommend the country increase or decrease trade barriers on athletic shoes? Write a report making your recommendation and explaining why you want to increase or decrease specific trade barriers.

Understanding Economic Concepts

Recall that comparative advantage is the ability to produce something relatively more efficiently, or at a relatively lower opportunity cost, than another country can.

ACTIVITY **Creating a Chart Showing Comparative Advantage** Suppose that you and a partner are starting a lawn-mowing business. You will each contribute an equal amount of money to buy a mower, a trimmer, gas, and other materials. Work with your partner to list the tasks associated with your business in a table like the one shown here. Keep in mind that some of these tasks, such as asking a sibling or parent to drive you and the mower to the lawn site, will not be related to physically mowing. Revise your list after an extended time frame to use what you have learned about comparative advantage to divide the tasks.

Task	Time It Took	Opportunity Cost of Doing the Task

 Making Connections

Consider the following questions: Why is the economic health of all nations important in a global economy? In addition to the obvious economic benefits to companies and consumers, what social benefits arise from a healthy economy?

ACTIVITY **Being an Active Citizen** Identify and research one developing country that has used specialization to boost its economy. Prepare a short speech that explains the country's choices and describes the impact these choices have had on the country's GDP, employment, and wages. Also describe the impact that economic growth has had on the country's political and social stability. Deliver your speech to your classmates. As you speak, make eye contact with your audience and speak clearly.

APPENDIX

World Religions Handbook   **A3**

Reference Atlas   **A25**

Glossaries

Geography Glossary   A74

Economics Glossary   A89

Indexes

Geography Index   A100

Economics Index   A110

World Religions Handbook

People across the world use different religions and spiritual traditions to explain the meaning of life.

Introduction	A4
Buddhism	A6
Christianity	A8
Confucianism	A10
Hinduism	A12
Islam	A14
Judaism	A16
Sikhism	A18
Indigenous Religions	A20
Assessment	A23

Introduction

A religion is a set of beliefs that helps people explain their lives and the world around them. In many cultures, religious beliefs help people answer basic questions about life's meaning. People use religion to help them lead a meaningful life and to explain the mysteries of life, such as how and why the world was created. Religions explain what happens to people when they die or why there is suffering. Religion is an important part of culture. Most religions have their own sacred texts, symbols, and sites. These beliefs and practices unite followers wherever they live in the world.

The religions described in this handbook have sacred elements, celebrations, and worship styles. We can gain a better understanding of these religions by examining what sets them apart.

Percentage of World Population

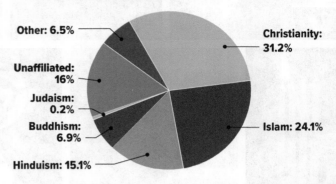

Other: 6.5%
Unaffiliated: 16%
Judaism: 0.2%
Buddhism: 6.9%
Hinduism: 15.1%
Christianity: 31.2%
Islam: 24.1%

Source: The CIA World Factbook, (2015 estimates)

World Religions Today

This map shows, on a large scale, where many people of each religion live. This map does not show the diversity of the religions practiced in each region throughout the world. In every region shown on the map, there are people who practice different religions than those shown.

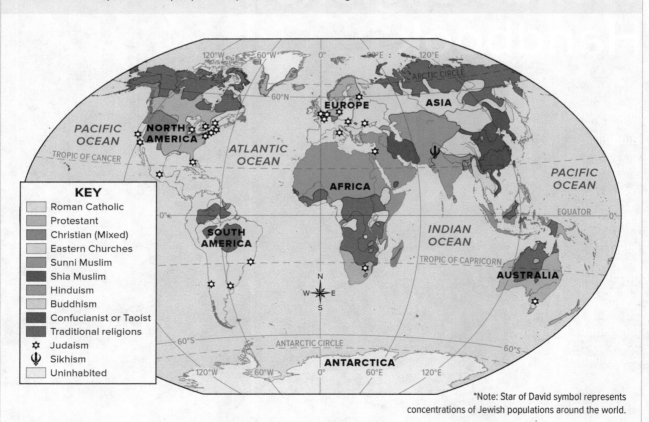

KEY
- Roman Catholic
- Protestant
- Christian (Mixed)
- Eastern Churches
- Sunni Muslim
- Shia Muslim
- Hinduism
- Buddhism
- Confucianist or Taoist
- Traditional religions
- ✡ Judaism
- ☬ Sikhism
- Uninhabited

*Note: Star of David symbol represents concentrations of Jewish populations around the world.

Early Diffusion of Major World Religions

The diffusion, or spreading, of religion throughout the world has been caused by a variety of factors including migration, missionary work, trade, and war.

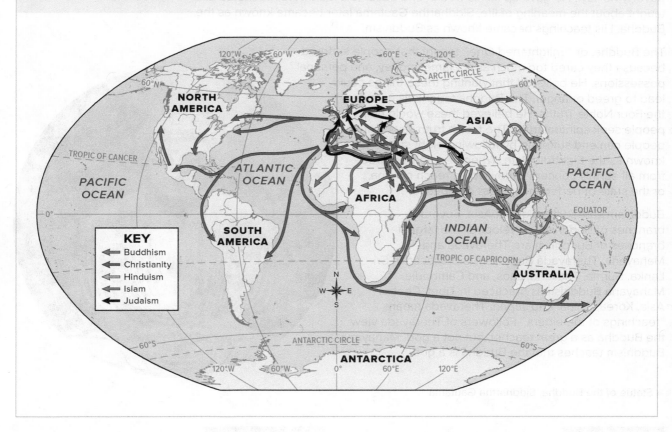

KEY
- Buddhism
- Christianity
- Hinduism
- Islam
- Judaism

We study religion because it is an important part of culture. Religion can determine how people interact with one another. Religion can also influence what people wear and eat.

- Buddhism, Christianity, and Islam are the three major religions that spread their religion through missionary activities.

- Religions such as Hinduism, Sikhism, and Judaism are associated with a particular culture group. Followers are usually born into these religions.

- Religion has spread throughout the world because of trade and war. As people trade, they interact with each other, sharing their beliefs. Conquerors bring their cultures, including their religions, to conquered areas.

animism belief that spirits inhabit natural objects and forces of nature

atheism disbelief in the existence of any god

monotheism belief in one God

polytheism belief in more than one god

secularism belief that life's questions can be answered apart from religious belief

sect a subdivision within a religion that has its own distinctive beliefs and/or practices

tenet a belief, doctrine, or principle believed to be true and held in common by members of a group

Buddhism

Buddhism was founded about 2,500 years ago by Siddhartha Gautama. He grew up as a prince near the Himalaya. Today, this area is in southern Nepal. One day, Siddhartha Gautama left his palace to explore the world. As he traveled, he was shocked at the poverty he saw. He gave up all his wealth and became a monk. He meditated and taught about the meaning of life. Siddhartha Gautama later became known as the Buddha. His teachings became known as Buddhism.

The Buddha, or "Enlightened One," taught that people suffered because they cared too much about fame, money, and personal possessions. He believed that wanting these things could lead to greed or anger. The Buddha taught his followers the Four Noble Truths. He believed these would help people seek spiritual truth. The fourth truth says that people can end suffering by following eight steps, known as the Eightfold Path. When people were free from all earthly concerns, they would reach nirvana, or the state of perfect happiness and peace.

Buddhism spread throughout Asia. Several branches of Buddhism developed. The largest branches of Buddhism are Theravada and Mahayana. Theravada Buddhism is practiced in Sri Lanka, Burma, Thailand, Laos, and Cambodia. Mahayana Buddhism is practiced in Tibet, Central Asia, Korea, China, and Japan. Theravada means "teachings of the elders." Followers of Theravada view the Buddha as a great teacher, but not a god. Mahayana Buddhism teaches that the Buddha is a god.

» Statue of the Buddha, Siddhartha Gautama

SACRED TEXT

For Theravada Buddhists, the sacred collection of Buddhist texts is the Tripitaka ("three baskets"). This excerpt from the *Dhammapada*, a famous text within the Tripitaka, urges responding to hatred with love:

> 66 *For hatred does not cease by hatred at any time: hatred ceases by love, this is an old rule.* 99
>
> —*Dhammapada* 1.5

SACRED SYMBOL

The *dharmachakra* ("wheel of the law") is an important Buddhist symbol. The eight spokes represent the Eightfold Path—right view, right intention, right speech, right action, right livelihood, right effort, right mindfulness, and right concentration.

PHOTO:(t)Serg Zastavkin/Shutterstock.com; (b)Armands/Pharyos/Alamy Stock Photo; TEXT:"The Dhammapada." Translated 1881 by F. Max Muller. In The Sacred Books of the East. Vol. X. Translated by Various Oriental Scholars and Edited by F. Max Muller. Delhi: Motilal Banarsidass, Reprinted 1988.

SACRED SITE

Buddhists believe that Siddhartha Gautama achieved enlightenment beneath the Bodhi Tree in Bodh Gayā, India. Today, Buddhists from around the world flock to Bodh Gayā in search of their own spiritual awakening.

WORSHIP AND CELEBRATION

The goal of Buddhists is to achieve nirvana, the enlightened state in which individuals are free from ignorance, greed, and suffering. Theravada Buddhists believe that monks are most likely to reach nirvana when they reject worldly objects, behave morally, and devote their lives to meditation.

Christianity

Christianity has more members than any of the other world religions. It began with the death of Jesus in 33 C.E. in what is now Israel. Christianity is based on the belief in one God and on the life and teachings of Jesus. Christians believe that Jesus, who was born a Jew, is the son of God and that Jesus was both God and man. Christians believe that Jesus is the messiah (Christ), or savior, who died for people's sins. Christians believe that when people accept Jesus as their savior, they will achieve eternal life.

The major forms of Christianity are Roman Catholicism, Eastern Orthodoxy, and Protestantism. In 1054, disputes over doctrine and the leadership of the Christian Church caused the Church to divide into the Roman Catholic Church, headed by the Bishop of Rome, also known as the pope, and the Eastern Orthodox Church, led by patriarchs. Protestant churches emerged in the 1500s in an era known as the Reformation. Protestants disagreed with some Catholic doctrines and questioned the pope's authority. Despite their different theologies, all three forms of Christianity are united in their belief in Jesus as savior.

» Stained glass window depicting Jesus

SACRED TEXT

The Christian Bible is the spiritual text for all Christians and is considered to be inspired by God.

> *Blessed are the poor in spirit, for theirs is the kingdom of heaven.*
>
> *Blessed are those who mourn, for they shall be comforted.*
>
> *Blessed are the meek, for they shall inherit the earth.*
>
> *Blessed are those who hunger and thirst for righteousness, for they shall be satisfied.*
>
> *Blessed are the merciful, for they shall obtain mercy.*
>
> *Blessed are the pure in heart, for they shall see God.*
>
> *Blessed are the peacemakers, for they shall be called sons of God.*
>
> *Blessed are those who are persecuted for righteousness' sake, for theirs is the kingdom of heaven.*
>
> *Blessed are you when men revile you and persecute you and utter all kinds of evil against you falsely on my account.*
>
> *Rejoice and be glad, for your reward is great in heaven, for so men persecuted the prophets who were before you.*

—This excerpt, from Matthew 5:3–12, is from Jesus's Sermon on the Mount.

SACRED SYMBOL

Christians believe that Jesus died for their sins. His death redeemed, or freed, people from their sins. The statue *Christ the Redeemer*, located in Rio de Janeiro, Brazil, symbolizes this important Christian belief.

The gospels are books in the Christian Bible that describe the life and teachings of Jesus. The gospels state that Bethlehem was the birthplace of Jesus. Because of this, Bethlehem is very important to Christians. The Church of the Nativity is located in Bethlehem. Christians believe that the church sits on the exact site where Jesus was born.

WORSHIP AND CELEBRATION

Christians participate in many events to celebrate the life and death of Jesus. Among the most widely known and observed events are Christmas, Good Friday, and Easter. People attend Christmas church services to celebrate the birth of Jesus. As part of the celebration, followers often light candles to symbolize their belief that Jesus is the light of the world.

Confucianism

Confucianism began more than 2,500 years ago in China. Although considered a religion, it is actually a philosophy. It is based upon the teachings of Confucius.

Confucius believed that people needed to have a sense of duty. Duty meant that a person must put the needs of the family and the community before his or her own needs. Each person owed a duty to another person. Parents owed their children love, and children owed their parents honor. Rulers had a duty to govern fairly. In return, the ruler's subjects had to be loyal to the ruler and obey the law. Confucius also promoted the idea that people should treat others the same way that they would like to be treated. Eventually, Confucianism spread from China to other East Asian societies.

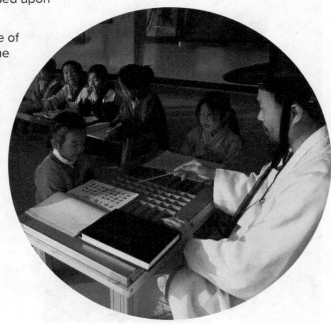

» Students study Confucianism, Chunghak-dong, South Korea

SACRED TEXT

Confucius was famous for his sayings. These teachings were gathered into a book called the *Analects* after Confucius's death.

Confucius said:

> ❝ To learn, and then, in its due season [appropriate time], put what you have learned into practice - isn't that still a great pleasure? And to have a friend visit from somewhere far away - isn't that still a great joy? When you're ignored by the world like this, and yet bear no resentment [anger] - isn't that great nobility [magnificence]? ❞

SACRED SYMBOL

Yin-yang symbolizes the harmony offered by Confucianism. Black and white represent two energies. These energies are different from each other, but they also cannot be separated from one another. An example of a *yin-yang* relationship is night and day. Night looks and is different from day, but you cannot have night without day. Both rely on each other to complete each other. It is believed that the *yin* and *yang* act together to balance one another.

PHOTO:(t)michel Setboun/The image Bank Unreleased/Getty Images; (b)Bettmann/Getty Images; TEXT:Confucius. The Analects of Confucius. Translated by David Hinton. Washington DC: Counterpoint, Member of Perseus Books Group, 1998.

The temple at Qufu is a group of buildings dedicated to Confucius. Confucius's family home was located here. It is one of the largest ancient architectural complexes in China. Every year, followers gather at Qufu to celebrate the birthday of Confucius.

WORSHIP AND CELEBRATION

Followers of Confucianism do not worship Confucius as a god. Confucius is their spiritual leader, and there are temples dedicated to him. Followers believe that Confucianism is a guide for living and effective governing.

Hinduism

Hinduism is one of the world's oldest religions. It is the world's third-largest religion, after Christianity and Islam. Hinduism does not have a single founder or founding date. Hinduism grew out of the religious customs of many people over thousands of years in India. Hindus think of all their deities as different parts of one universal spirit. This universal spirit is called Brahman.

According to Hindu sacred texts, every living thing has a soul that is part of Brahman. The body is part of life on Earth. At death, the soul leaves the body and joins with Brahman. However, Hindus believe that the soul is not joined to Brahman immediately after a person dies. Instead, a soul must pass through many lives, or be reincarnated, before it is united with Brahman. According to the idea of karma, the life into which a soul is reborn is determined by the good and evil acts performed in past lives. Dharma is the divine law that requires people to do their duties in the life that they are living in order to achieve a better life.

» Statue of Vishnu

SACRED TEXT

The Vedas are the entire collection of Hindu sacred writings. The Vedas are written in ancient Sanskrit. They are the oldest religious texts in an Indo-European language. The Vedas include Hindu hymns, prayers, and descriptions of rituals.

> 66 *Now, whether they perform a cremation for such a person or not, people like him pass into the flame, from the flame into the day, from the day into the fortnight of the waxing moon from the fortnight of the waxing moon into the six months when the sun moves north, from these months into the year, from the year into the sun, from the sun into the moon, and from the moon into the lightning. Then a person who is not human—he leads them to Brahman. This is the path to the gods, the path to Brahman. Those who proceed along this path do not return to this human condition.* 99
>
> —The Chandogya Upanishad 4:15.5

SACRED SYMBOL

One important symbol of Hinduism is actually a symbol for a sound. "Om" is a sound that Hindus often chant during prayer and rituals.

Hindus believe that when a person dies, his or her soul is reborn. This is known as reincarnation. Many Hindus bathe in the Ganges and other sacred rivers to purify their soul and to be released from rebirth.

WORSHIP AND CELEBRATION

Holi is an important North Indian Hindu festival celebrating the triumph of good over evil. As part of the celebration, men, women, and children splash colored powders and water on each other. In addition to its religious importance, Holi also celebrates the beginning of spring.

Islam

Followers of Islam, known as Muslims, believe in one God, whom they call Allah. The word *Allah* is Arabic for "God." The founder of Islam, Muhammad, began his teachings in Makkah (Mecca) in 610 C.E. Eventually, the religion spread throughout much of Asia, including parts of India to the borders of China and much of Africa. According to Muslims, the Quran, their holy book, contains the direct word of God, revealed to their prophet Muhammad. Muslims believe that God created nature. Without God, Muslims believe, there would be nothingness.

Muslims are expected to fulfill the Five Pillars of Islam, or acts of worship. Muslims must believe and declare that there is no God but Allah and that Muhammad is his prophet. Muslims must pray five times a day facing toward Makkah. Muslims must tithe, or give a portion of their income to the poor. Muslims must fast during the holy month of Ramadan, meaning that they must not eat from dawn to dusk. Finally, Muslims, if possible, should visit Makkah at least once in their lifetime.

Muhammad died in 632 C.E., and disagreement broke out about who should be the caliph, or the successor to Muhammad. Two groups with different opinions emerged. The Shia believed that the rulers should descend from Muhammad. The Sunni believed that the rulers need only to be followers of Muhammad. Most Muslims today are Sunni.

» The Dome of the Rock, Jerusalem

SACRED TEXT

The Quran instructs Muslims about how they should live and treat others. The Quran also contains rules that affect Muslims' daily lives. For example, Muslims are not allowed to eat pork, drink liquor, or gamble. The Quran also has rules about marriage, divorce, and business practices. The excerpt below is a verse repeated by all Muslims during their five daily prayers.

66 *In the name of the merciful and compassionate God. Praise belongs to God, the Lord of the worlds, the merciful, the compassionate, the ruler of the day of judgment! Thee we serve and Thee we ask for aid. Guide us in the right path, the path of those Thou art gracious to, not of those Thou art wroth with, nor of those who err.* 99

—The Quran

SACRED SYMBOL

Islam is often symbolized by the crescent moon. It is an important part of Muslim rituals, which are based on the lunar calendar.

PHOTO:(t)Ralph Curtin/Natural Selection/Design Pics; (b)Fred de Noyelle/Stone/Getty Images; TEXT:The Qur'an, Part I, translated by E. H. Palmer, Oxford at the Clarendon Press, 1880.

Makkah is a sacred site for all Muslims. One of the Five Pillars of Islam states that all who are physically and financially able must make a hajj, or pilgrimage, to the holy city once in their life. Practicing Muslims are also required to pray facing Makkah five times a day.

WORSHIP AND CELEBRATION

Muslims believe that Muhammad received the Quran from Allah. They celebrate this event during the holy month of Ramadan. Muslims fast from dawn until sunset during the month. Muslims believe that fasting helps them focus on one's spiritual needs rather than one's physical needs. They also believe that fasting makes them aware of the needs of others. Ramadan ends with a feast known as Eid-al-Fitr, or Feast of the Fast.

Judaism

Judaism was the first major religion based on monotheism—the belief in one God. Jews trace their national and religious origins to Abraham. According to the Hebrew Bible— the Tanakh—God made a covenant, or agreement, with Abraham around 1800 B.C.E. If Abraham moved to the land of Canaan, which today is Lebanon, Israel, and Jordan, Abraham and his descendants would be blessed. Abraham's descendants would continue to be blessed as long as they followed God's laws. Jews believe that if they strive for justice and live moral lives, they will help create a new era of universal peace.

» El Ghriba Synagogue, Jerba, Tunisia

SACRED TEXT

Jews believe that God gave Moses the writings and teachings found in the Torah. The Torah includes the first five books of Moses in the Hebrew Bible. These books include Genesis, Exodus, Leviticus, Numbers, and Deuteronomy. They tell the story of the origins of the Jews and explain Jewish laws. The remainder of the Hebrew Bible contains the writings of the prophets, Psalms, and ethical and historical works.

» The Torah scroll

> 66 *I am the Lord thy [your] God, who brought thee out of the land of Egypt, out of the house of bondage [slavery]. Thou [you] shalt have no other gods before me.* 99
>
> —Exodus 20:2–3

SACRED SYMBOLS

One of the oldest Jewish symbols is the menorah. The menorah is lit to celebrate Hanukkah. Hanukkah is an eight-day celebration. During Hanukkah, Jews celebrate the rededication of the Temple of Jerusalem following the victory over the Syrian-Greeks around 167 B.C.E. Another important Jewish symbol is the Star of David, also known as the Magen David, or Shield of David.

» The Magen David

» The Menorah

The Second Jerusalem Temple was a sacred building to Jews after they were freed from slavery in Babylon around 538 B.C.E. It was located in Jerusalem, the holiest city for Judaism. The Temple was destroyed by the Romans in 70 C.E. The Western Wall is all that remains of the structure that surrounded the Second Jerusalem Temple. The wall quickly became a sacred site in Jewish religious tradition. Jews throughout the world pray facing Jerusalem in the morning, afternoon, and evening. Jews living in Jerusalem pray facing toward the Western Wall.

WORSHIP AND CELEBRATION

The day-long Yom Kippur service ends with the blowing of the ram's horn (shofar). Yom Kippur is the holiest day in the Jewish calendar. During Yom Kippur, Jews do not eat or drink for 25 hours. The purpose is to reflect on the past year, repent for one's sins, and gain forgiveness from God. It falls in September or October, 10 days after Rosh Hashanah, the Jewish New Year.

Sikhism

Sikhism emerged in the mid-1400s in the Punjab, in northwest India. Sikhism is a religion that arose out of the teachings of Guru Nanak. Sikh tradition says that Guru Nanak's teachings were revealed directly to him by God.

Sikhs believe in one God who is formless, all-powerful, and all-loving. One can achieve unity with God by helping others, meditating, and working hard. Although about 76 percent of the world's 27 million Sikhs live in the Punjab, Sikhism has spread widely as many Sikhs have migrated to different regions throughout the world.

» Sikh men often wear long beards and cover their hair with turbans.

» Guru Nanak

SACRED TEXT

The primary sacred text for Sikhs is the *Guru Granth Sahib*. Collected from the mid-1500s through the 1600s, it includes contributions from Sikh Gurus, along with others also claimed as saints by Hindus and Muslims, such as Namdev, Ravidas, and Kabir.

> 66 *Enshrine the Lord's Name within your heart. The Word of the Guru's Bani prevails throughout the world; through this Bani, the Lord's name is obtained.* 99
>
> —Guru Amar Das

SACRED SYMBOL

One of the sacred symbols of the Sikhs is the *khanda*. It is composed of four traditional Sikh weapons.

CENTRAL SYMBOL

Ek Onkar is one of the central Sikh symbols. It represents the belief that there is one God for all people, regardless of religion, gender, race, or culture.

PHOTO: (l)Aloysius Patrimonio/Alamy Stock Photo (r)PhotosIndia.com RM 9/Alamy Stock Photo; TEXT:Sri Guru Granth Sahib: Translation of the Sikh Religion Holy Scriptures. Brooklyn: Sukari Publishing Universe.

Darbar Sahib, also known as the Golden Temple, is located in Amritsar, Punjab. It is one of the most popular Sikh houses of worship. It is believed that Guru Nanak meditated along the lake that surrounds the temple.

WORSHIP AND CELEBRATION

Vaisakhi is a significant Punjabi and Sikh festival in April. Sikhs celebrate Vaisakhi as the day Guru Gobind Singh, the 10th Guru, established the Khalsa, the community of people who have been initiated into the Sikh religion. In Punjab, Vaisakhi is celebrated as the New Year and the beginning of the harvest season.

Indigenous Religions

There are many types of religions that are limited to certain ethnic groups. These local religions are found in Africa, as well as isolated regions of Japan, Australia, and the Americas.

Most local, or indigenous, religions reflect a close relationship between humans and the environment. Some groups teach that people are a part of nature, not separate from it. Animism is characteristic of many indigenous religions. Natural features are sacred, and stories about how nature developed are an important part of religious heritage. Although many of these stories have been written down in modern times, they were originally transmitted orally.

Africa

Many people living in Africa practice a variety of local religions. Despite their differences, most African religions recognize the existence of one creator, in addition to spirits that inhabit all forms of life. Religious ceremonies are often celebrated with music and dance.

» Rituals are an important part of African religions. These Masai boys are wearing ceremonial dress as part of a ritual.

» These masked dancers from Mali are performing a funeral ritual.

» Masks are a component of ritual and ceremony in many African religions.

Japan

Shinto, founded in Japan, is the world's largest indigenous religion. It developed in prehistoric times and has no formal teachings. The gods are known as *kami*. Ancestors are also worshipped. Shinto's 4 million followers often practice Buddhism, too.

» This Shinto priest performs a ceremony at the annual fish harvest blessing to ensure a good harvest for fishers.

» Shinto shrines, like this one, are usually built in places of great natural beauty to emphasize the relationship between people and nature.

Australia

The Australian Aboriginal religion does not have any gods. It is based upon a belief known as the Dreaming, or Dreamtime. Followers believe that ancestors sprang from the Earth and created people, plants, and animals. Followers also believe that ancestors continue to control the natural world.

» These Aborigine men are taking part in a traditional smoking ceremony. The smoke from burning plants is thought to keep bad spirits away.

» Aborigines often paint their faces with the symbols of their clan or family group.

Native Americans

The beliefs of most Native Americans focus on the spirit world; however, the rituals and practices of individual groups vary. Most Native Americans believe in a Great Spirit who, along with other spirits, influences life on Earth. These spirits make their presence known primarily through acts of nature.

Good health, a productive harvest, and successful hunting often serve as the reasons for Native American rituals, prayers, and ceremonies. Native Americans also observe a person's passage into different stages of life. Rituals celebrate birth, adulthood, and death. Prayers, in the form of songs and dances, are offered to spirits.

» Rituals are passed down from generation to generation. This Native American woman is performing a ritual dance.

» There are many different Native American groups throughout the United States and Canada. This Pawnee is wearing ceremonial dress during a celebration in Oklahoma.

» Totem poles, like this one in Alaska, were popular among the Native American peoples of the Northwest Coast. They were often decorated with mythical beings, family crests, or other figures. They were placed outside homes.

Assessment

Reviewing Vocabulary

Match the following terms with their definitions.

1. sect
2. monotheism
3. polytheism
4. animism
5. atheism

a. belief that spirits inhabit natural objects and forces of nature
b. belief in one God
c. a subdivision within a religion that has its own distinctive beliefs and/or practices
d. belief in more than one god
e. disbelief in the existence of any god

Reviewing the Main Ideas

World Religions

6. Which religion has the most followers worldwide?
7. On a separate sheet of paper, make a table of the major world religions. Use the chart below to get you started.

Name	Founder	Geographic distribution	Sacred sites
Buddhism			
Christianity			
Confucianism			
Hinduism			
Islam			
Judaism			
Sikhism			
Indigenous			

Buddhism

8. According to Buddhism, how can the end of suffering in the world be achieved?
9. What is nirvana? According to Buddhists, who is most likely to achieve nirvana and why?

Christianity

10. In what religion was Jesus raised?
11. Why do Christians accept Jesus as their savior?

Confucianism

12. What is Confucianism based on?
13. What does yin-yang symbolize?

Hinduism

14. Where did Hinduism develop?
15. What role do Hindus believe karma plays in reincarnation?

Islam

16. What are the two branches of Islam? What is the main difference between the two groups?
17. What role does Makkah play in the Islamic faith?

Judaism

18. What is the Torah?
19. What is the purpose of Yom Kippur?

Sikhism

20. Where do most Sikhs live? Why?
21. What is Vaisakhi?

Indigenous Religions

22. Why would local religions feature sacred stories about the creation of people, animals, and plant life?
23. Which of the indigenous religions has the largest membership?

Problem-Solving Activity

24. **Research Project** Use library and Internet sources to research the role of food and food customs in one of the world's major religions. Create a presentation to report your findings to the class.

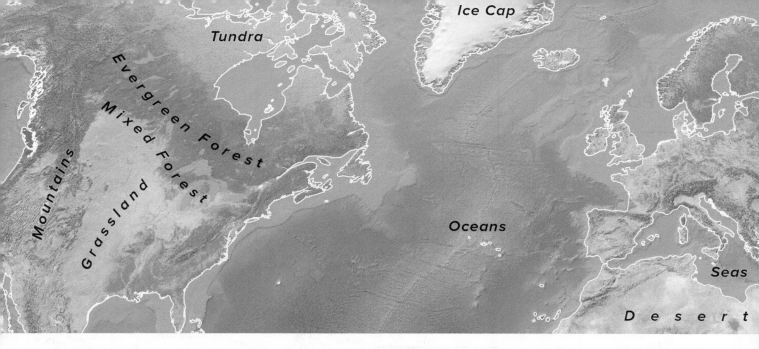

Tundra

Evergreen Forest

Mixed Forest

Mountains

Grassland

Ice Cap

Oceans

Seas

Desert

Atlas and Symbol Key

········ Claimed boundary
———— International boundary (political map)
———— Internatonal boundary (physical map)

✪ National capital
○ State/Provincial capital
● Towns

▼ Depression
▲ Elevation

Dry salt lake
Lake
Rivers
Canal

Reference Atlas

World Political	A26	Africa Political	A50
World Physical	A28	Africa Physical	A51
North America Political	A30	Middle East Physical/Political	A52
North America Physical	A31	Oceania Physical/Political	A54
United States Political	A32	Pacific Rim Physical/Political	A56
United States Physical	A34	Ocean Floor	A58
Canada Physical/Political	A36	Polar Regions Physical	A60
South America Political	A38	World GDP Per Capita Cartogram	A62
South America Physical	A39	World Population Growth Rate Cartogram	A64
Middle America Physical/Political	A40	World Population Dot Density	A66
Europe Political	A42	World Population Choropleth	A68
Europe Physical	A44	Time Zones	A70
Asia Political	A46	Geographic Dictionary	A72
Asia Physical	A48		

WORLD
POLITICAL

0   2,000 miles at Equator
0   2,000 kilometers at Equator
Equal Earth projection

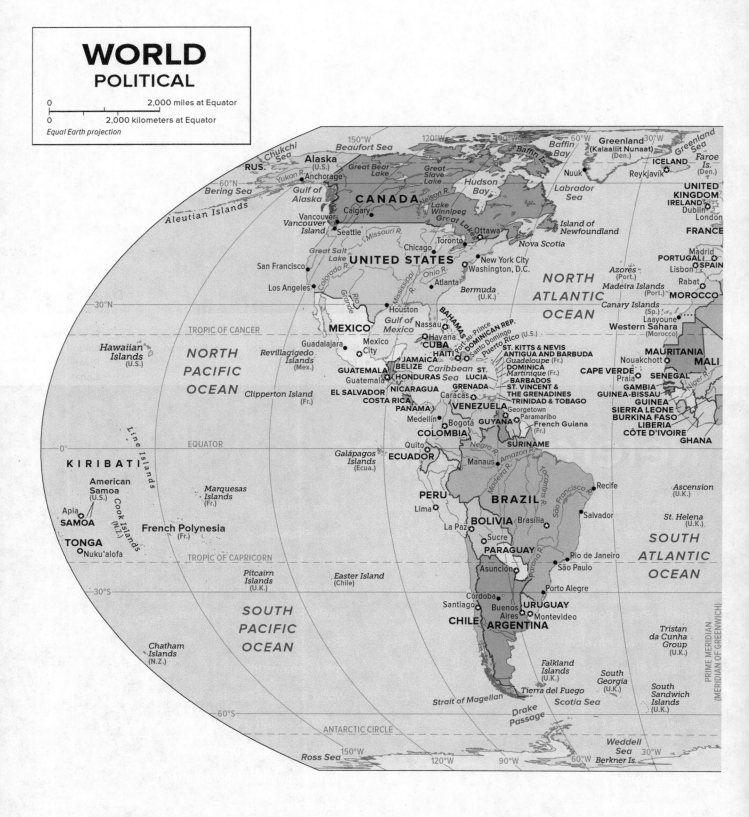

Chukchi Sea
Beaufort Sea
Baffin Bay
Greenland (Kalaallit Nunaat) (Den.)
ICELAND
Greenland Sea
Faroe Is. (Den.)
RUS.
Alaska (U.S.)
Great Bear Lake
Nuuk
Reykjavik
UNITED KINGDOM
IRELAND
Bering Sea
Anchorage
60°N
Yukon R.
Gulf of Alaska
Great Slave Lake
Hudson Bay
Labrador Sea
Dublin
London
Aleutian Islands
Vancouver
Vancouver Island
Calgary
CANADA
Nelson R.
Lake Winnipeg
Great Lakes
Island of Newfoundland
FRANCE
Seattle
Missouri R.
Ottawa
Nova Scotia
Madrid
PORTUGAL
SPAIN
Chicago
Toronto
San Francisco
Great Salt Lake
UNITED STATES
New York City
Washington, D.C.
NORTH ATLANTIC OCEAN
Azores (Port.)
Lisbon
Los Angeles
Colorado R.
Ohio R.
Atlanta
Bermuda (U.K.)
Madeira Islands (Port.)
Rabat
MOROCCO
Rio Grande
Mississippi R.
Houston
Gulf of Mexico
Canary Islands (Sp.)
Laayoune
30°N
TROPIC OF CANCER
Hawaiian Islands (U.S.)
NORTH PACIFIC OCEAN
Revillagigedo Islands (Mex.)
MEXICO
Nassau
BAHAMAS
Port-au-Prince
Havana
CUBA
Santo Domingo
DOMINICAN REP.
Puerto Rico (U.S.)
Western Sahara (Morocco)
MAURITANIA
MALI
Guadalajara
Mexico City
ST. KITTS & NEVIS
ANTIGUA AND BARBUDA
Nouakchott
GUATEMALA
BELIZE
HAITI
Guadeloupe (Fr.)
DOMINICA
Martinique (Fr.)
CAPE VERDE
Praia
SENEGAL
Clipperton Island (Fr.)
JAMAICA
Caribbean Sea
ST. LUCIA
BARBADOS
GAMBIA
GUINEA-BISSAU
Guatemala
HONDURAS
GRENADA
ST. VINCENT & THE GRENADINES
GUINEA
EL SALVADOR
NICARAGUA
Caracas
TRINIDAD & TOBAGO
SIERRA LEONE
BURKINA FASO
COSTA RICA
VENEZUELA
Georgetown
LIBERIA
PANAMA
Paramaribo
CÔTE D'IVOIRE
Medellín
Bogotá
GUYANA
French Guiana (Fr.)
GHANA
COLOMBIA
SURINAME
Niger R.
Quito
Negro R.
EQUATOR
Galápagos Islands (Ecua.)
ECUADOR
Manaus
Amazon R.
KIRIBATI
Line Islands
PERU
Madeira R.
BRAZIL
Recife
Ascension (U.K.)
American Samoa (U.S.)
Marquesas Islands (Fr.)
Lima
Tocantins R.
São Francisco R.
St. Helena (U.K.)
Apia
SAMOA
Cook Islands (N.Z.)
French Polynesia (Fr.)
BOLIVIA
La Paz
Brasília
Salvador
TONGA
Nuku'alofa
TROPIC OF CAPRICORN
Sucre
PARAGUAY
Paraná R.
Rio de Janeiro
SOUTH ATLANTIC OCEAN
Asunción
São Paulo
Pitcairn Islands (U.K.)
Easter Island (Chile)
Porto Alegre
30°S
Córdoba
Santiago
Buenos Aires
URUGUAY
Montevideo
Tristan da Cunha Group (U.K.)
SOUTH PACIFIC OCEAN
Chatham Islands (N.Z.)
CHILE
ARGENTINA
Falkland Islands (U.K.)
South Georgia (U.K.)
South Sandwich Islands (U.K.)
Tierra del Fuego
Strait of Magellan
Scotia Sea
PRIME MERIDIAN (MERIDIAN OF GREENWICH)
60°S
Drake Passage
ANTARCTIC CIRCLE
Ross Sea
Weddell Sea
Berkner Is.

ARCTIC OCEAN

Svalbard (Nor.) · Severnaya Zemlya · Laptev Sea · New Siberian Is. · East Siberian Sea · Kara Sea · Barents Sea · Novaya Zemlya · Norwegian Sea

ARCTIC CIRCLE

RUSSIA

Oslo · NORWAY · SWEDEN · FINLAND · St. Petersburg · Yekaterinburg · Ob' R. · Yenisey R. · Angara R. · Lena R. · Yakutsk · Kamchatka Peninsula · Bering Sea · Aleutian Is.

EST. · LATV. · LITH. · DEN. · NETH. · POLAND · BELARUS · Moscow · Samara · Omsk · Novosibirsk · Lake Baikal · Amur R. · Sea of Okhotsk

GERMANY · BELG. · CZECHIA · Paris · SWITZ. · AUST. · HUNG. · SLO. · MOLD. · UKRAINE · Kyiv (Kiev) · KAZAKHSTAN · Nur-Sultan · Ulaanbaatar · MONGOLIA · Harbin · Shenyang · NORTH KOREA · Sapporo · Hokkaidō · Honshū

ITALY · B.&H. · SERB. · BULG. · ROMANIA · Black Sea · Aral Sea · Tashkent · Bishkek · Almaty · KYRGYZSTAN · Beijing · Tianjin · P'yŏngyang · SOUTH KOREA · Seoul · Osaka · JAPAN · Tokyo

Rome · MONT. · ALB. · KOS. · N. MAC. · GREECE · GEORGIA · ARMENIA · AZERBAIJAN · UZBEKISTAN · TURKMENISTAN · TAJIKISTAN · Dushanbe · Huang He (Yellow R.) · Chengdu · Shanghai · Wuhan · East China Sea · Kyūshū

Mediterranean Sea · TUNISIA · Ankara · TURKEY · Caspian Sea · Ashgabat · AFGHANISTAN · Kābul · Islamabad · CHINA · Chang Jiang (Yangtze R.) · Guangzhou · Taipei · NORTH PACIFIC OCEAN

Algiers · CYPRUS · LEBANON · SYRIA · Tehran · Baghdad · Delhi · Lahore · NEPAL · Thimphu · BHUTAN · Hanoi · Hong Kong · TAIWAN

ALGERIA · LIBYA · ISRAEL · IRAQ · JORDAN · IRAN · KUWAIT · New Delhi · Karachi · Dhaka · BANGLADESH · MYANMAR (BURMA) · LAOS · Hainan · Philippine Sea · Northern Mariana Is. (U.S.)

Cairo · EGYPT · BAHRAIN · QATAR · Riyadh · U.A.E. · PAKISTAN · Kolkata (Calcutta) · Nay Pyi Taw · VIETNAM · South China Sea · Luzon · Manila · Guam (U.S.)

Nile R. · Red Sea · SAUDI ARABIA · OMAN · Mumbai (Bombay) · INDIA · Hyderabad · Vientiane · THAILAND · Bangkok · PHILIPPINES · MARSHALL ISLANDS

NIGER · CHAD · SUDAN · ERITREA · YEMEN · Arabian Sea · Bay of Bengal · Chennai (Madras) · CAMBODIA · Mindanao · PALAU · FEDERATED STATES OF MICRONESIA

Niamey · TOGO · N'Djamena · Khartoum · Sanaa · DJIBOUTI · Socotra (Yemen) · Bengaluru (Bangalore) · Phnom Penh · Ho Chi Minh City · BRUNEI · KIRIBATI

BENIN · NIGERIA · Lagos · CENTRAL AFRICAN REP. · SOUTH SUDAN · ETHIOPIA · Addis Ababa · SRI LANKA · Colombo · Sri Jayewardenepura Kotte · MALAYSIA · SINGAPORE · NAURU

EQ. GUINEA · CAMEROON · Bangui · Juba · SOMALIA · Male · MALDIVES · Kuala Lumpur · Borneo · Celebes

SÃO TOMÉ & PRÍNCIPE · GABON · Brazzaville · UGANDA · KENYA · Nairobi · Mogadishu · Sumatra · INDONESIA · New Guinea · PAPUA NEW GUINEA · SOLOMON ISLANDS · TUVALU

Cabinda (Angola) · DEM. REP. OF THE CONGO · Kinshasa · RWANDA · BURUNDI · Dodoma · SEYCHELLES · Chagos Archipelago (U.K.) · Jakarta · Java · Surabaya · Dili · Port Moresby

Luanda · ANGOLA · ZAMBIA · TANZANIA · Dar es Salaam · MALAWI · COMOROS · EAST TIMOR (TIMOR-LESTE) · Darwin · Arafura Sea · Coral Sea · VANUATU · FIJI

Lusaka · Harare · ZIMBABWE · MOZAMBIQUE · MADAGASCAR · Antananarivo · Port Louis · MAURITIUS · Réunion (Fr.) · Christmas Island (Australia) · Cocos Islands (Australia) · New Caledonia (Fr.)

NAMIBIA · BOTSWANA · Windhoek · Gaborone · Tshwane (Pretoria) · Johannesburg · Maputo · ESWATINI · INDIAN OCEAN · AUSTRALIA · Brisbane · SOUTH PACIFIC OCEAN

Bloemfontein · LESOTHO · SOUTH AFRICA · Cape Town · Perth · Darling R. · Sydney · Canberra · Melbourne · Murray R. · Tasman Sea · North Island · Auckland · NEW ZEALAND · Wellington · South Island · Tasmania

Crozet Islands (Fr.) · Kerguelen Islands (Fr.) · Prince Edward Islands (S. Af.) · Auckland Islands (N.Z.)

The Atlantic, Indian, and Pacific Oceans merge around Antarctica. Some define this as an ocean, calling it the Antarctic Ocean, Austral Ocean, or Southern Ocean. While most accept four oceans (including the Arctic Ocean), there is little international agreement on the name and extent of a fifth ocean.

SOUTHERN OCEAN

ANTARCTICA

Ross Sea

ABBREVIATIONS

ALB.	ALBANIA
AUST.	AUSTRIA
B.&H.	BOSNIA & HERZEGOVINA
BELG.	BELGIUM
BULG.	BULGARIA
CRO.	CROATIA
DEM. REP. OF THE CONGO	DEMOCRATIC REPUBLIC OF THE CONGO
DEN.	DENMARK
EQ. GUINEA	EQUATORIAL GUINEA
EST.	ESTONIA
HUNG.	HUNGARY
KOS.	KOSOVO
LATV.	LATVIA
LITH.	LITHUANIA
MOLD.	MOLDOVA
MONT.	MONTENEGRO
NETH.	NETHERLANDS
N. MAC.	NORTH MACEDONIA
SERB.	SERBIA
SLO.	SLOVAKIA
SLOV.	SLOVENIA
SWITZ.	SWITZERLAND
U.A.E.	UNITED ARAB EMIRATES

WORLD PHYSICAL

WORLD
PHYSICAL

0 _____ 2,000 miles at Equator
0 _____ 2,000 kilometers at Equator

Equal Earth projection

North Magnetic Pole (2021)
Queen Elizabeth Islands
Ellesmere Is.
GREENLAND
ARCTIC CIRCLE
Chukchi Sea
Beaufort Sea
Victoria Island
Baffin Bay
Baffin Is.
Iceland
Faroe Is.
Brooks Range
ALASKA
Great Bear Lake
Hudson Bay
Labrador Sea
British Isles
Ireland
Great Britain
60°N
Bering Sea
Denali (Mt. McKinley) 20,310 ft., 6,190 m
Alaska Ra.
Great, Slave Lake
Nelson R.
Canadian Shield
Labrador
Alexander Archipelago
Great Lakes
Lake Winnipeg
Island of Newfoundland
Iberian Peninsula
Haida Gwaii (Queen Charlotte Is.)
Vancouver Island
Coast Mts.
Cascade Mts.
ROCKY MOUNTAINS
Missouri R.
Nova Scotia
Azores
Madeira Islands
Atlas
NORTH AMERICA
Central Lowland
Appalachian Mts.
NORTH ATLANTIC OCEAN
S
30°N
Great Salt Lake
Great Plains
Death Valley -282 ft. -86 m
Colorado R.
Mississippi R.
Baja California
Gulf of Mexico
Bahama Islands
Canary Islands
S
TROPIC OF CANCER
Rio Grande
Cuba
WEST INDIES
NORTH PACIFIC OCEAN
Revillagigedo Islands
Greater Antilles
Hispaniola
Jamaica
Lesser Antilles
Cape Verde Islands
Hawaiian Islands
Hawai'i
Caribbean Sea
Upper
CENTRAL AMERICA
Clipperton Island
Trinidad
Llanos
Orinoco R.
Guiana Highlands
P O L Y N E S I A
EQUATOR
Galápagos Islands
ANDES
AMAZON BASIN
Negro R.
Amazon R.
Ascension
Line Islands
Marquesas Islands
Madeira R.
Tocantins R.
São Francisco R.
SOUTH AMERICA
Samoa Islands
Tuamotu Archipelago
Brazilian Highlands
St. Helena
Tonga Is.
Fiji Is.
Cook Islands
Society Islands
Tahiti
Lake Titicaca
Mato Grosso Plateau
SOUTH ATLANTIC OCEAN
Austral Islands
TROPIC OF CAPRICORN
Atacama Desert
Gran Chaco
Paraguay R.
Paraná R.
Pitcairn Islands
Easter Island
ANDES
Pampas
Tristan da Cunha Group
30°S
Aconcagua 22,834 ft. 6,960 m
SOUTH PACIFIC OCEAN
Chatham Islands
Chiloé Island
Chonos Archipelago
Valdés Peninsula -131 ft. -40 m
Patagonia
Queen Adelaide Archipelago
Falkland Islands
South Georgia
South Sandwich Islands
Tierra del Fuego
Strait of Magellan
Scotia Sea
60°S
Drake Passage
South Shetland Islands
South Orkney Islands
ANTARCTIC CIRCLE
Vinson Massif 16,066 ft. 4,897 m
Antarctic Peninsula
Weddell Sea
Ross Sea
Marie Byrd Land
Berkner Is.
PRIME MERIDIAN (MERIDIAN OF GREENWICH)

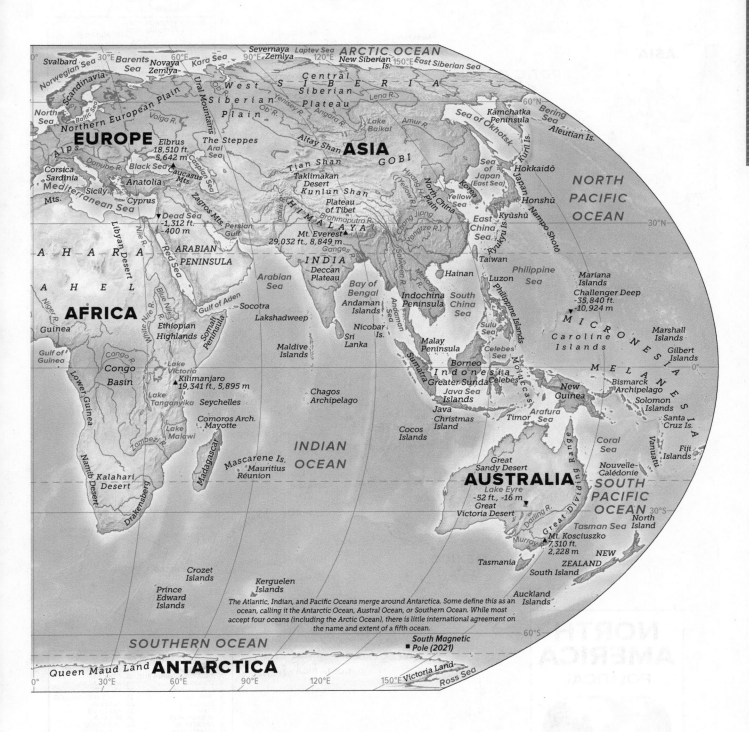

ARCTIC OCEAN

Svalbard
Norwegian Sea
Scandinavia
North Sea
Baltic Sea
Barents Sea
Novaya Zemlya
Kara Sea
Ural Mountains
Severnaya Zemlya
Laptev Sea
New Siberian Is.
East Siberian Sea

West Siberian Plain
Central Siberian Plateau
Lena R.
Bering Sea
Kamchatka Peninsula
Aleutian Is.

EUROPE
Alps
Corsica
Sardinia
Sicily
Mediterranean Sea
Mts.
Cyprus
Anatolia
Black Sea
Caucasus Mts.
Caspian Sea
Elbrus 18,510 ft. 5,642 m
The Steppes
Aral Sea
Danube R.
Volga R.
Ob R.
Yenisey R.
Angara R.
Altay Shan
Lake Baikal
Amur R.
Sea of Okhotsk
Hokkaidō
Sea of Japan (East Sea)
Honshū
Kuril Is.
Japan
Nampo Shotō

NORTH PACIFIC OCEAN

ASIA
GOBI
Tian Shan
Taklimakan Desert
Kunlun Shan
Plateau of Tibet
HIMALAYA
Zagros Mts.
Persian Gulf
Dead Sea -1,312 ft. -400 m
Red Sea
Nile R.
Libyan Desert
ARABIAN PENINSULA
Mt. Everest 29,032 ft., 8,849 m
Ganges
Brahmaputra R.
INDIA
Deccan Plateau
Indus R.
Huang He (Yellow R.)
North China Plain
Chong Jiang
Yongtze R.)
Yellow Sea
North China Sea
Korea
East China Sea
Ryukyu Is.
Kyūshū
Taiwan
Hainan

SAHARA
SAHEL
AFRICA
Guinea
Gulf of Guinea
Lower Guinea
Niger R.
White Nile R.
Blue Nile R.
Congo R.
Congo Basin
Ethiopian Highlands
Somali Peninsula
Gulf of Aden
Socotra
Lakshadweep
Arabian Sea
Maldive Islands
Bay of Bengal
Andaman Islands
Andaman Sea
Nicobar Is.
Sri Lanka
Salween R.
Indochina Peninsula
Mekong R.
South China Sea
Malay Peninsula
Philippine Islands
Luzon
Philippine Sea
Mariana Islands
Challenger Deep -35,840 ft. -10,924 m
MICRONESIA
Caroline Islands
Marshall Islands
Gilbert Islands

Lake Victoria
Kilimanjaro 19,341 ft., 5,895 m
Lake Tanganyika
Seychelles
Lake Malawi
Comoros Arch.
Mayotte
Madagascar
Zambezi R.
Namib Desert
Kalahari Desert
Drakensberg
Mascarene Is.
Mauritius
Réunion
Chagos Archipelago
INDIAN OCEAN
Cocos Islands
Christmas Island
Java
Sumatra
Indonesia
Greater Sunda Islands
Java Sea
Borneo
Celebes Sea
Sulu Sea
Celebes
Moluccas
Timor
Arafura Sea
New Guinea
Bismarck Archipelago
Solomon Islands
Santa Cruz Is.
Vanuatu
Fiji Islands
MELANESIA
Coral Sea
Nouvelle-Calédonie

Crozet Islands
Prince Edward Islands
Kerguelen Islands

Great Sandy Desert
Great Victoria Desert
AUSTRALIA
Lake Eyre -52 ft., -16 m
Darling R.
Murray R.
Great Dividing Range
Mt. Kosciuszko 7,310 ft. 2,228 m
Tasman Sea
Tasmania
SOUTH PACIFIC OCEAN
North Island
NEW ZEALAND
South Island
Auckland Islands

The Atlantic, Indian, and Pacific Oceans merge around Antarctica. Some define this as an ocean, calling it the Antarctic Ocean, Austral Ocean, or Southern Ocean. While most accept four oceans (including the Arctic Ocean), there is little international agreement on the name and extent of a fifth ocean.

SOUTHERN OCEAN
South Magnetic Pole (2021)
Queen Maud Land
ANTARCTICA
Victoria Land
Ross Sea

0° 30°E 60°E 90°E 120°E 150°E

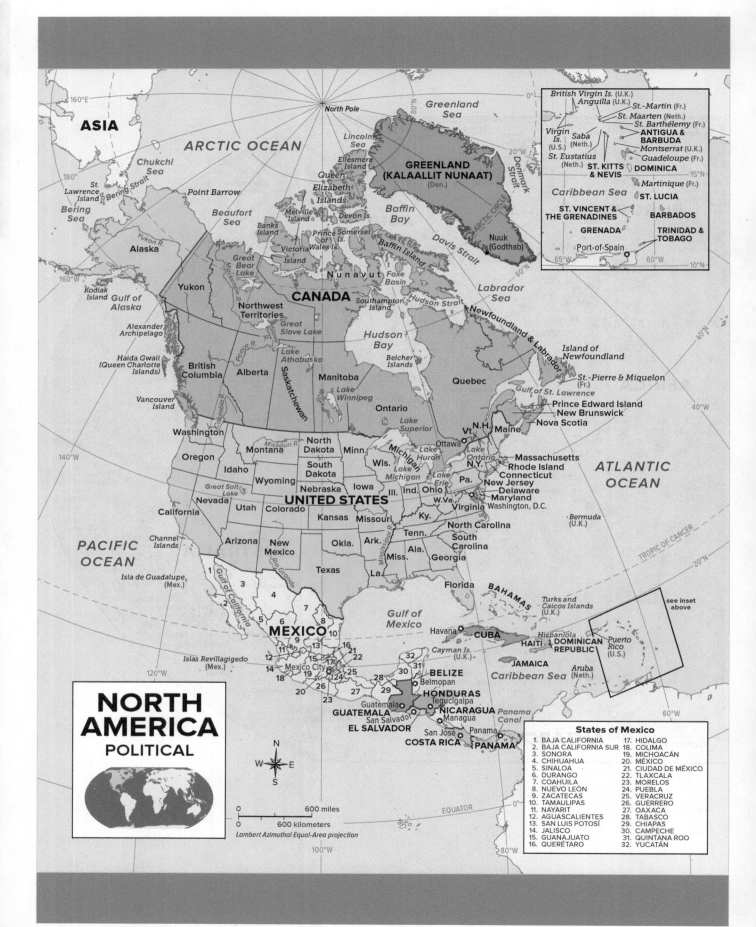

NORTH AMERICA POLITICAL

ASIA

ARCTIC OCEAN

Chukchi Sea

St. Lawrence Island

Bering Strait

Bering Sea

Yukon R.

Alaska

Kodiak Island

Gulf of Alaska

Alexander Archipelago

Haida Gwaii (Queen Charlotte Islands)

Vancouver Island

Point Barrow

Beaufort Sea

Banks Island

Melville Island

Prince Somerset Is.

Victoria Island

Wales Is.

Great Bear Lake

Mackenzie R.

Yukon

Northwest Territories

Great Slave Lake

British Columbia

Alberta

Saskatchewan

Peace R.

Lake Athabasca

Manitoba

Lake Winnipeg

North Pole

Queen Elizabeth Islands

Ellesmere Island

Lincoln Sea

Devon Is.

Greenland Sea

GREENLAND (KALAALLIT NUNAAT) (Den.)

ARCTIC CIRCLE

Baffin Bay

Nuuk (Godthab)

Denmark Strait

Davis Strait

Baffin Island

Foxe Basin

Nunavut

CANADA

Southampton Island

Hudson Strait

Hudson Bay

Belcher Islands

Labrador Sea

Newfoundland & Labrador

Island of Newfoundland

St.-Pierre & Miquelon (Fr.)

Gulf of St. Lawrence

Ontario

Quebec

Prince Edward Island

New Brunswick

Nova Scotia

PACIFIC OCEAN

Channel Islands

Isla de Guadalupe (Mex.)

Islas Revillagigedo (Mex.)

Washington

Oregon

Idaho

Nevada

California

Arizona

Montana

Wyoming

Utah

Colorado

New Mexico

North Dakota

South Dakota

Nebraska

Kansas

Oklahoma

Texas

UNITED STATES

Minn.

Wis.

Iowa

Missouri

Ark.

La.

Miss.

Ala.

Tenn.

Ky.

Ill.

Ind.

Ohio

Michigan

Lake Superior

Lake Huron

Lake Michigan

Lake Erie

Lake Ontario

W.Va.

Virginia

North Carolina

South Carolina

Georgia

Florida

Pa.

N.Y.

Vt. N.H. Maine

Massachusetts

Rhode Island

Connecticut

New Jersey

Delaware

Maryland

Washington, D.C.

Ottawa

Great Salt Lake

Missouri R.

Rio Grande

Columbia R.

Snake R.

Mississippi R.

ATLANTIC OCEAN

Bermuda (U.K.)

TROPIC OF CANCER

MEXICO

Mexico City

Gulf of California

Gulf of Mexico

Havana

CUBA

BAHAMAS

Turks and Caicos Islands (U.K.)

Cayman Is. (U.K.)

JAMAICA

Hispaniola

HAITI

DOMINICAN REPUBLIC

Puerto Rico (U.S.)

see inset above

Aruba (Neth.)

Caribbean Sea

BELIZE

Belmopan

GUATEMALA

Guatemala

San Salvador

EL SALVADOR

HONDURAS

Tegucigalpa

NICARAGUA

Managua

COSTA RICA

San José

Panama Canal

PANAMA

Panama

EQUATOR

Caribbean inset

British Virgin Is. (U.K.)

Anguilla (U.K.)

St.-Martin (Fr.)

St. Maarten (Neth.)

St. Barthélemy (Fr.)

Virgin Is. (U.S.)

Saba (Neth.)

St. Eustatius (Neth.)

ANTIGUA & BARBUDA

Montserrat (U.K.)

Guadeloupe (Fr.)

ST. KITTS & NEVIS

DOMINICA

Martinique (Fr.)

Caribbean Sea

ST. LUCIA

ST. VINCENT & THE GRENADINES

BARBADOS

GRENADA

TRINIDAD & TOBAGO

Port-of-Spain

States of Mexico

1. BAJA CALIFORNIA	17. HIDALGO
2. BAJA CALIFORNIA SUR	18. COLIMA
3. SONORA	19. MICHOACÁN
4. CHIHUAHUA	20. MÉXICO
5. SINALOA	21. CIUDAD DE MÉXICO
6. DURANGO	22. TLAXCALA
7. COAHUILA	23. MORELOS
8. NUEVO LEÓN	24. PUEBLA
9. ZACATECAS	25. VERACRUZ
10. TAMAULIPAS	26. GUERRERO
11. NAYARIT	27. OAXACA
12. AGUASCALIENTES	28. TABASCO
13. SAN LUIS POTOSÍ	29. CHIAPAS
14. JALISCO	30. CAMPECHE
15. GUANAJUATO	31. QUINTANA ROO
16. QUERÉTARO	32. YUCATÁN

0 — 600 miles
0 — 600 kilometers
Lambert Azimuthal Equal-Area projection

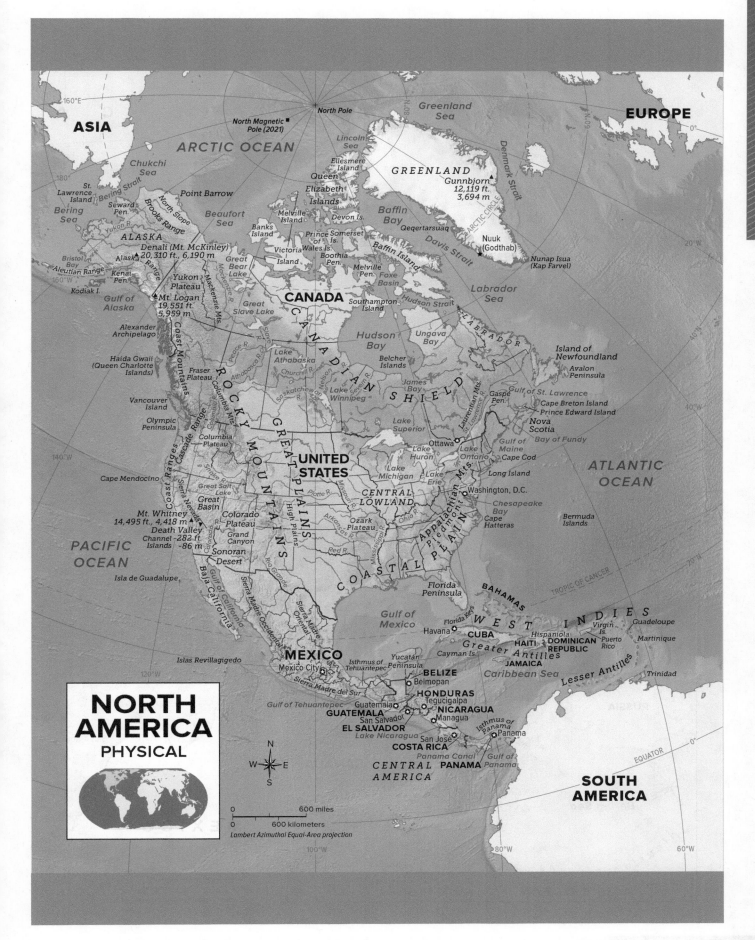

NORTH AMERICA
PHYSICAL

0   600 miles

0   600 kilometers

Lambert Azimuthal Equal-Area projection

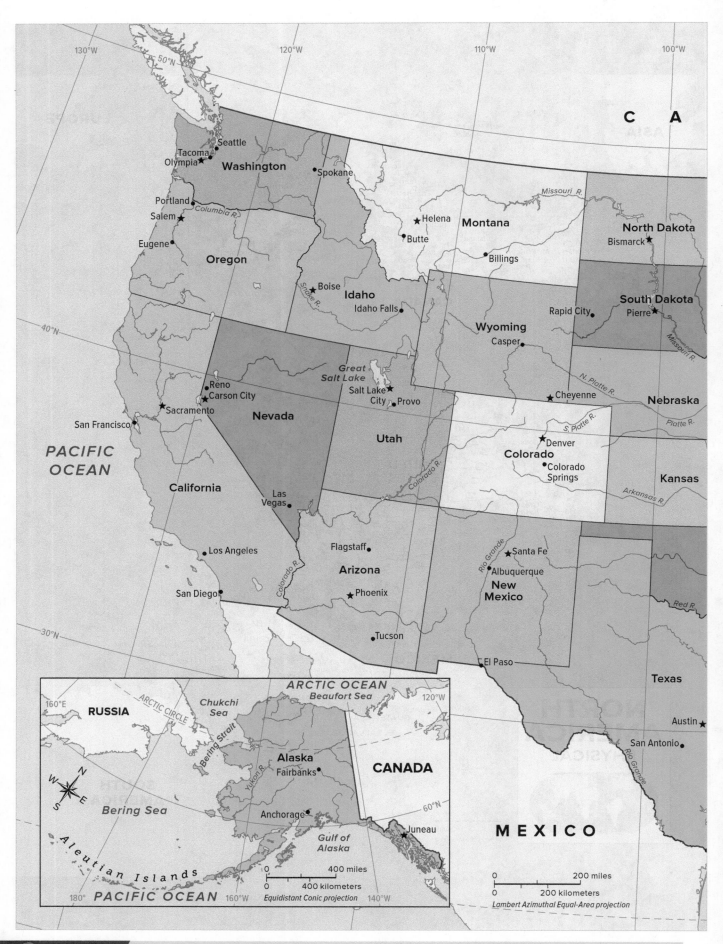

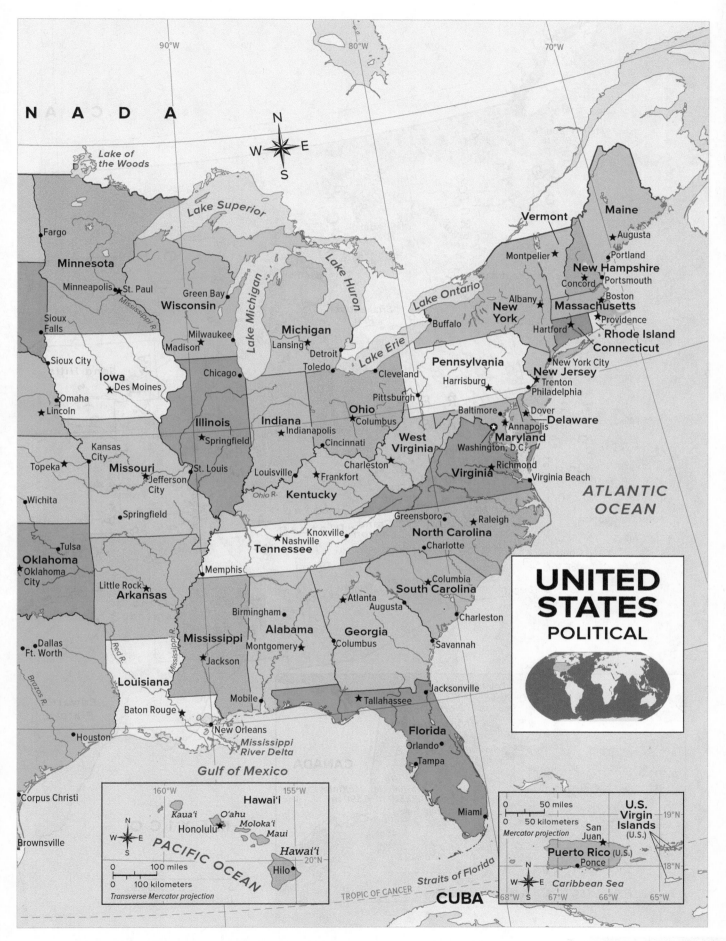

UNITED
STATES
POLITICAL

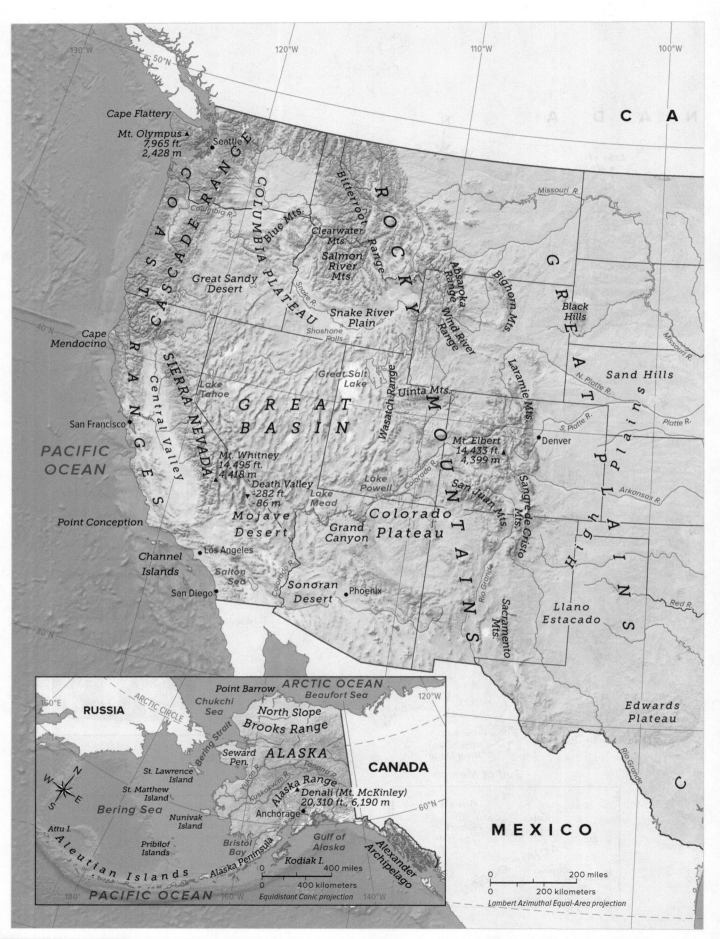

CANADA

130°W · 120°W · 110°W · 100°W

50°N

Cape Flattery

Mt. Olympus ▲
7,965 ft.
2,428 m   Seattle

COAST RANGE

COLUMBIA PLATEAU

Columbia R.

Blue Mts.

Bitterroot Range

R O C K Y

Clearwater
Mts.

Salmon
River
Mts.

Great Sandy
Desert

Snake R.

Snake River
Plain

Shoshone
Falls

40°N

Cape
Mendocino

Absaroka
Range

Wind River
Range

Bighorn Mts.

G R E A T

Black
Hills

Missouri R.

Missouri R.

Sand Hills

SIERRA NEVADA

Central Valley

Lake
Tahoe

GREAT
BASIN

Great Salt
Lake

Wasatch Range

Uinta Mts.

M O U N T A I N S

Laramie Mts.

S. Platte R.

N. Platte R.

Platte R.

San Francisco

Mt. Whitney
14,495 ft.
4,418 m

Death Valley
-282 ft.
-86 m

Lake
Powell

Lake
Mead

Colorado R.

Mt. Elbert
14,433 ft.
4,399 m   Denver

San Juan Mts.

Sangre de Cristo Mts.

Arkansas R.

PACIFIC
OCEAN

Point Conception

Mojave
Desert

Grand
Canyon

Colorado
Plateau

Rio Grande

High   Plains

Channel
Islands

Los Angeles

Salton
Sea

Colorado R.

Sonoran
Desert   Phoenix

Sacramento
Mts.

Llano
Estacado

Red R.

San Diego

30°N

Edwards
Plateau

Rio Grande

C

MEXICO

Alaska inset

ARCTIC OCEAN

160°E

RUSSIA

ARCTIC CIRCLE

Point Barrow

Chukchi
Sea

Beaufort Sea

120°W

North Slope

Brooks Range

Bering Strait

Seward
Pen.

ALASKA

St. Lawrence
Island

St. Matthew
Island

Bering Sea

Nunivak
Island

Yukon R.

Kuskokwim R.

Tanana R.

Alaska Range

▲ Denali (Mt. McKinley)
20,310 ft., 6,190 m

Anchorage

CANADA

60°N

N
W   E
S

Attu I.

Pribilof
Islands

Aleutian   Islands

Bristol
Bay

Alaska Peninsula

Kodiak I.

Gulf of
Alaska

Alexander
Archipelago

0   400 miles
0   400 kilometers
Equidistant Conic projection

180°   160°W   PACIFIC OCEAN   140°W

0   200 miles
0   200 kilometers
Lambert Azimuthal Equal-Area projection

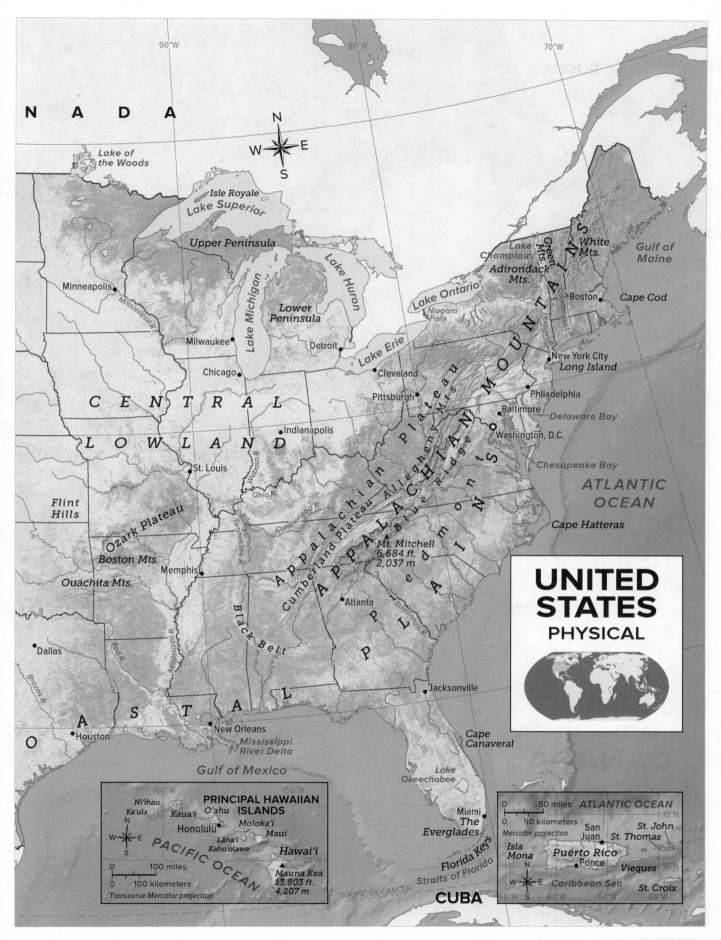

N A D A

Lake of
the Woods

Isle Royale
Lake Superior

Upper Peninsula

Minneapolis

Lower
Peninsula

Milwaukee

Lake Michigan

Lake Huron

Detroit

Chicago

Lake Erie

Cleveland

Pittsburgh

Lake Ontario

Niagara
Falls

Lake
Champlain

Adirondack
Mts.

Green
Mts.

White
Mts.

Gulf of
Maine

Boston

Cape Cod

New York City
Long Island

Philadelphia

Baltimore

Delaware Bay

Washington, D.C.

Chesapeake Bay

C E N T R A L

L O W L A N D

Indianapolis

Mississippi R.

Wabash R.

Ohio R.

St. Louis

Flint
Hills

Ozark Plateau

Boston Mts.

Ouachita Mts.

Memphis

Tennessee R.

Cumberland R.

Appalachian Plateau

Allegheny Mts.

Cumberland Plateau

Blue Ridge

A P P A L A C H I A N   M O U N T A I N S

Connecticut R.

Hudson R.

ATLANTIC
OCEAN

Cape Hatteras

Mt. Mitchell
6,684 ft.
2,037 m

Atlanta

Savannah R.

P
i
e
d
m
o
n
t

Black Belt

Dallas

Red R.

Brazos R.

Mississippi R.

Houston

New Orleans

Mississippi
River Delta

Gulf of Mexico

C O A S T A L   P L A I N

Jacksonville

Cape
Canaveral

Lake
Okeechobee

Miami
The
Everglades

Florida Keys

Straits of Florida

TROPIC OF CANCER

CUBA

**UNITED
STATES
PHYSICAL**

**PRINCIPAL HAWAIIAN
ISLANDS**

Ni'ihau
Ka'ula

Kaua'i

O'ahu

Honolulu

Moloka'i

Lāna'i

Kaho'olawe

Maui

Hawai'i

Mauna Kea
13,803 ft.
4,207 m

PACIFIC OCEAN

0   100 miles

0   100 kilometers

Transverse Mercator projection

160°W

155°W

20°N

0   50 miles   ATLANTIC OCEAN

0   50 kilometers

Mercator projection

Isla
Mona

San
Juan

St. John

St. Thomas

Puerto Rico

Ponce

Vieques

Caribbean Sea

St. Croix

19°N

18°N

68°W   67°W   66°W   65°W

90°W

80°W

70°W

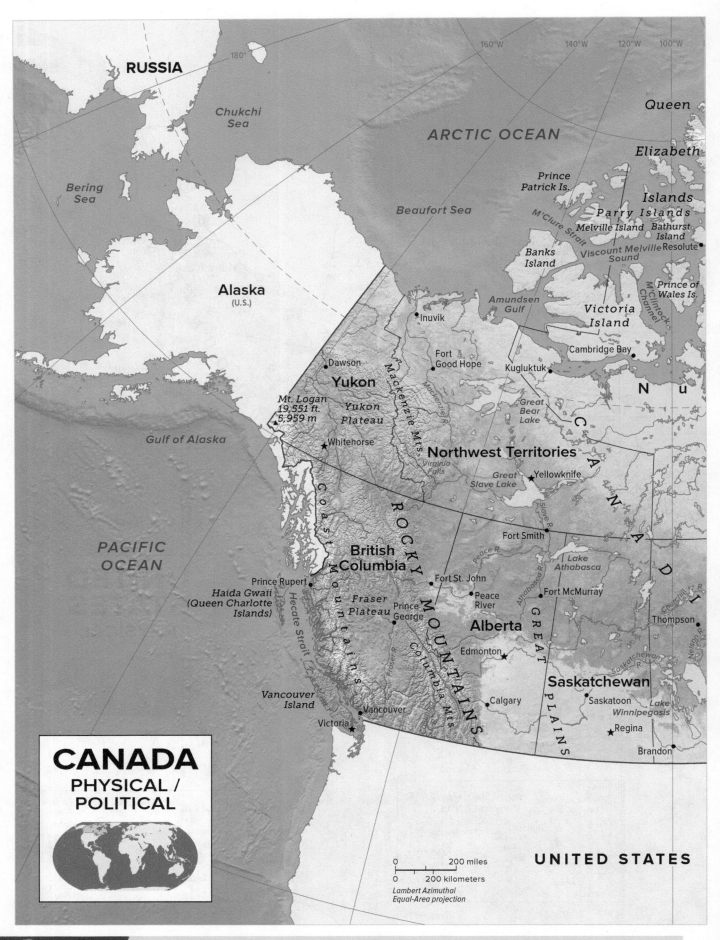

RUSSIA

*Chukchi
Sea*

ARCTIC OCEAN

*Bering
Sea*

Beaufort Sea

Queen

Elizabeth

*Prince
Patrick Is.*

M'Clure Strait

Islands

Parry Islands

Melville Island

Bathurst
Island

Viscount Melville
Sound

Resolute

*Banks
Island*

*Amundsen
Gulf*

*Prince of
Wales Is.*

Alaska
(U.S.)

Inuvik

Fort
Good Hope

Kugluktuk

*Victoria
Island*

Cambridge Bay

M'Clintock Channel

N
u

Dawson

Yukon

Mackenzie Mts.

Mackenzie R.

*Mt. Logan
19,551 ft.
5,959 m*

*Yukon
Plateau*

*Great
Bear
Lake*

C
A

Gulf of Alaska

Whitehorse

*Virginia
Falls*

Northwest Territories

*Great
Slave Lake*

Yellowknife

N

PACIFIC
OCEAN

Coast Mountains

**British
Columbia**

R
O
C
K
Y

Fort Smith

Slave R.

A

D

Prince Rupert

*Haida Gwaii
(Queen Charlotte
Islands)*

Hecate Strait

*Fraser
Plateau*

Prince
George

M
O
U
N
T
A
I
N
S

Fort St. John

Peace
River

Peace R.

*Lake
Athabasca*

Fort McMurray

Athabasca R.

I

Thompson

Churchill R.

Columbia Mts.

Fraser R.

Alberta

Edmonton

G
R
E
A
T

Saskatchewan R.

Nelson R.

*Vancouver
Island*

Vancouver

Calgary

P
L
A
I
N
S

Saskatchewan

Saskatoon

*Lake
Winnipegosis*

Victoria

Regina

Brandon

0   200 miles

0   200 kilometers

*Lambert Azimuthal
Equal-Area projection*

UNITED STATES

160°W   140°W   120°W   100°W

180°

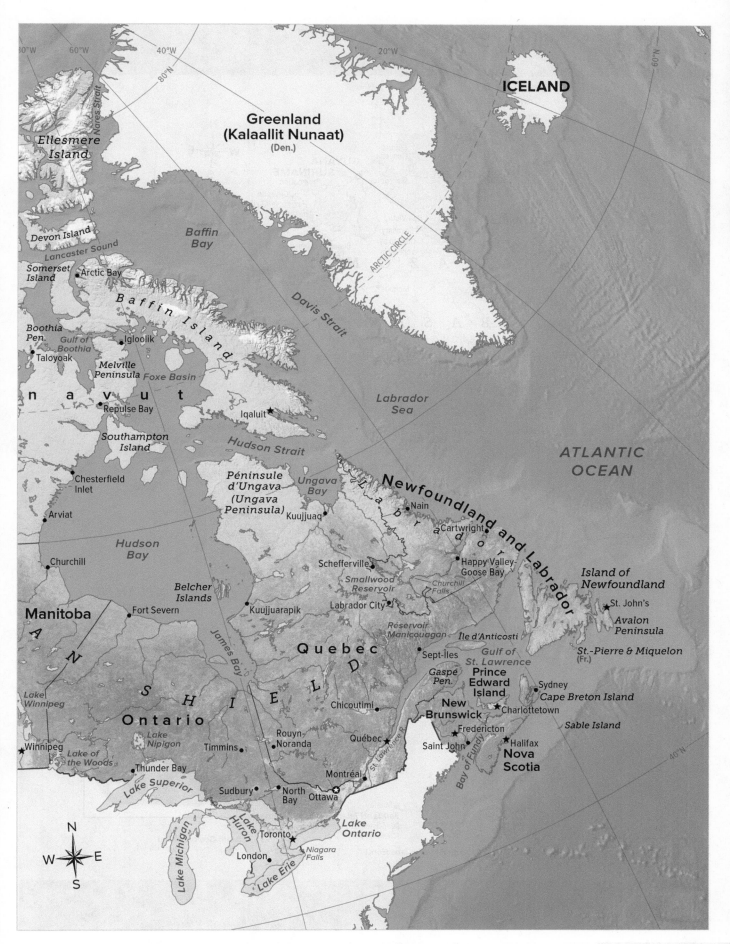

ICELAND

Greenland
(Kalaallit Nunaat)
(Den.)

*Ellesmere
Island*

Devon Island

*Baffin
Bay*

Lancaster Sound

*Somerset
Island* • Arctic Bay

*Boothia
Pen.* *Baffin Island*
• Taloyoak *Gulf of
Boothia* • Igloolik

*Melville
Peninsula* *Foxe Basin*

Davis Strait

ARCTIC CIRCLE

n a v u t • Repulse Bay *Iqaluit* ★

*Labrador
Sea*

*Southampton
Island*

Hudson Strait

ATLANTIC
OCEAN

• Chesterfield
Inlet

*Péninsule
d'Ungava
(Ungava
Peninsula)* *Ungava
Bay* Nain •

Newfoundland and Labrador

• Arviat

*Hudson
Bay* • Kuujjuaq *L a b r a d o r* Cartwright •

*Belcher
Islands* • Schefferville • Happy Valley-
Goose Bay

• Churchill *Island of
Newfoundland*

Manitoba • Fort Severn • Kuujjuarapik *Smallwood
Reservoir* *Churchill
Falls* St. John's ★

A Labrador City • *Avalon
Peninsula*

*Réservoir
Manicouagan* *Île d'Anticosti* *St.-Pierre & Miquelon*
(Fr.)

S Q u e b e c Sept-Îles • *Gulf of
St. Lawrence*

N *Gaspé
Pen.* **Prince
Edward
Island** Sydney •
Cape Breton Island

*Lake
Winnipeg* I H E L D • Chicoutimi **New
Brunswick** Charlottetown ★ *Sable Island*

Ontario *Lake
Nipigon* Rouyn-
Noranda • Québec • ★ Fredericton

★ Winnipeg *Lake of
the Woods* • Timmins Québec • Saint John • ★ Halifax
**Nova
Scotia**

• Thunder Bay *St. Lawrence R.* *Bay of Fundy*

Lake Superior • Sudbury • North
Bay Montréal •

• London Toronto ★ *Niagara
Falls* *Lake
Ontario* ★ Ottawa

Lake Michigan *Lake
Huron* *Lake Erie*

N
W ✦ E
S

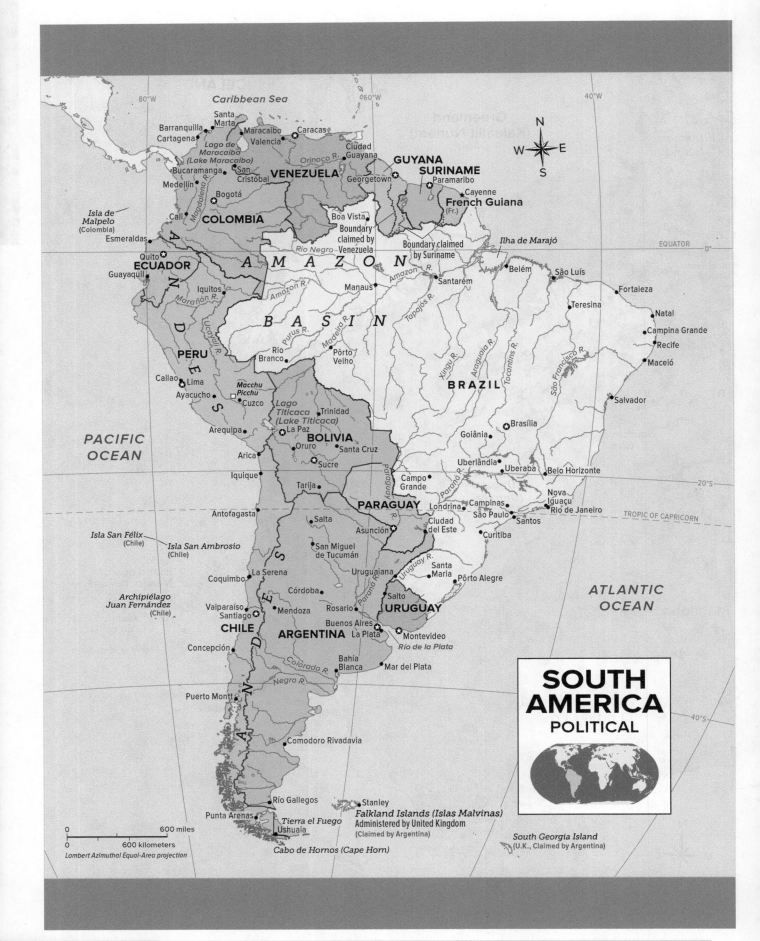

Caribbean Sea

80°W · 60°W · 40°W

Santa Marta
Barranquilla
Cartagena
Maracaibo · Caracas
Valencia
Lago de Maracaibo (Lake Maracaibo)
Ciudad Guayana
GUYANA
SURINAME
Bucaramanga · San Cristóbal
VENEZUELA
Georgetown
Paramaribo
Medellín
Magdalena R.
Orinoco R.
Cayenne
Bogotá
French Guiana (Fr.)
Cali
COLOMBIA
Boa Vista
Isla de Malpelo (Colombia)
Boundary claimed by Venezuela
Boundary claimed by Suriname
Ilha de Marajó
Esmeraldas
Rio Negro
EQUATOR · 0°
Quito
ECUADOR
A M A Z O N
Amazon R.
Belém
São Luís
Guayaquil
Iquitos
Manaus
Santarém
Fortaleza
Amazon R.
Marañón R.
Teresina
B A S I N
Tapajós R.
Natal
Purus R.
Madeira R.
Campina Grande
PERU
Rio Branco
Pôrto Velho
Xingu R.
Araguaia R.
Tocantins R.
São Francisco R.
Recife
Ucayali R.
Maceló
Callao · Lima
Macchu Picchu
BRAZIL
Ayacucho
Cuzco
Salvador
Lago Titicaca (Lake Titicaca)
Trinidad
Arequipa
La Paz
Brasília
Oruro
BOLIVIA
Santa Cruz
Goiânia
Arica
Sucre
Uberlândia
PACIFIC OCEAN
Uberaba
Belo Horizonte
Iquique
20°S
Tarija
Campo Grande
Nova Iguaçu
Antofagasta
Paraguay R.
Paraná R.
Londrina
Campinas
Rio de Janeiro
PARAGUAY
São Paulo
Santos
TROPIC OF CAPRICORN
Salta
Ciudad del Este
Isla San Félix (Chile)
Asunción
Curitiba
Isla San Ambrosio (Chile)
San Miguel de Tucumán
Uruguaiana
Paraná R.
Santa Maria
Uruguay R.
Pôrto Alegre
Coquimbo
La Serena
ATLANTIC OCEAN
Archipiélago Juan Fernández (Chile)
Córdoba
Salto
Valparaíso
Mendoza
Rosario
URUGUAY
Santiago
ARGENTINA
Buenos Aires
La Plata
Montevideo
CHILE
Río de la Plata
Concepción
Bahía Blanca
Mar del Plata
Colorado R.
Negro R.
Puerto Montt

A N D E S

Comodoro Rivadavia

40°S

SOUTH AMERICA POLITICAL

Río Gallegos
Stanley
Punta Arenas
Tierra el Fuego
Falkland Islands (Islas Malvinas)
Administered by United Kingdom
(Claimed by Argentina)
Ushuaia

South Georgia Island
(U.K., Claimed by Argentina)

Cabo de Hornos (Cape Horn)

0 — 600 miles
0 — 600 kilometers
Lambert Azimuthal Equal-Area projection

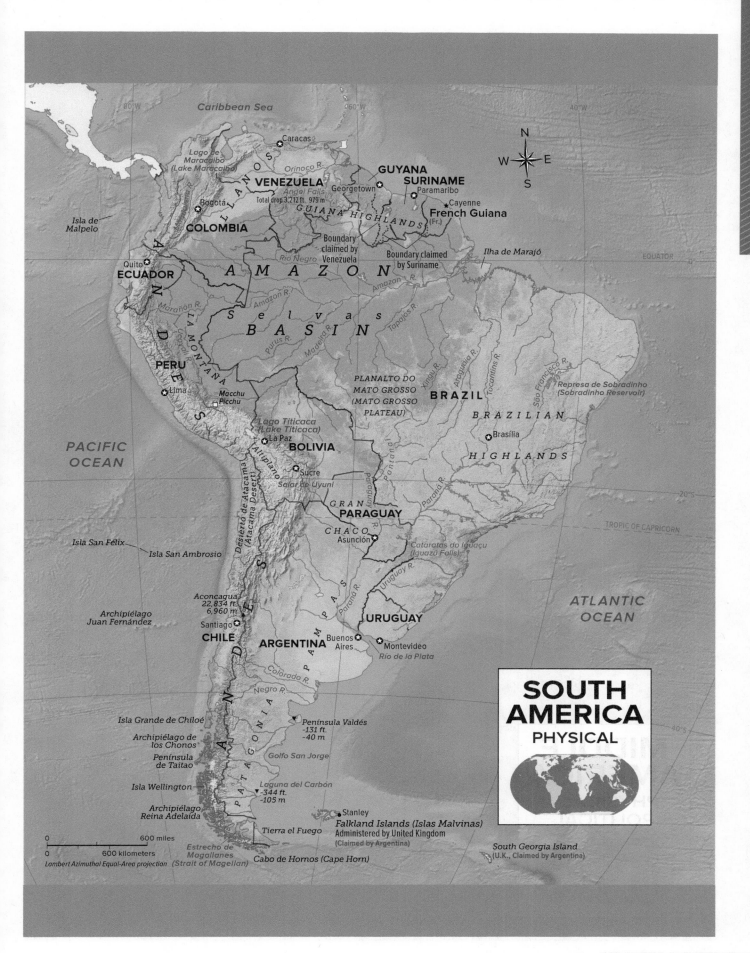

Caribbean Sea

Caracas

Lago de
Maracaibo
(Lake Maracaibo)

Orinoco R.

VENEZUELA

GUYANA
SURINAME

Georgetown

Paramaribo

Cayenne
French Guiana
(Fr.)

Angel Falls
Total drop 3,212 ft., 979 m

Bogotá

COLOMBIA

GUIANA HIGHLANDS

Isla de
Malpelo

Boundary
claimed by
Venezuela

Boundary claimed
by Suriname

Ilha de Marajó

EQUATOR

Rio Negro

A M A Z O N

Quito

ECUADOR

Marañón R.

Amazon R.

Amazon R.

Selvas

B A S I N

Topajos R.

purus R.

Madeira R.

Xingu R.

Araguaia R.

Tocantins R.

São Francisco R.

PERU

Lima

Macchu
Picchu

PLANALTO DO
MATO GROSSO
(MATO GROSSO
PLATEAU)

BRAZIL

B R A Z I L I A N

Represa de Sobradinho
(Sobradinho Reservoir)

Lago Titicaca
(Lake Titicaca)

La Paz

BOLIVIA

Sucre

Brasília

H I G H L A N D S

PACIFIC
OCEAN

Altiplano

Salar de Uyuni

Desierto de Atacama
(Atacama Desert)

GRAN

CHACO

PARAGUAY

Asunción

Paraguay R.

Paraná R.

Cataratas do Iguaçu
(Iguazú Falls)

20°S

TROPIC OF CAPRICORN

Isla San Félix

Isla San Ambrosio

Uruguay R.

Aconcagua
22,834 ft.
6,960 m

Archipiélago
Juan Fernández

Santiago

CHILE

ARGENTINA

Buenos
Aires

URUGUAY

Montevideo

Río de la Plata

ATLANTIC
OCEAN

Colorado R.

Negro R.

Isla Grande de Chiloé

Archipiélago de
los Chonos

Península
de Taitao

Isla Wellington

Archipiélago
Reina Adelaida

Península Valdés
-131 ft.
-40 m

Golfo San Jorge

Laguna del Carbón
▼-344 ft.
-105 m

Stanley

Falkland Islands (Islas Malvinas)
Administered by United Kingdom
(Claimed by Argentina)

40°S

0   600 miles

0   600 kilometers

Lambert Azimuthal Equal-Area projection

Tierra el Fuego

Estrecho de
Magallanes
(Strait of Magellan)

Cabo de Hornos (Cape Horn)

South Georgia Island
(U.K., Claimed by Argentina)

SOUTH
AMERICA
PHYSICAL

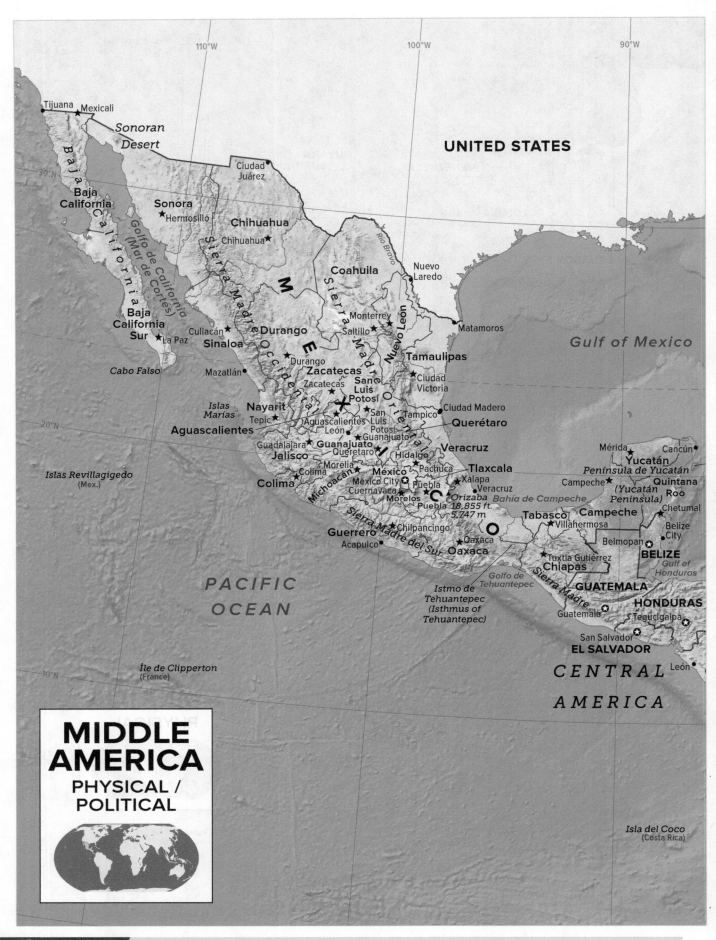

MIDDLE
AMERICA
PHYSICAL /
POLITICAL

Labels visible on the map:

UNITED STATES

Tijuana
Mexicali
Sonoran Desert
30°N
Baja California
Baja California (Mar de Cortés)
Golfo de California
Sonora
Hermosillo
Chihuahua
Chihuahua
Ciudad Juárez
Coahuila
Nuevo Laredo
Monterrey
Saltillo
Nuevo León
Matamoros
Gulf of Mexico
Baja California Sur
La Paz
Cabo Falso
Culiacán
Sinaloa
Durango
Durango
Mazatlán
Zacatecas
Zacatecas
San Luis Potosí
Tamaulipas
Ciudad Victoria
Islas Marías
Nayarit
Tepic
Aguascalientes
Aguascalientes
León
San Luis Potosí
Ciudad Madero
Tampico
Querétaro
Islas Revillagigedo (Mex.)
Guadalajara
Jalisco
Guanajuato
Querétaro
Guanajuato
Hidalgo
Pachuca
Veracruz
Mérida
Cancún
Yucatán
Peninsula de Yucatán
20°N
Colima
Morelia
Michoacán
México
Mexico City
Puebla
Xalapa
Veracruz
Campeche
Quintana Roo
Colima
Cuernavaca
Morelos
Puebla
Orizaba 18,855 ft. 5,747 m
Bahía de Campeche
(Yucatán Peninsula)
Chetumal
Guerrero
Chilpancingo
Tabasco
Villahermosa
Campeche
Belize City
Acapulco
Sierra Madre del Sur
Oaxaca
Oaxaca
Tuxtla Gutiérrez
Chiapas
Belmopan
BELIZE
Gulf of Honduras
Istmo de Tehuantepec (Isthmus of Tehuantepec)
Golfo de Tehuantepec
Sierra Madre
GUATEMALA
HONDURAS
PACIFIC OCEAN
Guatemala
Tegucigalpa
San Salvador
EL SALVADOR
León
CENTRAL AMERICA
Île de Clipperton (France)
10°N
Sierra Madre Occidental
Sierra Madre Oriental
Río Bravo

Isla del Coco (Costa Rica)

110°W   100°W   90°W

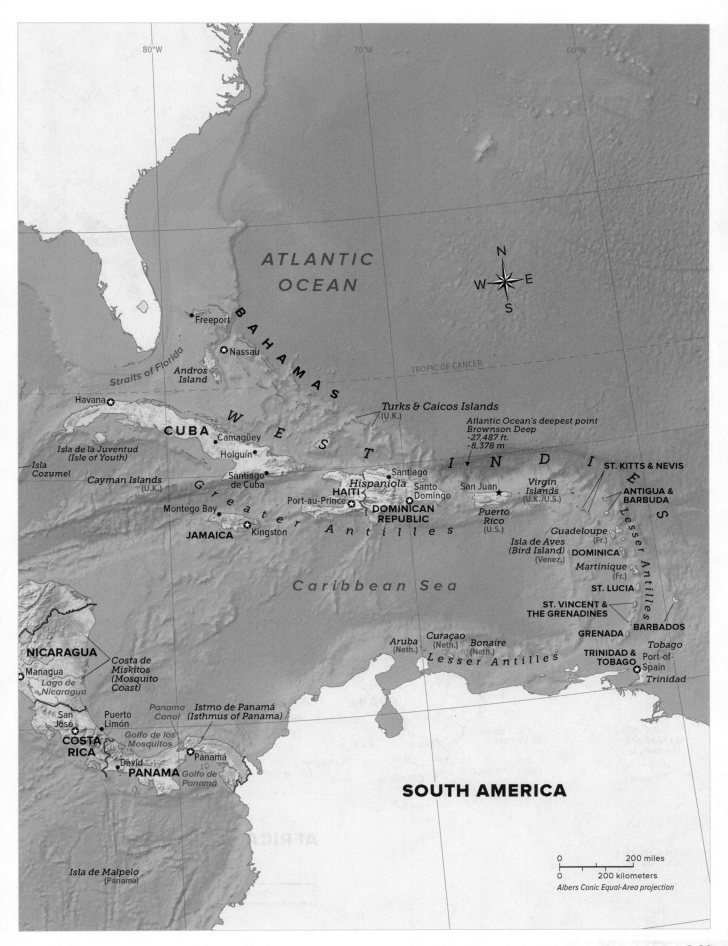

ATLANTIC
OCEAN

N
W E
S

Freeport

B
A
H
A
M
A
S

Nassau

Straits of Florida

Andros
Island

TROPIC OF CANCER

Turks & Caicos Islands
(U.K.)

Atlantic Ocean's deepest point
Brownson Deep
-27,487 ft.
-8,378 m

Havana

CUBA

Camagüey

Holguín

Isla de la Juventud
(Isle of Youth)

Isla
Cozumel

Cayman Islands
(U.K.)

W
E
S
T

Santiago
de Cuba

Montego Bay

JAMAICA

Kingston

G
r
e
a
t
e
r

A
n
t
i
l
l
e
s

Hispaniola

HAITI

Port-au-Prince

Santiago

Santo
Domingo

DOMINICAN
REPUBLIC

San Juan

Puerto
Rico
(U.S.)

I
N
D
I
E
S

Virgin
Islands
(U.K./U.S.)

ST. KITTS & NEVIS

ANTIGUA &
BARBUDA

Isla de Aves
(Bird Island)
(Venez.)

Guadeloupe
(Fr.)

DOMINICA

Martinique
(Fr.)

ST. LUCIA

L
e
s
s
e
r

A
n
t
i
l
l
e
s

Caribbean Sea

ST. VINCENT &
THE GRENADINES

BARBADOS

GRENADA

Aruba
(Neth.)

Curaçao
(Neth.)

Bonaire
(Neth.)

Lesser Antilles

TRINIDAD &
TOBAGO

Tobago

Port-of-
Spain

Trinidad

NICARAGUA

Managua

Costa de
Miskitos
(Mosquito
Coast)

Lago de
Nicaragua

San
José

Puerto
Limón

COSTA
RICA

David

PANAMA

Panamá

Panama
Canal

Istmo de Panamá
(Isthmus of Panama)

Golfo de los
Mosquitos

Golfo de
Panamá

SOUTH AMERICA

Isla de Malpelo
(Panama)

0   200 miles
0   200 kilometers
Albers Conic Equal-Area projection

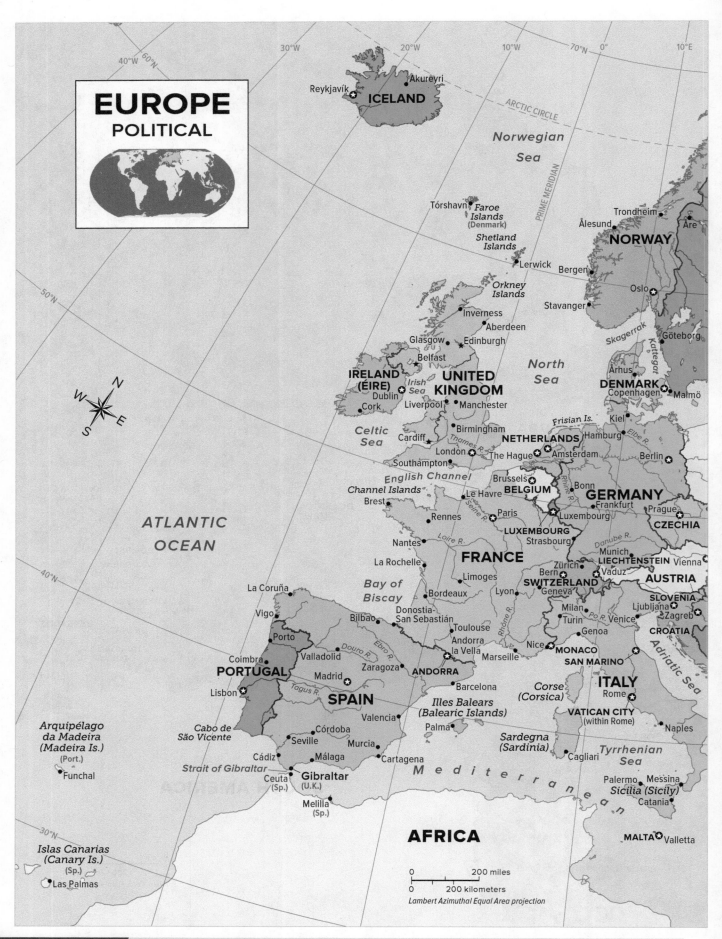

EUROPE
POLITICAL

ICELAND
Reykjavík • Akureyri

Norwegian Sea

ARCTIC CIRCLE

Tórshavn *Faroe Islands (Denmark)*

Shetland Islands

Lerwick

Orkney Islands

Inverness
Aberdeen
Glasgow • Edinburgh

NORWAY
Trondheim
Ålesund
Åre
Bergen
Stavanger
Oslo
Skagerrak
Göteborg
Kattegat
Århus
DENMARK
Copenhagen
Malmö

IRELAND (ÉIRE)
Dublin
Cork
Irish Sea

UNITED KINGDOM
Belfast
Liverpool • Manchester
Birmingham
Cardiff
London
Southampton
North Sea
Frisian Is.
Kiel
Hamburg
Elbe R.
Berlin

Celtic Sea

English Channel
Channel Islands
Brest
Le Havre
Rennes
Paris
Seine R.
NETHERLANDS
The Hague
Amsterdam
Brussels
BELGIUM
Bonn
GERMANY
Frankfurt
Prague
CZECHIA
Rhine R.
Luxembourg
LUXEMBOURG

ATLANTIC OCEAN

Nantes
Loire R.
La Rochelle
Limoges
FRANCE
Bordeaux
Strasbourg
Danube R.
Munich
LIECHTENSTEIN Vienna
Zürich
Bern
SWITZERLAND
Geneva
Vaduz
AUSTRIA
Lyon
Rhône R.
Milan
Turin
Po R. Venice
SLOVENIA
Ljubljana
Zagreb
CROATIA
Genoa
Nice
MONACO
SAN MARINO
Adriatic Sea

La Coruña
Vigo
Porto
Coimbra
Bilbao
Donostia-San Sebastián
Toulouse
Andorra la Vella
Marseille
Valladolid
Douro R.
Ebro R.
Zaragoza
ANDORRA
Corse (Corsica)
ITALY
Rome
PORTUGAL
Madrid
SPAIN
Barcelona
Tagus R.
VATICAN CITY (within Rome)
Lisbon
Valencia
Illes Balears (Balearic Islands)
Naples
Palma
Sardegna (Sardinia)
Tyrrhenian Sea

Bay of Biscay

Arquipélago da Madeira (Madeira Is.) (Port.)

Córdoba
Seville
Murcia
Cartagena
Cagliari
Cabo de São Vicente
Cádiz
Málaga
Funchal
Strait of Gibraltar
Ceuta (Sp.)
Gibraltar (U.K.)
Palermo Messina
Sicilia (Sicily)
Catania
Melilla (Sp.)

M e d i t e r r a n e a n

AFRICA

MALTA Valletta

Islas Canarias (Canary Is.) (Sp.)
Las Palmas

0 ————— 200 miles
0 ————— 200 kilometers
Lambert Azimuthal Equal Area projection

20°E 30°E 40°E 50°E 60°E 70°E

Barents Sea

Tromsø

Murmansk

Ivalo

Kirovsk

Kiruna

Umba

Beloye More (White Sea)

Tobseda

Pechora R.

Pechora

Kemi

Luleå

Oulu

Kem

Arkhangel'sk

Severodvinsk

Severnaya Dvina R.

Syktyvkar

URAL MOUNTAINS

A commonly accepted division between Europe and Asia is formed by the Ural Mountains, Ural River, Caspian Sea, Caucasus Mountains, the Black Sea with its outlets, the Bosporus and the Dardanelles.

Europe/Asia boundary

ASIA

Umeå

FINLAND

Kuopio

SWEDEN

Vaasa

Sundsvall

Tampere

Onezhskoye Ozero (Lake Onega)

Perm'

Kirov

Gulf of Bothnia

Pori

Turku

Helsinki

Ladozhskoye Ozero (Lake Ladoga)

RUSSIA

Kama R.

Uppsala

Stockholm

Tallinn

St. Petersburg

Gulf of Finland

ESTONIA

Novgorod

Yaroslavl'

Ufa

Gotland

LATVIA

Tver'

Nizhniy Novgorod

Kazan'

Öland

Riga

Volga R.

Baltic Sea

Daugavpils

Moscow

Samara

Orenburg

LITHUANIA

Vitsyebsk

Smolensk

Ryazan'

Penza

Ural R.

Kaliningrad

Kaunas

RUSSIA

Vilnius

Saratov

Oral

Gdańsk

Minsk

Bryansk

Volga R.

Bydgoszcz

BELARUS

Homyel'

Kursk

KAZAKHSTAN

POLAND

Warsaw

Łódź

Chernihiv

Sumy

Volgograd

Wrocław

Vistula R.

Oder R.

Kyiv (Kiev)

Kharkiv

Don R.

Kraków

L'viv

Dnieper R.

Poltava

Dniester R.

Vinnytsya

Dnipropetrovs'k

Astrakhan

SLOVAKIA

UKRAINE

Donets'k

Don R.

Bratislava

Rostov

Caspian Sea

Budapest

MOLDOVA

Chişinău

HUNGARY

Tisza R.

Stavropol'

Drava R.

ROMANIA

Odessa

Sea of Azov

GrozNyy

GEORGIA

Crimea

Kerch

Caucasus

BOSNIA & HERZEGOVINA

Beograd (Belgrade)

Bucureşti (Bucharest)

Simferopol'

Sevastopol'

Yalta

Mountains

Baku

Sarajevo

SERBIA

Danube R.

Constanţa

Black Sea

AZERBAIJAN

Pristina

Sofia

Varna

MONT.

KOSOVO

Skopje

BULGARIA

Podgorica

NORTH MACEDONIA

Bosporus

Tiranë

Istanbul

ALBANIA

Thessaloníki

TURKEY

Dardanelles

Sea of Marmara

Ionian Sea

GREECE

Aegean Sea

Athína (Athens)

ASIA

Ionian Is.

Cyprus includes a Greek Cypriot south with an internationally recognized government. The government of the Turkish Cypriot north is recognized only by Turkey. The UN patrols a buffer zone between the two and works toward reunification of the island.

Ródos (Rhodes)

Nicosia

Iraklíon

CYPRUS

Kríti (Crete)

Sea

EUROPE
PHYSICAL

Reykjavík ★ **ICELAND**
Surtsey —

ARCTIC CIRCLE

Lofoten

Norwegian Sea

PRIME MERIDIAN

NORWAY

Oslo ★

Rockall —

Faroe Islands

Shetland Islands

Vänern

Outer Hebrides
Isle of Lewis

Orkney Islands

Skagerrak

Vättern

Inner Hebrides

Highlands

British Isles

Edinburgh ★

Kattegat

Jutland

Sjælland

Belfast ★

North Sea

DENMARK
Copenhagen ★

IRELAND (ÉIRE) ★
Dublin

Irish Sea

UNITED KINGDOM

Frisian Is.

Elbe R.

Cardiff ★

Great Britain

NETHERLANDS

Berlin ★

N O R

Celtic Sea

Thames R.

London ★

The Hague ★ ★ Amsterdam

Brussels ★

Land's End

English Channel

BELGIUM

Rhine R.

GERMANY

Channel Islands

Seine R.

Paris ★

★ Luxembourg

Prague ★

N O R

Brittany

LUXEMBOURG

CZECHIA

ATLANTIC OCEAN

Loire R.

Danube R.

LIECHTENSTEIN

Vienna ★

FRANCE

Vaduz ★

Bay of Biscay

Bern ★

AUSTRIA

Mont Blanc 15,771 ft. 4,807 m

SWITZERLAND

A L P S

SLOVENIA

Ljubljana ★

Cordillera Cantabrica

Pyrenees

Massif Central

Po R.

Rhône R.

Andorra la Vella

Zagreb ★

Douro R.

Ebro R.

MONACO ★

Riviera

SAN MARINO ★

Adriatic Sea

CROATIA

I B E R I A N

ANDORRA

Corse (Corsica)

ITALY

Appennines

PORTUGAL

Madrid ●

Rome ★

Lisbon ★

Tagus R.

SPAIN

Illes Balears (Balearic Islands)

VATICAN CITY (within Rome)

Arquipélago da Madeira (Madeira Is.)
(Port.)

P E N I N S U L A

Sardegna (Sardinia)

Tyrrhenian Sea

Cabo de São Vicente

Sistemas Béticos

Strait of Gibraltar

M e d i t e r r a n e a n

Sicilia (Sicily)

Ceuta (Sp.) ●

Gibraltar (U.K.)

Etna ▲
10,902 ft.
3,323 m

Melilla (Sp.) ●

Islas Canarias (Canary Is.)
(Sp.)

AFRICA

MALTA ★ Valletta

0 _____ 200 miles
0 _____ 200 kilometers
Lambert Azimuthal Equal Area projection

A commonly accepted division between Europe and Asia is formed by the Ural Mountains, Ural River, Caspian Sea, Caucasus Mountains, the Black Sea with its outlets, the Bosporus and the Dardanelles.

20°E Nordkapp 30°E 40°E Ostrov 50°E 60°E 70°E
Kolguyev

Vesterålen Barents Sea

Poluostrov
Kanin

U R A L

Kol'skiy
Poluostrov

Beloye More (White Sea)

Pechora R.

Europe/Asia
boundary

ASIA

D I N A V I A N D
L A P L A N D

Severnaya Dvina R.

M O U N T A I N S

FINLAND
Järvi-Suomi
(Finnish
Lakeland)

Onezhskoye Ozero
(Lake Onega)

SWEDEN

Gulf of Bothnia

RUSSIA

Kama R.

Helsinki

Ladozhskoye
Ozero
(Lake Ladoga)

Gulf of Finland
Stockholm   Tallinn

ESTONIA

Volga R.

Kuybyshevskoye
(Samara Reservoir)

Gotland

Baltic Sea

LATVIA
Riga

Moscow

Öland

C E N T R A L

LITHUANIA
Vilnius

RUSSIA

R U S S I A N

Ural R.

Minsk

KAZAKHSTAN

BELARUS

U P L A N D

Warsaw

T H E R N

Vistula R.

Volga R.

Oder R.

E U R O P E A N   P L A I N

Caspian Depression

POLAND

Kyiv (Kiev)

Don R.

Dnieper R.

Caspian Sea

SLOVAKIA
Bratislava

Dniester R.

UKRAINE

Tsimlyanskoye
Vodokhranilishche
(Tsimlyansk
Reservoir)

Volga R.

Budapest

Carpathian Mts

MOLDOVA
Chișinău

Don R.

HUNGARY

Tisza R.

Sea of
Azov

ROMANIA

Crimea

Elbrus
18,510 ft.
5,642 m

Drava R.

Beograd (Belgrade)

București
(Bucharest)

Caucasus Mountains

**BOSNIA &
HERZEGOVINA**

Caspian Sea

Sarajevo

SERBIA
MONT.

Danube R.

GEORGIA

Baku

Podgorica

KOSOVO
Pristina

BALKAN
Sofia

Black Sea

AZERBAIJAN

Skopje

BULGARIA

Bosporus

Tiranë

**NORTH
MACEDONIA**

P E N I N S U L A

ALBANIA

T U R K E Y

Ionian
Sea

Dardanelles

Sea of
Marmara

GREECE

Aegean

ASIA

Ionian
Is.

Athína
(Athens)

Peloponnísos

Sea

Ródos
(Rhodes)

Nicosia

Kríti (Crete)

CYPRUS

Sea

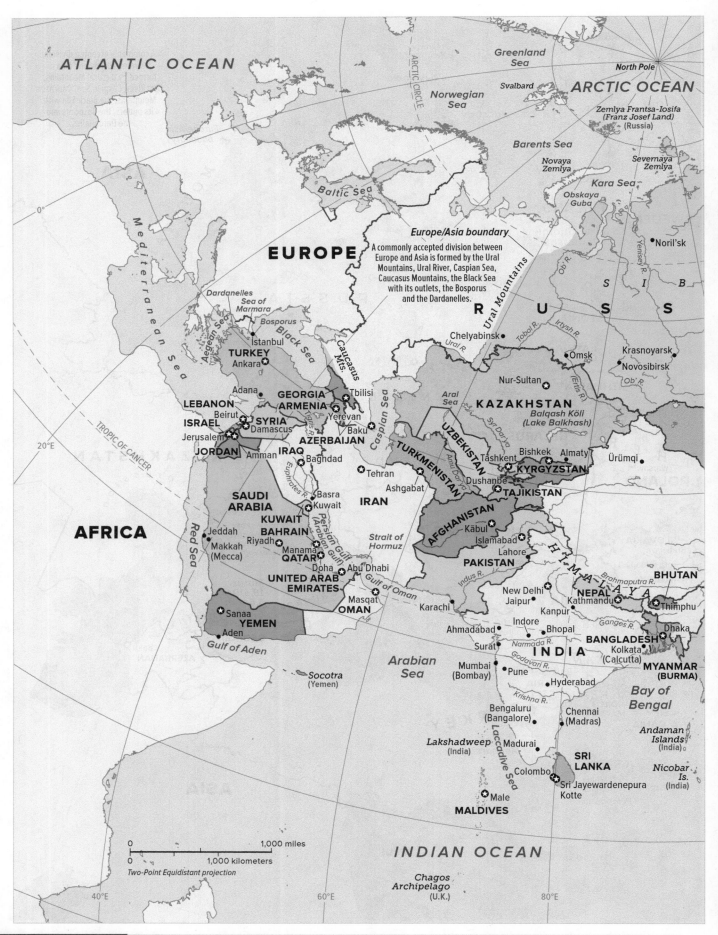

ATLANTIC OCEAN

Greenland Sea

ARCTIC OCEAN

North Pole

ARCTIC CIRCLE

Norwegian Sea

Svalbard

Zemlya Frantsa-Iosifa
(Franz Josef Land)
(Russia)

Barents Sea

Novaya Zemlya

Severnaya Zemlya

Kara Sea

0°

Baltic Sea

Obskaya Guba

Noril'sk

EUROPE

Europe/Asia boundary

A commonly accepted division between
Europe and Asia is formed by the Ural
Mountains, Ural River, Caspian Sea,
Caucasus Mountains, the Black Sea
with its outlets, the Bosporus
and the Dardanelles.

Ural Mountains

Yenisey R.

Ob R.

S I B
R U S S

Dardanelles
Sea of
Marmara

Bosporus

Black Sea

Caucasus Mts.

Chelyabinsk

Tobol R.

Ural R.

Irtysh R.

Ertis R.

Omsk

Krasnoyarsk

Novosibirsk

Ob' R.

Mediterranean Sea

Aegean Sea

Istanbul

TURKEY
Ankara

Adana

GEORGIA
ARMENIA

Tbilisi

Nur-Sultan

KAZAKHSTAN

Balqash Köli
(Lake Balkhash)

LEBANON
Beirut
ISRAEL
Jerusalem
JORDAN
Amman

SYRIA
Damascus

Yerevan

AZERBAIJAN

Baku

Caspian Sea

IRAQ

Baghdad

Tigris R.

Euphrates R.

20°E

TROPIC OF CANCER

Aral Sea

Syr Dar'ya

Tashkent

UZBEKISTAN

Amu Dar'ya

TURKMENISTAN

Tehran

Bishkek

Almaty

Ürümqi

KYRGYZSTAN

Dushanbe

TAJIKISTAN

Ashgabat

IRAN

SAUDI
ARABIA

Basra
Kuwait

KUWAIT

BAHRAIN

Persian Gulf
(Arabian Gulf)

Strait of
Hormuz

AFGHANISTAN

Kābul

Islamabad

Lahore

HIMALAYA

Indus R.

Brahmaputra R.

BHUTAN

Thimphu

AFRICA

Red Sea

Jeddah

Makkah
(Mecca)

Riyadh

Manama

QATAR

Doha

UNITED ARAB
EMIRATES

Abu Dhabi

Gulf of Oman

Masqat

OMAN

PAKISTAN

New Delhi
Jaipur

Kanpur

Indore

Bhopal

Ganges R.

NEPAL
Kathmandu

Dhaka

BANGLADESH

Kolkata
(Calcutta)

Karachi

Ahmadabad

Surat

Narmada R.

INDIA

Godavari R.

MYANMAR
(BURMA)

YEMEN

Sanaa

Aden

Gulf of Aden

Socotra
(Yemen)

Arabian
Sea

Mumbai
(Bombay)

Pune

Hyderabad

Krishna R.

Bay of
Bengal

Bengaluru
(Bangalore)

Chennai
(Madras)

Andaman
Islands
(India)

Lakshadweep
(India)

Madurai

Laccadive Sea

SRI
LANKA

Nicobar
Is.
(India)

Colombo

Sri Jayewardenepura
Kotte

Male

MALDIVES

0   1,000 miles
0   1,000 kilometers
Two-Point Equidistant projection

INDIAN OCEAN

Chagos
Archipelago
(U.K.)

40°E   60°E   80°E

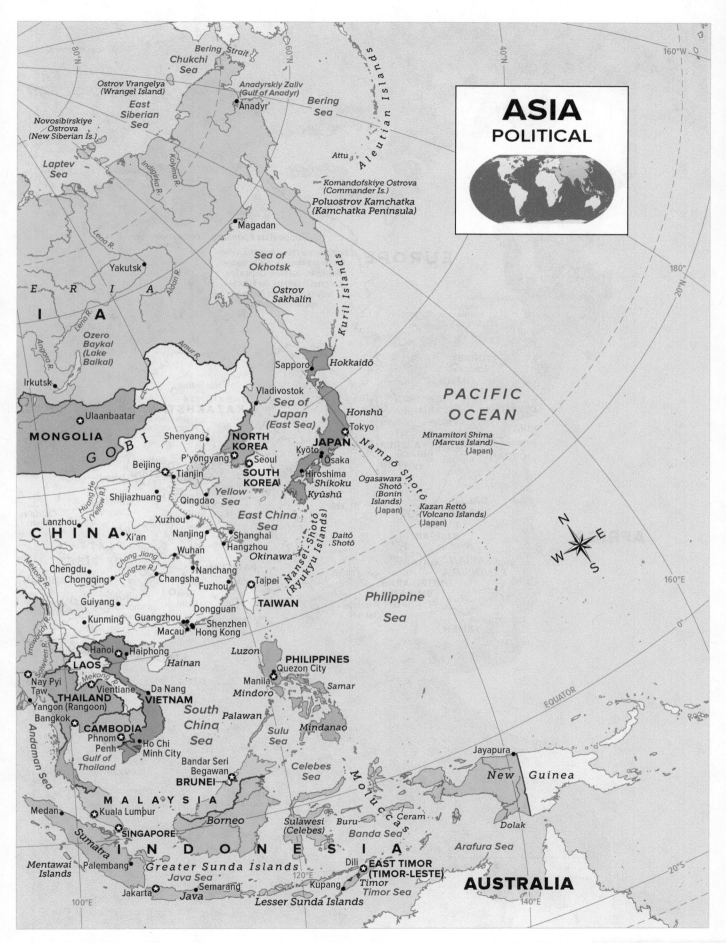

ASIA
POLITICAL

Chukchi Sea
Ostrov Vrangelya (Wrangel Island)
East Siberian Sea
Novosibirskiye Ostrova (New Siberian Is.)
Laptev Sea
Bering Strait
Anadyrskiy Zaliv (Gulf of Anadyr)
Anadyr'
Bering Sea
Attu
Aleutian Islands
Komandofskiye Ostrova (Commander Is.)
Poluostrov Kamchatka (Kamchatka Peninsula)
Magadan
Yakutsk
Sea of Okhotsk
Ostrov Sakhalin
Lena R.
Indigirka R.
Kolyma R.
Aldan R.
Amur R.
Lena R.
Angara R.
Ozero Baykal (Lake Baikal)
Irkutsk
Ulaanbaatar
MONGOLIA
GOBI
Shenyang
Beijing
Tianjin
Shijiazhuang
Huang He (Yellow R.)
Lanzhou
CHINA
Xi'an
Xuzhou
Nanjing
Wuhan
Chang Jiang (Yangtze R.)
Chengdu
Chongqing
Changsha
Nanchang
Fuzhou
Mekong R.
Guiyang
Kunming
Guangzhou
Dongguan
Macau
Shenzhen
Hong Kong
Sapporo
Hokkaidō
Vladivostok
Sea of Japan (East Sea)
Honshū
Tokyo
JAPAN
Kyōto
Osaka
Hiroshima
Shikoku
Kyūshū
NORTH KOREA
P'yŏngyang
Seoul
SOUTH KOREA
Yellow Sea
Qingdao
Shanghai
Hangzhou
Okinawa
Nansei-Shotō (Ryukyu Islands)
East China Sea
Daitō Shotō
Taipei
TAIWAN
Kuril Islands
PACIFIC OCEAN
Minamitori Shima (Marcus Island) (Japan)
Nampō Shotō
Ogasawara Shotō (Bonin Islands) (Japan)
Kazan Rettō (Volcano Islands) (Japan)
Philippine Sea
Hanoi
Haiphong
LAOS
Nay Pyi Taw
Vientiane
THAILAND
Yangon (Rangoon)
Bangkok
VIETNAM
Da Nang
CAMBODIA
Phnom Penh
Ho Chi Minh City
Gulf of Thailand
Hainan
Luzon
PHILIPPINES
Quezon City
Manila
Mindoro
Samar
Palawan
South China Sea
Mindanao
Sulu Sea
Irrawaddy R.
Salween R.
Mekong R.
Andaman Sea
Medan
Bandar Seri Begawan
BRUNEI
MALAYSIA
Kuala Lumpur
SINGAPORE
Borneo
Celebes Sea
Sulawesi (Celebes)
Buru
Ceram
Moluccas
Banda Sea
Jayapura
New Guinea
Dolak
Arafura Sea
Sumatra
Mentawai Islands
Palembang
INDONESIA
Greater Sunda Islands
Java Sea
Jakarta
Semarang
Java
Dili
EAST TIMOR (TIMOR-LESTE)
Timor
Kupang
Timor Sea
Lesser Sunda Islands
AUSTRALIA
EQUATOR

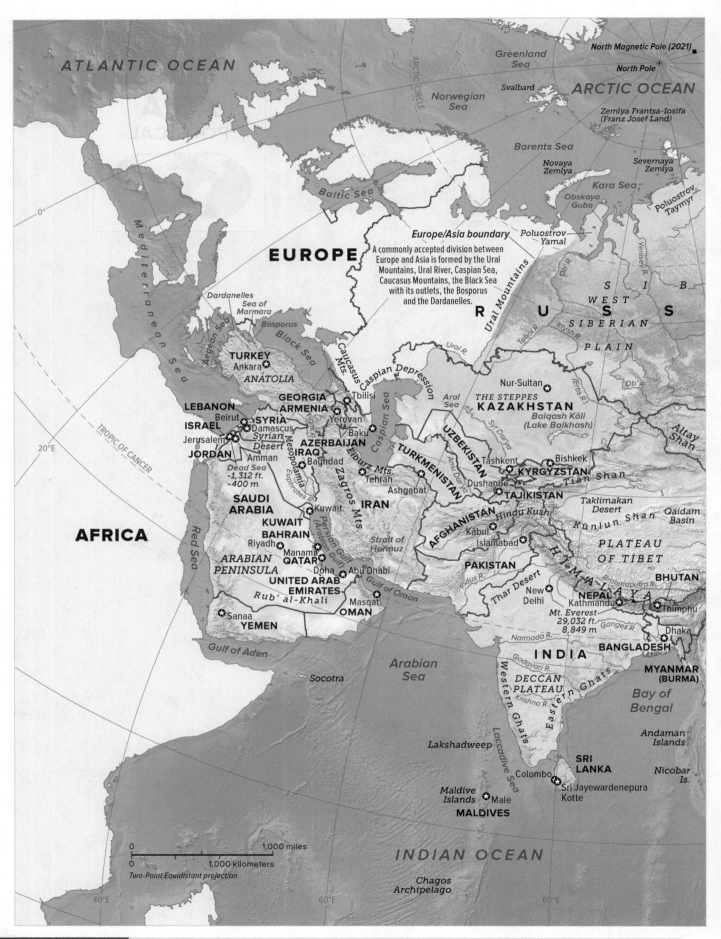

ATLANTIC OCEAN

ARCTIC OCEAN

North Magnetic Pole (2021)

North Pole

Greenland Sea

Norwegian Sea

Svalbard

Zemlya Frantsa-Iosifa (Franz Josef Land)

Barents Sea

Novaya Zemlya

Severnaya Zemlya

Kara Sea

Poluostrov Taymyr

Baltic Sea

Obskaya Guba

Poluostrov Yamal

EUROPE

Europe/Asia boundary
A commonly accepted division between Europe and Asia is formed by the Ural Mountains, Ural River, Caspian Sea, Caucasus Mountains, the Black Sea with its outlets, the Bosporus and the Dardanelles.

Ural Mountains

RUSSIA

WEST SIBERIAN PLAIN

Ob' R.

Yenisey R.

Ob' R.

Tobol R.

Irtysh R.

Ural R.

Mediterranean Sea

Dardanelles
Sea of Marmara
Bosporus

Aegean Sea

Black Sea

Caucasus Mts.

Caspian Depression

TURKEY
Ankara
ANATOLIA

GEORGIA
Tbilisi

ARMENIA
Yerevan

Caspian Sea

Aral Sea

THE STEPPES
KAZAKHSTAN
Nur-Sultan
Balqash Köli (Lake Balkhash)

Altay Shan

LEBANON
Beirut
SYRIA
Damascus
Syrian Desert

ISRAEL
Jerusalem

Baku

AZERBAIJAN
IRAQ
Baghdad
Mesopotamia
Euphrates R.

Elburz Mts.
Tehran

TURKMENISTAN
Ashgabat

UZBEKISTAN
Tashkent

Amu Darya
Syr Darya

Bishkek
KYRGYZSTAN

Dushanbe
TAJIKISTAN

Tian Shan

JORDAN
Amman

TROPIC OF CANCER

20°E

Dead Sea
-1,312 ft.
-400 m

Zagros Mts.

IRAN

Strait of Hormuz

AFGHANISTAN
Kabul
Islamabad

Hindu Kush

Kunlun Shan

Taklimakan Desert

Qaidam Basin

SAUDI ARABIA

KUWAIT
Kuwait

BAHRAIN
Riyadh
Manama
QATAR
Doha

Persian Gulf (Arabian Gulf)

Abu Dhabi

PAKISTAN

Indus R.

PLATEAU OF TIBET

HIMALAYA

BHUTAN
Thimphu

AFRICA

Red Sea

ARABIAN PENINSULA

UNITED ARAB EMIRATES
Rub' al-Khali

Gulf of Oman

Masqat
OMAN

Thar Desert

New Delhi

NEPAL
Kathmandu

Mt. Everest
29,032 ft.
8,849 m

Brahmaputra R.

Ganges R.

Dhaka

BANGLADESH

Sanaa
YEMEN

Gulf of Aden

Arabian Sea

Socotra

Narmada R.

Godavari R.

INDIA

DECCAN PLATEAU

Western Ghats

Eastern Ghats

Krishna R.

MYANMAR (BURMA)

Bay of Bengal

Andaman Islands

Lakshadweep

Laccadive Sea

SRI LANKA
Colombo
Sri Jayewardenepura Kotte

Nicobar Is.

Maldive Islands
Male
MALDIVES

0   1,000 miles
0   1,000 kilometers
Two-Point Equidistant projection

INDIAN OCEAN

Chagos Archipelago

40°E   60°E   80°E

0°

ARCTIC CIRCLE

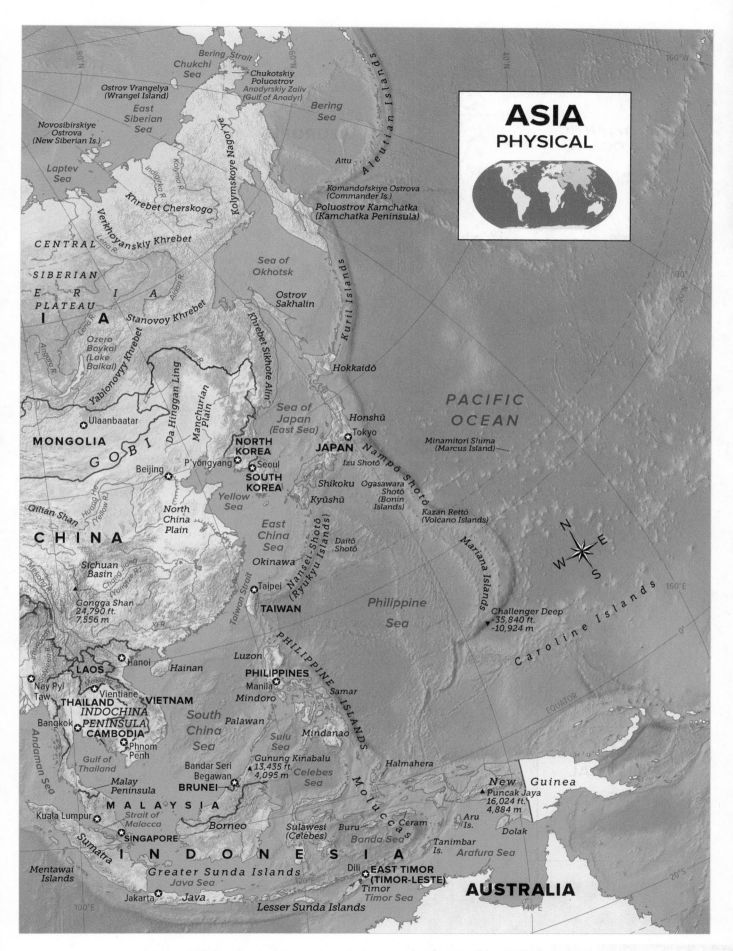

ASIA
PHYSICAL

Ostrov Vrangelya (Wrangel Island)
Chukchi Sea
Bering Strait
Chukotskiy Poluostrov
Anadyrskiy Zaliv (Gulf of Anadyr)
Bering Sea

Novosibirskiye Ostrova (New Siberian Is.)
East Siberian Sea
Laptev Sea

Kolymskoye Nagor'ye
Khrebet Cherskogo
Indigirka R.
Kolyma R.

Attu
Aleutian Islands

Verkhoyanskiy Khrebet
Lena R.

CENTRAL
SIBERIAN
PLATEAU

Komandofskiye Ostrova (Commander Is.)
Poluostrov Kamchatka (Kamchatka Peninsula)

Sea of Okhotsk

Lena R.
Aldan R.
Stanovoy Khrebet

Ostrov Sakhalin

Angara R.
Ozero Baykal (Lake Baikal)
Yablonovyy Khrebet

Amur R.
Khrebet Sikhote Alin'

Kuril Islands

PACIFIC
OCEAN

Da Hinggan Ling
Manchurian Plain

Hokkaidō

Ulaanbaatar

MONGOLIA
G O B I

Sea of Japan (East Sea)

Honshū
Tokyo

Minamitori Shima (Marcus Island)

Beijing
Huang He (Yellow R.)

NORTH KOREA
P'yŏngyang
Seoul
SOUTH KOREA

JAPAN

Izu Shotō

Nampo Shotō

Qilian Shan
North China Plain

Yellow Sea

Shikoku
Kyūshū

Ogasawara Shotō (Bonin Islands)

Kazan Rettō (Volcano Islands)

C H I N A

East China Sea

Daitō Shotō

Sichuan Basin
Chang Jiang (Yangtze R.)
Mekong R.

Gongga Shan
24,790 ft.
7,556 m

Okinawa

Nansei-Shotō (Ryukyu Islands)

Taipei
TAIWAN

Taiwan Strait

Mariana Islands

Philippine Sea

Caroline Islands

Challenger Deep
-35,840 ft.
-10,924 m

N
E
S
W

160°E

0°

Irrawaddy R.

Salween R.
Mekong R.

LAOS
Hanoi

Hainan

Luzon

PHILIPPINES
Manila
Mindoro

Samar

PHILIPPINE ISLANDS

Nay Pyi Taw

Vientiane
THAILAND
INDOCHINA
PENINSULA

VIETNAM

South China Sea

Palawan

Mindanao

CAMBODIA

Bangkok

Phnom Penh

Sulu Sea

Gulf of Thailand

Bandar Seri Begawan

Gunung Kinabalu
13,435 ft.
4,095 m
Celebes Sea

Halmahera

Moluccas

New Guinea
Puncak Jaya
16,024 ft.
4,884 m

Andaman Sea

Malay Peninsula

BRUNEI

Kuala Lumpur
M A L A Y S I A
Strait of Malacca

Borneo

Sulawesi (Celebes)

Buru
Banda Sea

Ceram

Aru Is.

Dolak

SINGAPORE

Sumatra

I N D O N E S I A

Tanimbar Is.

Arafura Sea

20°S

Mentawai Islands

Greater Sunda Islands

Dili
EAST TIMOR (TIMOR-LESTE)

Jakarta
Java
Java Sea
Lesser Sunda Islands

Timor
Timor Sea

AUSTRALIA

100°E

120°E

140°E

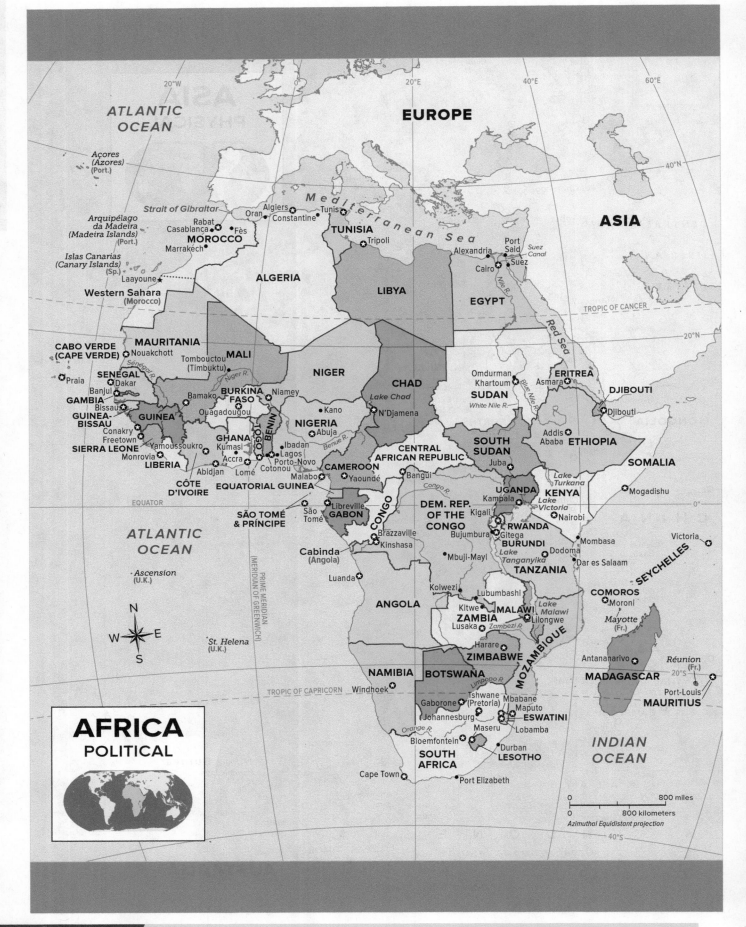

ATLANTIC
OCEAN

EUROPE

ASIA

Mediterranean Sea

Açores
(Azores)
(Port.)

Arquipélago
da Madeira
(Madeira Islands)
(Port.)

Islas Canarias
(Canary Islands)
(Sp.)

Strait of Gibraltar
Oran ● Algiers ● Tunis
● Constantine

Rabat ●
Casablanca ● ● Fès
MOROCCO
Marrakech ●

TUNISIA
● Tripoli

Alexandria ● Port
Said
● Cairo ● Suez

Suez
Canal

Laayoune ★- - - -

Western Sahara
(Morocco)

ALGERIA

LIBYA

EGYPT

TROPIC OF CANCER

CABO VERDE
(CAPE VERDE)

MAURITANIA
Nouakchott ●

MALI

Tombouctou
(Timbuktu) ●

NIGER

CHAD

Omdurman ●
Khartoum ●

ERITREA
Asmara ●

DJIBOUTI

20°N

Senegal R.
SENEGAL
Dakar ●
Banjul ●
GAMBIA
Bissau ●
GUINEA-
BISSAU
Conakry ●
Freetown ●
SIERRA LEONE
Monrovia ●
LIBERIA

● Praia

Niger R.
Bamako ●

BURKINA
FASO
Ouagadougou ●

Niamey ●

● Kano

Lake Chad

N'Djamena ●

SUDAN

White Nile R.

Blue Nile R.

Addis
Ababa ●
Djibouti ●

ETHIOPIA

NIGERIA
● Abuja

Benue R.

CENTRAL
AFRICAN REPUBLIC

SOUTH
SUDAN
Juba ●

SOMALIA

GHANA
Yamoussoukro ●
Kumasi ●
Accra ●
Abidjan ●
CÔTE
D'IVOIRE

TOGO
BENIN
Ibadan ●
Lagos ●
Porto-Novo ●
Lomé Cotonou

CAMEROON
Malabo ●
● Yaoundé

● Bangui

Lake
Turkana

EQUATOR

SÃO TOMÉ
& PRÍNCIPE

São
Tomé ●
Librevílle ●
GABON

EQUATORIAL GUINEA

Congo R.

UGANDA
Kampala ●

KENYA

Lake
Victoria

Mogadishu ●

0°

DEM. REP.
OF THE
CONGO

Kigali ●
RWANDA
Gitega ●
BURUNDI

Nairobi ●

CONGO
Brazzaville ●
Kinshasa ●

Bujumbura ●

Mombasa ●

SEYCHELLES

Cabinda
(Angola)

Luanda ●

Mbuji-Mayi ●

Lake
Tanganyika

Dodoma ●

Dar es Salaam ●

Victoria ●

ATLANTIC
OCEAN

Kolwezi ●

Lubumbashi ●

TANZANIA

COMOROS
Moroni ●

Mayotte
(Fr.)

PRIME MERIDIAN
(MERIDIAN OF GREENWICH)

Ascension
(U.K.)

ANGOLA

Kitwe ●
ZAMBIA
Lusaka ●

MALAWI
Lilongwe ●

Lake
Malawi

Zambezi R.

St. Helena
(U.K.)

Harare ●

ZIMBABWE

MOZAMBIQUE

Antananarivo ●

Réunion
(Fr.)

N
W ✦ E
S

NAMIBIA

BOTSWANA

Limpopo R.

MADAGASCAR

20°S

Port-Louis ●
MAURITIUS

TROPIC OF CAPRICORN
Windhoek ●

Gaborone ●
Johannesburg ●

Tshwane
(Pretoria) ●
Maseru ●

Mbabane ●
Maputo ●
ESWATINI
Lobamba

AFRICA
POLITICAL

Orange R.
Bloemfontein ●

SOUTH
AFRICA

Durban ●

LESOTHO

INDIAN
OCEAN

Cape Town ●

● Port Elizabeth

40°S

0   800 miles
0   800 kilometers
Azimuthal Equidistant projection

20°W   0   20°E   40°E   60°E
40°N

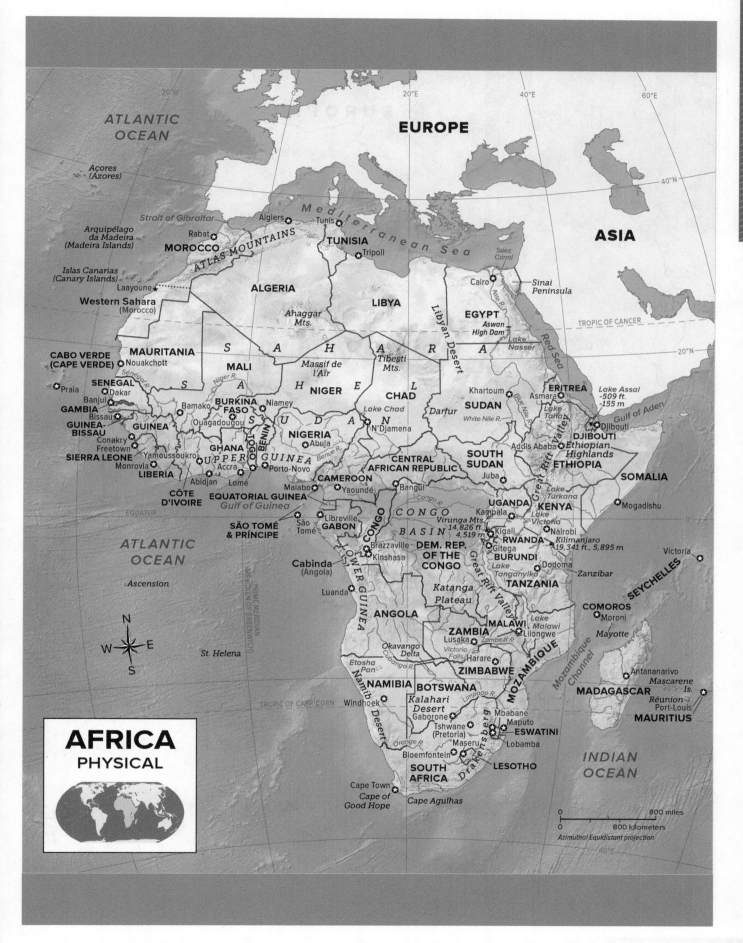

AFRICA
PHYSICAL

ATLANTIC OCEAN

ASIA

EUROPE

Açores (Azores)

Arquipélago da Madeira (Madeira Islands)

Islas Canarias (Canary Islands)

Strait of Gibraltar

ATLAS MOUNTAINS

Mediterranean Sea

Algiers · Tunis

Rabat ·

MOROCCO

TUNISIA

Tripoli ·

Suez Canal

Cairo ·

Sinai Peninsula

Laayoune ★

Western Sahara (Morocco)

ALGERIA

LIBYA

EGYPT

Ahaggar Mts.

Aswan High Dam

Lake Nasser

TROPIC OF CANCER

CABO VERDE (CAPE VERDE)

Nouakchott ·

MAURITANIA

S A H

MALI

Massif de l'Aïr

Tibesti Mts.

A R A

Libyan Desert

Red Sea

Lake Assal -509 ft. -155 m

20°N

· Praia

SENEGAL

Dakar ·

Senegal R.

Niger R.

BURKINA

NIGER

CHAD

Khartoum ·

ERITREA

Asmara ·

Gulf of Aden

Banjul ·

GAMBIA

Bamako ·

FASO

Niamey ·

S A

Darfur

SUDAN

Blue Nile R.

Lake Tana

Bissau ·

GUINEA-BISSAU

Ouagadougou ·

H E L

N'Djamena

White Nile R.

DJIBOUTI

Djibouti ·

Conakry ·

GUINEA

S U

D A N

CENTRAL AFRICAN REPUBLIC

SOUTH SUDAN

Addis Ababa ·

Ethiopian Highlands

Freetown ·

GHANA

NIGERIA

Abuja ·

Benue R.

ETHIOPIA

SIERRA LEONE

Yamoussoukro ·

Accra ·

TOGO

BENIN

GUINEA

Porto-Novo

Bangui ·

Juba ·

SOMALIA

Monrovia ·

UPPER

Lomé ·

LIBERIA

Abidjan ·

CÔTE D'IVOIRE

EQUATORIAL GUINEA

Gulf of Guinea

Malabo ·

CAMEROON

Yaoundé ·

UGANDA

Kampala ·

KENYA

Lake Turkana

· Mogadishu

SÃO TOMÉ & PRÍNCIPE

São Tomé ·

Libreville ·

GABON

CONGO

CONGO BASIN

Virunga Mts. 14,826 ft. 4,519 m

Kigali ·

RWANDA

Lake Victoria

Nairobi ·

Kilimanjaro 19,341 ft., 5,895 m

EQUATOR

ATLANTIC OCEAN

Ascension

PRIME MERIDIAN MERIDIAN OF GREENWICH

LOWER GUINEA

Brazzaville ·

Kinshasa ·

DEM. REP. OF THE CONGO

Congo R.

Great Rift Valley

Gitega ·

BURUNDI

Dodoma ·

Zanzibar

Victoria ·

SEYCHELLES

Cabinda (Angola)

Lake Tanganyika

TANZANIA

St. Helena

Luanda ·

Katanga Plateau

COMOROS

Moroni ·

Mayotte

N W E S

ANGOLA

Okavango Delta

Cubango R.

MALAWI

Lusaka ·

ZAMBIA

Zambezi R.

Lake Malawi

Lilongwe ·

MOZAMBIQUE

Mascarene Is.

Antananarivo ·

MADAGASCAR

Réunion

Port-Louis ·

MAURITIUS

Etosha Pan

Victoria Falls

Harare ·

Mozambique Channel

TROPIC OF CAPRICORN

Namib Desert

NAMIBIA

Windhoek ·

BOTSWANA

Kalahari Desert

Gaborone ·

ZIMBABWE

Limpopo R.

Mbabane ·

Maputo ·

ESWATINI

Orange R.

Tshwane (Pretoria) ·

Drakensberg

Lobamba ·

Bloemfontein ·

Maseru ·

LESOTHO

INDIAN OCEAN

Cape Town ·

SOUTH AFRICA

Cape of Good Hope

Cape Agulhas

0 ____ 800 miles
0 ____ 800 kilometers

Azimuthal Equidistant projection

20°W · 20°E · 40°E · 60°E · 40°N · 20°S · 40°S

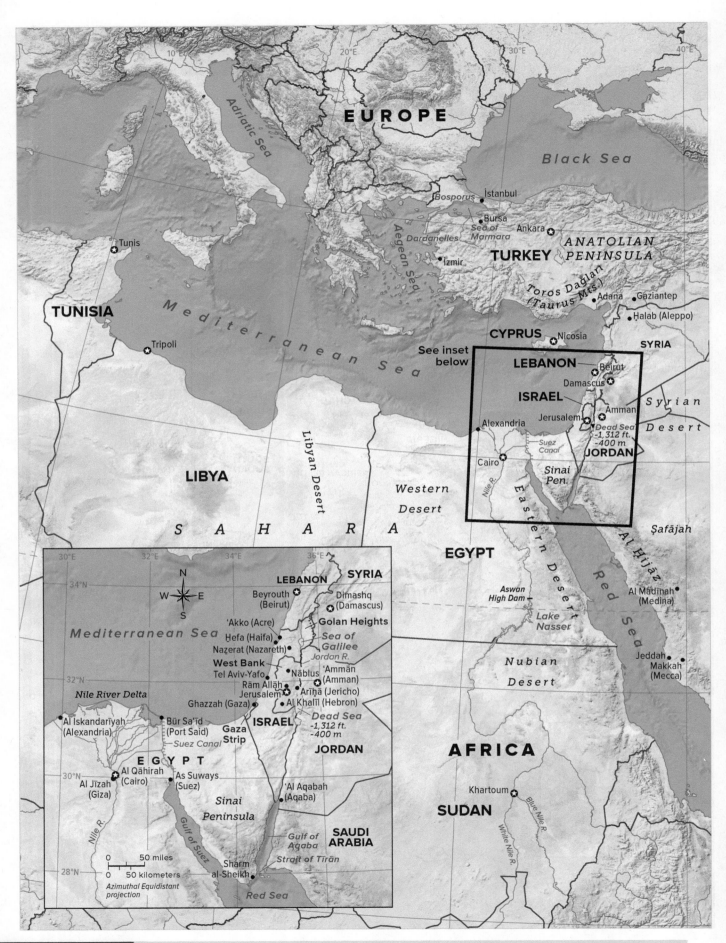

EUROPE

Black Sea

Adriatic Sea

İstanbul
Bosporus
Bursa
Sea of
Marmara
Ankara ✪
Dardanelles
Izmir

ANATOLIAN
PENINSULA

TURKEY

Toros Dağları
(Taurus Mts.)
Adana   Gaziantep
Halab (Aleppo)

CYPRUS   Nicosia   SYRIA

See inset
below

LEBANON   Beirut
Damascus ✪
ISRAEL
Jerusalem ✪   Amman ✪
Alexandria   *Dead Sea*
-1,312 ft.
-400 m   *Syrian*
Desert
Cairo ✪   JORDAN

Aegean Sea

Tunis ✪

M e d i t e r r a n e a n   S e a

TUNISIA

Tripoli ✪

Libyan Desert

Western
Desert

LIBYA

Suez
Canal

Sinai
Pen.

Nile R.

Eastern Desert

Al Hijāz

Red Sea

Şafājah

S A H A R A

EGYPT

Aswan
High Dam

Al Madīnah
(Medina)

Lake
Nasser

Nubian

Desert

Jeddah
Makkah
(Mecca)

AFRICA

Khartoum ✪

Blue Nile R.

SUDAN

White Nile R.

Inset map

30°E   32°E   34°E   36°E

34°N

N
W ✦ E
S

LEBANON   SYRIA

Beyrouth   Dimashq
(Beirut)   (Damascus)

M e d i t e r r a n e a n   S e a

'Akko (Acre)   Golan Heights
Ḥefa (Haifa)   *Sea of*
Galilee
Naẓerat (Nazareth)   *Jordan R.*

32°N

West Bank
Tel Aviv-Yafo   Nāblus
Rām Allāh   'Ammān
Jerusalem ✪   (Amman)
Arīḥā (Jericho)
Ghazzah (Gaza)   Al Khalīl (Hebron)

Nile River Delta

Al Iskandarīyah
(Alexandria)

Būr Saʻīd
(Port Said)   ISRAEL   *Dead Sea*
-1,312 ft.
-400 m

Suez Canal   Gaza
Strip   JORDAN

30°N

Al Qāhirah
(Cairo) ✪

As Suways
(Suez)

Nile R.

EGYPT

Al Jīzah
(Giza)

Sinai
Peninsula

'Al Aqabah
(Aqaba)

SAUDI
ARABIA

Gulf of Suez

Gulf of
Aqaba

Strait of Tīrān

28°N

0   50 miles
0   50 kilometers
Azimuthal Equidistant
projection

Sharm
al-Sheikh

Red Sea

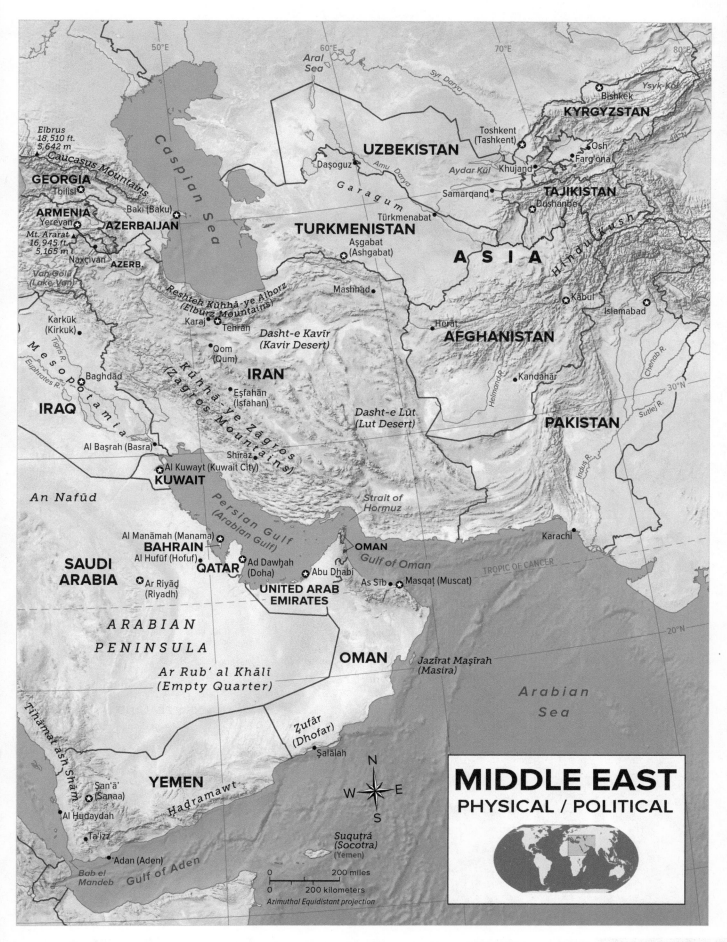

Elbrus
18,510 ft.
5,642 m

Caucasus Mountains

Aral Sea

Syr Darya

Bishkek ★

Ysyk-Köl

KYRGYZSTAN

Toshkent (Tashkent) ★

Osh ●

Farg'ona ●

UZBEKISTAN

Daşoguz ●

Amu Darya

Khujand ●

Aydar Kül

TAJIKISTAN

GEORGIA

Tbilisi ★

Caspian Sea

ARMENIA

Yerevan ★

Baki (Baku) ★

AZERBAIJAN

Mt. Ararat
16,945 ft.
5,165 m

Naxçivan ●

AZERB.

Van Gölü (Lake Van)

Samarqand ●

Dushanbe ★

TURKMENISTAN

Garagum

Türkmenabat ●

A S I A

Aşgabat (Ashgabat) ★

Mashhad ●

Hindu Kush

Kabul ★

Islamabad ★

Reshteh Kūhhā-ye Alborz (Elburz Mountains)

Karaj ● ● Tehrān

Herāt ●

AFGHANISTAN

Karkūk (Kirkuk) ●

Dasht-e Kavīr (Kavir Desert)

Tigris R.

Qom (Qum) ●

IRAN

Mesopotamia

Euphrates R.

Baghdād ●

Eşfahān (Isfahan) ●

Kūhhā-ye Zāgros (Zagros Mountains)

Dasht-e Lūt (Lut Desert)

Kandahār ●

Helmand R.

30°N

Chenab R.

IRAQ

PAKISTAN

Indus R.

Al Başrah (Basra) ●

Shīrāz ●

Sutlej R.

An Nafūd

Al Kuwayt (Kuwait City) ★

KUWAIT

Persian Gulf (Arabian Gulf)

Strait of Hormuz

Karachi ●

Al Manāmah (Manama) ★

BAHRAIN

Al Hufūf (Hofuf) ●

OMAN

Gulf of Oman

TROPIC OF CANCER

SAUDI ARABIA

QATAR

Ad Dawḩah (Doha) ★

Abu Dhabi ●

As Sīb ●

Masqaţ (Muscat) ★

Ar Riyāḑ (Riyadh) ★

UNITED ARAB EMIRATES

ARABIAN PENINSULA

20°N

OMAN

Jazīrat Maşīrah (Masira)

Ar Rub' al Khālī (Empty Quarter)

Arabian Sea

Ẓufār (Dhofar)

Tihāmat ash Shām

Şalālah ●

YEMEN

San'ā' (Sanaa) ★

Hadramawt

N
W ✦ E
S

Suquṭrá (Socotra) (Yemen)

Al Hudaydah ●

Ta'izz ●

'Adan (Aden) ●

Bab el Mandeb

Gulf of Aden

0   200 miles
0   200 kilometers

Azimuthal Equidistant projection

MIDDLE EAST
PHYSICAL / POLITICAL

50°E   60°E   70°E   80°E

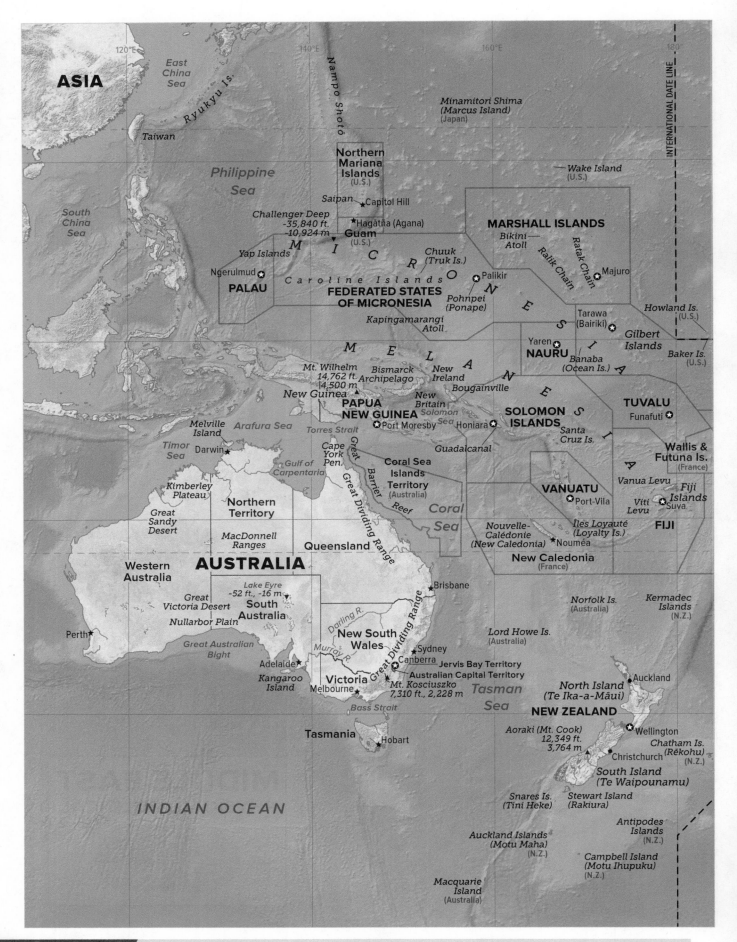

ASIA

East
China
Sea

120°E   140°E   160°E   180°

Ryukyu Is.

Taiwan

Philippine
Sea

South
China
Sea

Nampo Shotō

INTERNATIONAL DATE LINE

Minamitori Shima
(Marcus Island)
(Japan)

Northern
Mariana
Islands
(U.S.)

Saipan
★ Capitol Hill

Challenger Deep
-35,840 ft.
-10,924 m

★ Hagåtña (Agana)
Guam
(U.S.)

Yap Islands

Ngerulmud ◉

PALAU

M I C

Caroline Islands

FEDERATED STATES
OF MICRONESIA

Chuuk
(Truk Is.)

R   O

Palikir
◉
Pohnpei
(Ponape)

Kapingamarangi
Atoll

N

Wake Island
(U.S.)

MARSHALL ISLANDS

Bikini
Atoll

Ratak Chain

Ralik Chain

Majuro
◉

Tarawa
(Bairiki)
◉

Howland Is.
(U.S.)

Gilbert
Islands

Baker Is.
(U.S.)

M   E   L   A

Mt. Wilhelm
14,762 ft.
4,500 m

Bismarck
Archipelago

New Guinea ▲

PAPUA
NEW GUINEA

New
Ireland

New
Britain
Solomon
Port Moresby ◉ Sea

Bougainville

E   S

Yaren
◉
NAURU

Banaba
(Ocean Is.)

I   A

N   E   S   I   A

SOLOMON
ISLANDS
Honiara ◉

Santa
Cruz Is.

TUVALU
Funafuti ◉

Melville
Island

Darwin ★

Arafura Sea

Timor
Sea

Torres Strait

Cape
York
Pen.

Great

Great Dividing Range

Barrier

Reef

Guadalcanal

Coral Sea
Islands
Territory
(Australia)

Coral
Sea

A

Wallis &
Futuna Is.
(France)

Vanua Levu

VANUATU
◉ Port-Vila

Fiji
Islands

Viti
Levu   Suva ◉

Gulf of
Carpentaria

Kimberley
Plateau

Great
Sandy
Desert

Northern
Territory

MacDonnell
Ranges

Nouvelle-
Calédonie
(New Caledonia)

Îles Loyauté
(Loyalty Is.)
★ Nouméa

FIJI

Western
Australia

AUSTRALIA

Queensland

New Caledonia
(France)

Great
Victoria Desert

Nullarbor Plain

Lake Eyre
-52 ft., -16 m

South
Australia

Brisbane

Norfolk Is.
(Australia)

Kermadec
Islands
(N.Z.)

Perth ★

Great Australian
Bight

Adelaide ★

Kangaroo
Island

Darling R.

Murray R.

New South
Wales

Victoria

Great Dividing Range

Melbourne ★

Sydney •
Canberra ◉

Mt. Kosciuszko
7,310 ft., 2,228 m

Lord Howe Is.
(Australia)

Jervis Bay Territory
Australian Capital Territory

Tasman
Sea

North Island
(Te Ika-a-Māui)

Auckland •

Bass Strait

Tasmania

Hobart ★

Aoraki (Mt. Cook)
12,349 ft.
3,764 m

NEW ZEALAND

◉ Wellington

Chatham Is.
(Rēkohu)
(N.Z.)

Christchurch •

INDIAN OCEAN

Snares Is.
(Tini Heke)

Auckland Islands
(Motu Maha)
(N.Z.)

Stewart Island
(Rakiura)

South Island
(Te Waipounamu)

Antipodes
Islands
(N.Z.)

Campbell Island
(Motu Ihupuku)
(N.Z.)

Macquarie
Island
(Australia)

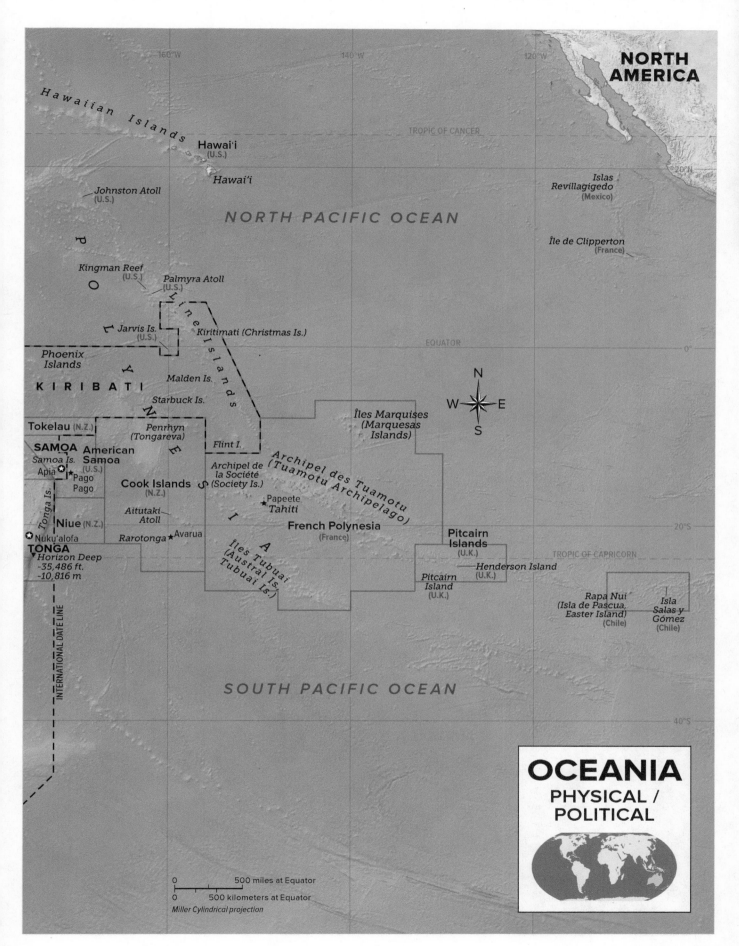

NORTH
AMERICA

160°W · 140°W · 120°W

TROPIC OF CANCER

20°N

Hawaiian Islands

Hawai'i
(U.S.)

Hawai'i

Johnston Atoll
(U.S.)

NORTH PACIFIC OCEAN

*Islas
Revillagigedo*
(Mexico)

Île de Clipperton
(France)

P

O

Kingman Reef
(U.S.)

Palmyra Atoll
(U.S.)

L

Jarvis Is.
(U.S.)

Kiritimati (Christmas Is.)

EQUATOR

0°

*Phoenix
Islands*

Line Islands

K I R I B A T I

Malden Is.

Starbuck Is.

N

*Îles Marquises
(Marquesas
Islands)*

N
W ✦ **E**
S

Tokelau (N.Z.)

*Penrhyn
(Tongareva)*

Flint I.

SAMOA

Samoa Is.

Apia ⊕ ★
Pago
Pago

**American
Samoa**
(U.S.)

Cook Islands
(N.Z.)

*Archipel de
la Société
(Society Is.)*

*Archipel des Tuamotu
(Tuamotu Archipelago)*

Tonga Is.

*Aitutaki
Atoll*

★ Papeete
Tahiti

Niue (N.Z.)

Rarotonga ★ Avarua

French Polynesia
(France)

**Pitcairn
Islands**
(U.K.)

20°S

★ Nuku'alofa

TONGA
*Horizon Deep
-35,486 ft.
-10,816 m*

*Îles Tubuai
(Austral Is.
Tubuai Is.)*

TROPIC OF CAPRICORN

*Pitcairn
Island*
(U.K.)

Henderson Island
(U.K.)

*Rapa Nui
(Isla de Pascua,
Easter Island)*
(Chile)

*Isla
Salas y
Gómez*
(Chile)

INTERNATIONAL DATE LINE

SOUTH PACIFIC OCEAN

40°S

0 ____ 500 miles at Equator
0 ____ 500 kilometers at Equator
Miller Cylindrical projection

OCEANIA
PHYSICAL /
POLITICAL

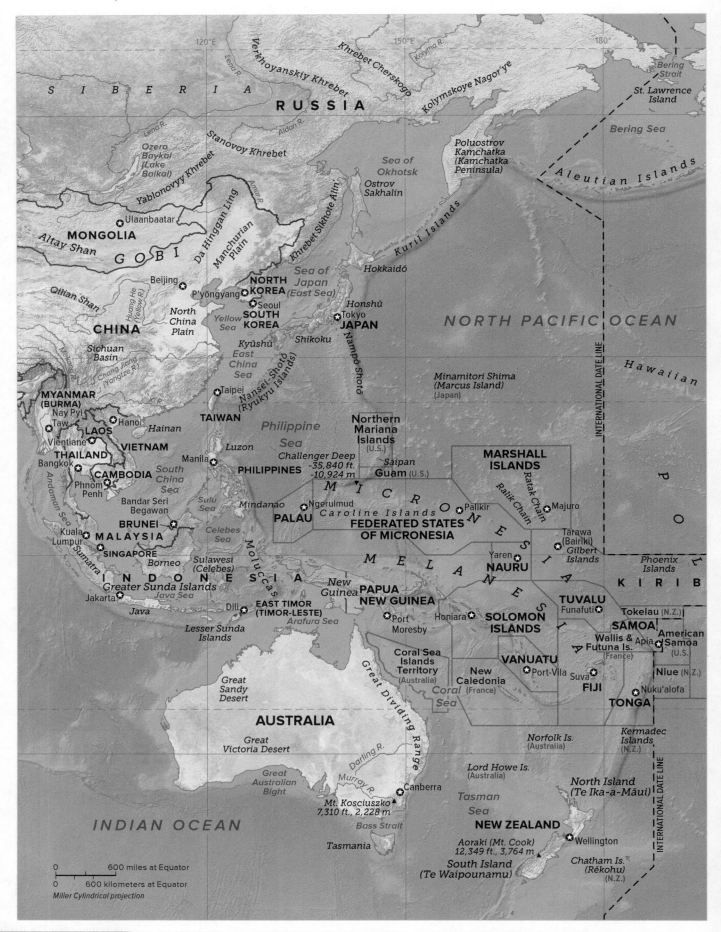

SIBERIA

RUSSIA

Verkhoyanskiy Khrebet

Khrebet Cherskogo

Kolyma R.

Kolymskoye Nagor'ye

Bering Strait

St. Lawrence Island

Bering Sea

Lena R.

Aldan R.

Sea of Okhotsk

Poluostrov Kamchatka (Kamchatka Peninsula)

Aleutian Islands

Ozero Baykal (Lake Baikal)

Stanovoy Khrebet

Ostrov Sakhalin

Yablonovyy Khrebet

Amur R.

Khrebet Sikhote Alin'

Kuril Islands

Ulaanbaatar

MONGOLIA

Da Hinggan Ling

Manchurian Plain

Hokkaidō

Altay Shan

GOBI

Sea of Japan (East Sea)

NORTH PACIFIC OCEAN

Qilian Shan

Beijing

NORTH KOREA

P'yŏngyang

Honshū

Huang He (Yellow R.)

North China Plain

Seoul

SOUTH KOREA

Tokyo

JAPAN

Hawaiian

CHINA

Yellow Sea

Minamitori Shima (Marcus Island) (Japan)

INTERNATIONAL DATE LINE

Sichuan Basin

Chang Jiang (Yangtze R.)

Kyūshū

Shikoku

East China Sea

Nampō Shotō

Taipei

Nansei-Shotō (Ryukyu Islands)

MYANMAR (BURMA)

Nay Pyi Taw

Hanoi

TAIWAN

Philippine Sea

Northern Mariana Islands (U.S.)

MARSHALL ISLANDS

P

LAOS

Hainan

O

Vientiane

VIETNAM

Luzon

Challenger Deep -35,840 ft. -10,924 m

Saipan

THAILAND

Manila

Guam (U.S.)

Ratak Chain

L

Bangkok

CAMBODIA

PHILIPPINES

M I C R O N E

Ralik Chain

Majuro

Y

Phnom Penh

South China Sea

Tarawa (Bairiki)

Andaman Sea

Bandar Seri Begawan

Sulu Sea

Mindanao

Ngerulmud

Palikir

S

Gilbert Islands

N

BRUNEI

PALAU

Caroline Islands

FEDERATED STATES OF MICRONESIA

I

Phoenix Islands

E

Kuala Lumpur

MALAYSIA

Celebes Sea

A

Yaren

KIRIB

SINGAPORE

Borneo

Sulawesi (Celebes)

M E L A N

NAURU

Sumatra

INDONESIA

Moluccas

TUVALU

Greater Sunda Islands

New Guinea

E

Funafuti

Tokelau (N.Z.)

Jakarta

Java Sea

Dili

EAST TIMOR (TIMOR-LESTE)

PAPUA NEW GUINEA

S

SAMOA

American Samoa (U.S.)

Java

Arafura Sea

Port Moresby

Honiara

SOLOMON ISLANDS

I

Wallis & Futuna Is. (France)

Apia

Lesser Sunda Islands

A

Niue (N.Z.)

Coral Sea Islands Territory (Australia)

VANUATU

New Caledonia (France)

Port-Vila

Suva

Nuku'alofa

Great Sandy Desert

FIJI

Coral Sea

TONGA

Great Victoria Desert

AUSTRALIA

Great Dividing Range

Norfolk Is. (Australia)

Kermadec Islands (N.Z.)

Darling R.

Lord Howe Is. (Australia)

North Island (Te Ika-a-Māui)

Great Australian Bight

Murray R.

Canberra

Tasman Sea

INDIAN OCEAN

Mt. Kosciuszko 7,310 ft., 2,228 m

INTERNATIONAL DATE LINE

Bass Strait

NEW ZEALAND

Tasmania

Aoraki (Mt. Cook) 12,349 ft., 3,764 m

Wellington

Chatham Is. (Rēkohu) (N.Z.)

South Island (Te Waipounamu)

0   600 miles at Equator

0   600 kilometers at Equator

Miller Cylindrical projection

Denali
(Mt. McKinley) ▲
20,310 ft.
6,190 m

Yukon R.

ARCTIC CIRCLE

Mackenzie R.

*Great Bear
Lake*

*Great Slave
Lake*

*Hudson
Bay*

60°N

Gulf of Alaska

*Kodiak
Island*

*Alexander
Archipelago*

*Haida Gwaii
(Queen Charlotte
Islands)*

*Lake
Athabaska*

CANADA

C A N A D I A N   S H I E L D

Coast Mountains

*Vancouver
Island*

R O C K Y

Cascade Range

G R E A T

M O U N T A I N S

P L A I N S

Missouri R.

*Lake
Winnipeg*

*Lake
Superior*

Lake Huron

Ottawa ✪
Lake Ontario

**UNITED
STATES**

*Lake
Michigan*

*CENTRAL
LOWLAND*

*Lake
Erie*

A p p a l a c h i a n   M t s .

*Colorado
R.*

Mississippi R.

Ohio R.

Washington, D.C.

C O A S T A L   P L A I N

**ATLANTIC
OCEAN**

30°N

Gulf of California

Sierra Madre Occidental

Sierra Madre Oriental

TROPIC OF CANCER

Islands

*Baja
California*

*Gulf of
Mexico*

BAHAMAS

Havana

Hawai'i
(U.S.)   *Hawai'i*

MEXICO

✪ Mexico City

*Islas
Revillagigedo
(Mexico)*

CUBA

**DOMINICAN
REPUBLIC**

BELIZE

Belmopan

Tegucigalpa
✪

HAITI

JAMAICA

Caribbean Sea

Guatemala ✪

HONDURAS

GUATEMALA

San Salvador

NICARAGUA

EL SALVADOR

Managua

*Île de Clipperton
(France)*

San José ✪

✪ Panamá

Caracas

COSTA RICA

VENEZUELA

PANAMA

✪ Bogotá

Kiritimati (Christmas Is.)

EQUATOR

COLOMBIA

Quito

*Archipiélago de Colón
(Galápagos Islands)
(Ecuador)*

A   T   I

Line Islands

*Îles Marquises
(Marquesas
Islands)*

ECUADOR

A M A Z O N

A N D E S

BASIN

BRAZIL

Y

N

**Cook
Islands**
(N.Z.)

*Society
Islands*

Tuamotu Archipelago

PERU

Lima

E

Tahiti

French Polynesia
(France)

S

I

A

La Paz

BOLIVIA

Austral Islands

Sucre ✪

**Pitcairn
Islands**
(U.K.)

TROPIC OF CAPRICORN

*Rapa Nui
(Isla de Pascua,
Easter Island)
(Chile)*

Aconcagua
22,834 ft.
6,960 m

A N D E S

Santiago ✪

SOUTH PACIFIC OCEAN

30°S

PACIFIC RIM
PHYSICAL / POLITICAL

CHILE

*Isla Grande
de Chiloé*

ARGENTINA

PATAGONIA

150°W   120°W   90°W

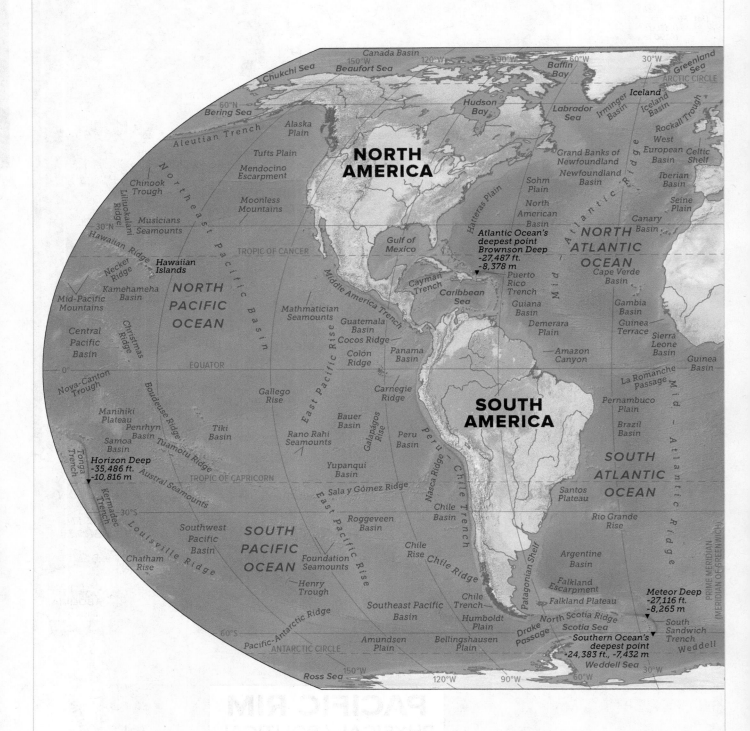

OCEAN FLOOR

0 — 2,000 miles at Equator
0 — 2,000 kilometers at Equator
Equal Earth projection

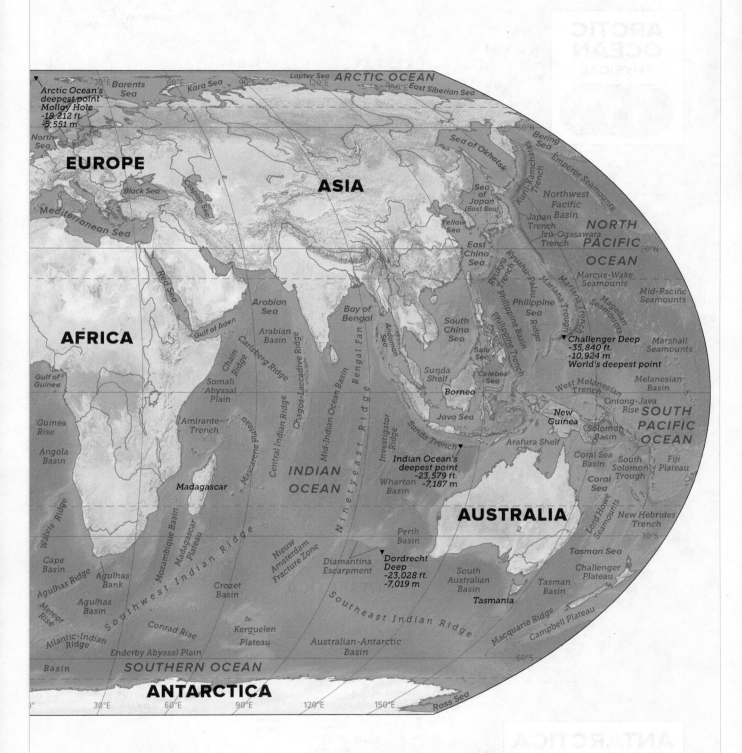

Arctic Ocean's
deepest point
Molloy Hole
-18,212 ft.
-5,551 m

North
Sea

Baltic Sea

EUROPE

Black Sea

Caspian Sea

Mediterranean Sea

ASIA

ARCTIC OCEAN

Laptev Sea

Barents
Sea

Kara Sea

East Siberian Sea

Bering
Sea

Emperor Seamounts

Sea of Okhotsk

Kuril-Kamchatka Trench

Sea
of
Japan
(East Sea)

Northwest
Pacific

Japan Basin

NORTH

Japan
Trench

Izu-Ogasawara
Trench

PACIFIC

Yellow
Sea

East
China
Sea

Ryukyu Trench

Kyushu-Palau Ridge

Philippine Trench

Philippine
Sea

OCEAN

Marcus-Wake
Seamounts

Mariana Trench

Magellan
Seamounts

Mid-Pacific
Seamounts

AFRICA

Red Sea

Gulf of Aden

Arabian
Sea

Arabian
Basin

Bay of
Bengal

Andaman
Sea

South
China
Sea

Sulu
Sea

Philippine Basin

Challenger Deep
-35,840 ft.
-10,924 m
World's deepest point

Marshall
Seamounts

Melanesian
Basin

Gulf of
Guinea

Chain Ridge

Carlsberg Ridge

Somali
Abyssal
Plain

Chagos-Laccadive Ridge

Bengal Fan

Mid-Indian Ocean Basin

Sunda
Shelf

Celebes
Sea

Borneo

Java Sea

New
Guinea

West Melanesian Trench

Ontong-Java
Rise

SOUTH

Guinea
Rise

Amirante
Trench

Macarene plateau

Central Indian Ridge

Ninetyeast Ridge

Investigator
Ridge

Sunda Trench

Arafura Shelf

Solomon
Basin

PACIFIC

OCEAN

Angola
Basin

Indian Ocean's
deepest point
-23,579 ft.
-7,187 m

Coral Sea
Basin

South
Solomon
Trough

Fiji
Plateau

Madagascar

INDIAN

OCEAN

Wharton
Basin

AUSTRALIA

Coral
Sea

Lord Howe
Seamounts

New Hebrides
Trench

Walvis Ridge

Mozambique Basin

Madagascar
Plateau

Nieuw
Amsterdam
Fracture Zone

Perth
Basin

Cape
Basin

Agulhas Ridge

Agulhas
Bank

Southwest Indian Ridge

Crozet
Basin

Diamantina
Escarpment

Dordrecht
Deep
-23,028 ft.
-7,019 m

South
Australian
Basin

Tasman Sea

Tasman
Basin

Challenger
Plateau

Meteor
Rise

Agulhas
Basin

Tasmania

Atlantic-Indian
Ridge

Conrad Rise

Kerguelen
Plateau

Australian-Antarctic
Basin

Southeast Indian Ridge

Macquarie Ridge

Campbell Plateau

Basin

Enderby Abyssal Plain

SOUTHERN OCEAN

ANTARCTICA

Ross Sea

30°E

60°E

90°E

120°E

150°E

The Atlantic, Indian, and Pacific Oceans
merge around Antarctica. Some define this as
an ocean, calling it the Antarctic Ocean,
Austral Ocean, or Southern Ocean. While
most accept four oceans (including the Arctic
Ocean), there is little international agreement
on the name and extent of a fifth ocean.

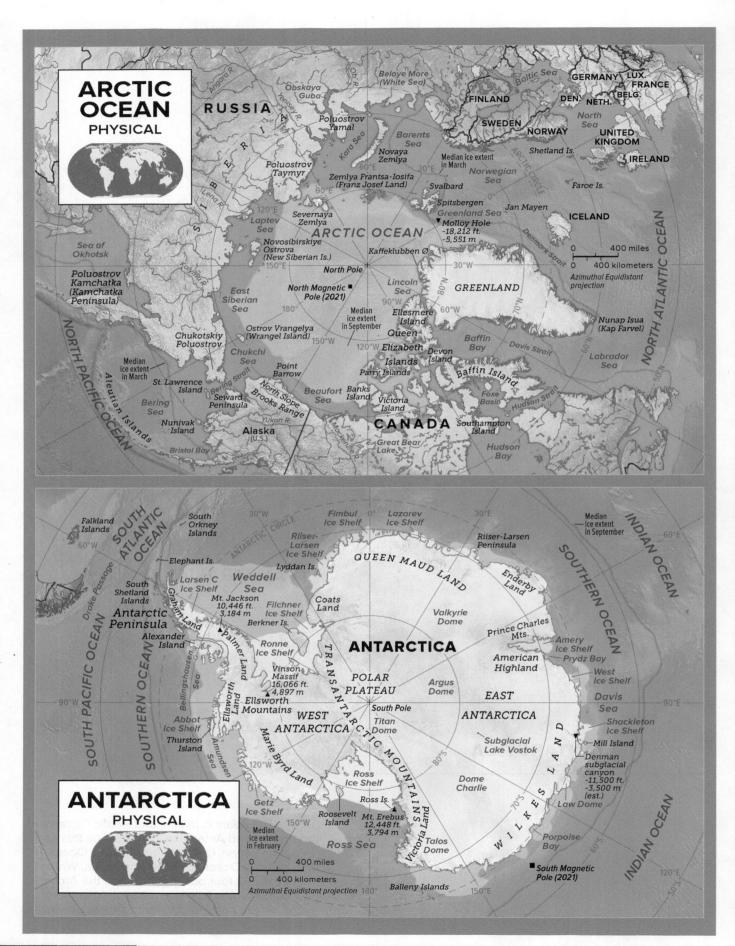

ARCTIC OCEAN
PHYSICAL

RUSSIA

Angara R.

Obskaya Guba

Beloye More (White Sea)

Baltic Sea

GERMANY LUX.
FINLAND FRANCE
DEN. NETH. BELG.

Yenisey R.

Poluostrov Yamal

Barents Sea

SWEDEN NORWAY UNITED KINGDOM

North Sea

Kara Sea

Novaya Zemlya

Median ice extent in March

Shetland Is.

IRELAND

Poluostrov Taymyr

Zemlya Frantsa-Iosifa (Franz Josef Land)

Svalbard

Norwegian Sea

Faroe Is.

SIBERIA

Lena R.

120°E

90°E

Severnaya Zemlya

Laptev Sea

ARCTIC OCEAN

Spitsbergen
Greenland Sea
▼ Molloy Hole
-18,212 ft.
-5,551 m

Jan Mayen

ARCTIC CIRCLE

ICELAND

Denmark Strait

NORTH ATLANTIC OCEAN

Sea of Okhotsk

Novosibirskiye Ostrova (New Siberian Is.)

150°E

Kaffeklubben Ø

30°W

80°N

GREENLAND

0   400 miles
0   400 kilometers

Azimuthal Equidistant projection

Poluostrov Kamchatka (Kamchatka Peninsula)

East Siberian Sea

180°

North Pole

North Magnetic Pole (2021) ■

Lincoln Sea

90°W 60°N

Nunap Isua (Kap Farvel)

Kolyma R.

Median ice extent in September

Ostrov Vrangelya (Wrangel Island)

150°W

Ellesmere Island
Queen
Elizabeth
Islands

Devon Island

Baffin Bay

Davis Strait

Labrador Sea

Chukotskiy Poluostrov

Chukchi Sea

120°W

Parry Islands

Baffin Island

NORTH PACIFIC OCEAN

Aleutian Islands

Point Barrow

Beaufort Sea

Banks Island

Foxe Basin

Median ice extent in March

Bering Strait

St. Lawrence Island

Bering Sea

North Slope
Brooks Range

Victoria Island

Hudson Strait

Seward Peninsula

Nunivak Island

Yukon R.

Alaska (U.S.)

Mackenzie R.

CANADA

Southampton Island

Hudson Bay

Bristol Bay

Great Bear Lake

ANTARCTICA
PHYSICAL

Falkland Islands

SOUTH ATLANTIC OCEAN

60°W

South Orkney Islands

30°W

ANTARCTIC CIRCLE

Fimbul Ice Shelf

0°

Lazarev Ice Shelf

30°E

Median ice extent in September

60°E

INDIAN OCEAN

Elephant Is.

Riiser-Larsen Ice Shelf

Lyddan Is.

Riiser-Larsen Peninsula

SOUTHERN OCEAN

South Shetland Islands

Larsen C Ice Shelf

Weddell Sea

QUEEN MAUD LAND

Enderby Land

Antarctic Peninsula

Graham Land

Mt. Jackson
10,446 ft.
3,184 m

Filchner Ice Shelf

Coats Land

Valkyrie Dome

Prince Charles Mts.

Amery Ice Shelf

Alexander Island

Berkner Is.

ANTARCTICA

American Highland

Prydz Bay

West Ice Shelf

SOUTHERN OCEAN

Palmer Land

Ronne Ice Shelf

Davis Sea

Bellingshausen Sea

Vinson Massif
16,066 ft.
▲ 4,897 m

TRANSANTARCTIC MOUNTAINS

POLAR PLATEAU

Argus Dome

EAST ANTARCTICA

90°E

SOUTH PACIFIC OCEAN

90°W

Ellsworth Mountains

South Pole

Shackleton Ice Shelf

Abbot Ice Shelf

Ellsworth Land

WEST ANTARCTICA

Titan Dome

Subglacial Lake Vostok

Mill Island

Denman subglacial canyon
-11,500 ft.
-3,500 m
(est.)

Thurston Island

Marie Byrd Land

Amundsen Sea

120°W

Ross Ice Shelf

Dome Charlie

Law Dome

WILKES LAND

Getz Ice Shelf

Roosevelt Island

Ross Is.

Victoria Land

Porpoise Bay

60°E

Median ice extent in February

150°W

Mt. Erebus
12,448 ft.
3,794 m

Talos Dome

70°S

■ South Magnetic Pole (2021)

0   400 miles
0   400 kilometers

Azimuthal Equidistant projection

Ross Sea

180°

Balleny Islands

150°E

INDIAN OCEAN

120°E

90°S

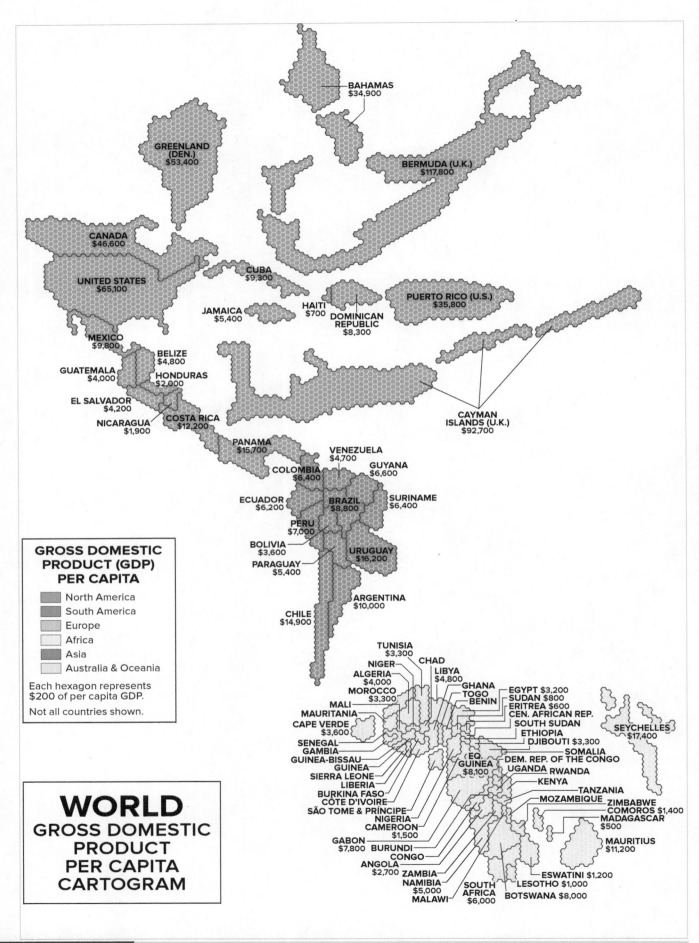

GROSS DOMESTIC PRODUCT (GDP) PER CAPITA

North America
South America
Europe
Africa
Asia
Australia & Oceania

Each hexagon represents $200 of per capita GDP.

Not all countries shown.

WORLD
GROSS DOMESTIC PRODUCT PER CAPITA CARTOGRAM

BAHAMAS
$34,900

GREENLAND
(DEN.)
$53,400

BERMUDA (U.K.)
$117,800

CANADA
$46,600

CUBA
$9,300

UNITED STATES
$65,100

PUERTO RICO (U.S.)
$35,800

JAMAICA
$5,400

HAITI
$700

DOMINICAN
REPUBLIC
$8,300

MEXICO
$9,800

BELIZE
$4,800

GUATEMALA
$4,000

HONDURAS
$2,000

EL SALVADOR
$4,200

NICARAGUA
$1,900

COSTA RICA
$12,200

CAYMAN
ISLANDS (U.K.)
$92,700

PANAMA
$15,700

VENEZUELA
$4,700

GUYANA
$6,600

COLOMBIA
$6,400

SURINAME
$6,400

ECUADOR
$6,200

BRAZIL
$8,800

PERU
$7,000

BOLIVIA
$3,600

URUGUAY
$16,200

PARAGUAY
$5,400

ARGENTINA
$10,000

CHILE
$14,900

TUNISIA
$3,300

NIGER

CHAD

ALGERIA
$4,000

LIBYA
$4,800

MOROCCO
$3,300

GHANA

TOGO

BENIN

EGYPT $3,200

SUDAN $800

ERITREA $600

CEN. AFRICAN REP.

MALI

MAURITANIA

SOUTH SUDAN

CAPE VERDE
$3,600

ETHIOPIA

DJIBOUTI $3,300

SENEGAL

SOMALIA

GAMBIA

DEM. REP. OF THE CONGO

GUINEA-BISSAU

EQ.
GUINEA
$8,100

UGANDA

RWANDA

SEYCHELLES
$17,400

GUINEA

KENYA

SIERRA LEONE

TANZANIA

LIBERIA

MOZAMBIQUE

ZIMBABWE

BURKINA FASO

COMOROS $1,400

CÔTE D'IVOIRE

MADAGASCAR
$500

SÃO TOMÉ & PRÍNCIPE

NIGERIA

MAURITIUS
$11,200

CAMEROON
$1,500

GABON
$7,800

BURUNDI

CONGO

ANGOLA
$2,700

ZAMBIA

ESWATINI $1,200

NAMIBIA
$5,000

SOUTH
AFRICA
$6,000

LESOTHO
$1,000

BOTSWANA $8,000

MALAWI

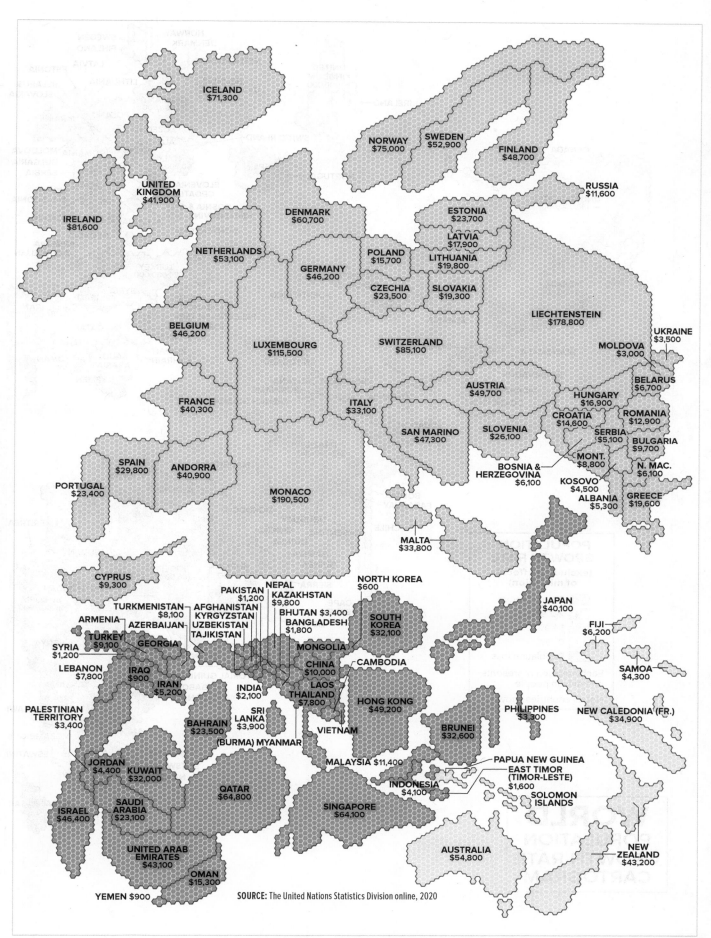

ICELAND
$71,300

NORWAY
$75,000

SWEDEN
$52,900

FINLAND
$48,700

RUSSIA
$11,600

UNITED
KINGDOM
$41,900

DENMARK
$60,700

ESTONIA
$23,700

LATVIA
$17,900

IRELAND
$81,600

NETHERLANDS
$53,100

POLAND
$15,700

LITHUANIA
$19,800

GERMANY
$46,200

CZECHIA
$23,500

SLOVAKIA
$19,300

LIECHTENSTEIN
$178,800

UKRAINE
$3,500

MOLDOVA
$3,000

BELGIUM
$46,200

LUXEMBOURG
$115,500

SWITZERLAND
$85,100

BELARUS
$6,700

AUSTRIA
$49,700

HUNGARY
$16,900

FRANCE
$40,300

ITALY
$33,100

CROATIA
$14,600

ROMANIA
$12,900

SAN MARINO
$47,300

SLOVENIA
$26,100

SERBIA
$5,100

BULGARIA
$9,700

SPAIN
$29,800

ANDORRA
$40,900

MONT.
$8,800

N. MAC.
$6,100

BOSNIA &
HERZEGOVINA
$6,100

KOSOVO
$4,500

PORTUGAL
$23,400

MONACO
$190,500

ALBANIA
$5,300

GREECE
$19,600

CYPRUS
$9,300

MALTA
$33,800

PAKISTAN
$1,200

NEPAL
$9,800

KAZAKHSTAN

NORTH KOREA
$600

JAPAN
$40,100

TURKMENISTAN
$8,100

AFGHANISTAN

KYRGYZSTAN

BHUTAN $3,400

FIJI
$6,200

ARMENIA

AZERBAIJAN

UZBEKISTAN

BANGLADESH
$1,800

SOUTH
KOREA
$32,100

TURKEY
$9,100

GEORGIA

TAJIKISTAN

MONGOLIA

SAMOA
$4,300

SYRIA
$1,200

LEBANON
$7,800

IRAQ
$900

INDIA
$2,100

CHINA
$10,000

CAMBODIA

IRAN
$5,200

LAOS

HONG KONG
$49,200

NEW CALEDONIA (FR.)
$34,900

THAILAND
$7,800

PHILIPPINES
$3,300

PALESTINIAN
TERRITORY
$3,400

BAHRAIN
$23,500

SRI
LANKA
$3,900

VIETNAM

BRUNEI
$32,600

(BURMA) MYANMAR

JORDAN
$4,400

KUWAIT
$32,000

MALAYSIA $11,400

PAPUA NEW GUINEA

EAST TIMOR
(TIMOR-LESTE)
$1,600

INDONESIA
$4,100

SOLOMON
ISLANDS

QATAR
$64,800

ISRAEL
$46,400

SAUDI
ARABIA
$23,100

SINGAPORE
$64,100

UNITED ARAB
EMIRATES
$43,100

OMAN
$15,300

AUSTRALIA
$54,800

NEW
ZEALAND
$43,200

YEMEN $900

SOURCE: The United Nations Statistics Division online, 2020

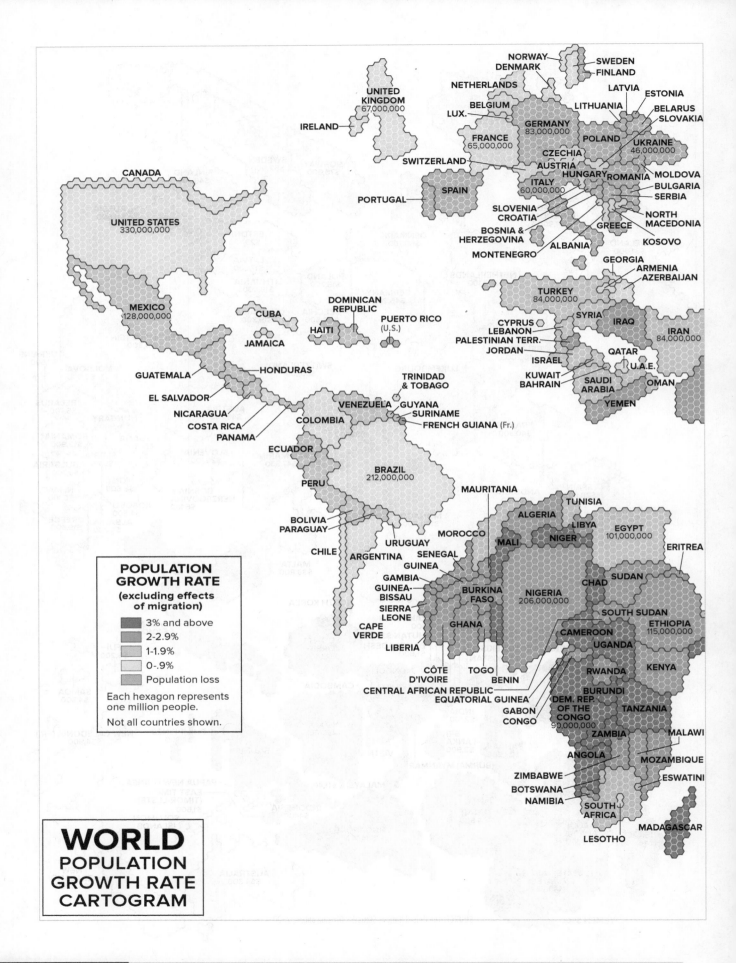

WORLD
POPULATION
GROWTH RATE
CARTOGRAM

POPULATION GROWTH RATE
(excluding effects of migration)

- 3% and above
- 2-2.9%
- 1-1.9%
- 0-.9%
- Population loss

Each hexagon represents one million people.

Not all countries shown.

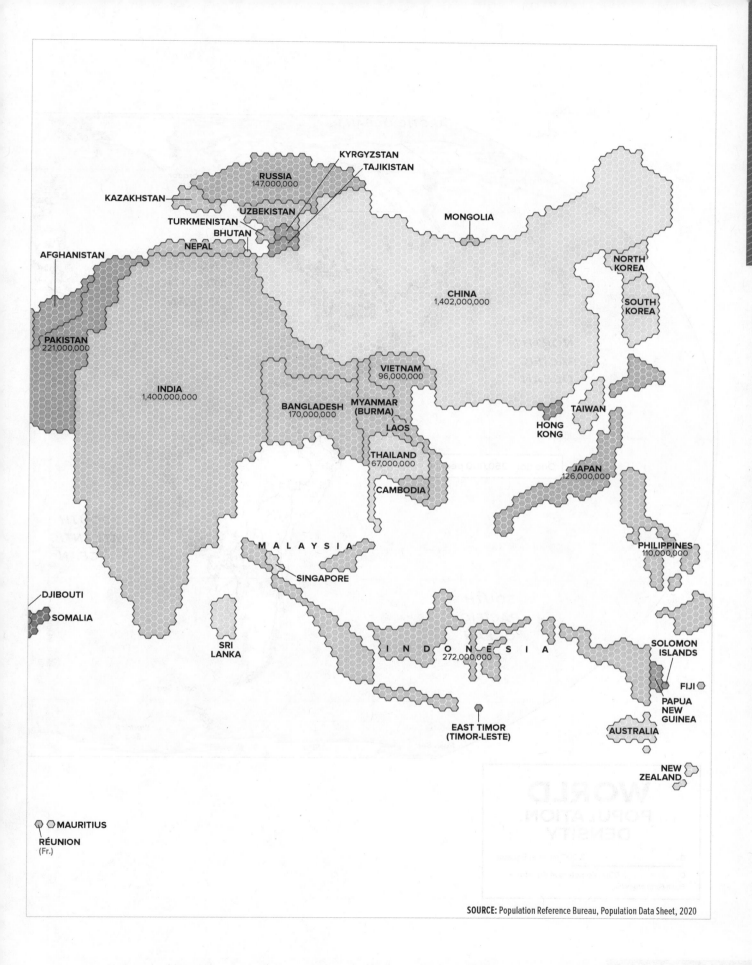

KYRGYZSTAN
TAJIKISTAN

RUSSIA
147,000,000

KAZAKHSTAN

MONGOLIA

UZBEKISTAN

TURKMENISTAN

BHUTAN

NEPAL

AFGHANISTAN

NORTH
KOREA

CHINA
1,402,000,000

SOUTH
KOREA

PAKISTAN
221,000,000

VIETNAM
96,000,000

INDIA
1,400,000,000

BANGLADESH
170,000,000

MYANMAR
(BURMA)

TAIWAN

LAOS

HONG
KONG

THAILAND
67,000,000

JAPAN
126,000,000

CAMBODIA

M A L A Y S I A

PHILIPPINES
110,000,000

SINGAPORE

DJIBOUTI

SOMALIA

SRI
LANKA

I N D O N E S I A
272,000,000

SOLOMON
ISLANDS

FIJI

EAST TIMOR
(TIMOR-LESTE)

PAPUA
NEW
GUINEA

AUSTRALIA

NEW
ZEALAND

MAURITIUS

RÉUNION
(Fr.)

SOURCE: Population Reference Bureau, Population Data Sheet, 2020

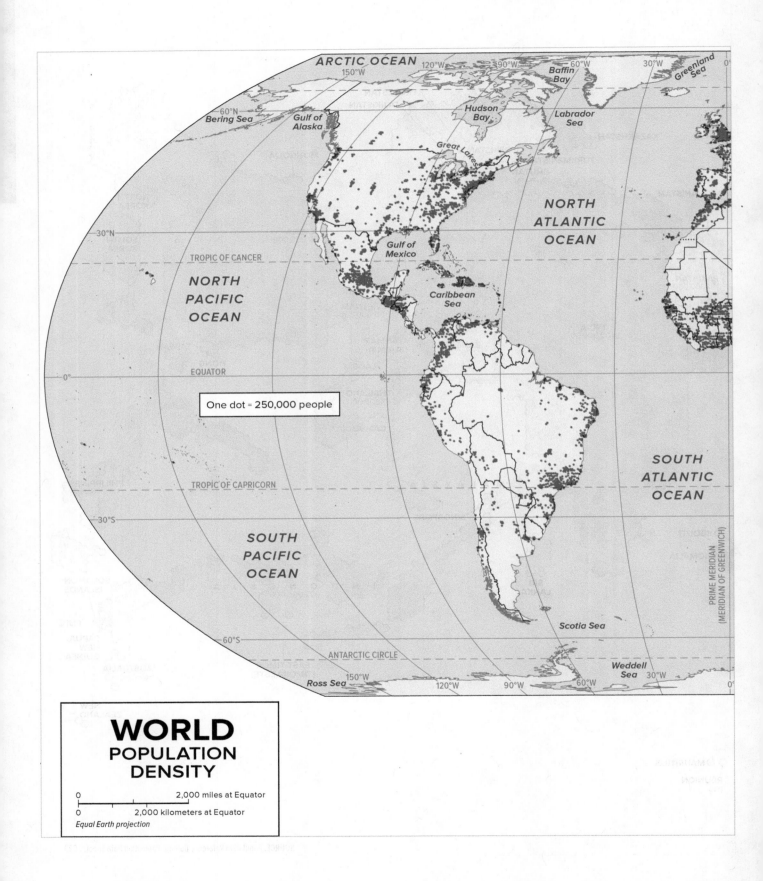

ARCTIC OCEAN

150°W   120°W   90°W   60°W   30°W   0°

Baffin Bay

Greenland Sea

60°N
Bering Sea

Gulf of Alaska

Hudson Bay

Labrador Sea

Great Lakes

NORTH ATLANTIC OCEAN

30°N

TROPIC OF CANCER

NORTH PACIFIC OCEAN

Gulf of Mexico

Caribbean Sea

0°   EQUATOR

One dot = 250,000 people

TROPIC OF CAPRICORN

SOUTH ATLANTIC OCEAN

30°S

SOUTH PACIFIC OCEAN

PRIME MERIDIAN (MERIDIAN OF GREENWICH)

60°S

ANTARCTIC CIRCLE

150°W

Ross Sea

120°W   90°W   60°W

Scotia Sea

Weddell Sea

30°W   0°

WORLD
POPULATION DENSITY

0   2,000 miles at Equator

0   2,000 kilometers at Equator

Equal Earth projection

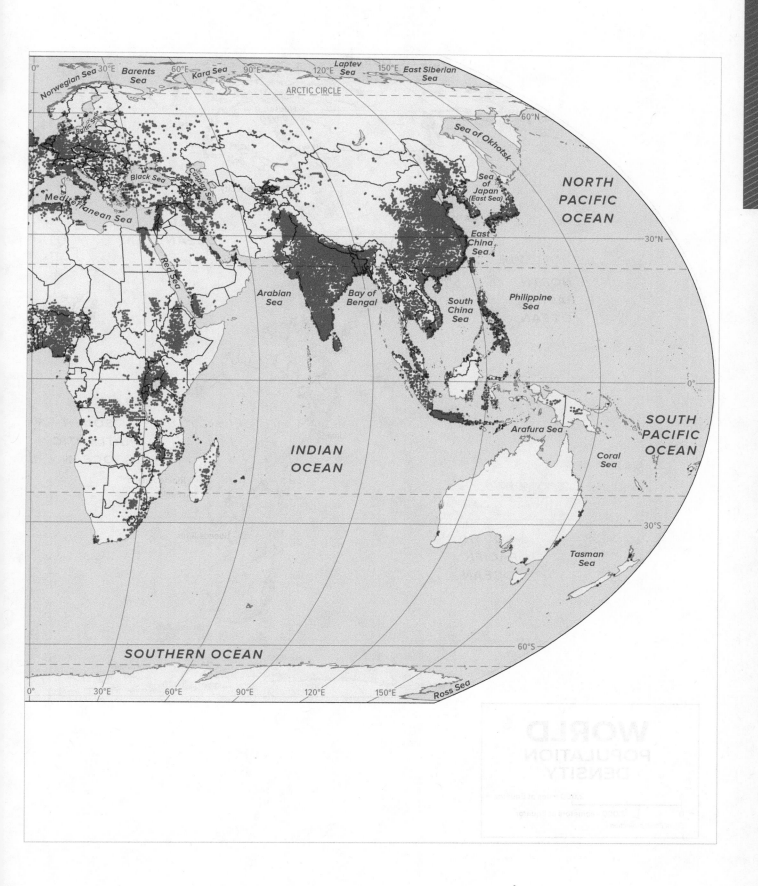

Norwegian Sea · Barents Sea · 30°E · 60°E · Kara Sea · 90°E · 120°E · Laptev Sea · 150°E · East Siberian Sea · 0°

ARCTIC CIRCLE

Baltic Sea

60°N

Black Sea

Caspian Sea

Sea of Okhotsk

Mediterranean Sea

NORTH PACIFIC OCEAN

Sea of Japan (East Sea)

30°N

Red Sea

East China Sea

Arabian Sea

Bay of Bengal

Philippine Sea

South China Sea

0°

INDIAN OCEAN

SOUTH PACIFIC OCEAN

Arafura Sea

Coral Sea

30°S

Tasman Sea

60°S

SOUTHERN OCEAN

0° · 30°E · 60°E · 90°E · 120°E · 150°E · Ross Sea

WORLD POPULATION DENSITY

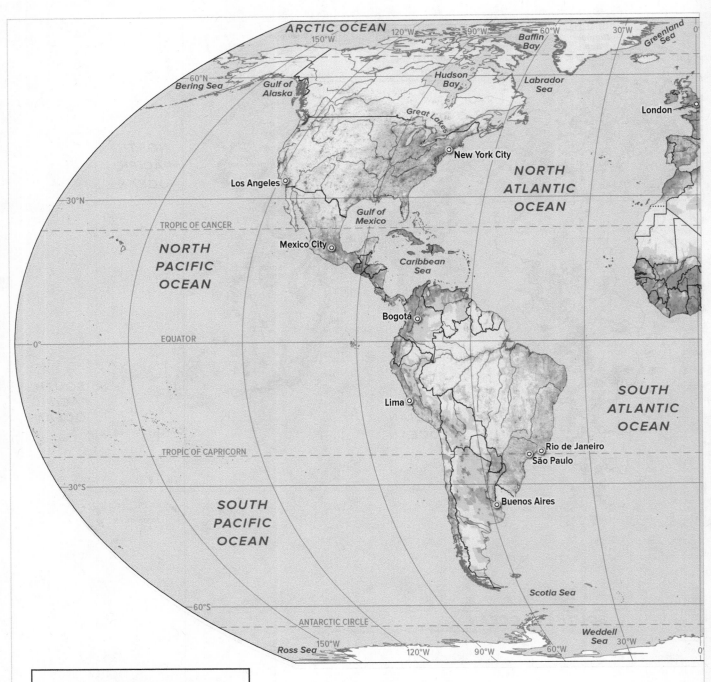

WORLD
POPULATION
DENSITY

0   2,000 miles at Equator

0   2,000 kilometers at Equator

Equal Earth projection

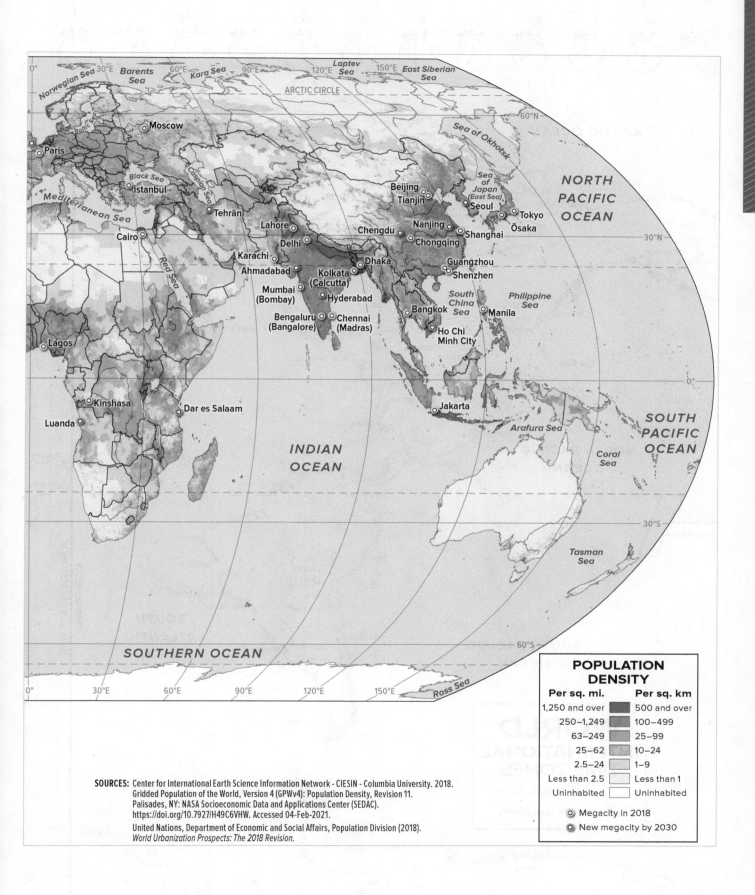

POPULATION
DENSITY

Per sq. mi.	Per sq. km
1,250 and over	500 and over
250–1,249	100–499
63–249	25–99
25–62	10–24
2.5–24	1–9
Less than 2.5	Less than 1
Uninhabited	Uninhabited

◎ Megacity in 2018
◉ New megacity by 2030

SOURCES: Center for International Earth Science Information Network - CIESIN - Columbia University. 2018.
Gridded Population of the World, Version 4 (GPWv4): Population Density, Revision 11.
Palisades, NY: NASA Socioeconomic Data and Applications Center (SEDAC).
https://doi.org/10.7927/H49C6VHW. Accessed 04-Feb-2021.

United Nations, Department of Economic and Social Affairs, Population Division (2018).
World Urbanization Prospects: The 2018 Revision.

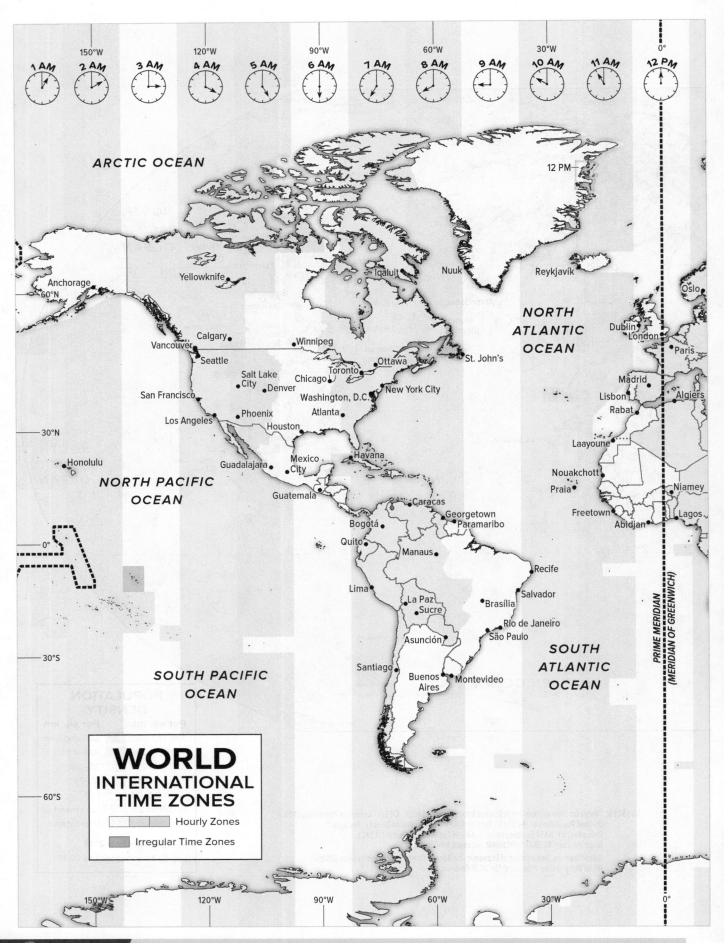

WORLD
INTERNATIONAL TIME ZONES

Hourly Zones

Irregular Time Zones

1 AM · 2 AM · 3 AM · 4 AM · 5 AM · 6 AM · 7 AM · 8 AM · 9 AM · 10 AM · 11 AM · 12 PM

150°W · 120°W · 90°W · 60°W · 30°W · 0°

ARCTIC OCEAN

NORTH ATLANTIC OCEAN

NORTH PACIFIC OCEAN

SOUTH PACIFIC OCEAN

SOUTH ATLANTIC OCEAN

PRIME MERIDIAN (MERIDIAN OF GREENWICH)

Anchorage
Yellowknife
Iqaluit
Nuuk
Reykjavík
Oslo
60°N
Calgary
Winnipeg
Dublin
London
Vancouver
Seattle
Ottawa
St. John's
Paris
Salt Lake City
Toronto
Chicago
Madrid
Denver
New York City
Lisbon
San Francisco
Washington, D.C.
Rabat
Algiers
30°N
Phoenix
Atlanta
Los Angeles
Houston
Laayoune
Mexico City
Havana
Honolulu
Guadalajara
Nouakchott
Niamey
Guatemala
Praia
Freetown
Abidjan
Lagos
Caracas
Georgetown
Paramaribo
Bogotá
Quito
Manaus
0°
Lima
Recife
La Paz
Sucre
Brasília
Salvador
Asunción
Río de Janeiro
São Paulo
30°S
Santiago
Buenos Aires
Montevideo
60°S

12 PM

150°W · 120°W · 90°W · 60°W · 30°W · 0°

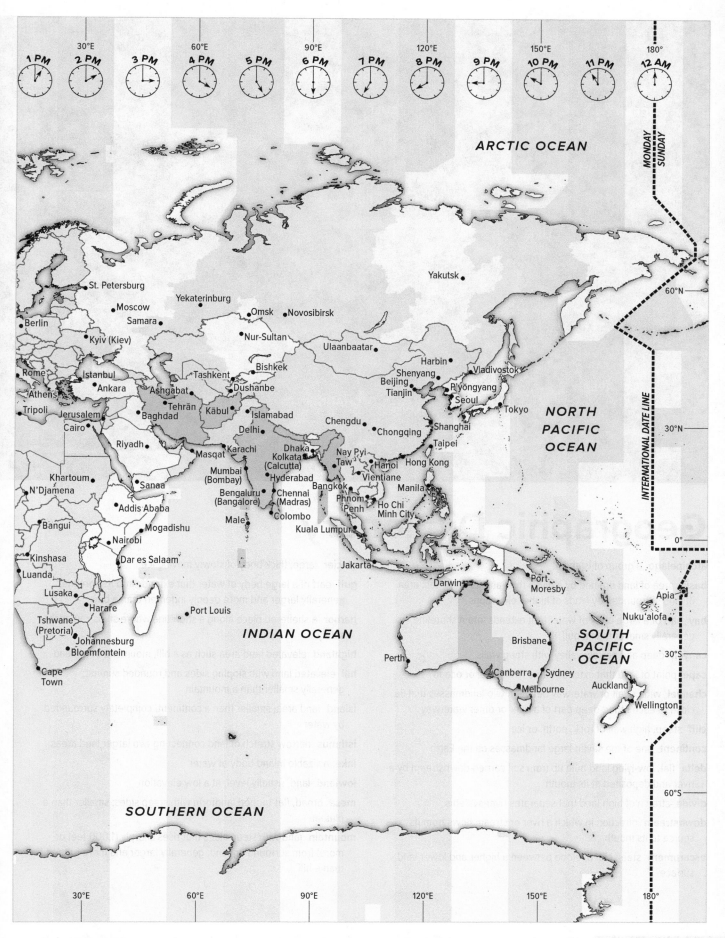

1 PM · 2 PM · 3 PM · 4 PM · 5 PM · 6 PM · 7 PM · 8 PM · 9 PM · 10 PM · 11 PM · 12 AM

30°E · 60°E · 90°E · 120°E · 150°E · 180°

ARCTIC OCEAN

MONDAY
SUNDAY

Yakutsk

60°N

St. Petersburg

Moscow
Samara
Berlin
Kyiv (Kiev)
Rome
Istanbul
Athens
Ankara
Tripoli
Jerusalem
Cairo

Yekaterinburg
Omsk
Novosibirsk
Nur-Sultan
Bishkek
Tashkent
Ashgabat
Dushanbe
Tehrān
Kābul
Baghdad
Islamabad

Ulaanbaatar

Harbin
Shenyang
Beijing
Tianjin

Vladivostok

P'yŏngyang
Seoul
Tokyo

NORTH
PACIFIC
OCEAN

30°N

Riyadh
Masqat
Karachi
Mumbai
(Bombay)
Khartoum
N'Djamena
Sanaa
Addis Ababa
Bangui
Nairobi
Mogadishu
Kinshasa
Dar es Salaam
Luanda
Lusaka
Harare
Tshwane
(Pretoria)
Johannesburg
Bloemfontein
Cape
Town

Delhi
Dhaka
Kolkata
(Calcutta)
Hyderabad
Bengaluru
(Bangalore)
Chennai
(Madras)
Male
Colombo

Chengdu
Chongqing

Shanghai

Taipei

Nay Pyi
Taw
Hanoi
Vientiane
Bangkok
Phnom
Penh
Kuala Lumpur

Hong Kong

Ho Chi
Minh City

Manila

Port Louis

INDIAN OCEAN

Jakarta

Darwin

Port
Moresby

SOUTH
PACIFIC
OCEAN

Apia
Nuku'alofa

INTERNATIONAL DATE LINE

0°

30°S

Perth

Brisbane

Canberra
Melbourne

Sydney

Auckland

Wellington

60°S

SOUTHERN OCEAN

30°E · 60°E · 90°E · 120°E · 150°E · 180°

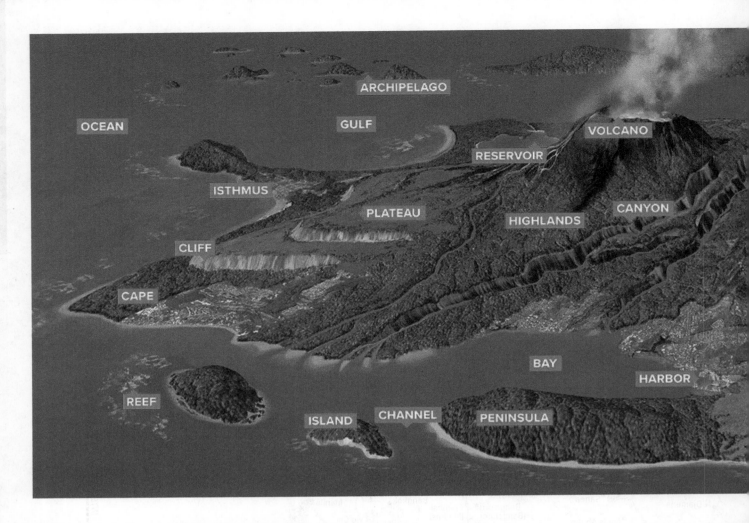

Geographic Dictionary

archipelago a group of islands

basin area of land drained by a given river and its branches; area of land surrounded by lands of higher elevations

bay part of a large body of water that extends into a shoreline, generally smaller than a gulf

canyon deep and narrow valley with steep walls

cape point of land that extends into a river, lake, or ocean

channel wide strait or waterway between two landmasses that lie close to each other; deep part of a river or other waterway

cliff steep, high wall of rock, earth, or ice

continent one of the seven large landmasses on the Earth

delta flat, low-lying land built up from soil carried downstream by a river and deposited at its mouth

divide stretch of high land that separates river systems

downstream direction in which a river or stream flows from its source to its mouth

escarpment steep cliff or slope between a higher and lower land surface

glacier large, thick body of slowly moving ice

gulf part of a large body of water that extends into a shoreline, generally larger and more deeply indented than a bay

harbor a sheltered place along a shoreline where ships can anchor safely

highland elevated land area such as a hill, mountain, or plateau

hill elevated land with sloping sides and rounded summit; generally smaller than a mountain

island land area, smaller than a continent, completely surrounded by water

isthmus narrow stretch of land connecting two larger land areas

lake a sizable inland body of water

lowland land, usually level, at a low elevation

mesa broad, flat-topped landform with steep sides; smaller than a plateau

mountain land with steep sides that rises sharply (1,000 feet or more) from surrounding land; generally larger and more rugged than a hill

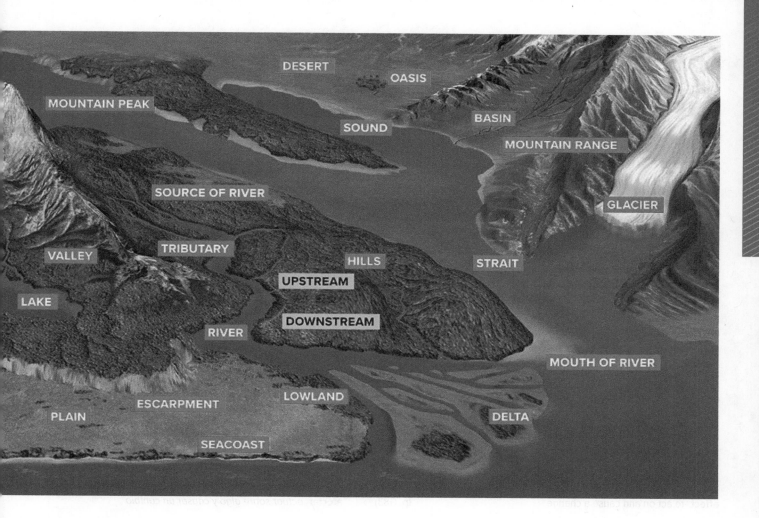

mountain peak pointed top of a mountain

mountain range a series of connected mountains

mouth (of a river) place where a stream or river flows into a larger body of water

oasis small area in a desert where water and vegetation are found

ocean one of the five major bodies of salt water that surround the continents

ocean current stream of either cold or warm water that moves in a definite direction through an ocean

peninsula body of land jutting into a lake or ocean, surrounded on three sides by water

physical feature characteristic of a place occurring naturally, such as a landform, body of water, climate pattern, or resource

plain area of level land, usually at low elevation and often covered with grasses

plateau area of flat or rolling land at a high elevation, about 300 to 3,000 feet (90 to 900 m) high

reef a chain of rocks, coral or sand at or near the surface of the water

reservoir a large natural or artificial lake used as a source of water supply

river large natural stream of water that runs through the land

sea large body of water completely or partly surrounded by land

seacoast land lying next to a sea or an ocean

sound broad inland body of water, often between a coastline and one or more islands off the coast

source (of a river) place where a river or stream begins, often in highlands

strait narrow stretch of water joining two larger bodies of water

tributary small river or stream that flows into a large river or stream; a branch of the river

upstream direction opposite the flow of a river; toward the source of a river or stream

valley area of low land usually between hills or mountains

volcano mountain or hill created as liquid rock and ash erupt from inside the Earth

Geography Glossary/
Glosario de geografía

All vocabulary words are **boldfaced** and **highlighted in yellow** in your textbook.

A		ESPAÑOL
abolition the end of slavery	(p. G16)	**abolición** *fin de la esclavitud*
absolute location the exact spot at which a place is found on Earth; the exact position of a place on Earth's surface, often determined by latitude and longitude	(pp. G8, G19)	**ubicación absoluta** *punto exacto en que se encuentra un lugar en la Tierra; ubicación exacta de un lugar en la superficie de la Tierra, a menudo, determinada por la longitud y la latitud*
absolute monarchy a form of autocracy with a hereditary king or queen exercising supreme power	(p. G234)	**monarquía absoluta** *forma de autocracia con un rey o reina que hace ejercicio del poder supremo hereditariamente*
access a way to get information	(p. G28)	**acceso** *modo de obtención de información*
accurate without mistakes or error	(p. G22)	**preciso** *sin errores*
acid rain rain that contains harmful amounts of poisons due to pollution; precipitation carrying large amounts of dissolved acids	(pp. G82, G283)	**lluvia ácida** *lluvia que contiene cantidades nocivas de venenos debido a la contaminación; precipitación que contiene grandes cantidades de ácidos disueltos*
acquire to gain	(p. G8)	**adquirir** *obtener*
affect to act on and cause a change	(p. G83)	**afectar** *actuar sobre algo y causar un cambio*
alluvial soil rich soil made up of sand and mud deposited by running water	(p. G266)	**suelo aluvial** *suelo fértil compuesto de arena y lodo depositados por una corriente de agua y tierra fértil que ocurre en la naturaleza*
authority the power to influence or command thought, opinion, or behavior	(p. G233)	**autoridad** *poder de influenciar o dominar el pensamiento, la opinión o el comportamiento*
autocracy a system of government in which one person rules with unlimited power and authority	(p. G233)	**autocracia** *sistema de gobierno en el cual una persona dirige con poder y autoridad ilimitados*
axis an imaginary line that runs through the center of Earth between the Poles	(p. G95)	**eje** *línea imaginaria que atraviesa el centro de la Tierra entre los Polos*

B		ESPAÑOL
benefit a good or helpful result	(p. 174)	**beneficio** *resultado bueno o útil*
biodiversity the wide variety of life on Earth	(p. G279)	**biodiversidad** *la gran variedad de vida en la Tierra*
biome a type of large ecosystem with similar life-forms and climates	(p. G99)	**bioma** *tipo de ecosistema extenso con formas de vida y climas similares*
birth rate the number of births per year for every 1,000 people	(p. G120)	**tasa de natalidad** *número de nacimientos al año por cada 1,000 personas*

A74

C		ESPAÑOL
boreal forest the forest in the far northern regions of Earth	(p. G279)	**bosque boreal** *bosque de las regiones más al norte de la Tierra*
brain drain migration of skilled or talented people out of a place	(p. G130)	**fuga de cerebros** *migración de personas calificadas o talentosas fuera de un lugar*
category a group of similar things	(p. G209)	**categoría** *grupo de cosas similares*
clearcutting the removal of all trees in a stand of timber	(p. G270)	**tala uniforme de árboles** *tala de todos los árboles de un rodal de madera*
clerical general office work	(p. G180)	**de oficina** *trabajo general de oficina*
climate change any significant change in the measures of climate lasting for an extended period of time	(p. G283)	**cambio climático** *cualquier cambio significativo en las medidas del clima en un período prolongado*
climate weather patterns typical for an area over a long period of time	(p. G95)	**clima** *patrones de tiempo atmosférico característicos de un área en un periodo extenso*
command economy an economy in which the means of production are owned by the government and are supposed to be managed for the benefit of the public	(p. G194)	**economía planificada** *economía en la que los medios de producción son propiedad del gobierno y se supone que deben administrarse en beneficio del público*
commercial farming growing large quantities of crops or livestock to sell for a profit	(pp. G204, G263)	**agricultura comercial** *cultivo de grandes cantidades de plantas o ganado para venderlos con fines de lucro*
commission an assignment of responsibility	(p. G15)	**comisión** *asignación de responsabilidad*
comparative advantage a place's ability to produce something more efficiently than another place	(p. G207)	**ventaja comparativa** *habilidad de un lugar de producir algo de forma más eficiente que otro*
complementarity the relationship between two places for the demand and supply of a product	(p. G206)	**complementariedad** *relación entre dos lugares por la demanda y la oferta de un producto*
coniferous evergreen trees that produce cones to hold seeds and that have needles instead of leaves; a cone-bearing tree with needle- or scale-like leaves	(pp. G101, G105)	**conífera** *árboles de hoja perenne que producen conos para almacenar semillas y que tienen agujas en lugar de hojas; árbol con forma de cono, con hojas en forma de agujas o escamas*
conservation the act of protecting Earth's natural resources	(p. G271)	**conservación** *acto de proteger los recursos naturales de la Tierra*
constant happening all the time	(p. G97)	**constante** *que sucede todo el tiempo*
context the situation in which something happens	(p. G47)	**contexto** *situación en la que algo acontece*
continental shelf the part of a continent that extends out underneath the ocean and is fairly flat, but then drops sharply to the ocean floor	(p. G79)	**plataforma continental** *parte de un continente que se extiende por debajo del océano y es bastante plana, pero luego cae bruscamente al fondo del océano*
contract to create a business arrangement between two parties	(p. G276)	**contrato** *crear un acuerdo de negocio entre dos partes*
contribute to give	(p. G129)	**contribuir** *dar*
convert to change from one thing to another	(p. G18)	**convertir** *cambiar de una cosa a otra*
core the innermost layer of Earth	(p. G83)	**núcleo** *la capa más interna de la Tierra*

GEOGRAPHY GLOSSARY

English		Español
crust outer layer of Earth, a hard rocky shell	(p. G84)	**corteza** *capa externa de la tierra, rocosa y sólida*
crux the most important or decisive part	(p. G214)	**punto** *crucial la parte más importante o decisiva*
cultural boundary a geographical boundary between two different cultures	(p. G236)	**frontera cultural** *límite geográfico entre dos culturas diferentes*
cultural diffusion the spread of new ideas from a source point	(pp. G129, G171)	**difusión cultural** *expansión de nuevas ideas desde un punto de origen*
culture hearth a center where cultures developed and from which ideas and traditions spread outward	(p. G171)	**centro cultural** *centro donde se desarrollaron las culturas y desde el que se difundieron las ideas y tradiciones*
culture region section of the Earth in which people share a similar way of life, including language, religion, economic systems, and types of government	(p. G161)	**región cultural** *sección de la Tierra en la cual las personas comparten una forma similar de vida, incluyendo la lengua, la religión, sistemas económicos y tipos de gobierno*
culture trait a characteristic of the culture that is shared by most members	(p. G161)	**característica cultural** *particularidad de una cultura que es compartida por la mayoría de sus miembros*
culture the set of beliefs, behaviors, and traits shared by a group of people	(p. G157)	**cultura** *conjunto de creencias, comportamientos y rasgos compartidos por un grupo de personas*
curry generic term used to describe many Indian dishes	(p. G168)	**curry** *término genérico utilizado para describir muchos platos de la India*
custom an action or behavior that is traditional among the people in a particular group or place	(p. G157)	**costumbre** *acción o comportamiento que es tradicional entre las personas de un grupo o lugar en particular*

D   ESPAÑOL

English		Español
death rate the number of deaths per year for every 1,000 people	(p. G119)	**tasa de mortalidad** *número de muertes al año por cada 1,000 personas*
deciduous trees that shed their leaves in the autumn	(p. G101)	**caducifolio** *árboles que pierden sus hojas en otoño*
deforestation the loss or destruction of forests, mainly for logging or farming	(p. G279)	**deforestación** *pérdida o destrucción de bosques, principalmente por la tala o la agricultura*
degrade to wear out and make lower quality	(p. G274)	**degradar** *desgastar o reducir la calidad*
democracy a type of government in which leaders rule with the consent of the citizens	(p. G232)	**democracia** *tipo de gobierno en el que los líderes dirigen con el consentimiento de los ciudadanos*
depositary a place for safe storage	(p. G275)	**depósito** *lugar de almacenamiento seguro*
desalination a process that makes salt water safe to drink; the removal of salt from seawater to make it usable for drinking and farming	(pp. G80, G267)	**desalinización** *proceso que hace que el agua salada sea segura para el consumo; proceso que elimina la sal del agua marina para que sea potable y apta para la agricultura*
desertification the process by which an area turns into a desert	(p. G280)	**desertización** *proceso por el cual un área se convierte en desierto*
despite without being prevented by	(p. G208)	**a pesar de** *sin ser impedido por*
developing having an economy mainly based on agriculture and a relatively low level of wealth	(p. G63)	**en vía de desarrollo** *tener una economía basada principalmente en la agricultura y un nivel de riqueza relativamente bajo*
dialect a regional variety of a language with unique features, such as vocabulary, grammar, or pronunciation	(p. G158)	**dialecto** *variación regional de una lengua con características únicas, tales como vocabulario, gramática o pronunciación*

diaspora the dispersal of a group of people from one location to many others	(p. G176)	**diáspora** dispersión de un grupo de personas de un lugar a muchos otros	
dictatorship a form of autocracy in which the one person has absolute power to rule and control the government, the people, and the economy	(p. G234)	**dictadura** forma de autocracia en la cual una sola persona tiene poder absoluto de dirigir y controlar el gobierno, la economía y a las personas	
dike a long wall or earthen bank used to prevent flooding from the sea; large bank of earth and stone that holds back water	(pp. G288, G267)	**dique** pared larga o muro de tierra usado para prevenir inundaciones del mar gran muro de tierra y piedra que retiene el agua	
diminish to make or become less	(p. G134)	**reducir** hacer o convertirse en menos	
diplomatic involving communication among countries to maintain relationships	(p. G253)	**diplomático** que involucra comunicación entre países para mantener relaciones	
distort to pull or change something so it is no longer accurate	(p. G18)	**distorsionar** quitar o cambiar algo para que ya no sea preciso	
diverse different	(p. G100)	**diverso** diferente	
domestic relating to a home country	(p. G60)	**nacional** relativo a un país de origen	
domesticity home and family life	(p. G277)	**domesticidad** vida hogareña y familiar	
doubling time the number of years it takes for a population to double in size	(p. G120)	**tiempo de duplicación** número de años que tarda una población en doblar su tamaño	

E — ESPAÑOL

earthquake an event in which the ground shakes or trembles, brought about by the collision of tectonic plates	(p. G86)	**terremoto** evento en el cual el suelo se sacude o tiembla como consecuencia del choque de las placas tectónicas	
economic system the way a society decides on the ownership and distribution of its economic resources	(p. G193)	**sistema económico** forma en que una sociedad decide acerca de la propiedad y distribución de sus recursos económicos	
ecosystem the complex community of interdependent living things in a given environment	(p. G279)	**ecosistema** comunidad compleja de seres vivos interdependientes en un entorno dado	
element a particular part of something	(p. G26)	**elemento** parte específica de algo	
elevation the height of a land surface above or below sea level; vertical position, usually measured in distance from sea level	(pp. G24, G34, G78)	**elevación** altura de una superficie de tierra por encima o por debajo del nivel del mar; posición vertical, por lo general, medida en distancia desde el nivel del mar	
emigrate to leave one's home to live in another place	(p. G127)	**emigrar** ejar el país natal para vivir en otro lugar	
emphasis importance	(p. G207)	**énfasis** importancia	
empire an extensive group of countries or places under a single supreme authority	(p. G163)	**imperio** grupo extensivo de países o lugares bajo una sola autoridad suprema	
enclave a piece of territory wholly enclosed within another territory	(p. G242)	**enclave** parte de territorio completamente encerrado dentro de otro	
enforce to make sure that people do what is required	(p. G231)	**imponer** asegurarse de que las personas hagan lo que se requiere	
enormous very great in size	(p. G79)	**enorme** de gran tamaño	
environment the natural world	(p. G172)	**medioambiente** mundo natural	

GEOGRAPHY GLOSSARY

English		Español
equinox one of two days each year when the sun is directly overhead at the Equator	(p. G96)	**equinoccio** *uno de los dos días de cada año en que el sol se encuentra directamente sobre el Ecuador*
erosion the process by which weathered bits of rock are moved elsewhere by water, wind, or ice	(p. G87)	**erosión** *proceso por el cual pedazos de roca erosionados se mueven a otra parte por el agua, el viento o el hielo*
ethnic group a group of people who share a common ancestry, language, religion, customs, or place of origin	(p. G160)	**grupo étnico** *grupo de personas que comparte un linaje, lengua, religión, costumbres o lugar de origen en común*
expand to increase in size	(p. G55)	**expandir** *aumentar de tamaño*
exploit to get value or use from	(p. G56)	**explotar** *obtener valor o uso de*
export to send a product from one country to another for purposes of trade	(pp. G58, G209)	**exportar** *enviar un producto de un país a otro con propósitos de comercio*
extract pull out; withdraw	(p. G270)	**extraído** *arrancado; retirado*

F		ESPAÑOL
fall line a boundary where a higher, upland area drops to the lower land of a plain	(p. G265)	**línea de descenso** *frontera en el oriente de Estados Unidos donde la meseta de Piedmont se une con la llanura litoral atlántica*
fault an extended break in a body of rock, marked by movement of rock on either side	(p. G86)	**falla** *fractura prolongada en un cuerpo de roca, marcada por un movimiento de roca en ambos lados*
fast to go without eating	(p. G275)	**ayunar** *abstenerse de comer*
federal system a form of government in which powers are divided between the national government and state or provincial government	(p. G232)	**sistema federal** *forma de gobierno en la cual los poderes están divididos entre el gobierno nacional y el gobierno provincial o estatal*
formal region a region defined by a common characteristic	(p. G54)	**región formal** *región definida por una característica común*
fossil fuel a resource formed in the Earth by plant and animal remains	(p. G269)	**combustible fósil** *recurso que se forma en la Tierra por restos de plantas y animales*
foundation a basis on which something stands	(p. G159)	**cimiento** *base sobre la cual algo se sostiene*
free trade arrangement whereby a group of countries decides to set little or no tariffs or quotas	(p. G210)	**libre comercio** *acuerdo por el cual un grupo de países decide establecer aranceles o cuotas escasas o nulas*
function purpose	(p. G59)	**función** *propósito*
functional region a central place and the surrounding territory linked to it	(p. G54)	**región funcional** *un lugar central y territorio circundante que se vincula a este*

G		ESPAÑOL
genocide the intentional destruction of a group of people that can include mass murder, imposing harsh conditions, and forcibly removing children	(p. G245)	**genocidio** *destrucción intencional de un grupo de personas que puede incluir asesinato en masa, imposición de condiciones severas y traslado forzoso de niños*
genre style	(p. G65)	**género** *estilo*
geographic information systems computer programs that process and organize details about places on Earth	(p. G26)	**sistemas de información** *geográfica programas informáticos que procesan y organizan detalles sobre lugares de la Tierra*
geography the study of Earth and its peoples, places, and environments	(p. G7)	**geografía** *estudio de la Tierra y sus habitantes, lugares y entornos*

geometric boundary a boundary that follows a geometric pattern (p. G236)

frontera geométrica *límite que sigue un patrón geométrico*

geothermal energy the electricity produced by natural, underground sources of steam (p. G270)

energía geotérmica *electricidad producida por fuentes de vapor naturales subterráneas*

glacier a large body of ice that moves slowly across land (p. G88)

glaciar *enorme masa de hielo que se desliza lentamente sobre la tierra*

global positioning system a navigational system that can determine absolute location by using satellites and receivers on Earth (p. G25)

sistema de posicionamiento global *sistema de navegación que puede determinar una localización absoluta mediante satélites y receptores en la Tierra*

globalization increasing interconnection of economic, political, and cultural processes to the point that they become global in scale and impact; the process by which economic, political, and cultural processes expand to reach across the world (pp. G60, G173, G211)

globalización *creciente interconexión de procesos económicos, políticos y culturales hasta el punto de tener una escala e impacto global; medio por el cual procesos económicos, políticos y culturales se expanden para llegar a todo el mundo*

gradient an increase or decrease in the amount of something (p. G81)

gradiente *aumento o descenso en la cantidad de algo*

Greenbushes the world's largest hard-rock lithium mine in Western Australia (p. G198)

Greenbushes *la mina de litio de roca dura más grande del mundo, en Australia Occidental*

gross domestic product (GDP) the dollar value of all final goods and services produced in a country during a single year (p. G203)

producto interno bruto (PIB) *valor en dólares de todos los bienes y servicios finales producidos en un país durante un solo año*

groundwater the water contained inside Earth's crust (p. G80)

agua subterránea *agua ubicada debajo de la corteza de la Tierra*

H

ESPAÑOL

harlequin a form of comic theater in which characters wear masks and bright colors (p. G15)

arlequín *forma de teatro cómico en el cual los personajes usan máscaras y colores brillantes*

headwater the source of a stream or river (p. G239)

cabecera *fuente de un arroyo o río*

hemisphere each half of Earth (p. G17)

hemisferio *cada mitad de la tierra*

heritage something of value to a group or society that can be passed to future generations (p. G16)

herencia *algo de valor para un grupo o sociedad que puede dejarse a futuras generaciones*

human rights the rights belonging to all individuals simply because we exist as human beings; these include freedoms and rights, such as freedom of speech, that all people should enjoy (p. G234)

derechos humanos *derechos que pertenecen a todos los individuos por el simple hecho de existir como seres humanos; estos incluyen libertades y derechos tales como la libertad de expresión, que todas las personas deberían disfrutar*

humidity moisture in the air (p. G101)

humedad *cantidad de agua o de vapor de agua en el aire*

hydroelectric power the electricity that is created by flowing water (p. G272)

energía hidroeléctrica *electricidad que es producida por agua en movimiento*

I

ESPAÑOL

immigrate to enter and live in a new place or country (p. G127)

inmigrar *entrar y vivir en un sitio o país nuevo*

impinging striking against (p. G90)

transgredir *quebrantar*

import to bring a product into one country from another country (pp. G58, G209)	**importar** *llevar un producto proveniente de un país a otro*
improvisational done without advanced preparation (p. G167)	**improvisado** *hecho sin preparación previa*
indigenous the earliest known inhabitants of a place (p. G163)	**indígena** *los habitantes más antiguos conocidos de un lugar*
induce to cause (p. G243)	**inducir** *causar*
industrialize to build factories and engage in manufacturing (p. G56)	**industrializar** *construir fábricas y dedicarse a la manufactura*
industrialized having an advanced economy and a relatively high level of wealth (p. G63)	**industrializado** *tener una economía avanzada y un nivel relativamente alto de riqueza*
infrastructure the set of systems that affect how well a place or organization operates, such as telephone or transportation systems, within a country (pp. G133, G141)	**infraestructura** *conjunto de sistemas que afectan el funcionamiento de un lugar u organización, como los sistemas telefónicos o de transporte en un país*
integrated with functions linked or coordinated (p. G252)	**integrado** *con funciones entrelazadas o coordinadas*
interdependence a condition in which people or groups rely on each other, rather than only relying on themselves (p. G59)	**interdependencia** *condición en la cual las personas o grupos dependen unos de otros, en lugar de ellos mismos*
Inuit an indigenous people of northern Canada (p. G13)	**inuit** *pueblo indígena del norte de Canadá*
irrigation the process of collecting and moving water for use in agriculture (p. G265)	**irrigación** *proceso de acumular y trasladar agua para uso agrícola*
issue an important subject (p. G163)	**asunto** *tema importante*
isthmus a narrow strip of land that connects two larger land areas (p. G78)	**istmo** *franja delgada de tierra que conecta dos territorios más grandes*

K	**ESPAÑOL**
kitch in poor taste (p. G176)	**kitsch** *de mal gusto*

L	**ESPAÑOL**
landscape a portion of Earth's surface that can be viewed at one time from a location (p. G46)	**paisaje** *porción de la superficie de la Tierra que se puede ver de una sola vez desde una ubicación*
language family a group of related languages that have all developed from one earlier language (p. G158)	**familia lingüística** *grupo de lenguas relacionadas que se han desarrollado de una anterior*
latitude the lines on a map or globe that run east to west but measure distance on Earth in a north to south direction (p. G19)	**latitud** *líneas de un mapa o globo que van de oriente a occidente, pero que miden distancias en la Tierra en dirección de norte a sur*
lava melted rock on Earth's surface (p. G84)	**lava** *roca fundida en la superficie de la Tierra*
less developed country a country with the lowest indicators of socioeconomic development (pp. G124, G204)	**país en vía de desarrollo** *país con los indicadores de desarrollo socioeconómico más bajos*
lichen tiny sturdy plants that grow in rocky areas (p. G102)	**liquen** *plantas diminutas y robustas que crecen en áreas rocosas*
lingua franca a common language among speakers whose native languages are different (p. G158)	**lengua franca** *lengua común entre hablantes cuyas lenguas nativas son distintas*

linguist a person who studies language (p. G169)

lingüista *persona que estudia el lenguaje*

longitude the lines on a map or globe that run north to south but measure distance on Earth in an east to west direction (p. G19)

longitud *líneas de un mapa o globo que van del norte al sur, pero que miden distancias en la Tierra en dirección oriente-occidente*

M	ESPAÑOL

magma melted rock below Earth's surface (p. G84)

magma *roca fundida bajo la superficie de la Tierra*

major greater in importance or interest (p. G159)

principal *de mayor importancia o interés*

mantle a thick layer of hot, dense rock surrounding the Earth's core (p. G83)

manto *capa gruesa de roca caliente densa que rodea el núcleo de la Tierra*

map projection the method used to represent the round Earth on a flat map (p. G22)

proyección cartográfica *método usado para representar la superficie curva de la Tierra en un mapa plano*

market economy an economy in which most of the means of production are privately owned and these owners compete for selling what they produce (p. G194)

economía de mercado *economía en la cual la mayoría de los medios de producción son de propiedad privada y estos propietarios compiten por vender aquello que producen*

mature to age; to grow up (p. G120)

madurar *envejecer; crecer*

megalopolis a huge city or cluster of cities with an extremely large population (p. G132)

megalópolis *ciudad grande o grupo de ciudades con una población extremadamente densa*

mental map a map that people have in their minds that is based on the understandings and perceptions they have of features on Earth's surface (p. G24)

mapa mental *mapa que las personas tienen en sus mentes y que está basado en el entendimiento y percepción que tienen sobre las características de la superficie de la Tierra*

metallurgy the science and technology of purifying and working with metallic elements (p. G278)

metalurgia *ciencia y tecnología de purificar y trabajar con elementos metálicos*

migrate to move from one place to another (p. G56)

migrar *mudarse de un lugar a otro*

migration the movement of people from one place to another (pp. G120, G127, G172)

migración *mudanza de las personas de un lugar a otro*

mixed economy an economy in which parts of the economy are privately owned and parts are owned by the government (p. G194)

economía mixta *economía en la cual una parte es propiedad privada y la otra del gobierno*

Moors northwest African Muslims who controlled territory that is part of present-day Spain from the 700s to the 1600s A.D. (p. G177)

Moros *musulmanes del noroeste de África que controlaron el territorio de lo que es parte de la actual España, desde el siglo VIII al XVII d. C.*

more developed country a country with a highly developed economy and advanced technological infrastructure (pp. G124, G203)

país desarrollado *país con una economía altamente desarrollada e infraestructura tecnológica avanzada*

munitions military weapons, ammunition, and equipment (p. G275)

municiones *armas militares, artillería pesada y equipamiento*

GEOGRAPHY GLOSSARY

N | ESPAÑOL

nationalism belief in the right of each people with a unique cultural identity based on common language, religion, and national symbols to be independent (p. G244)

nacionalismo *creencia en que cada pueblo con una identidad cultural única basada en la lengua, la religión y los símbolos nacionales comunes tiene derecho a ser independiente*

natural boundary a boundary created by a physical feature, such as a mountain, river, or strait (p. G236)

frontera natural *límite fijado por una formación física como una montaña, río o estrecho*

natural disaster a sudden and extreme event in nature that usually results in serious damage and loss of life (p. G268)

desastre natural *evento repentino y extremo de la naturaleza que por lo general resulta en daños graves y pérdidas de vidas*

natural increase the growth rate of a population; the difference between birthrate and death rate (p. G120)

crecimiento natural *tasa de crecimiento de una población; diferencia entre la tasa de natalidad y la tasa de mortalidad*

natural resource materials or substances such as minerals, forests, water, and fertile land that occur in nature (p. G191, G269)

recurso natural *materiales o sustancias tales como minerales, bosques, agua y tierra fértil que se encuentran en la naturaleza*

natural vegetation plant life that grows in a certain area if people have not changed the natural environment (p. G99)

vegetación natural *vida vegetal que crece en un área determinada si las personas no han cambiado el entorno natural*

newly industrialized country a country transitioning from primarily agricultural to primarily manufacturing and industrial activity (p. G204)

país recientemente industrializado *país en transición de actividades principalmente agrícolas a actividades de manufactura e industria*

nonrenewable resource a resource that cannot be replaced (pp. G192, G269)

recurso no renovable *recurso que no puede ser reemplazado*

nuclear energy the energy created by splitting the nucleus of an atom (p. G270)

energía nuclear *energía que surge de la división del núcleo de un átomo*

O | ESPAÑOL

obtain to gain (p. G191)

obtener *ganar*

oligarchy a system of government in which a small group holds power (p. G234)

oligarquía *sistema de gobierno en el cual un grupo pequeño ejerce el poder*

ongoing continuing (p. G121)

en curso *continuo*

outsourcing obtaining goods or services from an outside supplier (p. G211)

subcontratación *obtener bienes o servicios de un proveedor externo*

ozone layer a layer around the Earth's atmosphere that blocks out many of the most harmful rays from the sun (p. G283)

capa de ozono *capa que rodea la atmósfera terrestre y bloquea muchos de los rayos más nocivos del Sol*

P | ESPAÑOL

pandemic the rapid spread of a disease around the world that affects a large number of people over a wide area (p. G174)

pandemia *rápida propagación de una enfermedad alrededor del mundo que afecta a un gran número de personas en un área amplia*

pastoralism the raising of animals for food (p. G264)

pastoreo *cría de animales para el consumo*

peninsula a portion of land nearly surrounded by water (p. G78)

península *porción de tierra que está prácticamente rodeada por agua*

per-capita GDP the average value of goods and services produced by one person (p. G63)

PIB per cápita *valor en promedio de bienes y servicios producidos por una persona*

perception the way you think about something	(p. G47)	**percepción** *forma en la que piensas acerca de algo*
perceptual region a region defined through common understanding developed over time	(p. G55)	**región perceptiva** *región definida a través de un entendimiento común desarrollado con el tiempo*
perishable likely to decay or spoil quickly	(p. G173)	**perecedero** *con probabilidad de deteriorarse o echarse a perder rápidamente*
permafrost permanently frozen layer of soil beneath the surface of the ground	(pp. G102, G284)	**permafrost** *capa de suelo permanentemente congelada bajo la superficie*
phenomena a rare or significant event	(p. G9)	**fenómeno** *evento raro o significativo*
Pilbara a region in Western Australia	(p. G198)	**Pilbara** *región en el occidente de Australia*
pilgrim a person who travels to a shrine or other holy place	(p. G275)	**peregrino** *persona que viaja a un templo u otro lugar sagrado*
place what a specific location is like; a location on Earth that has distinctive characteristics that makes it meaningful to people	(p. G8, G45)	**lugar** *aspecto de un sitio específico; sitio en la Tierra que tiene características distintivas que lo hacen significativo para las personas*
plankton plants or animals that ride along with water currents	(p. 284)	**plancton** *plantas o animales que se desplazan con las corrientes de agua*
polder low-lying area from which seawater has been drained to create new land	(p. G267)	**pólder** *área baja en la cual se ha drenado el agua del mar para crear tierra nueva*
policy an overall plan that establishes goals and determines procedures, decisions, and actions	(p. G244)	**política** *plan general que establece metas y determina procedimientos, decisiones y acciones*
population density the average number of people living in a square mile or a square kilometer	(p. G125)	**densidad de población** *número promedio de personas que viven en una milla cuadrada o un kilómetro cuadrado*
population distribution the geographic pattern of where people live	(p. G123)	**distribución de población** *patrón geográfico de donde viven las personas*
population pyramid a diagram that shows the distribution of a population by age and gender	(p. G121)	**pirámide demográfica** *diagrama que muestra la distribución de una población por edad y género*
porter a person hired to carry luggage and other loads	(p. G214)	**botones** *persona contratada para llevar equipaje y otras cargas*
portion a part of the whole	(p. G123)	**porción** *parte de un todo*
prairie an inland grassland area with no trees except along rivers	(p. G101)	**pradera** *área de pastizal en el interior sin árboles, excepto a lo largo de los ríos*
precipitation the water that falls on the ground as rain, snow, sleet, hail, or mist	(p. G96)	**precipitación** *agua que cae al suelo en forma de lluvia, nieve, granizo o niebla*
process a series of actions that produces something	(p. G77)	**proceso** *serie de acciones que producen algo*
prohibit to forbid somebody from doing something through law or rule	(p. G283)	**prohibir** *imposibilitar que alguien haga algo por ley o regla*
promote advance	(p. G210)	**promover** *avanzar*
promote to help	(p. G130)	**fomentar** *ayudar*
pull factor factors that attract people to a place	(p. G128)	**factor de atracción** *factor que atrae a las personas a un lugar*
push factor factors that drive people from a place		**factor de expulsión** *factor que aleja a las personas de un lugar*

GEOGRAPHY GLOSSARY

R		ESPAÑOL
rain shadow an area that receives reduced rainfall because it is on the side of a mountain facing away from the ocean	(p. G98)	**sombra pluviométrica** *área que recibe poca lluvia porque está en la ladera de una montaña de espaldas al océano*
recover to regain	(p. G268)	**recuperar** *recobrar*
reforestation the planting and cultivating of new trees in an effort to restore a forest where the trees have been cut down or destroyed	(p. G280)	**reforestación** *plantación y cultivo de árboles como esfuerzo para recuperar un bosque donde los árboles han sido talados o destruidos*
refugee a person who flees a country because of violence, war, persecution, or disaster	(pp. G128, G172, G245)	**refugiado** *persona que huye de un país debido a la violencia, guerra, persecución o desastre*
region a group of places with similar characteristics; an area of Earth's surface that has similar physical or human characteristics	(pp. G8, G53)	**región** *grupo de lugares con características similares; área de la superficie de la Tierra que tiene características físicas o humanas similares*
relative location a place's location in relation to other places	(pp. G8, G19)	**ubicación relativa** *localización de un lugar en relación con otros sitios*
relief the vertical elevation change that exists in a landscape	(p. G24)	**relieve** *cambio de elevación vertical que existe en un paisaje*
remittance money a migrant worker sends back to people in his or her home country	(pp. G122, G130, G218)	**remesa** *dinero que un trabajador migrante envía a personas en su país de origen*
remote sensing the method of getting information from far away, usually with aircraft or satellites	(p. G27)	**detección remota** *método de obtención de información desde muy lejos, por lo general con aeronaves o satélites*
renewable resource a resource that can be replenished in a relatively short period of time	(pp. G192, G269)	**recurso renovable** *recurso que puede ser repuesto en un período relativamente corto*
representative democracy a form of democracy in which citizens elect government leaders to represent the people	(p. G233)	**democracia representativa** *forma de democracia en la cual los ciudadanos eligen a los líderes del gobierno para representar al pueblo*
retain to keep or hold	(p. G96)	**retener** *mantener o contener*
reveal to make known	(p. G164)	**revelar** *hacer algo conocido*
revolution a complete trip of Earth around the sun; a sudden or complete change	(pp. G95, G195)	**revolución** *un giro completo de la Tierra alrededor del sol; cambio repentino o completo*
Ring of Fire a path along the Pacific Ocean characterized by volcanoes and earthquakes	(p. G86)	**Cinturón de Fuego** *patrón a lo largo del Océano Pacífico caracterizado por volcanes y terremotos*
rural an area that is lightly populated	(p. G124)	**rural** *área poco poblada*

S		ESPAÑOL
savanna a large area of land with grass and scattered trees	(p. G100)	**sabana** *gran área de tierra con pasto y árboles dispersos*
scale the relationship between distances on the map and on Earth; the geographic size of the area being studied	(pp. G23, G54)	**escala** *relación entre las distancias en un mapa y en la Tierra; tamaño geográfico del área que está siendo estudiada*
scarcity the situation in which there are limited resources to satisfy unlimited needs	(p. G191)	**escasez** *situación en la cual hay recursos limitados para satisfacer necesidades ilimitadas*

series a set of things arranged in order (p. G265) — **serie** *conjunto de cosas arregladas en orden*

shift to change the place, position, or direction of (p. G280) — **mover** *cambiar de sitio, posición o dirección*

similar comparable (p. G99) — **similar** *comparable*

site the characteristics of a place, including its physical setting (p. G46) — **sitio** *características de un lugar, incluido su ámbito físico*

situation the geographic position of a place in relation to other places or features of a larger region (p. G46) — **situación** *posición geográfica de un lugar en relación con otros o características de una región más grande*

slash-and-burn agriculture a method of farming that involves cutting down trees and underbrush and burning the area to create a field for crops (p. G267) — **agricultura de tala y quema** *método de agricultura que incluye la tala de árboles y maleza, y la quema del área para crear un campo para los cultivos*

smog haze caused by chemical fumes from automobile exhausts and other pollution sources (p. G283) — **esmog** *niebla ocasionada por gases químicos de escapes y otras fuentes de contaminación*

solstice one of two days of the year when the sun reaches its northernmost or southernmost point (p. G95) — **solsticio** *uno de los dos días del año en que el sol alcanza su punto más al norte o al sur*

sovereignty the supreme and absolute authority within territorial boundaries; independence; right to self-determination (pp. G231, G250) — **soberanía** *autoridad suprema y absoluta dentro de los límites territoriales; independencia; derecho a la autodeterminación*

spatial interaction the movement and flow of people, products, and ideas related to human activity (p. G57) — **interacción especial** *movimiento y flujo de personas, productos e ideas relacionadas con la actividad humana*

spatial Earth's physical and human features in terms of their places, shapes, relationships to one another, and the patterns they form (p. G7) — **espacial** *características físicas y humanas de la Tierra en términos de sus lugares, formas, relaciones entre sí y los patrones que forman*

specific special or particular (p. G22) — **específico** *especial o particular*

standard of living the level at which a person, group, or country lives as measured by the extent to which it meets its needs (p. G203) — **estándar de vida** *nivel en el que vive una persona, grupo o país, medido por el grado en que satisface sus necesidades*

structure the arrangement of parts that gives something its basic form (p. G84) — **estructura** *disposición de las partes que le da a algo su forma básica*

subducted forced under another layer of crust (p. G90) — **subducción** *forzado debajo de otra capa de corteza*

subsistence farming farming that only provides the basic needs of a family (pp. G205, G263) — **agricultura de subsistencia** *agricultura que solo provee las necesidades básicas de una familia*

subsistence relating to production for one's own use or consumption (p. G213) — **subsistencia** *relativo a la producción para uso o consumo propio*

suffrage voting rights (p. G16) — **sufragio** *voto*

sustain to give support to (p. G269) — **sostener** *dar soporte a*

sustainability the idea that a country works to create conditions where all natural resources for meeting the needs of society are available (pp. G212, G272) — **sostenibilidad** *idea de que un país trabaja para crear condiciones donde todos los recursos naturales están disponibles para suplir las necesidades de la sociedad*

GEOGRAPHY GLOSSARY

T		**ESPAÑOL**
tariff a tax on trade	(p. G254)	**arancel** *impuesto sobre el comercio*
technology any way that scientific discoveries are applied to practical use	(p. G25)	**tecnología** *cualquier forma en la que los descubrimientos científicos son aplicados a un uso práctico*
tectonic plate one of the large pieces of Earth's crust	(p. G85)	**placa tectónica** *una de las grandes piezas de corteza de la Tierra*
temperance avoidance of alcoholic drinks	(p. G16)	**abstinencia** *evasión del consumo de bebidas alcohólicas*
territorial of or relating to land belonging to a government	(p. G243)	**territorial** *de o relativo a la tierra perteneciente a un gobierno*
terrorism violence committed in order to frighten people or governments into granting demands	(p. G245)	**terrorismo** *violencia cometida con el propósito de atemorizar a las personas o gobiernos en concesión de demandas*
thematic map a map that shows specialized information	(p. G24)	**mapa temático** *mapa que muestra información especializada*
theocracy a system of government in which the officials are regarded as divinely inspired	(p. G235)	**teocracia** *sistema de gobierno en el cual los funcionarios son considerados como inspirados por una divinidad*
topography the shape of the land on Earth's surface	(p. G24)	**topografía** *forma del terreno de la superficie de la Tierra*
traditional economy an economy in which goods are produced by individuals for themselves and are distributed mainly through families	(p. G194)	**economía tradicional** *economía en la cual los bienes son producidos por los individuos para sí mismos y son distribuidos principalmente a través de las familias*
treaty an official agreement, negotiated and signed by each party	(p. G243)	**tratado** *acuerdo oficial, negociado y firmado por cada parte*
trench a long, narrow, steep-sided cut on the ocean floor	(p. G79)	**fosa** *corte largo, angosto y empinado en el fondo del océano*
tributary a stream or river that feeds into a larger stream or river	(p. G239)	**tributario** *arroyo o río que desemboca en un arroyo o río más grande*
tsunami a sea wave caused by an undersea earthquake	(p. G87)	**tsunami** *ola del mar causada por un terremoto bajo el mar*

U		**ESPAÑOL**
unitary system a form of government in which all powers reside with national government	(p. G232)	**sistema unitario** *forma de gobierno en la cual todos los poderes residen en el Gobierno nacional*
urban sprawl spreading of urban developments on undeveloped land near a city	(p. G134)	**expansión urbana** *diseminación de los desarrollos urbanos hacia las áreas sin desarrollo cercanas a una ciudad*
urban an area that is densely populated	(p. G124)	**urbano** *área densamente poblada*
urbanization the growth of cities as people migrate from rural areas to urban areas	(p. G131)	**urbanización** *crecimiento de las ciudades a medida que las personas migran de las áreas rurales a las urbanas*

V | ESPAÑOL

vary to make different (p. G126)

variar *hacer diferente*

vector-borne spread to humans through insect bites (p. G33)

transmisión por vectores *propagación a los humanos a través de picaduras de insectos*

vehemently in a forceful or passionate manner (p. G254)

vehementemente *de forma forzada o apasionada*

volcano a vent or rupture in Earth's crust from which hot rock, steam, and gases erupt (p. G86)

volcán *abertura o ruptura en la corteza terrestre de la que emergen rocas calientes, vapor y gases*

W | ESPAÑOL

water cycle the process in which water is used and reused on Earth, including precipitation, collection, evaporation, and condensation (p. G81)

ciclo del agua *proceso en el cual el agua se usa y reutiliza en la Tierra, incluyendo la precipitación, recolección, evaporación y condensación*

weather condition of the atmosphere in one place during a short time (p. G95)

clima *condición de la atmósfera en un lugar durante un tiempo corto*

weathering the process of wearing away or breaking down rocks into smaller pieces (p. G87)

meteorización *proceso de desgastar o romper rocas en pedazos más pequeños*

wetland land flooded with water permanently or seasonally (p. G282)

humedal *tierra inundada con agua de forma permanente o estacional*

whimsy playfulness (p. G15)

capricho *determinación que se toma por antojo o humor*

world city a city generally considered to play an important role in the global economic system (p. G133)

ciudad global *ciudad que generalmente se considera que juega un papel importante en el sistema económico global*

Y | ESPAÑOL

yak a type of long-haired domesticated cattle found in the Himalaya region (p. G214)

yak *tipo de ganado domesticado, de pelo largo, que se encuentra en la región del Himalaya*

Economics Glossary/ Glosario de economía

All vocabulary words are **boldfaced** and **highlighted in yellow** in your textbook.

A		ESPAÑOL
absolute advantage the ability to produce something cheaper than anyone else can	(p. E166)	**ventaja absoluta** *habilidad para producir algo más económico de lo que pueden los demás*
aggregate total	(p. E118)	**acumulado** *total*
antitrust law legislation to prevent monopolies from forming and to preserve and promote competition	(p. E92)	**ley antimonopolio** *legislación que evita la formación de monopolios y protege y promueve la competencia*
appropriations bill legislation that sets spending on particular programs for the coming year	(p. E132)	**ley de apropiaciones** *legislación que destina rubros presupuestales a programas específicos para el año siguiente*
arbitration situation in which union and company officials submit the issues they cannot agree on to a neutral third party for a final decision	(p. E78)	**arbitraje** *situación en la que representantes de sindicatos y compañías someten a un tercero neutral los asuntos que no pueden concertar para que este tome la decisión final*
attain to achieve	(p. E95)	**lograr** *conseguir*
automatic stabilizer program that works to preserve income without additional government action during economic downturns	(p. E145)	**estabilizador automático** *programa que trabaja para ajustar los ingresos, sin necesidad de acción gubernamental adicional, durante recesiones económicas*

B		ESPAÑOL
balance of trade the difference between the value of a nation's exports and its imports	(p. E171)	**balanza comercial** *diferencia entre el valor de las exportaciones e importaciones de un país*
balanced budget annual budget in which expenditures equal revenues	(p. E146)	**presupuesto equilibrado** *presupuesto anual en el que los gastos son iguales a los ingresos*
barter to trade a good or service for another good or service	(p. E79)	**trueque** *intercambio de un bien o servicio por otro bien o servicio*
bear market period during which stock prices decline for a substantial period	(p. E124)	**mercado bajista** *periodo durante el cual los precios de las acciones caen durante largo tiempo*
benefit-cost analysis economic decision-making model that divides the total benefits by the total costs	(p. E11)	**análisis costo-beneficio** *modelo de elección económica que divide los beneficios totales entre los costos totales*
black market market where illegal goods are bought and sold	(p. E181)	**mercado negro** *mercado en el que se venden y compran bienes ilegales*
board of directors the people elected by the shareholders of a corporation to act on their behalf	(p. E66)	**junta directiva** *personas elegidas por los accionistas de una corporación para que actúen en su nombre*
budget deficit situation that occurs when a government spends more than it collects in revenue	(p. E146)	**déficit presupuestario** *situación que se presenta cuando un Gobierno gasta más de lo que recibe por concepto de ingresos*

budget surplus situation that occurs when a government collects more revenues than it spends (p. E146)

superávit presupuestario *situación que se presenta cuando los ingresos de un Gobierno superan sus gastos*

bull market period during which stock prices steadily increase (p. E124)

mercado alcista *periodo durante el cual los precios de las acciones suben constantemente*

business cycle alternating periods of real economic decline and growth (p. E120)

ciclo económico *fluctuaciones periódicas de recesión y crecimiento económico reales*

C	ESPAÑOL

capital factories, tools, and equipment that manufacture goods or help work go more quickly (p. E8)

capital *fábricas, herramientas y equipos que producen bienes o agilizan el trabajo*

capitalism system in which private citizens own most, if not all, of the means of production and decide how to use them within legal limits (p. E57)

capitalismo *sistema en el que ciudadanos privados poseen la mayoría de los bienes de producción, si no todos, y deciden cómo utilizarlos dentro de los límites legales*

central bank a banker's bank that can lend to other banks in time of need and can regulate the money supply; a bankers' bank that lends money to other banks in difficult times (pp. E84, E125)

banco central *banco bancario que presta a otros bancos en épocas de necesidad y regula la oferta monetaria; banco bancario que presta dinero a otros bancos en épocas difíciles*

certificate of deposit (CD) a timed consumer loan to a bank that states the amount of the loan, maturity, and rate of interest being paid (p. E81)

certificado de depósito (CD) *crédito de consumo a término en el que consta el monto del préstamo, el vencimiento y la tasa de interés que se paga a un banco*

charter state government document granting permission to organize a corporation (p. E66)

acta constitutiva *documento gubernamental estatal que concede permiso para crear una corporación*

checking account an account from which deposited money can be withdrawn at any time by writing a check, using a debit card, or swiping a cell phone; also known as demand deposit accounts (p. E81)

cuenta corriente *cuenta de la cual se puede retirar en cualquier momento el dinero depositado escribiendo un cheque, utilizando una tarjeta débito o haciendo una transacción telefónica; se conoce también como cuenta de depósito a la vista*

circular flow model a model showing how goods, services, resources, and money flow among sectors and markets in the American economy (p. E37)

modelo de flujo circular *modelo que muestra cómo fluyen los bienes, servicios, recursos y el dinero en los sectores y mercados de la economía estadounidense*

civilian labor force group made up of "all persons 16 years of age or over who are working or not working but are able and willing to work" and who are not in the military (p. E73)

fuerza laboral civil *grupo conformado por "todas las personas de 16 años de edad o más que están trabajando o no están trabajando, pero que pueden y quieren trabajar" y no están en las fuerzas militares*

clarify explain (p. E65)

aclarar *explicar*

coin metallic form of money, such as a penny (p. E80)

moneda *dinero acuñado en metal, como un centavo*

collective bargaining process by which unions and employers negotiate the conditions of employment (p. E77)

negociación colectiva *mecanismo mediante el cual sindicalistas y empleadores negocian las condiciones de empleo*

command economy economic system in which the government owns and directs the majority of a country's land, labor, and capital resources (p. E14)

economía planificada *sistema económico en el que el Gobierno posee y controla la mayoría de las tierras, del trabajo y de los recursos de capital de un país*

commercial bank a financial institution that offers the most banking services to individuals and businesses (p. E82)

banco comercial *institución financiera que presta la mayoría de los servicios bancarios a individuos y empresas*

English		Español
communism theoretical state where all property is publicly owned, and everyone works according to their abilities and is paid according to their needs	(p. E15)	**comunismo** *Estado teórico en el que toda propiedad es pública y las personas trabajan según sus habilidades y reciben un pago acorde con sus necesidades*
comparative advantage a country's ability to produce a good relatively more efficiently than other countries can (or, a country's ability to produce a good at a lower opportunity cost than another country)	(p. E166)	**ventaja comparativa** *capacidad de un país para producir un bien de manera más eficiente que otros países (o capacidad de un país para producir un bien a menor costo de oportunidad que otro país)*
compensation payment to unemployed or injured workers to make up for lost wages	(p. E98)	**indemnización** *pago que se hace a trabajadores desempleados o lesionados para compensar el salario perdido*
competition efforts by different businesses to sell the same good or service; the struggle that goes on between buyers and sellers to get the best products at the lowest prices	(pp. E34, E60)	**competencia** *esfuerzo que hacen las distintas empresas para vender los mismos bienes y servicios; pugna entre compradores y vendedores por obtener los mejores productos al precio más bajo*
confirmed approved	(p. E126)	**confirmado** *aprobado*
consumer person who buys goods and services	(p. E25)	**consumidor** *persona que compra bienes y servicios*
consumer sovereignty role of the consumer as ruler of the market that determines the types and quantities of goods and services produced	(p. E58)	**soberanía del consumidor** *rol del consumidor como gobernante del mercado que determina las clase y cantidades de los bienes y servicios producidos*
contribute to provide or give	(p. E70)	**contribuir** *aportar, dar*
corporation type of business organization owned by many people but treated by law as though it were a person	(p. E66)	**corporación** *tipo de empresa u organización de la que son dueñas muchas personas, pero que es considerada legamente como una sola entidad*
credit union nonprofit service cooperative that accepts deposits, makes loans, and provides other financial services to its members	(p. E82)	**cooperativa de crédito** *servicio cooperativo sin ánimo de lucro que acepta depósitos, hace préstamos y presta otros servicios financieros a sus miembros*
cryptocurrency electronic money not issued or managed by any country or central bank	(p. E80)	**criptomoneda** *dinero electrónico que no es emitido ni administrado por un país o banco central*
currency money, both coins and paper bills; system of money in general use in a country	(pp. E80, E169)	**moneda** *dinero en moneda y en papel; sistema monetario en general que utiliza un país*

D		**ESPAÑOL**
deflation a prolonged decrease in the general level of prices	(p. E124)	**deflación** *disminución prolongada del nivel general de los precios*
demand amount of a good or service that consumers are willing and able to buy over a range of prices	(p. E25)	**demanda** *cantidad de bienes y servicios que los consumidores quieren y pueden comprar entre un rango de precios*
deposit the money that customers put into a financial institution	(p. E81)	**depósito** *dinero que consignan los consumidores en una institución financiera*
deposit insurance program government-backed program that protects bank deposits up to a certain amount if a financial institution fails	(p. E83)	**programa de seguro de depósitos** *programa respaldado por el Gobierno que cubre los depósitos bancarios hasta un monto específico en caso de que una institución financiera quiebre*
depression state of the economy with high unemployment, severely depressed real GDP, and general economic hardship	(p. E120)	**depresión** *estado de la economía caracterizado por una tasa de desempleo alta, una caída drástica del PIB real y recesión económica*

ECONOMICS GLOSSARY

English		Español
developed country country with a high standard of living, a high level of industrialization, and a high per capita income	(p. E179)	**país desarrollado** *país que tiene un nivel de vida alto, un nivel de industrialización y un ingreso per cápita altos*
developing country nonindustrial country with a low per capita income in which a large number of people have a low standard of living	(p. E179)	**país en desarrollo** *país no industrializado cuyo ingreso per cápita es bajo y en el que numerosas personas tienen un nivel de vida bajo*
discount rate interest rate the Fed charges on its loans to financial institutions	(p. E129)	**tasa de descuento** *tasa de interés que la Reserva Federal cobra por sus préstamos a instituciones financieras*
discretionary spending spending for federal programs that must receive approval each year	(p. E132)	**gasto discrecional** *gasto asignado a programas federales que debe aprobarse anualmente*
dispose get rid of	(p. E61)	**desechar** *descartar*
distortion misleading impression	(p. E119)	**distorsión** *impresión falsa*
division of labor the breaking down of a job into separate, smaller tasks to be performed individually	(p. E40)	**división del trabajo** *fragmentación de un trabajo en tareas pequeñas e independientes que se realizan individualmente*

E / ESPAÑOL

English		Español
economic growth the increase in a country's total output of goods and services over time	(p. E39)	**crecimiento económico** *aumento progresivo en la producción total de bienes y servicios de un país*
economic interdependence the reliance of people and countries around the world on one another for goods and services	(p. E173)	**interdependencia económica** *dependencia mundial mutua de personas y países para obtener bienes y servicios*
economic system a nation's way of producing and distributing things its people want and need	(p. E13)	**sistema económico** *modo en el que un país produce y distribuye cosas que su población desea y necesita*
economics study of how individuals and nations make choices about ways to use scarce resources to fulfill their needs and wants	(p. E8)	**economía** *estudio del modo en que individuos y naciones toman decisiones relacionadas con la utilización de recursos escasos para satisfacer sus necesidades y deseos*
electronic money money in the form of a computer entry at a bank or other financial institution	(p. E80)	**dinero electrónico** *dinero de un banco u otra institución financiera al que se accede por computadora*
ensure to make certain of an outcome	(p. E61)	**asegurar** *garantizar un resultado*
entitlement program a government program that makes payments to people who meet certain requirements in order to help them meet minimum health, nutrition, and income needs	(p. E136)	**programa de subsidios** *programa gubernamental que hace pagos a personas con requerimientos específicos para ayudarlas a satisfacer necesidades básicas en salud, nutrición e ingreso*
entrepreneurs risk-taking individuals who start a new business, introduce a new product, or improve a method of making something	(p. E8)	**emprendedores** *individuos que se arriesgan a crear una nueva empresa, introducir un producto nuevo o perfeccionar la manera de hacer algo*
equilibrium price market price where quantity demanded and quantity supplied are equal	(p. E34)	**precio de equilibrio** *precio en el mercado en el que la cantidad demandada es igual a la cantidad ofrecida*
equilibrium quantity quantity of output supplied that is equal to the quantity demanded at the equilibrium price	(p. E34)	**cantidad de equilibrio** *cantidad de producción ofrecida que es igual a la cantidad en demanda al precio de equilibrio*
evolve to progress or develop gradually	(p. E14)	**evolucionar** *progresar o desarrollarse gradualmente*
exchange rate the value of a nation's currency in relation to another nation's currency	(p. E169)	**tasa de cambio** *valor que tiene la moneda de un país en relación con la moneda de otro país*

export to sell goods to other countries	(p. E165)	**exportar** *vender bienes a otros países*
externality economic side effect that affects an uninvolved third party	(p. E92)	**externalidad** *efecto secundario económico que afecta a una tercera parte no implicada*

F		ESPAÑOL
factor market a market where productive resources (land, labor, capital) are bought and sold	(p. E38)	**mercado de factores** *mercado en el que los recursos productivos (tierra, trabajo, capital) se compran y venden*
factors of production four categories of resources used to produce goods and services: natural resources, capital, labor, and entrepreneurs	(p. E8)	**factores de producción** *cuatro categorías de recursos utilizados para producir bienes y servicios: recursos naturales, capital, trabajo y emprendedores*
federal debt money the government has borrowed and not yet paid back	(p. E146)	**deuda federal** *dinero que el Gobierno ha pedido en préstamo y no ha pagado aún*
Federal Open Market Committee (FOMC) powerful committee of the Fed that makes decisions affecting the economy by managing the money supply in order to affect interest rates	(p. E126)	**Comité Federal de Mercado Abierto (FOMC)** *poderoso comité de la Reserva Federal que toma decisiones de incidencia económica, como regular la oferta monetaria a fin de modificar las tasas de interés*
financial capital the money used to run or expand a business	(p. E64)	**capital financiero** *dinero utilizado para operar o expandir una empresa*
fiscal policy government use of taxes and spending to stimulate the economy and reach economic goals	(p. E144)	**política fiscal** *uso que hace el Gobierno de los impuestos y gastos para estimular la economía y alcanzar metas económicas*
fiscal year any 12-month period chosen for keeping accounts	(p. E131)	**año fiscal** *periodo de 12 meses fijado para llevar la contabilidad*
fixed income an income that remains the same each month and does not have the potential to go up when prices are going up	(p. E123)	**ingreso fijo** *ingreso que permanece igual mes a mes y no tiene el potencial de subir cuando los precios están en alza*
foreign exchange the domestic currencies that are used to finance foreign trade	(p. E169)	**divisas** *las monedas nacionales que se utilizan para financiar el comercio exterior*
foundation organization established by a company or an individual to provide money for a particular purpose, especially for charity or research	(p. E70)	**fundación** *organización establecida por una compañía o un individuo que suministra dinero para un propósito en particular, especialmente de carácter benéfico o investigativo*
franchise company that has permission to sell the supplier's goods or services in a particular area in exchange for payment	(p. E68)	**franquicia** *compañía autorizada para vender los bienes y servicios del proveedor en un lugar específico a cambio de un pago*
free enterprise economic system in which individuals and businesses are allowed to compete for profit with a minimum of government interference	(p. E57)	**libre empresa** *sistema económico en el que se permite a individuos y empresas competir por lucro con mínima intervención del Estado*

G		ESPAÑOL
GDP per capita GDP on a per-person basis; GDP divided by population	(p. E118)	**PIB per cápita** *PIB por persona; PIB dividido entre la población*
goods things we can touch or hold	(p. E7)	**bienes** *cosas que podemos tocar o asir*
Gross Domestic Product (GDP) total market value of all final goods and services produced in a country during a single year	(p. E115)	**producto interno bruto (PIB)** *valor total en el mercado del conjunto de los bienes y servicios producidos en un país durante un año*

ECONOMICS GLOSSARY

H		ESPAÑOL
human capital the sum of people's knowledge and skills that can be used to create products	(p. E40)	**capital humano** *conjunto de conocimientos y habilidades de una persona que pueden utilizarse para producir productos*

I		ESPAÑOL
import to buy goods from another country	(p. E165)	**importar** *comprar bienes a otro país*
incentive motivation or reward	(pp. E16, E27)	**incentivo** *motivación o recompensa*
inflation a long-term increase in the general level of prices	(p. E123)	**inflación** *aumento a largo plazo en el nivel general de precios*
infrastructure highways, levees, bridges, power, water, sewage, and other public goods needed to support a growing economy	(p. E146)	**infraestructura** *carreteras, diques, puentes, energía, agua, alcantarillado y otros bienes públicos necesarios para sostener el crecimiento de la economía*
injunction a court order to stop some kind of action	(p. E78)	**medidas cautelares** *orden emanada de un tribunal para impedir algún tipo de acción*
integrate combine	(p. E175)	**integrar** *aunar*
intergovernmental revenue funds that one level of government receives from another level of government	(p. E135)	**ingreso intergubernamental** *fondos que un nivel del Gobierno recibe de otro nivel del Gobierno*

L		ESPAÑOL
labor workers and their abilities	(p. E8)	**fuerza de trabajo** *los trabajadores y sus habilidades*
labor union association of workers organized to improve wages and working conditions	(p. E75)	**sindicato** *asociación de trabajadores organizados para mejorar los salarios y las condiciones laborales*
laissez-faire economics belief that government should not interfere in the marketplace	(p. E61)	**laissez faire** *política económica según la cual el Gobierno no debe intervenir en el mercado*
liability the legal responsibility for something, such as an action or a debt	(p. E64)	**responsabilidad legal** *obligación legal por algo, como una acción o una deuda*
likewise in the same way	(p. E32)	**igualmente** *de la misma manera*
lockout when management closes a workplace to prevent union members from working	(p. E78)	**cierre patronal** *cuando una empresa cierra el lugar de trabajo para impedir que los miembros del sindicato trabajen*

M		ESPAÑOL
maintain to preserve	(p. E93)	**mantener** *conservar*
mandatory spending federal spending required by law that continues without the need for congressional approval each year	(p. E132)	**gasto obligatorio** *gasto federal requerido por ley que continúa sin necesidad de que el Congreso lo apruebe cada año*
market place or arrangement where a buyer and a seller voluntarily exchange money for a good or service	(pp. E16, E33)	**mercado** *lugar o sistema en el que compradores y vendedores intercambian voluntariamente dinero por bienes o servicios*

market economy economic system in which individuals and businesses have the freedom to use their resources in ways they think best (p. E16)

economía de mercado *sistema económico en el que individuos y empresas gozan de libertad para utilizar sus recursos de la manera que consideren más conveniente*

mediation situation in which union and company officials bring in a third party to try to help them reach an agreement (p. E78)

mediación *situación en la que los representantes de un sindicato y una compañía recurren a una tercera parte para que los ayude a llegar a un acuerdo*

medium a means of doing something (p. E79)

medio *recurso que permite hacer algo*

mercantilism economic system in which government controls the products that come into and go out of a country (p. E61)

mercantilismo *sistema económico en el que el Gobierno controla los productos que ingresan al país y salen de él*

merger a combination of two or more companies to form a single business (p. E93)

fusión *unión de una o más compañías para formar una sola empresa*

mixed market economy economic system in which markets, government, *and* tradition each answer some of the WHAT, HOW, and FOR WHOM questions (p. E17)

economía de mercado mixta *sistema económico en el que los mercados, el Gobierno y la tradición responden cada cual algunas de las preguntas de QUÉ, CÓMO y PARA QUIÉN*

monetary policy Fed's management of the money supply to affect the cost and availability of credit (p. E126)

política monetaria *control de la oferta monetaria, por parte de la Reserva Federal, que afecta el costo y la disponibilidad de crédito*

monopoly exclusive control of a good or service (p. E92)

monopolio *control exclusivo de un bien o servicio*

multilateral agreement treaty in which multiple nations agree to give each other the same benefits (p. E174)

acuerdo multilateral *tratado en el que múltiples países acuerdan concederse mutuamente los mismos beneficios*

multinationals corporations that have manufacturing or service operations in several different countries (p. E176)

multinacionales *corporaciones que tienen operaciones industriales o de servicios en varios países*

N | ESPAÑOL

natural monopoly a market situation in which the costs of production are minimized by having a single firm produce the product (p. E93)

monopolio natural *situación económica en la que los costos de producción se minimizan teniendo una sola firma que produzca el producto*

natural resources land and all of the materials nature provides that can be used to make goods or services (p. E8)

recursos naturales *la tierra y todos los materiales proporcionados por la naturaleza que pueden utilizarse para producir bienes o servicios*

need basic requirement for survival (p. E7)

necesidad *requisito básico para la supervivencia*

nonprofit organization business that does not intend to make a profit for the goods and services it provides (p. E68)

organización sin ánimo de lucro *empresa que no se lucra de los bienes y servicios que suministra*

O | ESPAÑOL

offset to counterbalance (p. E145)

compensar *contrapesar*

open market operations (OMO) Fed's purchase or sale of U.S. government securities—bond notes and Treasury bills (p. E128)

operaciones de mercado abierto (OMO) *compra o venta, por parte de la Reserva Federal, de valores del Gobierno estadounidense (bonos y letras del Tesoro)*

opportunity cost cost of the *next* best use of time or money when choosing to do one thing rather than another (p. E10)

costo de oportunidad *costo de la siguiente mejor alternativa de utilizar tiempo o dinero cuando se opta por hacer una cosa en lugar de otra*

ECONOMICS GLOSSARY

option alternative, choice	(p. E9)	**opción** *alternativa, elección*
outsource to move factories to other countries with lower labor costs	(p. E177)	**tercerizar** *trasladar fábricas a otros países a menor costo laboral*

P		**ESPAÑOL**
partnership a business owned by two or more people	(p. E64)	**sociedad** *agrupación comercial de la que son dueñas dos o más personas*
picketing a union tactic in which striking workers walk with signs that express their grievances	(p. E77)	**piquete** *táctica sindical en la que los huelguistas caminan portando pancartas con sus reclamaciones*
potential possible	(p. E96)	**potencial** *posible*
price the monetary value of a product	(p. E31)	**precio** *valor monetario de un producto*
private good economic good that, when consumed by one person, cannot be used by another	(p. E91)	**bien privado** *bien económico que, cuando es consumido por una persona, no puede ser utilizado por otra*
private property rights the freedom to own and use our property as we choose as long as we do not interfere with the rights of others	(p. E61)	**derechos de propiedad privada** *libertad para poseer y utilizar la propiedad como se quiera siempre y cuando no se interfiera con los derechos de los demás*
privatization process of changing from state-owned businesses, factories, and farms to ones owned by private citizens	(p. E181)	**privatización** *proceso de traspasar empresas, fábricas y granjas de propiedad del Estado al sector privado*
producer person or business that provides goods and services	(p. E25)	**productor** *persona o empresa que suministra bienes y servicios*
product anything that is produced; goods and services	(p. E115)	**producto** *todo lo que es producido; bienes y servicios*
product market a market where goods and services are bought and sold	(p. E38)	**mercado de productos** *mercado en el que se compran y venden bienes y servicios*
productivity the degree to which resources are being used efficiently to produce goods and services	(p. E40)	**productividad** *grado de utilización eficiente de los recursos para producir bienes y servicios*
profit money a business receives for its products over and above what it cost to make the products	(pp. E16, E27, E59)	**ganancia** *dinero que recibe una empresa por sus productos por encima de lo que cuesta producirlos*
profit motive the driving force that encourages individuals and organizations to improve their material well-being	(p. E59)	**ánimo de lucro** *fuerza impulsora que alienta a individuos y organizaciones a mejorar su bienestar material*
promote to support or encourage	(p. E12)	**promover** *apoyar o alentar*
property tax tax on the value of land and property that people own	(p. E136)	**impuesto al patrimonio** *impuesto sobre el valor de las tierras y propiedades que poseen las personas*
protectionism use of tactics that make imported goods more expensive than domestic goods	(p. E167)	**proteccionismo** *tácticas empleadas para hacer que los bienes importados sean más costosos que los bienes nacionales*
public good economic good that is used collectively, such as a highway and national defense	(p. E91)	**bien público** *bien económico que es utilizado colectivamente, como las carreteras y la defensa nacional*

Q   ESPAÑOL

quota limit on the amount of foreign goods imported into a country   (p. E168)

cuota *límite impuesto a los bienes extranjeros que importa un país*

R   ESPAÑOL

rationing system of distributing goods and services without prices   (p. E32)

racionamiento *sistema de distribución de bienes y servicios sin precios*

real GDP GDP after adjustments for price changes   (p. E117)

PIB real *PIB resultante de los ajustes a las fluctuaciones de los precios*

recall government action that causes an unsafe product to be removed from consumer contact   (p. E94)

retiro *acción gubernamental para evitar el acceso de los consumidores a productos peligrosos*

recession period of declining economic activity lasting about six or more months   (p. E120)

recesión *periodo de depresión de las actividades económicas que dura seis o más meses*

rely to depend on   (p. E167)

depender *necesitar de*

reserve requirement percentage of a deposit that banks have to set aside as cash in their vaults or as deposits in their Federal Reserve district bank   (p. E128)

encaje bancario obligatorio *porcentaje del depósito que los bancos deben reservar en efectivo en sus bóvedas o depositar en el banco de la Reserva Federal de su distrito*

resources things used to make goods or services: natural resources, capital, labor, entrepreneurs   (p. E8)

recursos *cosas que se utilizan para producir bienes y servicios: recursos naturales, capital, trabajo, empresarios*

revenue money a government collects to fund its spending   (p. E131)

ingresos *dinero recaudado por el Gobierno para financiar sus gastos*

right-to-work laws state laws forbidding unions from forcing workers to join   (p. E77)

leyes de derecho al trabajo *leyes estatales que les prohíben a los sindicatos que obliguen a los trabajadores a afiliarse a ellos*

S   ESPAÑOL

sales tax tax paid by consumers at the time they buy goods or services   (p. E135)

impuesto a las ventas *impuesto que pagan los consumidores cuando compran bienes o servicios*

savings account an account that pays interest on deposits but allows only limited withdraws   (p. E81)

cuenta de ahorros *cuenta en la que el dinero depositado genera intereses pero de la cual solo se pueden retirar montos limitados*

scarcity situation of not having enough resources to satisfy all of one's wants and needs   (p. E9)

escasez *situación en la que no hay suficientes recursos para satisfacer los propios deseos o necesidades*

schedule table listing items or events   (p. E26)

calendario *tabla en la que se listan asuntos o actividades*

sector part or category distinct from other parts   (p. E38)

sector *parte o categoría distinta de otra*

services work that is done for us   (p. E7)

servicios *trabajos que se hacen por nosotros*

shortage situation in which the quantity of a good or service supplied at a certain price is less than the quantity demanded for it   (p. E35)

desabastecimiento *situación en la que la cantidad de un bien o servicio ofrecido a cierto precio es menor que la cantidad demandada por él*

social responsibility the obligation businesses have to pursue goals that benefit society as well as themselves   (p. E69)

responsabilidad social *obligación que tienen las empresas de alcanzar metas que beneficien tanto a la sociedad como a ellas mismas*

English		Español
Social Security anti-poverty program that taxes working people and pays benefits to retirees	(p. E98)	**Seguridad Social** *programa contra la pobreza que grava a los trabajadores y paga beneficios a los pensionados*
socialism economy in which government owns some factors of production so it can distribute products and wages more evenly among its citizens; a type of command economy	(p. E15)	**socialismo** *economía en la que el Gobierno posee algunos de los factores de producción para poder distribuir productos y salarios de manera más equitativa entre los ciudadanos; tipo de economía planificada*
sole proprietorship a business owned and operated by a single person	(p. E63)	**empresa unipersonal** *empresa que posee y opera una sola persona*
specialization when people, businesses, regions, and/or nations concentrate on goods and services that they can produce better than anyone else	(p. E40)	**especialización** *cuando personas, empresas, regiones o países se concentran en bienes y servicios que pueden producir mejor que otros*
standard of living the material well-being of an individual, a group, or a nation as measured by how well needs and wants are satisfied	(pp. E39, E118, E179)	**nivel de vida** *bienestar material de un individuo, grupo o país que se mide según lo bien que se satisfagan los deseos y las necesidades*
stocks shares of ownership in a corporation; same as stock certificates	(p. E66)	**acciones** *participaciones de propiedad en una corporación; lo mismo que certificados de acciones*
strike when workers deliberately stop working in order to force an employer to give in to their demands	(p. E77)	**huelga** *cuando los trabajadores dejan de trabajar deliberadamente para obligar al empleador a ceder a sus exigencias*
subsidize to aid or support a person, business, institution, or undertaking with money or tax breaks	(p. E136)	**subsidiar** *ayudar o subvencionar a una persona, un negocio, una institución o una empresa con dinero o mediante exenciones de impuestos*
subsidy payment or other benefit given by the government to help a domestic producer	(p. E168)	**subsidio** *dinero u otro auxilio dado por el Gobierno para ayudar a un productor nacional*
Supplemental Nutrition Assistance Program (SNAP) welfare program that provides nutritional benefits to supplement the food budgets of needy families	(p. E98)	**Programa de Asistencia Nutricional Suplementaria (SNAP)** *programa de asistencia social que proporciona auxilios nutricionales para complementar los presupuestos alimentarios de familias en situación de necesidad*
supply amount of a good or service that producers are willing and able to sell over a range of prices	(p. E27)	**oferta** *cantidad de un bien o servicio que los productos quieren y pueden vender a determinado precio*
surplus situation in which the amount of a good or service supplied by producers at a certain price is greater than the amount demanded by consumers	(p. E35)	**excedente** *situación en la que la cantidad de un bien o servicio ofrecido por los productores a determinado precio es mayor que la cantidad demandada por los consumidores*
survey examination, review	(p. E73)	**encuesta** *examen, sondeo*

T		ESPAÑOL
tariff tax on an imported good	(p. E167)	**arancel** *impuesto sobre bienes importados*
Temporary Assistance for Needy Families (TANF) welfare program paid for by the federal government and administered by the individual states	(p. E97)	**Ayuda Temporal para Familias Necesitadas (TANF)** *programa de asistencia social subvencionado por el Gobierno federal y administrado por cada estado*
trade bloc group of countries, usually located in a particular region, that cooperate to expand free trade among one another	(p. E174)	**bloque comercial** *grupo de países, ubicados por lo general en una misma región, que cooperan para ampliar entre sí el libre comercio*
trade-off alternative you face when you decide to do one thing rather than another	(p. E9)	*trade-off* *alternativa a la que te enfrentas cuando optas por hacer una cosa en vez de otra*

traditional economy economic system in which the decisions of WHAT, HOW, and FOR WHOM to produce are based on traditions or customs (p. E13)

economía tradicional *sistema económico en el que las decisiones de QUÉ, CÓMO y PARA QUIÉN producir se basan en tradiciones o costumbres*

transfer to shift or reassign (p. E181)

transferir *trasladar o reasignar*

transmit send (p. E132)

transmitir *hacer llegar*

transparency the process of making business deals or conditions more visible to everyone (p. E71)

transparencia *proceso de hacer más visibles para todos los tratos o condiciones comerciales*

U · ESPAÑOL

unemployment rate percentage of people in the civilian labor force who are not working but are looking for jobs (pp. E74, E121)

tasa de desempleo *porcentaje de personas de la fuerza laboral civil que no están trabajando pero están buscando empleo*

V · ESPAÑOL

voluntary exchange the act of buyers and sellers freely and willingly engaging in market transactions (p. E59)

intercambio voluntario *acción de compradores y vendedores que participan libre y voluntariamente en transacciones mercantiles*

W · ESPAÑOL

want desire for a good or a service (p. E7)

deseo *apetencia de un bien o servicio*

welfare aid given to the poor in the form of money or necessities (p. E97)

asistencia *ayuda dada a los pobres en forma de dinero o artículos de primera necesidad*

workfare programs that require welfare recipients to exchange some of their labor for benefits (p. E98)

workfare *programas en los que los beneficiarios de la asistencia social deben intercambiar parte de su trabajo por auxilios*

Geography Index

Note: Italicized page numbers refer to illustrations. The following abbreviations are used in the index: *m* = map; *c* = chart; *p* = photograph or picture; *g* = graph; *ptg* = painting; *crt* = cartoon; *q* = quote.

A

abolition movement, G16
absolute location, G8, G19, G29
absolute monarchy, G232
Abyei, G243
accessibility, G59–60
acid rain, G81–82, G283
Aditya, Arys, *qG286*
Afghanistan, *pG56*, G90, G128
Africa, G55–56, *pG58*, *G63*, *pG73*, *pG80*, *G84*, *G99*, *G100*, *G108*; Berlin Conference and, G234, G236, *crtG236*; birth rate in, G120; geometric boundaries, G234; migration from, G127, *pG127*; political borders of, G236, *crtG236*; population in, *gG115*, G120, *gG124*, G143; rainfall in, *mG31*; urbanization in, G132
Africa culture region, G163
African Americans, G51, G160, G178, *mG178*
African ethnic groups in Darfur, G243
Africans, enslaved, G172
Afsluitdijk, G288, *pG288*
agglomeration, G196, *pG196*
Agricultural Revolution, G172
agriculture, G188, *pG188*, G195, G197; alluvial soil and, G266, G267; in ancient times, G274; commercial farming and, G204–5, G263–64; comparative advantage and, G207; corn production in United States and, G199, *pG199*, G206–7; economic development and, *mG204*, G204–5; factors affecting, G206–7; gross domestic product and, *mG204*; irrigation and, G265–66; in less developed countries, G204–5; slash-and-burn, G267–68; subsistence farming and, G205, G263; water for, G192
air, on Earth, G74
airline route map, *mG30*
Alaska, *pG88*
Alberta Culture and Tourism, *qG278*
alluvial soil, G266, G267
altitude, G91, G96, G102, G108
Amazon, G207
Amazon River basin, G99

Andes mountains, *pG75*, *pG89*
Antarctica, G78, G80, G84, G88, *pG103*
Antarctic Treaty, *mG227*, G246
Appalachian Mountains, G87
aquifers, G80
Arab-Israeli conflict, G244–45
arable land, G126
Araya, Johnny, G251
Arches National Park, *pG74*
Arctic Circle, G102
Arctic Ocean, G80, G272
Argentina, *pG89*
Armenia, G242
Arnold, Matthew, *qG153*, *pG153*
Arsenault, Chris, *qG287*
art forms, culture and, G164
Articles of Confederation, G230
Asia, *pG78*, G84, G90, G99, G100; birth rate in, G120; immigration from, G129; migration from, G127; population in, *gG115*, G120, *gG124*, G125; urbanization in, G132
Asians in U.S. population, G129
Askew, Katy, *qG216*
Association of Southeast Asian Nations (ASEAN), G212
Atlantic Ocean, G80
atmosphere, *cG77*, G77–78, G97–98
Attal, Gabriel, *qG252*
Austin, Texas, G117
Australia, G78, G84, G92, G100, G101, G163; administrative boundaries in, *mG233*; government system in, G230; mining in, G198, *pG198*; population in, G124
Austria, *gG144*, G239
authoritarian government, G231–32, *cG232*
autocracy, G231
"Avoiding a water crisis: how Cape Town avoided 'Day Zero'" (*Resilience News*), *qG289*
axis, G95
Azerbaijan, G242

B

Baarle-Nassau, political borders of, G240, *pG240*
Badaga language, *pG174*
ballgame, G164
Bangkok, *pG131*

Bangladesh, G125, G126, G201, *pG201*
baseball, G179
Basque Country, G165
bay, G80
beavers, G290
Beijing, G251
Belarus, G125
Belgium, G125, G240, *pG240*
Bengaluru, India, *pG141*
Benin, West Africa, G160
Berlin Conference, G234, G236, *crtG236*
Bernstein, William J., *qG187*
Bhutan, *pG78*, G90
Billock, Jennifer, *qG92*
biodiversity, G279
biomes, G99, *mG101*
biosphere, G78
birth rate, G120, G137, *gG137*
Bish, Joseph J., *qG143*
Bismarck, Otto von, G236
Black Sea, G144
Bloomberg CityLab, *qG239*
"Bogotá, Colombia as a Cut-Flower Exporter for World Markets" (Cheever), G200
Bonnett, Alastair, *qG40*
border-crossing cards, G217
boreal forests, G105, *mG105*, G279
Bosporus Strait, *pG133*
"Bountiful Benefits of Bringing Back the Beavers, The" (Runyon), *qG290*
brain drain, G130
Brazil, G101, G230
Britain, G155, G163, G233
British Isles, G101
Broken Abroad (Massoud), *qG115*
Brown, Lester R., *qG114*
Brownsville, Texas, G217
Bryce Canyon, G88
Buddhism, *cG159*, G163
buildings, culture and, G177, *pG177*
Bulgaria, G144, *gG144*, G250

C

Cairo, Egypt, G197
California, *pG49*, G56, G101, G104, G132, G206, G241, G277
Campbell County, Wyoming, *pG272*

Canada, G13, *mG13*, G146; Canada-United States Air Quality Agreement and, G245; culture region of, G162, *mG162*; global food supply in, G287, *pG287*; government system in, G230; territories of, G13, *mG13*; traditional economy in, G186; world trade and, G212; WTO and, G246
Canada-United States Air Quality Agreement, G245
Canadian Rockies, *pG73*
Cape Town, South Africa, *pG4*, G289, *pG289*
capital, G195, G211
capitalism, *pG186*, G194
Cardiff Bay Wetlands Reserve, *pG281*
cardinal directions, G21
Caribbean, G99, *gG124*, G218
"Caribbean Nurse Migration-a Scoping Review" (Sands, Ingraham, and Salami), *qG218*
Carstensen, George, *qG15*
cartographers (mapmakers), G22
Cascade Mountains, G107
Cassedy, Ellen, G180
census, G135, *pG135*
Census Bureau, G138, G140, G145
Center for a Livable Future, *qG274*
Central America, G78, G100, G212
Central Asia culture region, G163
central business district (CBD), G208
central place theory, G132, *cG132*
Chad, G126, G243
Chang Jiang river valley, G125
Charles II, King, *pG20*
Charleston, West Virginia, *pG40*
Cheever, David, G200
Chicago, Illinois, *pG59*
Chile, G101
China, G90, G101; animal protein consumption and, G199; athletes in, G164; boycott of 2022 Winter Olympic Games in, G246; command economy in, G186; conflict with Uyghurs in, G243; culture hearths in, *mG171*; in East Asia cultural region, G163; ethnic groups in, G160; fast food in, G183; Fushan mine in, G278; in global economy, G251; Kashmir conflict and, *mG242*; lithium and, G198;

oligarchy, G232; population in, G125; Silk Road routes from, G275; world trade and, G211; WTO and, G246

"China's Stadium Diplomacy" (Will), *qG251*

Chinatown in Chicago, *pG161*

Chinese Americans, G129

Chinese New Year, G129

Christianity, *cG159,* G161, G163

cities, growth of, G115, G116, *pG117,* G131–34; central place theory and, G132, *cG132;* connectivity and, G131; factors leading to, G132; functions of cities and, *pG133,* G133–34; infrastructure and, G133–34; migration and, G115, *gG115,* G117; slums and, G121. *See also* urbanization.

"Cities and the Fall Line" (Virginia Studies), *qG276*

Civil Rights movement in 1966, G167

clearcutting, G270

climate: atmosphere and, G75, *cG77,* G77–78, G97–98; change, G55, *pG55,* G102, *cG282,* G283–84, G285, *pG285,* G288; definition of, G95, G103; Earth's, G73, *pG73;* economic activity and, G205–6; factors affecting, G95–98; forests and, G105, *mG105;* global warming and, G259, G261, G283, G284, G287; of Hawaii, G104, *pG104;* of Rwanda, G108. *See also* climate regions, temperature.

"Climate of Hawai'i" (Price), *qG104*

"Climate of Rwanda" (Climatestotravel.com), *qG108*

climate regions, G73, *pG73,* G99–102; dry, G100; high-latitude, *pG73,* G99, G101–2; midlatitude, G75, *pG75, mG97,* G99, G100, G101–2; tropical, G75, *pG99,* G99–100

Climatestotravel.com, *qG108*

coal, G192, G195, G198, *pG208,* G265, G278

"Coal" (Alberta Culture and Tourism), *qG278*

cocoa, G63, *pG63*

colliding boundaries, G85

Colombia, G200

colonialism, G162, G163

Colorado, G93, *pG93*

Colorado River, *pG87*

Columbian Exchange, G155, *mG155*

Columbia River, *pG192*

combines, *pG287*

command economy, G186, G194, G195

commercial farming, G204–5, G263–64

communication: cultural change and, G155; Information Revolution and, G172–73; language and, G158; systems, G58–59

communism, G194

Communist Party, G232

commute, G62, *mG62*

comparative advantage, G207, G211

compass rose, G21, *pG21*

complementarity, G206

condensation, G81, *cG82*

confederation, G230

conflict, G227, G241–45, G247–52; causes of, *mG225,* G227, G241–45, G247–52; examples of, G248–52; regional changes and, G55–56

Confucius Institute, G251

coniferous, G101, G105

connectivity, G131

conservation, G271

constitutional monarchies, G230–31

consumption, factors of, G195–96

continental shelf, G79, *cG79*

continents, G78

contour lines, on maps, G24

cooperation, G227, G245–52; alliances and, *mG245,* G245–46; Antarctic Treaty and, *mG227,* G246; effects of, G227, G247–52; examples of, G248–52

COP 26 Conference, *pG285*

Copenhagen, G15

Corbet, Sylvie, *qG252*

Corder, Mike, *qG288*

Corn Belt of United States, G199, G206–7

Corn Palace, *pG199*

Cortés, Hernán, *mG273*

Costa Rica, G251

Costa Rica National Stadium, G251, *pG251*

cottage industries, G205

COVID-19 pandemic, G145, G174, G212

Crimea, G244

Croatia, G250

crust, Earth's, *cG83,* G84, G89–94. *See also* forces of change, in Earth.

cultural boundary, G233

cultural change, G171–74; Agricultural Revolution and, G172; causes of, G171;

environments and, G172; globalization and, G173–74; in history, G171–72; Industrial Revolution and, G172; Information Revolution and, G172–73; migration and, G172

cultural convergence, G129, G171

cultural diffusion, G129, G171

cultural divergence, G129, G171

culture: buildings and, G177, *pG177;* changes in, G155; Columbian Exchange and, G155, *mG155;* continuity and change in, G175–80; definition of, G152–53, G157; elements of, G154, G157–60; expressions of, G154, G161–64; hula in Hawaiian culture, G176, *pG176;* identity and, G165–69; indigenous, decline of, G155; population distribution and, G124–25; study of, G154, G164–65. *See also* cultural change; culture regions; elements of culture.

culture hearth, *mG171,* G171–72

culture regions, G161–63; definition of, G161; languages and, G169; world, *mG162,* G162–63

culture trait, G161

currents, G97

curry houses, G168, *pG168*

customs, G157, G164

Cypriot, G249

Cyprus, G249, G250

Cyprus Orthodox Church, G249

Czechia, *gG144*

D

dance, culture and, G164

Day of the Dead, G170, *pG170*

Dead Sea, G80

death rate, G119, G120

Decatur, Illinois, *pG59*

deciduous, G101

deforestation, G56, G278, *pG279,* G279–81

degradation, G274

delta, G81

democracy, G224, G230–31, *cG232*

demographics, G135

densely populated. *See* population density.

Department of the Interior (DOI), *qG93*

depository, G275

desalination, G80, G267

desertification, G279–81, *mG280*

deserts, G100, *mG101*

Devils Tower National Monument, Wyoming, G50, *pG50*

dialect, G158

dictatorship, G224, G231

dikes, G267, *mG267*

direct democracy, G230

disaster drills, G268

discrimination, G180

distance decay, G59

Doctors Without Borders, *pG58*

domesticity, G277

Dominican Republic, G212

Dominican Republic-Central America Free Trade Agreement (CAFTA-DR), G212

doubling time, G120

Douglass, Frederick, G16

dry climates, G100

Dunhuang, China, G275, *pG275*

"Dunhuang" (Silk Roads Programme), *qG275*

Dutch, G155

"Dutch reinforce major dike as seas rise, climate changes" (Corder and Furtula), *qG288*

E

Earth: biomes, G99, *mG101;* change in, forces of; Equator, G17, *mG17,* G19, *mG19,* G20, *pG20,* G22, *mG22;* grid system on, *mG19,* G20; hemispheres, G17, *mG17;* layers of, *cG83,* G83–84; location on, determining, G19–20 (*See also* global positioning system (GPS)); maps of (*See* map); physical and human features of, spatial aspects of, G7–8; Prime Meridian, G17, *mG17,* G20, *pG20;* satellites for gathering data about (*See* satellites); seasons, G75, G95, *cG96,* G104; structure of, G72; sun, climate and, G75; surface of, G84; topography of, G24; water, land and air, *pG72,* G74.

earthquakes, G55, G59, G86, *pG86,* G87, G128, *pG258,* G261, G268

East Asia, G125, G163, *mG189*

East Jerusalem, G244

economic activities, G195–208; aspects of, G203–5; complementarity and, G206; Corn Belt of United States, G199, *pG199,* G206–7; cut flowers from Colombia, G200; factors of consumption in, G195–96; factors of production in, G195;

location of, G195–96, G197–202; manufacturing (*See also* manufacturing industry); mining in Australia, G198, *pG198;* overview of, G188; patterns of, G205–8; physical geography and, G1205–206; primary, G195, G197; quaternary, G195; resources and, G188, *pG188;* secondary, G195, G197; services (*See also* service industry); tertiary, G195, G197; textile manufacturing in Bangladesh, G201, *pG201;* topography and, G206; tourism in the Maldives, G202; types of, G187, *pG187,* G188

economic categories, G209

economic consumption, G195–96

economic development, G203–8; agriculture and, G204, *mG204,* G205, G206–7; gross domestic product and, G203; industry and, G204, *mG205,* G207–8, *pG208;* in less developed countries, G204–5; in more developed countries, G203–4; in newly industrialized countries, G204, *mG205;* services and, G205, *mG206;* standard of living and, G203; urban economies and, G208

economic geography, G185–222; economic development, G203–8; economic systems and activities, G191–202, G213–18; economies and world trade, G209–12; introducing, G186–90; reviewing, G219–22

economic interdependence, G187, G211–12

economic organizations, G212

economic performance, measures of, G203, G204

"Economics of Chocolate, The" (Gross), G63

economic systems, G186, *pG186,* G193–95; command economy in, G186, G194, G195; consumer preferences and behaviors in, G216; cross-border shopping and, G217, *pG217;* definition of, G193; market economy in, *pG186,* G194; migrating workers and, G218; mixed economy in, G188, G194–95; participation in, G213–18; Sherpas and tourism in Nepal, G214, *pG214;* traditional economy in, *pG186,* G194; women in the workplace, G215, *pG215*

economy: Agricultural Revolution and, G172; choices in, G193, *cG193;* culture and, G160; in culture regions, G161; definition

of, G197; economic activities in, G188, G195–208; economic systems in, G193–95; foreign labor and, G122; functions of cities and, G133; Industrial Revolution and, G172; migration and, G117, G129–30; negative population growth and, G121–22; population distribution and, G124–25; push-pull factors and, G128; remittances and, G122, G130, *gG130;* trade for connecting, G189, *mG189;* types of, G186; urbanization and, G134; wants and resources in, G191–92; world cities and, G133; world trade and, G189, *mG189,* G209–12. See also income.

ecosystems, G279

eco-tourism, G202

"Eco-Tourism: Encouraging Conservation or Adding to Exploitation?" (Nash), *qG202*

Egypt, *pG126, mG171,* G172, 197, *pG274*

elderly, cultural traditions and, G160

elements of culture, G154, G157–60; economy, G160; government, G160; language, G158, *mG158;* religion, G159, *cG159;* social systems, G160

elevation, G24, G34, *cG34,* G78, G96

Elizabeth II, Queen, *pG224*

El Jorullo volcano, G91

El Paso, Texas, G217

emigrate, G127

empires, G163

employment rate, G205

enclaves, G240

energy, G192, *pG192,* G270–72

Enosis, G249

enslaved Africans, migration of, G128

entrepreneurship, G195

environment, G9; globalization and, G212; population growth and, G122; waste products, exporting, G64, *cG64*

environment, human impact on, G259, G279–84; deforestation and desertification, G279–81; pollution and, G281–83; water availability and, G281

environmental change, G283–90; adapting to, G285–90; beavers and, G290; climate change in the Netherlands and, G288, *pG288;* global food supply in Canada and, G287, *pG287;* in Jakarta, Indonesia, G286, *pG286;* water supply in Cape

Town, South Africa and, G289, *pG289*

environment and human settlement, G263–68; adapting to the environment, G265–68; natural disasters and, G268; physical geography and, influences of, G263–65

equal rights in workplace, G180

Equator, G17, *mG17,* G19, *mG19,* G20, *pG20,* G22, *mG22, pG73, cG73,* G96, *cG96,* G97

equinoxes, G96

erosion, G87–88, *pG88*

Ese'eja Indian community, G202

estuary, G81

ethnic enclave, G54

ethnic groups, G160

ethnicity, in U.S. population, G129, G140, *gG140*

Eurasian Plate, *pG84, mG85, mG90*

Europe, G84; Berlin Conference and, G234, G236, *crtG236;* colonialism and, G162, G163; Columbian Exchange and, G155; cultural influences, G163; culture region, G162–63; expressions of culture in, G164; immigrants from, G163; immigration from, G129; migration from, G127; natural boundary in, G233, *mG234;* negative population growth, G121–22; polite greetings in, G164; political borders of, shifting, G239; population cluster in, G125; population in, *gG115,* G124, *gG124,* G125; rural areas in, G124; Schengen Agreement and, G250, *mG250;* urbanization in, G131; world trade in, *mG189,* G212; WTO and, G246

European Union (EU), *pG57,* G58, G144, G212, G244, G246

evaporation, G81, *cG82*

"Examining How K-pop Helps Drive Korea's Global Growth" (Williams), G65

exports, G58, G200, G205, *gG209,* G209–10

fall line, *mG264,* G265, G276

farming. See agriculture

fasting, G275

faults, *pG84,* G86–87, *pG86–87*

federal system, G230

Finland, "Winter War" against, G248

Fletcher, Kenneth R., G52

Floramérica, G200

flow resources, G192, G193

Fodor's Essential Scandinavia, *qG15*

Fonseca, Rae, *qG176*

food: Agricultural Revolution and, G172, G263; American fast-food cuisine, G183; culture and, G164, G166, *pG166;* curry houses and, G168, *pG168;* perishable, in freight cargo, G173; surplus, G172

forces of change, in Earth, G74, G83–88; buildup and movement, G88; erosion, G87–88, *pG88;* inside Earth, G83–84; weathering, G87. See also plate movements.

formal regions, *cG53,* G54

fossil fuels, G192, G269, G271

France, G125, G230, G233, *mG234,* G252

"France to announce sanctions amid fishing dispute with UK" (Corbet), *qG252*

Frederick Douglass National Historic Site, G16

Freedom Quilting Bee, G167

free trade, G210

freight cargo, G173, *pG173*

French, G155, G163

freshwater, G80, G81

Frey, William H., *qG145*

"From Tibet Trading to the Tourist Trade" (Stevens), *qG214*

fuel, advancements in civilization and, G278

Fulani people, *pG127*

fulfillment centers, G207

functional regions, *cG53,* G54–55

Furtula, Aleksandar, *qG288*

Fushan mine, G278

G

Gall-Peters projection, G23

Gandhi, Mohandas K., *qG152, pG152*

Ganges River, *pG42,* G125

gas, G79–80

Gásadalur, Faroe Islands, Denmark, *pG72*

Gateway to the Americas International Bridge, *pG217*

Gaza Strip, G244

Gee's Bend, Alabama, G167, *pG167*

gender, social status and, G180

general-purpose map, G24

genocide, G243

Genuine Progress Indicator (GPI), G203

geographer: career path of, G37; location determined by, G19–20; maps used by, G3, G5, pG5; perspective of, G4; skills and knowledge of, G8, G10; spatial perspective of, G7–8; technology used by, G3, pG3, G5; thinking like, G2–10; tools of, G3, G17–20

Geographer (Vermeer), ptgG29

geographers, tools of, G3, G18–20; absolute location, G8, G19, G29; for determining location, G19–20; globes, mG17, G17–G18, mG19; maps, G18, mG18; satellite images and photographs, pG3; technology, G3, pG3, G5

geography: definition of, G7; environment and society, G9; essential elements of, G8–10; knowledge of, G10; other subjects related to, G10; physical and human, G9; places and regions, G8–9; reasoning in, G37; skills, G10; uses of, G9–10; world in spatial terms, G8

Geologic Section of the New York Academy of Sciences, qG91

geometric boundaries, G234

geospatial technologies, G25–28; geographic information systems (GIS), pG26, G26–27; global positioning system (GPS), G3, pG3, G5, G25–26; limitations of, G28; satellites (*See* satellites)

geothermal energy, G270

Germany, G125, G144, gG144, G231, G233, mG234

ghost towns, G132

glacier, G88

Glasgow, Scotland, pG285

global grid, mG19 G20

globalization, G60, G65, G173–74, G211–12

global positioning system (GPS), G3, pG3, G5, G25–26, G38

global warming, G259, G261, G283, G284, G287

globes, G3, mG17, G17–G18, mG19

God, G232

Gold Rush, G56, G132, pG276, G277

Goode's Interrupted Equal-Area projection, G22, mG22

goods, G193, G195, G197, G213; agglomeration and, G207; agricultural, G207; capital and, G195; consumer preferences and behaviors and, G216; cross-border shopping and, G217; economic development and, G205–6, mG206; factors of production and, cG194; globalization and, G211–12; gross domestic product and, G203; in mixed economy, G194–95; outsourcing and, G211; sustainability and, G212; as tertiary economic activity, G195; in traditional economy, G194; in urban economies, G208; world trade in, G187, G189, mG189, G209–10, mG210, G211

government: around the world, map of, mG231; authoritarian, G231–32, cG232; culture and, G160; in culture regions, G161; democratic, G224, G230–31, cG232; features of, G229–34; functions of, cG229; geography and, G225–26, G233–40 (*See also* political borders); levels of, G229–30; types of, G224, G230–32, mG231, cG232

Grand Canyon, pG87, G88

Grand Coulee Dam, pG192

Great Britain, G125, G230, G265, G278

Great Lakes, G245, mG245, G265

Great Migration, G51, G178, mG178

Great Recession, G145

Great Salt Lake, G80

Great Sand Dunes, G93, pG93

Greece, G144, G164, 249

Greek Cypriot, G249

Greenbushes, G198

greenhouse effect, G97

Greenland, G84, G88

Grivas, George, G249

Gross, Daniel A., G63

Gross, Rebecca, qG167

gross domestic product (GDP): agriculture and, mG204; definition of, G203; goods and, G203; industry and, mG205; money and, G203; per capita, G203; services and, mG206; world trade and, G209

gross domestic product (GDP) per capita, G142, gG142

groundwater, G80, pG80

Gulf of Mexico, G94

gulfs, G80–81

Guo Jinlong, G251

Hall, Edward T., qG152, pG152

Han Chinese, G160

Harlem Renaissance, G51

Hart, Chad, qG199

Hawaii, rainfall in, G104

headwater, G237

hemispheres, G17, mG17

heritage, G16

heroes and heroines, honoring, G164

Hibbing, Minnesota, pG191

high-altitude ice masses, G88

high-latitude climates, pG73, G99, G101–2

Himalayas, G90, mG90, G125, G214

"Himalayas: Two continents collide, The" (USGS), qG90

Hinduism, cG159, G163

Hindus, G233

"Hip Tradition, A" (Kirk), qG176

Hiroshi Hiraoka, G179

Hispanic Americans, G129, G160, G161

"History of Agriculture" (Center for a Livable Future), qG274

"History of Women's Work and Wages and How It Has Created Success For Us All, The" (Yellen), qG215

Hitler, Adolf, G231

Holi, pG2

holiday celebrations, G164, G170, pG170

hoop dance, pG154

Houston, Texas, G125

"How Australia's 'White Gold' Could Power the Global Electric Vehicle Revolution" (Opray), qG198

"How gold rushes helped make the modern world" (Mountford and Tuffnell), qG276

"How to Slow Down the World's Fastest-Shrinking Country" (Hruby), qG144

Hruby, Denise, qG144

Huang He River valley, G125

Hughes, Langston, qG51

hula in Hawaiian culture, G176, pG176

human activities and natural resources, G269–72

Human Development Index (HDI), G203

human-environment interactions, G257–94; adapting to environmental change, G285–90; agriculture in ancient times and, G274; Dunhuang and the Mogao Caves and, G275, pG275; environmental impact on humans, G258, G260–61; environment and human settlement, G263–68; fuel advancements in civilization and, G278; human impact on the environment, G259, G261, G279–84; introducing, G258–62; mining of precious metals and, G277; natural resources and human activities, G269–72; reviewing, G291–94; throughout history, G723–278; Virginia's cities and, location of, G276

human geography, G9

human resources. *See* resources

human rights, G231

Humid continental climate, G101

humidity, G101

humid subtropical climates, G101

Hungary, G121, G244

hurricanes, G102, G261

Hurricane Sandy, pG28

Hussein, Saddam, G231

hydroelectric energy, G192, pG192, G272

hydrosphere, G77, cG77

I

Ice Age, G88

ice caps, G79, G80, G88

Iceland, G250

ice sheets, G88

Illinois, G199

immigrants: from Europe, G163; Italian, in New York City, G54; Japanese, G179; Nisei, G179; Somali, pG157

immigrate, G127

Imperial War Museums, G249

impinging plates, G90

imports, G58, gG209, G209–10, G211

Inca civilization, G155

"In Canada, climate change could open new farmland to the plow" (Arsenault), qG287

Inca Trail Marathon, G34, cG34

income: fertility and, G142, gG142; population growth and, G121–22; poverty and, G121, G122, G130, G134. *See also* economy.

India, G90, mG90, G101; closing of border between Pakistan and, pG235; cultural boundaries in, G233; government system in, G230; Kashmir conflict and, G242, mG242; population in, G125, pG141; religion in, G125

Indiana, G199

Indian Ocean, G80

indigenous peoples, G155, G163

Indo-European language family, G128

Indonesia, *pG86, pG99,* G286, *pG286*

"Indonesia Sets 2024 Deadline to Move Its New Capital to Borneo" (Aditya), *qG286*

Indus River, G125

industrialization, G56

Industrial Revolution, G124, G134, G172, *mG266,* G278

industry, G195; complementarity in, G206; cottage, G205; economic development and, *mG205;* factors affecting, G207–8, *pG208;* globalization and, G211; in newly industrialized countries, G204

information, processing and managing, *pG187,* G188, G195

Information Revolution, G172–73, G195

information technology (IT), G195, G211

infrastructure, G133–34, G141, *pG141*

Ingraham, Kenchera, *qG218*

inland deltas, G81

inner core, G83, *cG83*

intermediate directions, G21

International Date Line, G20

International Monetary Fund (IMF), G212

International Olympic Committee (IOC), G246

Internet, economic consumption and, G196

Inuit communities in northern Canada, *pG186*

Inuit people, G13

Iowa, G199

Iran, G242, G243

Iraq, *mG171,* G172, G231, G242

Ireland, *pG72,* G125

iron ore, *pG191,* G265, G278

irrigation, G265–66

Islam, *cG159,* G161, G163

Islamic law, G232

Islamic mosque, G177

islands, G78, G79

Israel, G244–45

İstanbul, Turkey, G133, *pG133*

isthmus, G78–79

Isthmus of Panama, G79

Italy, *pG71,* G122, G125, *gG144,* G239

J

Jakarta, Indonesia, G286, *pG286*

Japan, G101; athletes in, G164; baseball in, G179; in East Asia cultural region, G163; government system in, G230; population in, G120, G125, G139, *gG139;* tsunami in Ōfunato, *pG258*

Japanese-American baseball leagues, G179

"Japanese-America's Pastime: Baseball" (Maloney), *qG179*

Jim Crow laws, G51

Jordan, G230

Judaism, *cG159,* G163

K

Kalman, Bobbie, *qG153, pG153*

Kashmir, G242, *mG242*

Kauai, Hawaii, *pG104*

Khumbu Valley of Nepal, G214, *pG214*

Kim Jong Un, *pG224,* G231

Kiniero, Guinea, *pG80*

Kirk, Mimi, *qG176*

Korea, G125

Korean pop culture (K-pop), G65

Koreas, in East Asia cultural region, G163

Kurdish nationalism, G243

Kurds, G242–43

Kyrgyzstan, *pG166*

L

labor, G188, G191; cost of, factors in, G207–8; employment rate and, G205; factors of production and, *cG194,* G195; globalization and, G211; outsourcing and, G211, *pG211;* in textile and clothing production, G201, *pG201;* world trade and, G189, G209, G210–11

landfill, G64

landforms, Earth's, G74, G78–79; characteristics of, *pG78,* G78–79, *cG79;* climate affected by, G75 G98; created by buildup and movement, G88; examples of, G77; on ocean floor, G79; plate movements and, G84

landscapes, G46–47

land use, G199, G207

land-use map, G24

language families, G158

languages: culture and, G158, *mG158;* culture regions and, G169, *mG169;* linguists and, G169; world, G158, *mG158*

Laredo, Texas, G217

large scale maps, G23–24

Latin America: birth rate in, G120; culture region in, G162, *mG162;* immigration from, G129; population in, *gG115,* G120, *gG124;* urbanization in, G131, G132

latitude, G19, G20, G21, G25; Hawaii and, G104; high-latitude climates, *pG73, pG75, mG97,* G99, G101–2; low-latitude climates, G75, G96, *mG97,* G99, G100; midlatitude climates, G75, *pG75, mG97,* G99, G100, G101–2; temperature and, *pG73*

lava, G84

laws of holy books, G159

Lebanon, G235

less developed countries, G124, G204–5

Libya, G244

lichen, G102

Lichtenstein, G250

limestone, G88

lingua franca, G158

"Link between Fertility and Income, The" (Vandenbroucke), G142

Linn, Steve, *qG107*

Lippmann, Walter, *qG153, pG153*

literature, culture and, G164

lithium, G198, *pG198*

lithosphere, G77, *cG77*

"Local Brands are Winning Hearts and Minds" (Askew), *qG216*

location, determining, G19–20

London, England, G168, *pG168, pG258*

longitude, G19–20, G21, G25

lower-order goods and services, G59

Lyme disease, *mG33*

M

Maathai, Wangari, *qG152, pG152*

Madre de Dios River, *pG277*

Maeda, Wayne, *qG179*

magma, G84, G85

maharajas, G242

Makarios, Archbishop, G249

Malaysia, *pG279*

Maldives, G202, *pG284*

Maloney, Wendi, *qG179*

manganese, G206

mantle, G83, *cG83*

manufacturing industry, G188, G195, G197; in Bangladesh, G201, *pG201;* globalization and, G211; location of, G207, G210; in more developed countries, G203–4; in newly industrialized countries, G204, *mG205;* outsourcing, G211, *pG211*

Maori of New Zealand, G160

map, G3, G21–24; airline route, *mG30;* contour lines on, G24; distortion problems, G18; elevation on, G24; globes distinguished from, G18; key, G21; large scale or small scale, G23–24; of Lyme disease in United States, *mG33;* mental, G24; multiple perspectives on, understanding, G38; parts of, G21; physical, G24; political, *mG23* G24; projections of Earth, *mG22,* G22–23, *mG23;* relief of land on, G24; scale on, G23–24; thematic, G24; title of, G21; topography on, G24; types of maps, G24

mapmakers (cartographers), G22

Mariana Trench, G79

marine west coast climates, G101

market economy, *pG186,* G194

"Massive Corn Belt Crops Form Backbone Of Meat Industry" (Wendle), *qG199*

Massoud, Rasmenia, *qG115*

Maya of Mexico and Central America, G164

McDonald's, G183, *pG183*

McPhail, Thomas L., *qG153, pG153*

McQuaid, John, *qG200*

Mediterranean, G101

megalopolis, G132

men, cultural traditions and, G160

mental maps, G24

Mercado Común del Sur, G212

Mercator projection, G22, *mG22*

MERCOSUR, G212

meridians, G19–20, G21, G25

Meskwaki people, *pG154*

metallurgy, G278

Mexican-American War, G238

Mexico, G91, *pG91;* cross-border shopping and, G217, *mG217, pG217;* culture hearths in, *mG171;* Day of the Dead in, G170, *pG170;* government system in, G230; Mexican-American War and, G238, G241; NAFTA and, G245; Treaty of Guadalupe Hidalgo and, G241, *mG241;* USMCA and, G245; world trade and, G211, G212

Mexico City, *pG117,* G131

Miami, Florida, *pG47*

Mid-Atlantic Ridge, G79

Middle East, G199

Migiro, Asha-Rose, *qG237*

migration, G172, *pG172;* causes of, G117, G127–28; to cities, G115, *gG115,* G117; definition of, G127;

effects of, G129–30, *gG130;* of enslaved Africans, G128; global patterns of, *mG128;* population growth and, G120; of refugees, G128; regional changes and, G56; spatial interactions and, G58; of workers, G218

migratory increase, G146

minerals, G265

mining, *pG191,* G198, *pG198,* G272, *pG272,* G277

Mississippi River, G233

Mississippi River Delta, G94, *pG94*

Missouri, G199

Mitchell, South Dakota, *pG199*

mixed economy, G188, G194–95

Mogao Caves, G275

Moldova, G244

monarchy, G224, G230–31; absolute, G232; constitutional, G230–31

money: euro, G212; gross domestic product and, G203; International Monetary Fund and, G212; remittances and, G122, G130, *gG130,* G218; return/profit and, G207; sustainable growth and, G205; in world trade, G209, G210

Monroe Marketplace, *pG196*

Moorish rule, G177

more developed countries, G124, G203–4

Moscow, Summer Olympic Games in, G246

mountains: Himalayas, G90, *mG90,* G125, G214; rainfall and, G89; underwater, G79

Mount Etna, *pG71*

Mount Everest, G79

Mountford, Benjamin Wilson, *qG276*

mouth of river, G81

multinationals, G210

music, G164, G178

Muslims, G161, G163, G233, G242, G243

Myanmar, G128

N

Na Pali coast, *pG104*

Nash, Jonathan, *qG202*

National Aeronautics and Space Administration (NASA), G283, *pG284*

National Football League, regions of, *mG41*

nationalism, G242, G246

National Oceanic and Atmospheric Administration (NOAA), *pG26*

National Organisation of Cypriot Fighters (EOKA), G249

National Park Service, *qG16, qG50*

National Weather Service, *qG104*

Native Americans, G43, G50, G160, G162

natural boundary, G233, *mG234*

natural disasters, G268

natural gas, G192, G272

natural increase, G120

natural resources, G187, *pG187,* G188; around the world, *mG270;* culture and, G160; definition of, G191; economic interdependence and, G189; factors of production and, *cG194;* flow, G192, G193; globalization and, G212; human activities and, *pG259,* G269–72; land and, G195; mining in Australia and, G198, *pG198;* in primary economic activities, G195, G197; renewable/nonrenewable, G192, G193; sustainability and, G212; unequal distribution of, G210; using, G271–72; wants and, *pG191,* G191–92; world trade and, G189, G210

natural vegetation, G99

Navajo Nation, political borders of reservation for, G238

Navajo Nation Treaty of 1868, *qG238*

Nazi Germany, G231, G248

negative population growth, G121–22

Nepal, G90, G214, *pG214*

Netherlands, G125, G126, *gG144,* G240, *pG240,* G288, *pG288*

"New African American Identity: The Harlem Renaissance, A" (Smithsonian National Museum of African American History and Culture), *qG51*

newly industrialized countries, G204, *mG205*

New Orleans, G43

New York, forests in, *pG75*

New Zealand, G163

Nicosia, G249

Nile River and delta, *pG126*

Nile Valley, *pG274*

9to5 (National Association of Working Women), G180

"9to5: The Story of a Movement" (Vintzileos), *qG180*

Nisei, G179

non-religion, *cG159*

nonrenewable resources, G192, G193, G269

North Africa, G122, G161, G163, G199, G232

North America: beavers in, G290; political borders of, *mG226;* political map of, *mG226;* population in, *gG115,* G124, *gG124;* rural areas in, G124; world trade in, *mG189*

North American Free Trade Agreement (NAFTA), G212, G245

North American Plate, *pG84, mG85*

North Atlantic Treaty Organization (NATO), G244, G246

Northern European Plain, G264

Northern Hemisphere, G75, G95–96, G101

North Korea, G186, *pG186,* G224, G231

North Pole, G75, G84, G95, *cG96*

Northwest Territories, G13, *mG13*

Norway, G250

nuclear energy, G270

Nunavut, G13, *mG13*

Nussbaum, Karen, G180

O

Obama, Barack, *pG2, qG2,* G246

ocean floor, G79

Oceania, *gG124,* G163

ocean waves, *pG72,* G74, G80, G87, G88

Ōfunato, Japan, *pG258*

Ohio, G199

oil, G192, G210, *mG210,* 212, G265, G272

Okpilak River, *pG88*

Old City of Sanaa, Yemen, G14, *pG14*

oligarchy, G232

Olympic Games, G246

online shopping, G196

opportunity cost, G193

Opray, Max, *qG198*

orographic effect, G107, *cG107*

outer core, G83, *cG83*

outsourcing, G210–11, *pG211*

ozone layer, G283

P

Pacific Ocean, G79, G80, G86

Pacific Plate, *mG85,* G86

Pakistan, G90, G125, *mG171,* G172; closing of border between India and, *pG235;* cultural boundaries in, G233; Kashmir conflict and, G242, *mG242;* water conflict in, political borders and, G237, *mG237*

"Pakistan and Water: New Pressures on Global Security and Human Health" (Pappas), *qG237*

Palestine, G244–45

Panama, G78–79

Panama Canal, G79

Pan-American Highway, G56

pandemics, G174

Pappas, Gregory, *qG237*

parallels (lines of latitude), G19, G20, G21, G25

Parícutin volcano, G91, *pG91*

Paris, France, *mG46*

pastoralism, G264

patriotism, G242, G246

peninsula, G78

Pennsylvania, *pG196,* G244

Pentagon, G244

per capita GDP, G203

perceptual regions, *cG53,* G54, *mG54,* G55

permafrost, G102, G284

Peru, G189, G202

Petra, Jordan, *pG92*

phenomena, natural, G9

Philadelphia, Pennsylvania, G124

physical characteristics, social status and, G180

physical geography, G9, G71–112; change, forces of, G74, G83–88; Earth, G77–82; earth, sun, and climate, G75; Earth's crust, forces shaping, G89–94; Earth's structure, G72; economic activities and, G1205–206; introducing, G72–73; reviewing, G109–12; water, land, and air, G74

physical maps, G24

Pilbara, G198

Pilgangoora, *pG198*

pilgrims, G275

Pine Bluff, Arkansas, G117

place: background information, G11; Canada's provinces and territories, G13, *mG13;* characteristics of, G45–46; cocoa and, G63, *pG63;* commuting to work in, G62, *mG62;* connections between, G59–60, *mG61,* G61–66; definition of, G8, G11, G43, G45–47; Devils Tower National Monument, Wyoming, G50, *pG50;* Frederick Douglass National Historic Site, G16, *mG16;* globalization in, G60, G65; Harlem Renaissance, G51; identity and, G49–52; international travel destinations, G66, *cG66;* landscapes of, G46–47; Old City of Sanaa, Yemen, G14, *mG14;*

of origin, social status and, G180; perceptions of, G47–48; site and situation of, G46; spatial interaction in, G57–59; Tangier Island, Virginia, G52; Tivoli Gardens, G15, mG15; Ushuaia, Argentina, G12, mG12; waste products in, exporting, G64, cG64

plains, G78

plankton, G284

plateaus, G78

plate movements, G74, G84–87, mG85, G90; earthquakes and, G86, pG86; Earth's surface and, G84; faults and, pG84, G86–87, pG86–87; sudden changes and, G86–87, pG86–87; tectonic plates and, G85, mG85; tsunami and, pG86, G87; volcanoes and, G86–87, pG91

Poland, G125, G244, G248

polar areas, G75

polders, G267

political borders, mG226, G235–40; of Africa, G236, crtG236; Antarctic claims, mG227; of Baarle-Nassau, G240, pG240; of cities and neighborhoods, mG233; defining, G235–40; of North America, mG226; reservation for Navajo Nation and, G238; shifting European, G239; sovereignty and, G235; water conflict in Pakistan and, G237, mG237

political geography, G223–56; conflict and, G227, G241–45, G247–52; cooperation and, G227, G245–52; government and, features of, G229–34; introducing, G224–28; political borders and, G235–40; reviewing, G253–56

political maps, mG23 G24

pollution, G281–83

population, definition of, G135

population change, G135–46; in Africa, G143; birth rates in decline, G137, gG137; in Bulgaria, G144, gG144; in Canada, G1465; census and, G135, pG135; demographics and, G135; infrastructure and, G134, G141, pG141; in Japan, G139, gG139; in race and ethnicity in U.S. population, G129, G140, gG140; total fertility rate and, G142, gG142; in U.S., G145, gG145; of U.S. states, G138, mG138; world population growth, 1750-2100, gG136

population clusters, G125

population density, mG116, G125–26, pG126

population distribution, G114, pG114, mG116, G123–25, gG124, mG125; in Asia, gG124, G125; culture and, G124–25; Earth's physical geography and, G123–24; in Europe, G125; resources and, G124; in rural areas, G124

population geography, G113–50; growth of cities and, G115, G116, pG117, G131–34; introducing, G114–18; population change and, G135–46; population distribution and, G114, pG114, mG116; population growth and, G119–22; population movement and, G127–30; population patterns and, G116, G123–26; reviewing, G147–50

population growth, gG119, G119–22; causes of, G119–20; doubling time and, G120; effects of, G120–22; environment and, effects on, G122; migration and, G120; negative, G121–22; population pyramids and, gG120, G121, gG139; projections of, gG119; society and, effects on, gG121, G121–22; in South Sudan, gG120; uneven distribution of, gG124; urban, G115, gG115

"Population Growth in Africa: Grasping the Scale of the Challenge" (Bish), qG143

"Population Growth in Canada: From 1851 to 2061" (Statistics Canada), qG146

population movement, G127–30. See also migration.

population patterns, G116, G123–26; population density and, G125–26, pG126; population distribution and, G114, pG114, mG116, G123–25, gG124, mG125

population pyramids, gG120-G121, G121, gG139

population structure, G121

prairie, G101

precipitation, G81, cG82, G96

Price, Saul, qG104

primary economic activities, G195, G197

Prime Meridian, G17, mG17, mG19, G20, pG20

processes, G77

production, G188; agricultural, G206–7; clothing, G201, pG201; in command economy, G194; factors of, cG194, G195, G211; globalization and, G212; industrial, G207; in market economy, G194; oil, mG210; subsistence and, G213, pG213

productivity, G205

provinces, Canada's, G13, mG13

pull factors, G128, G172

Punakha valley, pG78

push factors, G128, G172

Putin, Vladimir, G244

Q

quaternary economic activities, G195

quilts, culture and, G167, pG167

"Quilts of Gee's Bend: A Slideshow, The" (Gross), qG167

quota, G210

R

race, in U.S. population, G140, gG140

rain, acid, G81–82, G283

rainfall: in Africa, mG31; climate and, G95; in dry climates, G100; in Hawaii, G104; influence of mountains on, G89; rain shadow and, G98, cG107; rivers and, G81; in Texas, G106, mG106; trees and, pG75, G101; in tropical climates, G99; in Washington state, G107, cG107

"Rainfall in Mountainous Areas" (Linn), qG107

Rainforest Expeditions, G202

rain shadow, G98, G107, cG107

reference map, G24

reforestation, G280

refugees, G128, G172, pG172, G243, pG243, G244

regions, G41, mG41, G43, mG43; changes in, G55–56; connections between, G59–60; definition of, G8–9, G41, G43, G53–55; globalization in, G60; identifying, G43; spatial interaction in, G57–59; types of, cG53, mG54, G54–55; of the United States, mG43

relative location, G8, G19

relief of land, on maps, G24

religion, G159, cG159, G164

remittances, G122, G130, gG130

remote sensing, G27, G32, pG32

renewable resources, G192, G193, G269

representative democracies, G230

republics, G230

Resilience News, qG289

resources, G58, G188, G197; capital, G195; definition of, G191; economic, G193 (See also economic systems); nonrenewable, G269; renewable, G269; sustainability and, G212; wants and, G191–92; world trade and, G189. See also natural resources.

revolution, G95

Rhine River, mG234

rice paddy fields in China, pG9

Ring of Fire, mG85, G86

road map, G24

Robinson projection, mG22, G23

Romania, G144, G244, G250

Rome, athletes in, G164

Romm, Joseph, qG259

Roosevelt, Franklin, G248

Royal Greenwich Observatory, pG20

Ruhr, Germany, pG208

Runyon, Luke, qG290

rural areas, G124

Rusizi River, G108

Russia, G125; culture region, G163; empires in, G163; government system in, G232; invasion of Ukraine by, G244, pG244; separatist rebellion against Ukrainian forces and, G243–44; Winter Olympic Games and, G246

Rwanda, G108

Ryman Auditorium in Nashville, Tennessee, pG45

S

"Sacred Site to American Indians, A" (National Park Service), qG50

Salami, Bukola Oladunni, qG218

saltwater, G80, G81

salt water oceans, G80

Sanaa, Yemen, G14, pG14

Sands, Shamel Rolle, qG218

sandstone, G88, G92

Sangre de Cristo Mountains, pG93

San José, G251

satellites: global positioning system and, G25; images and photographs, pG3, G27, pG27, pG32; remotes sensing, G27, G32, pG32

Saudi Arabia, G210, G232, G246

savannas, G100

scale, G23–24

scale bar, G21

scarcity, G191

Schengen Agreement, G250, mG250

"Schengen Area - The World's Largest Visa Free Zone" (SchengenVisaInfo.com), qG250

SchengenVisaInfo.com, qG250

Schengen Zone, G250

seasons, G75, G95, cG96, G104

Seattle, Washington, G107

secondary economic activities, G195, G197

"Secrets Behind Your Flowers, The" (McQuaid), qG200

Seine River, pG46

seismograph, pG86

September 11, 2001 terrorist attacks, G244

service industry, G187, pG187, G188, G193, G195, G213; agglomeration and, G207; economic development and, G205–6, mG206; factors of production and, cG194; gross domestic product and, G203, mG206; in mixed economy, G195; outsourcing and, G211; as tertiary economic activity, G195, G197; in urban economies, G208; world trade and, G187, G189, mG189, G209, G211

shari'ah, G232

Sherman, William Tecumseh, G238

Sherpas, G214, pG214

Shimbashi Athletic Club, G179

"Shimmy Through the World's Most Spectacular Slot Canyons" (Billock), qG92

shopping centers, G195–96, pG196, G208

Sikhism, cG159, G163

Silk Roads Programme, qG275

slash-and-burn agriculture, G267–68

slot canyons, G92, pG92

Slovak culture, pG164

Slovakia, G244

slums, G121

small scale maps, G23–24

Smithsonian National Museum of African American History and Culture, qG51

smog, G102, G283

social groups, in culture regions, G161

socialism, G194–95

social-networking sites, G173

social status in world cultures, G180

social systems, G160, G172

society, G9, gG121, G121–22

solstice, G95–96

Solvang, California, pG165

Somali immigrants, pG157

sources, analyzing, G11–17, G29–34

South Africa, G246, G289, pG289

South America, pG75, G79, G84, G99, G100; Mercado Común del Sur and, G212; population in, G124; world trade in, G189, mG189

South Asia, pG78, G125, G163

South Dakota, pG199

Southeast Asia, G163, G212

Southern Hemisphere, G75, G95–96

Southern Ocean, G80

South Korea, G56

South Pole, G75, cG96

South Sudan, gG120, G128, G243

Southwest Asia, G122, G125, G161, G163, G232

Southwest Louisiana as Cajun country, G165

sovereignty, G229, G235, G248

Soviet bloc, G144

Soviet Union, G244, G246, G248, pG248

Spain, G122, gG144, G155

Spanish Harlem in New York City, G161

spatial interaction, G57–59

spatial relationships, G8

spiritual teachings, G159

spodumene, G198

sports, culture and, G164

spreading boundaries, G85

Sri Lanka, G125

Stalin, Joseph, G248

standard of living, G203, G271

Statistics Canada, qG146

steppes, G100

Stevens, Stanley F., qG214

strike-slip boundaries, G85

subarctic climate, G102

subducted plates, G90

subsistence economy, G213

subsistence farming, G205, G263

suburbs, G208

Sudan, G243, pG243

Sumatra, Indonesia, pG86

summer solstice, G95

sun, Earth and, G75

sustainability, G272

sustainable growth, G205

Sutter's Mill, G277

Switzerland, G230, G239, G250

Syria, G128, G242, G243

Syrian refugees, pG172

Table Mountain, pG4

Taiwan, G211, G251

Taliban, pG56

Tambopata Candamo Reserved Zone, G202

Tangier Island, Virginia, G52

"Tangier Island and the Way of the Watermen" (Fletcher), G52

tariffs, G210, G252

technology: Agricultural Revolution and, G172; cultural change and, G155, G172; definition of, G25; geospatial, G25–28; globalization and, G174; remotes sensing, G32, pG32; used by geographers, G3, pG3, G5

tectonic plates, G85, mG85

temperature, pG73. See also climate.

terrorism, G244

tertiary economic activities, G195, G197

Texas, rainfall in, G106, mG106

Thailand, pG131, G230

Thames River, pG258

thematic maps, G24

theocracy, G232

Thingvellir National Park, Iceland, pG84

thumper trucks, G84

Tibet, G90, G214

Times Square in New York City, pG48

tin mining, G56

Tivoli Gardens, G15, pG15

topographic maps, G24

topography, G206

tornadoes, G102

total fertility rate, G142, gG142

totalitarian dictatorships, G231

tourism, G202, G214, pG214

trade, G58; Agricultural Revolution and, G172; cultural change and, G171; cultural identity and, pG175; in culture regions, G162; economy and, G160; international, pG173, G174; world economies and, G189, mG189, G209–12 See also world trade.

traditional economy, pG186, G194

transportation: costs, G207–8; networks, G189, G195, G206, pG208, G211; technologies, advances in, G173

Treaty of Guadalupe Hidalgo, G241, mG241

trees: in boreal forests, G105, mG105, G279; coniferous, G101, G105; deciduous, G101; deforestation and, G278, pG279, G279–81; rainfall and, pG75, G101; reforestation and, G280

trench, G79

tributary, G237

Troodos Mountains, G249

tropical climates (the Tropics), G75, pG99, G99–100

tropical rain forest climates, G75, G99–100

tropical wet/dry climates, G100

Tropic of Cancer, G75, G95, cG96

Tropic of Capricorn, G75, G95, cG96

tsunami, pG86, G87, G128, pG258, G268

Tuffnell, Stephen, qG276

tundra climate, G102

Turkey, G144, gG144; ethnicities in, G249; Kurdish populations in, G242m G243; Turkish Republic of Northern Cyprus, G249, pG249

Turkish Cypriot, G249

Turkish Northern Cyprus, G249

"12 Things You Didn't Know About Great Sand Dunes National Park and Preserve" (DOI), qG93

Tylor, Sir Edward B., qG152, pG152

typhoons, G268

Ukraine, G125, G206; refugees from, G244; Russian invasion of, G244, pG244; separatist rebellion against Ukrainian forces and, G243–44

underwater cliffs, G79

UN High Commission on Refugees (UNHCR), pG244

unitary system, G230

United Kingdom (UK), gG144, G224, G230, G252, pG281

United Nations (UN), G121, G246, pG247

United Nations Educational, Scientific and Cultural Organization (UNESCO), G174, G246

United Nations Population Division, G143

UN Refugee Agency, G128

United States (U.S.): agriculture in, G206–7; baseball and, G179; Berlin Conference and, G236, crtG236; boycotts of Olympic

T

U

Games by, G246; Canada-United States Air Quality Agreement and, G245; Corn Belt in, G199, *pG199*, G206–7; cross-border shopping and, G217, *mG217*, *pG217*; culture region of, G162, *mG162*; equal rights in the workplace in, G180; ethnic groups in, G160; fall line in, *mG264*, G265, G276; global positioning system in, G25–26; government system in, G230; Great Migration and, G178, *mG178*; hydroelectric energy in, G192, *pG192*; immigration to, G129, gG144, *pG157*; language and, *mG158*, G169; lithium and, G198; Lyme disease in, *mG33*; Mexican-American War and, G238, G241; NAFTA and, G245; natural boundaries in, G233; political map of, *mG23*; population in, G145, *gG145*; population in states, G138, *mG138*; regions of, *mG43*; Soviet Union and, G248, *pG248*; Treaty of Guadalupe Hidalgo and, G241, *mG241*; urbanization in, G131; USMCA and, G245; women in the workplace and, G215, *pG215*; world trade and, G211, G212; WTO and, G246

United States-Mexico-Canada Agreement (USMCA), G212, G245

U.S. Capitol switchboard, *pG215*

U.S. Constitution, G230

U.S. Department of Agriculture (USDA), G199

U.S. Department of State, *qG248*

U.S. Geological Survey, *qG90*

U.s. railroad system, *mG2*

"U.S.-Soviet Alliance, 1941-1945" (U.S. Department of State), *qG248*

urban areas, G124

urban economies, G208

urban heat islands, G102

urbanization: causes of, G131; challenges of, G134; definition of, G131; factors increasing/decreasing, G132; migration and, G130; patterns of, *pG131*, G131–34; world urban population and, G115, *gG115*. See also cities, growth of.

urban sprawl, G134

Ushuaia, Argentina, G12, *mG12*

Utah, *pG74*, G88

Uyghurs, G243

V

valley, G78

Vandenbroucke, Guillaume, G142

vector-borne disease, G33

Venezuela, G128

Vermeer, Johannes, *ptgG29*

Vimercate, Italy, *pG213*

Vintzileos, Jennifer, *qG180*

Virginia, G276

Virginia Studies, *qG276*

visual arts, culture and, G164

volcanoes, G55, G59, G79, G86–87, G91, *pG91*, G261, G268

W

Wadden Sea, *pG288*

Wagah Attari, *pG235*

Wang Medina, Jenny, *qG65*

wants and resources, G191–92

wars, regional changes and, G55–56

Washington state, rainfall in, G107, *cG107*

waste products, exporting, G64, *cG64*

water, Earth's, G74, G79–81; availability of, G281, G289, *pG289*; bodies of, G80–81; essentials of life provided by, G81; forms of, G79–80; freshwater, G80, G81; groundwater, G80, *pG80*; saltwater, G80, G81

water cycle, G81–82, *cG82*

water vapor, G79–80

weather, G95. See also climate.

weathering, G87

Wendle, Abby, *qG199*

West Africa, *pG80*

West Bank, G244

wetlands, *pG281*, G282, *cG282*

"What Caused the Division of the Island of Cyprus?" (Imperial War Museums), G249

"What the 2020 Census Will Reveal About America" (Frey), *qG145*

"When Borders Melt" (Bloomberg CityLab), *qG239*

wildfires, G261

Will, Rachel, *qG251*

Williams, Trevor, G65

winds, G96–97, *mG97*

Winkel Tripel projection, G22, *mG22*, G23

winter solstice, G95–96

women: cultural expressions and, G164; cultural traditions and, G160; suffrage and temperance of, G16; in the workplace, G215, *pG215*

Woolacombe, England, G123

world, in spatial terms, G8, G35–38

World Bank, G117, G199, G212

world city, G133

world culture regions, *mG162*, G162–63

World Heritage Sites, G246

world languages, G158, *mG158*

world trade, G187, G189, *mG189*, G209–12; agreements, G187, G212; definition of, G209; domestic and international trade, compared, G211; economic organizations and, G212; globalization and, G211–12; in goods, G187, G189, *mG189*, G209–10, *mG210*, G211; import and export of goods and services, G209, *gG209*; labor costs and education and, G210–11; outsourcing and, G210–11; patterns of, G209–12; unequal distribution of natural resources and, G210

World Trade Center, G244

World Trade Organization (WTO), G212, G246

Y

Yanukovych, Victor, G243

Yellen, Janet L., *qG215*

young people, cultural traditions and, G160

yurt, *pG166*

Z

Zelensky, Volodymr, G244

Economics Index

Note: Italicized page numbers refer to illustrations. The following abbreviations are used in the index: *m* = map; *c* = chart; *p* = photograph or picture; *g* = graph; *ptg* = painting; *crt* = cartoon; *q* = quote.

A

Aboriginal peoples, E14
absolute advantage, E166
ACH transfers, E88, *gE88*
acid rain, E188
advisory councils, E126, *cE127*
agency shops, E77
Amazon, E66
America. *See* United States (U.S.).
"America First" policy, E177
American Express, E70
American Federation of Labor and Congress of Industrial Organizations (AFL-CIO), E76
Americans with Disabilities Act of 1990, E97
American Taxpayer Relief Act of 2012, E138
American Telephone and Telegraph (AT&T), E93
Andean Community trading bloc, E176
antitrust laws, E92–93
Apple, E66, E70, E176
appropriations bill, E132
arbitration, E78
articles of partnership, E64–65
Asia, ASEAN and, E176
assembly lines, *pE4, pE40,* E44, *pE44*
Association of Southeast Asian Nations (ASEAN), E176
Automated Clearing House (ACH) transfers, E88, *gE88*
automated teller machines (ATMs), E84
automatic stabilizers, E145, *pE145*
automobile factories, *pE4, pE40,* E44, *pE44*
Aztec market scene, E20, *pE20*

B

balanced budget, E146
balanced budget amendment, E152
balance of trade, E171–72, *cE172;* imbalances, self-correcting, E172, *cE172;* negative, E171–72; positive, E171
banking, E83–84; changes in, over the years, E83; first banks, E83; Great Depression and, E84; national banks and the Fed, E83–84; technology and, E84
Banking Act of 1933, E84
bankruptcy, E151, E152
banks: commercial, E82; district, E126, *pE126, cE127;* first, E83; monetary policy for regulating and supervising, E129
bar codes on products, *pE173*
Barral, Miquel, *qE21*
barter system, E13, E19–21, E79
bear market, E112, *pE112,* E124
Bell, Alexander Graham, E43
benefit-cost analysis, E11
Berlin Wall, E49
Bessemer, Sir Henry, E43
Bessemer Converter, *pE43*
Bessemer process, E43
Beyoncé, E60
Biden, Joe, E144
black market, E181–82, *pE181*
Board of Governors, E126, *cE127*
bonds, E146, *pE146*
Brazil: soybean exports from, E166
British Petroleum, E176
Breedlove, Sarah, E53
budget, E113, E146–48; balanced, E146; changes in process, E132; deficits, *gE133,* E134, E147–48, *gE147;* expenditures, E113, E131, E133–34, *gE133;* infrastructure and, E146; for local governments, *cE135,* E136; process, steps in, *cE131,* E132; revenues, E113, E131, E133, *gE133;* for state governments, E134–36, *cE135;* surplus, E146, E147; understanding, E133–34. *See also* debt.
budget deficits, E113, *gE133,* E134, E146–49, *gE147, gE150*
budget resolution, E132
budget surplus, E146, E147
bull market, E112, *pE112,* E124, *pE124*
Bureau of Engraving and Printing, E129
Bureau of Labor Statistics survey, E73
Bush, George W., E78; on financial markets and world economy, *qE49*
business: competition in, maintaining, E92–93; government rules and regulations in, E55, E95; owning, E55; responsibilities to consumers, E71; responsibilities to employees, E71–72, *pE72;* responsibilities to owners, E71; social responsibilities of, E69–70, *pE69*
business cycles, *gE119,* E120
business organization, E63–68, *cE64;* corporations, E66–68, *pE66;* franchises, E68, *pE68;* nonprofit organizations, E68; partnerships, E64–65, *pE65;* sole proprietorship, E63–64, *pE63*
business sector, E38

C

capital, E23; as part of GDP, E116
capital equipment, E23
capitalism, E24, E54, E57; competition in, E60, *gE60;* economic freedom in, E58; identifying features of, E108, *pE108;* laissez-faire economics in, E61, *pE62;* markets in, E58–59; origins of, E61–62; private property rights in, E61; profit motive in, E59–60; in United States, E57–61; voluntary exchange in, E59, *pE61*
capital resources, E8, *pE8,* E38
Carnegie: steel factories, *pE46*
Castro, Fidel, E15
Centers for Disease Control and Prevention (CDC), E94
central bank, E84, E125
certificate of deposit (CD), E81–82
Change to Win, E76
charitable contributions, E70
charter, E66, E82, E83
Chavez, Hugo, E15
checking account, E81
checks, E88, *gE88*
child labor, *pE46*
China: Gross Domestic Product of, E116, E118; market economy in, transitioning to, E182, *cE182;* outsourcing to, *pE178;* pollution in, E118; Silk Road and, *pE19,* E21; tea exports from, E166; trade deficit with United States, E171, *pE171*
circular flow model, E37–39, *pE37;* business sector in, E38; circular flow in, E38–39; consumer sector in, E38; factor market in, E38; foreign sector in, E39; government sector in, E39; product market in, E38
civilian labor force (CLF), E73–74, *cE73*
Civil Rights Act of 1964, E97
Clayton Antitrust Act of 1914, E93
climate change, E186
Coast Guard, *pE111*
Coca-Cola, E124
coins, E80, *pE80*
collective bargaining, E77
command economies, E14–15, *pE14;* transitioning to market economy, E181, E182, *cE182*
commercial, E21
commercial banks, E82
commodity, E21
communism, E15, *pE15,* E49
Communist Party, E182
comparative advantage, E162, E166–67; factors of production and, E167; protectionist policies and, E163; specialization and, E166–67
compensation, unemployment, E98
competition, E34, E92–93; foreign, blocking, *crtE160;* in market economy, E17, *pE17;* price setting and, *pE33,* E34. *See also* monopoly.
Comptroller of the Currency, E83–84
Congress: Affordable Care Act of 2010 passed by, E72; appropriations bills and, E132; balanced budget amendment and, E152; Banking Act of 1933 passed by, E84; Clayton Antitrust Act of 1914 passed by, E93; Employment Act of 1946 passed by, E144; Fed created by, E84, E126, E130; minimum wage set by, E99, *pE99;* stimulus programs passed by, E144; taxes raised by, E138; writing a letter to, E154. *See also* government.
Congressional Budget Office (CBO), E132
Constitution, U.S., E54, E62, E152
consumer, E25; demand affected by, E28
consumer price index (CPI), E123

Consumer Product Safety Commission (CPSC), *pE70*, E94, *cE94*

consumers: monetary policy for protecting, E130; protecting, E94; responsibilities of businesses to, E71

consumer sector, E38

consumer sovereignty, E58

contactless purchases, *pE86*

continuing resolution, E132

cooling-off period, E78

cooperatives (co-ops), E68

corporations, E66–68, *pE66*; advantages of, E66, E67; disadvantages of, E68; organization of, E66; structure of, *cE67*; taxes paid by, E113

corruption, E163, E183, E184

cost-benefit analysis. See benefit-cost analysis.

cottage industries, *pE41*

cotton mill in Macon, Georgia, *pE46*

COVID-19 Recession of 2020, E84, E112, E113, E120, *gE121*, E122, E133, E144

credit cards, *pE86*, E88–89, *gE88*

credit union, E82

cryptocurrency, E80-81, *pE81*

Cuba: command economy in, E15, *pE15*, E49; Constitution of 1976 with Amendments through 2002, *qE49*

Cuba, E15, *pE15*; GDP per capita in, *mE180*, E181

cultural obstacles to economic development, E182–83

currency, E80, *pE80*, E83, *pE83*; in global economy, *pE159*; monetary policy for maintaining, E129–30; for trade, E169

D

Daily Evening Traveller, *qE43*

debit cards, E88, *gE88*

debt, E146, E149–54; balanced budget amendment, E152; deficits created by, E146, E147–48, *gE147*, E149, *gE150*; national debt clock and, E151, *pE151*; ratio of spending to income by fiscal year, E154, *gE154*; significance of, E153; of undeveloped countries, E183–84

deficits. See budget deficits.

deflation, E124

Dell Technologies, E176, *pE176*

demand, E25–30; change in, E28, *gE29*, E30; factors affecting, E28, *pE28*; graphing, E26, *gE26*, E27; quantity demanded, change in, *cE25*, E26, *gE26*, E30; schedule, E26, *gE26*, E30, *gE34*, *gE36*

demand curve, E26, *gE26*, E27

democracy: benefit of, E112; based on economic principles, E62

demand deposit account, E81. See *also* checking account.

deposit insurance program, E83

deposits, E81–82

depression, E120

deregulation, E93

developed/developing countries: classifying, E179–80, *cE179*; debt and, E183–84; standard of living and, E163, E179

direct deposit, E88

discount rate, E129

discretionary spending, E132, E133–34, *pE134*, *gE135*

discrimination: income inequality and, E96–97; laws passed to end, E96–97; in workplace, E71–72

diseases: in developing countries, E183

Disney, Walter Elias, E53, *pE53*

disproportionate wealth, E142

district banks, E126, *pE126*, *cE127*

diversified economies, E167, E183

division of labor, E40, *pE40*

Dobbs, Lou, E193, *qE193*

domestically, E23

Dow Jones Industrial Average (DJIA), E124

due process of law, E54

Durst, Seymour, E151

E

earnings gap, E102

economic choices, E5, E11–12; questions answered by, E11–12; values of a society and, E12; writing report about, E49

economic cooperation: global and regional, E173–76; military conflicts and, E178

economic decisions, E9–11; benefit-cost analysis in, E11; opportunity costs in, E9–11; trade-offs in, E9

economic flow, E37–39

economic growth, E39–40; cultural obstacles to, E182–83; in developing countries, challenges to, E163; economic obstacles to, E183–84;

environmental obstacles to, E183; increased productivity needed for, E40; political obstacles to, E183; productive resources needed for, E39. See *also* Industrial Revolution.

Economic Growth and Tax Relief and Reconciliation Act of 2001, E138

economic indicators, E123–24; Leading Economic Index, E124; price stability, E123–24, *cE123*; stock indexes, E124

economic interdependence, E163; consequences of, E176–78

economic performance, E119–24; business cycles and, *gE119*, E120; Great Depression and, E120, *pE120*; measures, E123–24; real GDP and, E119; recessions and unemployment, E120–22, *gE121*, *gE122*. See *also* economic indicators.

Economic Recovery Tax Act of 1981, E138

economics: choices in, *crtE2*, E5, E8; concepts of, graphing, E50, *gE50*; goods and services in, E4; markets in, E5; prices in, E5; resources in, E4

economic systems, E2, E13–24; barter system and early trade routes, E19–21; capitalism, E24; command, E14–15, *pE14*; manorialism, E22, *pE22*; market, E16–17; mercantilism, E23; mixed, E17–18; perspectives on, comparing and contrasting, E49; traditional, E13–14, *pE13*; WHAT, HOW, and FOR WHOM questions in, E13, E19, E49

economy: budget, E113; fixing, E113; government involvement in, reasons for, E112; Gross Domestic Product, E112, E115–18; health of, E110, E112; measuring, E119–24; recessions and, E120–22; unemployment and, E121–22. See *also* economic indicators; economic performance.

Ecuador: banana exports from, E167, *pE167*

education: income inequality and, E95–96, *pE95*, *gE96;* as human capital, E40

electronic money, E80

electronic transactions, E86

employees: responsibilities of businesses to, E71–72, *pE72*

employment, E73–74; civilian labor force in, E73–74, *cE73*; organized labor, E75–78; right-to-work laws, *pE76*, E77; by sector, then and now, E74, *cE74*; unemployment rate, E74, *cE74*. See *also* labor unions.

Employment Act of 1946, E144

entitlement programs, E136

entrepreneurs: in America, E53, *cE58*; creating a new product, *crtE52*, E108; GDP and, E116; piracy and, E24; as resources, E8, E38

environmental balance, E185–90; background information, E185; by consuming less, E190; economic growth affected by, E186; gaming and energy, E187, *pE187*; perception of energy use and economic progress, change in, *crtE189*; pollution tax, E188

environmental obstacles to economic development, E183

Environmental Protection Agency (EPA), *cE94*, E188

Equal Employment Opportunity Act of 1972, E97

Equal Pay Act of 1963, E96

equilibrium, E34

equilibrium price, E34

equilibrium quantity, E34–35

euro, E175

Europe: Industrial Revolution in, E42; manorial system in medieval, E22; mercantilism in, E23; Silk Road and, *pE19*

European Union (EU): free-trade zone, E163, E175; trade bloc, E163, E174–76, *cE175*, *mE175*; United Kingdom leaves, E176, E177

Everglades National Park, *pE111*

exchange rates, E162, E169–70, *pE169*

exempt, E23

expenditures, E113, E131, E133–34, *gE133*

exports, E23, E162, E165–66, *gE165*

externalities, E92, *pE92*

F

F-16 Fighting Falcon, *pE39*

factor market, E38, *pE39*

factors of production, E4, E8, E38

family wealth, E96

federal budget. See budget.

ECONOMICS INDEX

federal debt. *See* debt.

Federal Deposit Insurance Corporation (FDIC), E83, E84

Federal Open Market Committee (FOMC), E126, *cE127*

Federal Reserve Notes, E84

Federal Reserve System (the Fed), E84; advisory councils in, E126, *cE127*; Board of Governors in, E126, *cE127*; district banks in, E126, *pE126*, *cE127*; Federal Open Market Committee in, E126, *cE127*; functions of, E126–30; location of, *pE125*; member institutions in, E126, *cE127*; monetary policy conducted by, E113, E126, E128; structure of, E125–26, *cE127*

Federal Trade Commission (FTC), E93, *cE94*

final good or service, E116

financial capital, E64

financial institutions, E81–83; deposits, E81–82; functions of, E81; loans, E82; regulating, E82–83; types, E82. *See also* banking.

First Bank of the United States, E83

First Nations, E14

fiscal policy, E113, *pE113*, E144–45; advantages of, E144; automatic stabilizers in, E145; problems with, E144–45

fiscal year, E131, E154, *gE154*

fixed income, E123

Florida, *pE111*, E167

Food and Drug Administration (FDA), E94, *cE94*

Ford, Henry, E44, *qE44*

Ford Motor Company, *pE44*

foreign exchange, E169

foreign sector, E39

fossil fuels, E186

franchises, E68, *pE68*

Franklin, Benjamin, E107, *qE107*, E110, *qE110*

fraud, E89

free enterprise capitalism. *See* capitalism.

free enterprise system: characteristics of, *gE56*, E57; freedoms in, overview of, E52; government involvement in, E55; labor in, E54; money in, E55; owning a business in, E55, *pE57*; private property in, E54

free markets, GDP per capita in, E184

free trade: blocking, *crtE160*; issues in, E177. *See also* global trade alliances.

free-trade zone, E163, E175

G

gaming and energy, E187, *pE187*

GAP, E70

Gates, Bill, E70

GDP per capita, E118; in free markets, E184; in market economy, E180–82; in selected countries, E180, *mE180*

General Agreement on Tariffs and Trade (GATT), E174

General Motors, *pE66*, E121, E176

general partnership, E65

geographical problems for economic growth, E183

Germany: auto exports from, E166

global economy: in developed *vs.* developing countries, E163, E179–80, *cE179*; economic interdependence and, E163; environmental balance in, E185–90; GDP per capita, in selected countries, E180, *mE180*; standard of living and, E163, E179; transitioning to market economy, E180–84

globalization, E21

global trade alliances, E173–76; for economic interdependence, E173–76; EU trade bloc, E163, E174–76, *mE175*; GATT, E174; NAFTA, E176; political advantage of, E178; USMCA, E176; WTO, E174

global warming, E186

goods, E4, E7, E16; externalities, E92, *pE92*; final, E116; intermediate, E116; producer, E117; public and private, E91–92

Google, E60, *pE72*

government: assistance, viewpoints on, E107; benefit-cost analysis and, E11; budget, E113; in command economies, E14–15, *pE14*; competition encouraged by, E92–93; economic growth promoted by, E39, *pE39*; factors of production and, E16; financing, E131–36; goods provided by, E91–92; involved in economy, reasons for, E112; in market economy, E17; in mercantilism, E23, E24; in mixed market economy, E17, E18; poverty guidelines, E97, *cE97*; price setting and, E5, E35; rationing and, E32–33, *pE32*; regulation in free enterprise system, E55; regulation of financial institutions, E82–83; regulatory agencies, E94, *cE94*; shutdowns and closings, E112, E157; taxes contributed by, E111; trade-offs and, E9;

writing a letter to, E46. *See also* Congress.

government sector, E39

government securities, E128, E146

Great Clips, E68

Great Depression, E84, E97, E143–44, *pE143*; economic performance and, E120, *pE120*; fiscal policy and, E144; income tax rates and, E138; tariffs during, E167

Great Recession of 2008–2009, E84, E112, E113, E120, E121, *gE121*, E124, E144

Griswold, Daniel, E193, *qE193*

Gross Domestic Product (GDP), E112, E115–18; activities not included in, E117, *pE117*; as aggregate, E118; estimating total annual output, E116, *cE116*, E117; during Great Depression, E120; during Great Recession of 208-2009, E121; income represented by, E116; real GDP, E112, E117–18, *gE118*, E119; standard of living and, E118; total output measured by, E115–16; of United States, E116, E118, *gE118*. *See also* GDP per capita.

guild, E23

H

Hamilton, Alexander, E62

health insurance benefits, E72

Holmes, Oliver Wendell, E110, *qE110*

Hong Kong: free markets in, E49

human capital, E40

human resources, E4, *pE8*

Hungary, E181

I

imports, E23, E162, E165–66, *gE165*

Inc. (incorporated), E66

incentive, E16–17, E27

income: GDP for representing, E116

income inequality, E95–97; discrimination and, E96–97; education and, E95–96, *pE95*, *gE96*; family wealth and, E96; minimum wage and, E102

India: Dell Technologies in, *pE176*; global trade in, *pE161*; overcoming obstacle to development in, *pE184*

individual income taxes, E113

industrialism: benefits of, E45; problems of, E46, *pE46*

Industrial Revolution, E41–46; background information on, E41; industrialism and, benefits of, E45; industrialization and, problems of, E46; industries in, E42; inventions in, E43; production methods in, E44

industrial union, E75

industries, first, E42

inflation, E123

infrastructure, E146, E179

injunctions, E78

An Inquiry Into the Nature and Causes of the Wealth Of Nations (Smith), *qE24*

interest rates: monetary policy and, E126, E128, *gE128*, E129, *pE129*; supply and demand and, E146

intergovernmental revenue, E135

intermediate goods, E116

International Value of the U.S. Dollar, E170, *gE170*

Inuit, E14

inventions, E43

"invisible hand" of capitalism, E24, E61

IRS (Internal Revenue Service), E140

J

Jamestown, Virginia, E68

Japan: exchange rates in, E169–70; free markets in, E49

Jefferson, Thomas, E62

Jiffy Lube, E68

Johnson, Lyndon B., E107, *qE107*

joint-stock company, E68

K

Keppler, Udo J., *pE93*

Krugman, Paul, E157, *qE157*

L

labor, E54; as part of GDP, E116

labor negotiations, E77–78; arbitration, E78; collective bargaining, E77; injunctions, E78; lockouts, E78, *pE78*; mediation, E78; picketing, E77, *pE77*; strike, E77, *pE77*

labor resources, E8, E38

labor unions, E75–78; organization of, E76; right-to-work laws and, *mE76*, E77; types of, E75–76; in workplace, E76–77

LaHaye, Laura, *qE23*

laissez-faire economics, E61, *pE62*
land: in command economy, E14; in factor market, E38; in manorial system, E22; as natural resource, E8, *pE8*; as part of GDP, E116; preserving, for economic growth, E39
Leading Economic Index (LEI), E124
liability, E64
Lilly Ledbetter Fair Pay Act of 2009, E97
limited liability partnership (LLP), E65
limited partnership, E65
liquidity crisis, E129
Liverpool and Manchester Railway, E42
loans, E82, E130, *pE130,* E146
local governments: budgeting for, *cE135,* E136
locals, union, E76
local taxes, E113
location: as business decision, E5
lockouts, E78, *pE78*
logs. *See* timber.

M

Macy's Department Store, *pE45*
Ma, Jack, E60
Madison, James, E62
mandatory spending, *pE132,* E133–34, *gE135*
manorialism, E22, *pE22*
market, *pE5,* E16, E33; in capitalism, E58–59; competition in, E34; equilibrium in, E34–35; prices set in, E33–36; shortages in, E35, E36; surplus in, E35, *pE35*
market economy, E180–84; advantages and disadvantages of, E17; characteristics of, E16–17; competition in, E17, *pE17*; cultural obstacles to development, E182–83; demand in, E25–27, E28–30; economic obstacles to development, E183–84; environmental obstacles to development, E183; GDP per capita in, E180–82, *mE180*; political obstacles to development, E183; prices in, E16, E31–36; pure, E17, E18; supply in, E27–30; transitioning economies, E181–82, *cE182*; wealth of nations and, E184; WHAT, HOW, and FOR WHOM questions in, E16, E17

McDonald's, *pE17,* E124, *pE163*
median wage, E102
mediation, E78
Medicaid, E98, E133, E134
Medicare, E98, E133
member institutions of the Fed, E126, *cE127*
Men of Invention and Industry **(Smiles),** *qE42*
mercantilism, E23, E61
"Mercantilism" (LaHaye), *qE23*
mergers, E93, E129
Mexico, E163, E176; Aztec market scene in, E20, *pE20*
Microsoft Corporation, E70
minimum wage, E99–104; from 2000 to 2020, E100, *gE100*; background information on, E99; increasing, arguments for/against, E102–03, *pE104*; state, E101, *cE101*
Mitsubishi, E176
mixed economy, E182
mixed market economy, E17–18
modified union shop, E77
monetary base, E126, E128
monetary policy, E113; banks borrowing money from, E129; conducting, Fed's role in, E113, E126, E128; discount rate in, E129; Fed acting as government's bank in, E130; interest rates and, E126, 128, *gE128*; for maintaining currency, E129–30, *pE130*; open market operations in, E128; for protecting consumers, E130; for regulating and supervising banks, E129; reserve requirement in, E128–29; tools of, E128–29
money, E55; background information, E85; cash payment percentages, E90, *gE90*; characteristics of, E80; forms of, E80–81; functions of, E79–80; as measure of value, E80; as medium of exchange, E79; as store of value, E79. *See also* noncash payments.
monopolist, E92
monopoly, E92–93, *pE93*; antitrust laws and, E92–93; mergers and, E93; natural, E93
Monroy, Caine, E60
multilateral agreement, E174
multinationals, E163, E176, E178
My Life and Work **(Ford),** *qE44*
Musk, Elon, E60

N

Nabisco, E176
National Bank Act, E83–84
National Credit Union Share Insurance Fund (NCUSIF), E83
national debt clock, E151, *pE151*
national defense: as discretionary federal expenditure, E134
National Education Association (NEA), E76
National Highway Traffic Safety Administration (NHTSA), E94
National Labor Relations Board (NLRB), E76
National Park Service, *pE111*
natural monopoly, E93
natural resources, E8, *pE8,* E38
near-field communication (NFC) technology, E86, *pE86*
needs, E7
net interest, E134
new products, creating, E108
noncash payments: ACH transfers, E88; credit cards, *pE86,* E88–89, *gE88*; electronic transactions, E86; online shopping, E87, *cE87*; trends in, E88, *gE88*
nonprofit organizations, E68
North American Free Trade Agreement (NAFTA), E176
North Korea: command economy in, E15, E181; GDP per capita in, E180, *mE180,* E181, *cE182*

O

Occupational Safety and Health Administration (OSHA), *cE94*
oil: demand and supply for, *gE36*; in factor market, E38; preserving, for economic growth, E39; shortages in 1970s, *pE3*
Old Faithful, *pE111*
online banking, E84
online shopping, E87, *cE87*
open market operations (OMO), E128
opportunity costs, E9–11
organized labor, E75–77. *See also* labor unions.
outsourcing, E163, E177, *pE178*

P

pandemic of 2020, *pE3,* E33
partnerships, E64–65, *pE65*; advantages of, E65;

disadvantages of, E65; structure of, E64–65
Patient Protection and Affordable Care Act of 2010, E72, E138
Pichai, Sundar, E60
picketing, E77, *pE77*
piracy, E24
Plato, E110, *qE110*
plutocracy, E140
point-of-sale terminal, *pE86*
political advantages in economic cooperation, E178
political obstacles to economic development, E183, *pE183*
pollution, E46, *pE46*
pollution permits, E188
pollution tax, E188
population: and the Civilian Labor Force, 2020, *cE73,* E74
pork, E138
poverty, E97–98; guidelines, E97, *cE97*; social insurance programs and, E98; unemployment insurance and, E98; welfare and, E97–98
poverty line, E102
power loom, E42
prepaid debit cards, E88, *gE88*
price, E31–36; economic questions answered by, E31–32; in economics, E5, E31–33; equilibrium, E34; of goods and services, E16; in market economy, E16, E31–36; as measures of value, E32; quantity demanded and, E26, E30; quantity supplied and, E26–28, E30; rationing and, E32, *pE32,* E33; setting, E33–36; as signals, E32; tracking changes on chart or graph, E50
price stability, E123–24, *cE123*
primary sector, E74
private goods, E91–92
private property, E54
private property rights, E61
privatization, E181
producer, E25
producer goods, E117
product, E115
production: factors of, E4, E8, E38, E39; methods, in Industrial Revolution, E44
productive resources, E39
productivity, E40
product market, E38, *pE38*
profit, E16, E27
profit motive, E59–60
progressive tax, E133, E137; as automatic stabilizer, E145
property tax, E136

ECONOMICS INDEX

proportional tax, E133, E137
protectionism, E167
public goods, E91–92
pure market economy, E17, E18

Q

quantity demanded, *cE25*, E26, *gE26*, E30
quantity supplied, *cE25*, E27–28, *gE27*, E30
quaternary sector, E74, *gE74*
quotas, E23, E168

R

radio frequency identification (RFID) technology, E86, *pE86*
railroads, E42
Raise the Wage Act of 2021, E102
rationing, E32, E33; for dealing with scarcity, E3; during oil shortages in 1970s, *pE3*; during pandemic of 2020-2021, *pE3*, E33; during World War I and II, E32, E33; ration cards, *pE32*
real GDP, E112, E117–18, *gE118*, E119
recalls, E94
recession: economic performance and, E120–22, *gE121*, *gE122*; unemployment rate and, E121–22, *gE122*
regressive tax, E133, E137
reserve requirement, E128–29
resources, E4, E8; factors of production in, E8, E38; needed for economic growth, E39
revenue, E113, E131, E133, *gE133*
Revenue Act of 1916, E138
Richthofen, Ferdinand von, E21
right-to-work laws, *pE76*, E77
Roosevelt, Franklin D., E84, E97, E143–44, E151, *qE151*
Russia: black market in, E181–82, *pE181*; GDP per capita in, *mE180*, E181; market economy in, transitioning to, E182, *cE182*

S

Saez, Emmanuel, E140
sales tax, E135–36
sales tax holidays, E135–36
savings account, E81
savings and loan associations (S&Ls), E82
savings and loan industry crisis of 1980s, E84

scarcity, E3, E7–9; economic choices made due to, E11–12, *pE12*; economic problem of, E3, *pE3*, E9; needs and, E7; rationing and, E3; wants and, E7
schedule: demand, E26, *gE26*, E30, *gE34*, *gE36*; supply, E27–28, *gE27*, *gE30*, *gE34*, *gE36*
secondary sector, E74, *gE74*
Second Bank of the United States, E83
sectors: business, E38; consumer, E38; foreign, E39; government, E39
serfs, E22
services, E4, E7, E16
Sherman Antitrust Act of 1890, E93
shipping: during mercantile period, E23
shortage, E35, E36
signals: prices as, E32
Silk Road, *pE19*, E21, E83
"The Silk Road: The Route for Technological Exchange that Shaped the Modern World" (Barral), *qE21*
Singapore, free markets in, E49
single-resource economies, E167, E183
Small Business Administration, E133
Smiles, Samuel, *qE42*
Smith, Adam, E24, E61–62, E166, E184
social insurance programs, E98
social media networks, *gE60*
socialism, E15
social responsibilities of businesses, E69–70, *pE69*
Social Security, E98; mandatory expenditures on, E132; payments, E113; taxes, E113, E133
societies: economic choices made by, E11–12, *pE12*; values of, E12
sole proprietorship, E63–64, *pE63*
Sony, E176
South America: Andean Community trading bloc in, E176
South Korea: free markets in, E49
Soviet Union: black market in, E181–82, *pE181*; collapse of, E181; command economy in, E15; rebuilding economy, after World War II, E15
specialization, E40, E166–67
spinning jenny, E42
spinning mule, E42, *pE42*
Standard and Poor's (S&P) 500, E124

standard of living, E39, E118, E163, E179, E184
Standard Oil Company, E93, *pE93*
state governments: budgeting for, E134–36, *cE135*
state taxes, E113
steam engine, E41, E42
steel, E43, *pE43*, E45, *pE46*
stimulus programs, E121, E144–45
stock certificates, E66
stockholders: in corporations, E66, *cE67*; responsibilities of businesses to, E71
stock indexes, E124
Strauss, Levi, E53, *pE53*
strike, E77, *pE77*
subsidized spending, E136
subsidy, E168
Supplemental Nutrition Assistance Program (SNAP), E98
supply, E27–30; change in, E28–30, *gE30*; graphing, E27–28, *gE27*; quantity supplied, change in, *cE25*, E27–28, *gE27*, E30; schedule, E27–28, *gE27*, *gE30*, *gE34*, *gE36*
supply and demand: foreign exchange and, E169; interest rates and, E146
supply curve, E27–28, *gE27*
Surowiecki, James, E157, *qE157*
surplus, E35, *pE35*, E146
Sweden: command economy in, E15

T

Taiwan: free markets in, E49
tap-and-go purchases, *pE86*
tariffs, E23, E162, E167–68
taxes: amount paid by wealthy, E141, *cE141*; background information, E137; categories of, E137; fiscal policy and, E144; government contributions to, E111; in government sector, E23; impact on workers, *crtE110*; income before and after, E139, *gE139*; in mercantilism, E23; in mixed market economy, E18; opinions about, E110; pollution, E188; progressive *vs.* proportional system of, E145; property, E136; raised by Congress, E138; sales, E135–36; Social Security, E113; sources of, E113; tariffs, E162, E167–68; things paid for by, E111, *pE111*, E113; time line of top federal income tax rates, E138; wealth, E140–42, *cE141*
Tax Reform Act of 1986, E138

Tax Relief, Unemployment Insurance Reauthorization, and Job Creation Act of 2010, E138
technology: in banking, E84, E86, E88; in employment, E74; in production, E40, E42–44
telephone: invention of, E43
Temporary Assistance for Needy Families (TANF), E97
tertiary sector, E74, *gE74*
Theranos, E71
Tide corporation, *pE69*
timber: in factor market, E38; as natural resource, E4, *pE4*; preserving, for economic growth, E39
T-Mobile, E93
trade, E160, *crtE160*, E161; balance of, E171–72, *cE172*; barriers to, E167–68; beginning of, E160; benefits of, E162; blocking, *crtE160*; blocs, E163; comparative advantage in, E162, E166–67; cooperation in, global and regional, E173–76; economic interdependence and, E163, E173, E176–78; evidence of international, *pE162*; exchange rates, E169–70, *pE169*; exports and imports, E165–66, *gE165*, *mE166*; free trade agreements, E173–78; free trade issues, E177; global issues, *pE161*, E176–78; goods, paying for, E162; between nations, E165–67; specialization, E166–67; tariffs, E167–68. *See also* global trade alliances.
trade bloc: comparisons of major, *cE175*; EU, E163, E174–76, *mE175*
trade deficit, E171
trade-offs, E9; in early societies and civilizations, E9
trade routes, early, E19–21
trade surplus, E171
trade union, E75
trade war, E167, E171
traditional economies, E13–14, *pE13*
transparency, E71
Treasury bills, E146
Treasury notes, E146
Trump, Donald, E167, E171, E176, E177

U

Uncle Sam, *crtE160*
unemployment: benefits, E113; compensation, E133; economic performance and, E120–22,

gE121, gE122; insurance, E145, *pE145*; rate, E121–22, *gE122*; recessions and, E121–22, *gE121, gE122*

unemployment insurance, E98, E145

unemployment rate, E74, *cE74*

union shop, E76–77

United Auto Workers (UAW), E75

United Kingdom (UK), E176, E177

United States (U.S.): capitalism in, E57–61; changes in social and economic conditions, E39, E44–45, E74; circular flow of economic activity in, *pE37*; comparative advantages of, E167; economic growth in, E39; exchange rates in, E169–70, *gE170*; government sector in, E39; Gross Domestic Product of, E116, E118, *gE118*; Industrial Revolution in, E42; mixed market economy in, E18; standard of living in, E39; trade balance of, E171–72; trade deficit with China, E171, *pE171*; trading partners of, E167,

mE168; traditional economy in, E18

United States-Mexico-Canada Agreement (USMCA), E163, *cE175,* E176

V

value: price as measure of, E32

values of society, E12

Venezuela: command economy in, E15

Virginia Company, E68

voluntary exchange, E59, *pE61*

W

Walker, C. J., E53, *pE53*

Walmart, E66, E124

Walmart Foundation, E70

Walt Disney Company, E124

wants, E7

Warren, Elizabeth, E140

War Revenue Act of 1917, E138

Watt, James, E41

The Wealth of Nations **(Smith),** E24, *qE24*, E61–62, E184

wealth tax, E140–42, *cE141*

welfare, E97–98

welfare payments, E113

Wendy's, *pE68*

WHAT, HOW, and FOR WHOM questions, E13, E16, E17, E19, E49

White Castle, E70

women: cultural obstacles to contributions, E183; economic roles in traditional economies, E14

workers' compensation, E98

workfare programs, E97–98

workplace: discrimination in, E71–72

World Trade Organization (WTO), E174

World War I and II: income tax rates and, E138; military spending in, E144; rationing in, E32, E33

X

x-axis, E26

Y

Yap: stone money, E79

y-axis, E26

Yellowstone National Park, *pE111*

yen, E169

Young Communist's Union, *pE15*

Z

Zhang Quian, E21

Zucman, Gabriel, E140

X

x-axis, E26

Y

Yap, stone money on, E26
y-axis, E26
Yellowstone National Park, petri, E65
yen, E65
Young Communist's Union, E68

Z

Zhang Qian, E21
Zucman, Gabriel, E10

War Revenue Act of 1917, E138
Weil, James, E41
The Wealth of Nations (Smith), E24, gE24, E61–62, E161
wealth tax, E140–42, gE141
welfare, E97–98
welfare payments, E113
Wendy's, gE68
WHAT, HOW, and FOR WHOM questions, E22, E45, E17, E18, E93
White Castle, E70
women, cultural obstacles to contributions, E183; economic roles in traditional economies, E14
workers' compensation, E98
welfare programs, E97–98
workplace discrimination in, E71–72
World Trade Organization (WTO), E74
World War I and U.S. income tax rates and, E138; military spending in, E141; rationing in E22, E33

international economy, gE16, E16
United States-Mexico-Canada Agreement (USMCA), E143, gE175, E176

V

value, price as measure of, E32
values of society, E72
Venezuela, command economy in, E15
Virginia Company, E68
voluntary exchange, E55, pE61

W

Walker, C. J., E53, pE53
Walmart, E66, E124
Walmart Foundation, E70
Walt Disney Company, E124
wants, E1
Warren, Elizabeth, E140

gE121, gE122; insurance, E145;
pE145; rate, E121–22, gE122;
recessions and, E121–22;
gE121, gE122
unemployment insurance, E98,
E115
unemployment rate, E14, gE14
union shop, E76–77
United Auto Workers (UAW), E76
United Kingdom (UK), E115, E117;
United States (U.S.): capitalism
in, E57–61; changes in social
and economic conditions,
E39, E44–45, E74; circular
flow of economic activity in,
pE37; comparative advantages
of, E161; economic growth
in, E39; exchange rates in
E109–10, gE110; government
sector in, E59; Gross Domestic
Product of, E116, E118, pE118;
Industrial Revolution in, E47;
mixed market economy in,
E58; standard of living in, E39;
trade balance of, E171–72;
trade deficit with China, E171,
pE171; trading partners of, E167